PERGAMON
MECHANICS TODAY
SERIES

Editor: S. Nemat-Nasser
Northwestern University Evanston. Illinois

Mechanics Today

Volume 4

Mechanics Today

Volume 4

Edited by

S. NEMAT-NASSER, *Professor*

Department of Civil Engineering,
The Technological Institute,
Northwestern University,
Evanston, Illinois

Published by Pergamon Press
on behalf of the
AMERICAN ACADEMY OF MECHANICS

PERGAMON PRESS INC.

New York · Toronto · Oxford · Paris · Sydney · Frankfurt

U.K.	Pergamon Press Ltd., Headington Hill Hall, Oxford OX3 0BW, England
U.S.A	Pergamon Press Inc., Maxwell House, Fairview Park, Elmsford, New York 10523, U.S.A.
CANADA	Pergamon of Canada Ltd., 75 The East Mall, Toronto, Ontario, Canada
AUSTRALIA	Pergamon Press (Aust.) Pty. Ltd., 19a Boundary Street, Rushcutters Bay, N.S.W. 2011, Australia
FRANCE	Pergamon Press SARL, 24 rue des Ecoles, 75240 Paris, Cedex 05, France
FEDERAL REPUBLIC OF GERMANY	Pergamon Press GmbH, 6242 Kronberg-Taunus, Pferdstrasse 1, Federal Republic of Germany

First edition 1978

Library of Congress Cataloging in Publication Data

Nemat-Nasser, S
Mechanics today.

(Pergamon mechanics today series)
Vol. 1 : 1972.
"Published ... on behalf of the American Academy of Mechanics."
Includes bibliographical references.
1. Mechanics, Applied. I. American Academy of Mechanics. II. Title.
TA350.N4 620.1 72-10430
ISBN 0-08-0217923 (v. 4) Hard Cover

Printed in Great Britain by Pitman Press, Bath

Contents

Contributors

The number that follows each author's address refers to the page where his contribution begins.

I F. Erdogan, *Department of Mechanical Engineering and Mechanics, Lehigh University, Bethlehem, Pennsylvania 18015*, 1.

II K. Palaniswamy, *Northrop Research and Technology Center, Hawthorne, California 90250*, and

W. G. Knauss, *Graduate Aeronautical Laboratories, California Institute of Technology, Pasadena, California 91109*, 87.

III Subhendu K. Datta, *Department of Mechanical Engineering, University of Colorado, Boulder, Colorado 80302*, 149.

IV Yih-Hsing Pao, *Department of Theoretical and Applied Mechanics, Cornell University, Ithaca, New York 14850*, 209.

V Francis C. Moon, *Department of Theoretical and Applied Mechanics, Cornell University, Ithaca, New York 14850*, 307.

VI S. Nemat-Nasser, *Department of Civil Engineering, The Technological Institute, Northwestern University, Evanston, Illinois 60201*, 391.

Preface

The present volume of *Mechanics Today* follows the tradition that has been established by the previous volumes in this series and introduces the reader to contributions from some of the most active researchers in the fields of solid mechanics and applied mathematics. The volume consists of six articles in areas of applied mechanics that are of current interest and have enjoyed a great deal of attention in the recent past. As in the first three volumes, each article begins with a discussion of fundamentals and proceeds with a presentation of analytical and experimental (where applicable) results. The subject matter is hence developed in such a manner that the article is useful to specialists, while at the same time it remains accessible to nonexperts with sufficient background.

I wish to express my gratitude to Mrs. Erika Ivansons who has assisted with the editorial tasks.

S. Nemat-Nasser
Wilmette, Illinois
October 1976

Contents of Volume 1

Contents of Volume 2

Contents of Volume 3

Summary

For the convenience of the reader, an abstract of each chapter of this volume is given below.

I Mixed Boundary-value Problems in Mechanics *by* F. Erdogan, *Lehigh University*

In this article after a general description of the basic features of mixed boundary-value problems and the relevant concepts (e.g. points of geometric and flux singularity, multiple series equations, and multiple integral equations), the methods which may be used in the solution of mixed boundary-value problems in mechanics are considered. First, the direct application of complex potentials in potential theory and solid mechanics is briefly discussed and the technique of reducing the problem to a Riemann–Hilbert boundary-value problem is described. Then the method of reducing the multiple series and multiple integral equations to singular integral equations is introduced. The main emphasis in the article is on this reduction and on the discussion of various methods, particularly the numerical methods of solution of the resulting singular integral equations. The two numerical methods discussed are the solution by Gaussian integration formulas and that based on using the series in the related orthogonal polynomials. These methods are developed in detail for the singular integral equations of the first and of the second kind with indexes 1, 0, and −1. The singular integral equations with a generalized Cauchy kernel is then introduced and the techniques for obtaining their fundamental functions and for their numerical solution are described. Throughout the article solutions of sample problems are included in order to demonstrate the methods described.

II On the Problem of Crack Extension in Brittle Solids Under General Loading *by* K. Palaniswamy, *Northrop Research and Technology Center, and* W. G. Knauss, *California Institute of Technology*

The problem of crack extension in a solid under arbitrary loads is discussed.

Emphasis is placed on crack growth which leads from a planar (or line) to a nonplanar configuration.

The line crack in the infinite elastic plane is considered first, where it is assumed that an energy balance criterion is appropriate to describe brittle fracture behavior. The requisite stress analysis is performed via the Kolosov–Muskhelishvili method, by mapping an "angled" crack, having nonsharp tips, onto the unit circle. After computing the energy release rate resulting from the angled crack extension, its value for infinitesimal length and sharp edges is computed by a numerical limit process.

The results, giving both the critical far field stress for incipient crack growth as well as the (initial) angle between the original line crack and its extension, are compared with the stress criterion of fracture and with new experiments. The relation of these results to those obtained by other investigators is discussed. The implications for crack extension by a finite length jump, or with infinitesimal extension and infinitesimal deviation angle (smooth crack path) are examined; the corresponding problem for mildly ductile solids is also explored.

Experiments on crack extension under Mode III deformation are reviewed. Drawing on these and on the pronounced similarity of results derived for the two-dimensional case from the stress- and the energy-criteria, the propagation from the smooth front of a planar crack in an arbitrarily loaded solid is investigated. Finally, results for triple mode (I, II, III) interaction which leads to crack instability are cited; these are based on the crack tip stress field (stress criterion).

III **Scattering of Elastic Waves** *by* Subhendu K. Datta, *University of Colorado*

This article reviews various methods of solving the scattering of elastic waves by inclusions. One of these methods is the method of eigenfunction expansions and is used in Section 2 to solve the scattering by a fluid sphere or circular cylinder. Particular attention is paid to the short wavelength limit in which the results obtained from eigenfunction expansions are identified with the ray theoretic predictions. In Sections 3 and 4 a method of matched asymptotic expansions is outlined, and it is shown to give meaningful results for diffraction by spheroids in an infinite space and circular cavities in a semi-infinite space. In the fifth and final section a method of integral equations is illustrated by its application to solve scattering by a rigid circular disc. Various applications of the results obtained are pointed out.

IV Electromagnetic Forces in Deformable Continua *by* Yih-Hsing Pao, *Cornell University*

The interactions of electromagnetic field with deformable bodies in motion are determined by specifying the stress, body force, body couple, and supply of energy of electromagnetic origin in the balance equations of mechanics. The electromagnetic forces and the electromagnetic energy are expressed in terms of field variables which satisfy the Maxwell equations. In the literature there existed several formulations of the Maxwell equations for moving media. With additional postulates and application of first principles, many theories of field–matter interactions have been developed.

The most widely known theories are Minkowski's relativistic theory and Lorentz's theory of electrons. The former is based on the special principle of relativity, whereas the latter postulates that all electric and magnetic phenomena are attributed to the presence or motion of electrical charges (called electrons) embedded in material particles. Applying the modern principles of statistical mechanics to the theory of electrons, a comprehensive theory of electrodynamics has been recently formulated.

Other theories of electrodynamics have been developed by postulating Maxwell–Lorentz force or Maxwell stresses, or by postulating macroscopic models to represent the electromagnetic behavior of materials. The electrical polarization is usually represented by electric dipoles, and the magnetic polarization by either magnetic dipoles or current circuits.

Since each theory is founded on a different set of principles and postulations, the final expressions of one theory for the electromagnetic forces and energy are quite different from those of the other. These theories are compared and critically reviewed in this article. At the end, two theories together with complete set of field equations, boundary conditions, and constitutive relations are summarized; one theory being based on the Lorentz theory of electrons and statistical principles and the second postulating a two-dipole model for electrical polarization and magnetization.

V Problems in Magneto–solid Mechanics *by* Francis C. Moon, *Cornell University*

The purpose of this article is to introduce mechanicians to problems involving solids carrying high electric currents, or being placed in high magnetic fields. Problems in which nontrivial deformations result include ferromagnetic plates and beams in static magnetic fields, the mechanics of high current superconducting solids, and deformations produced by pulsed magnetic fields. The last two areas are related to the design of reactors for magnetic containment of fusion and to magnetic forming of metals using pulsed fields. Particular emphasis is given to instabilities of solid structures in magnetic fields. A review of the stability of ferromagnetic plates and beams in a magnetic field is presented including both theoretical and experimental work. Another problem discussed concerns current-carrying conductors which exhibit instabilities analogous to those found in plasmas. Both theoretical and experimental evidence is presented for the buckling of a toroidal set of superconducting rings. Other instabilities of high current-carrying elastic solids are reviewed. Finally a review of the mechanics of solids in pulsed magnetic fields is given. New experimental data on magnetically generated stress waves in an elastic ferromagnetic bar is presented. These experiments show that both tensile and compressive waves can be generated through the interaction of eddy current and magnetization forces. It is shown that these waves can be predicted using the measured Maxwell stress tensor outside the bar.

It is hoped that the article will demonstrate the rich source of nontrivial problems in magneto-solid mechanics. Such problems should have increasing application to the electric energy and magnetic fusion energy fields.

Areas in this field which should receive increased attention from mechanicians include the interaction of lasers, arcs, and electron beams with elastic solids and the development of analytical and numerical methods for handling the nonlinear mathematics which is intrinsic to this field.

VI **On Nonequilibrium Thermodynamics of Continua: Addendum** *by* S. Nemat-Nasser, *Northwestern University*

This article is supplementary to the article which appeared as Chapter II of Volume 2 of *Mechanics Today*. The implications of the second law of thermodynamics, as stated by Eqs. (4.4) and (4.5) in the above-mentioned article, are considered, and by means of an example it is shown that these statements are too restrictive and should be relaxed in order to apply to many commonly used models for the macroscopic behavior of materials. In addition, the consequences of (thermodynamic) *material stability* are studied, and the corresponding restrictions that this may place on the evolutionary equations for internal variables, are explored. In particular, it is shown that when the material is thermodynamically stable, and if there are only eight scalar-valued internal variables, then under additional rather mild assumptions, the rate of change of the internal variables (i.e. the fluxes) must necessarily be proportional to the gradient of a *scalar function of state,* taken with respect to the internal forces. This result is then used to show that the *inelastic* rate of change of strain, the (inelastic) rate of stress relaxation, the rate of entropy production due to internal dissipation, and the rate of change of temperature due to internal relaxation are proportional to the gradient of a suitable potential function (an *inelastic* potential depending on *state variables* only) taken with respect to the stress, strain, temperature, and entropy, respectively (i.e. a general set of normality rules).

I

Mixed Boundary-value Problems in Mechanics

F. Erdogan

Lehigh University, Bethlehem, Pennsylvania

1 INTRODUCTION

In attempting to formulate a given "equilibrium-type" of continuous system in mechanics one may either use some kind of variational principle and reduce the problem to a minimization problem subject to certain constraints, or, as is more commonly the case, one may directly apply the equilibrium principles and reduce the problem to a boundary value problem which consists of a (system of) differential equation(s) subject to certain boundary conditions. Even though in most practical applications the minimization problem is further reduced to a boundary-value problem, it may also be solved directly by using an appropriate technique such as the Ritz's method. To facilitate the definition of certain concepts consider the following boundary value problem in two dimensions:

$$L_{2m}(u) = f(x_1, x_2), \qquad (x_1, x_2) \in D, \tag{1.1}$$

$$B_i(u) = g_i(s), \qquad (i = 1, \ldots, m), \qquad s \in S, \tag{1.2}$$

where L_{2m} is a differential operator of order $2m$, x_1, x_2 are the spatial coordinates, D is the domain of definition for the unknown function u, S is the boundary of D, B_i $(i = 1, \ldots, m)$ is a differential operator (containing u and its normal derivatives) of order (at most) $2m - 1$, f and g are known functions, and s is any convenient coordinate defining the point on the boundary (say, the arc length). The domain D may contain the point at infinity and may be multiplyconnected. Contours forming the boundary are assumed to consist of piecewise smooth arcs. The points on S at which the tangent has a discontinuous slope will be called the *points of geometric singularity.*

There is another type of singular point on the boundary which results from the change in the nature of the homogeneous operators $B_i(u)$ specifying the boundary conditions. Such a point on a smooth boundary either side of which at least one of the operators B_i $(i = 1, \ldots, m)$ has different behavior is called a *point of flux singularity.* Note that if the behavior of an operator B_i changes at a "corner point" of the boundary, this point is then a point of both a geometric as well as a flux singularity. If the homogeneous operators B_i remain unchanged on each closed contour (but not necessarily the same on all contours), then the corresponding problem is called an *ordinary boundary value problem.* On the other hand, if there are points of flux singularity on the boundary, the problem is called a *mixed boundary value problem.*

In working with mixed boundary value problems it is often advantageous to keep in mind that the physical system has generally two types of quantities, namely, the *potential* and the *flux* type quantities. In potential theory the meaning of these concepts is unambiguous and clear. They are, for example, identified by temperature, velocity potential, electrostatic potential, mass concentration, or displacement (in anti-plane shear problems) as the potential type quantities, and heat flux, velocity, electrostatic charge, mass rate of diffusion, or stress, as the corresponding flux type quantities. Similarly, in solid mechanics one may classify the displacements and the stresses (or the strains) as respectively the potential and flux type quantities. The physics of the problem requires that the "potential" be bounded and continuous everywhere in $D + S$, including the points of both geometric and flux singularity.

To fix the ideas, consider the following simple problem in a wedge-shaped domain:

$$\nabla^2 u(r, \theta) = 0, \qquad 0 < r < \infty, \qquad 0 < \theta < \theta_0, \tag{1.3}$$

$$\frac{1}{r}\frac{\partial}{\partial\theta}u(r, 0) = f_1(r), \qquad \frac{1}{r}\frac{\partial}{\partial\theta}u(r, \theta_0) = f_2(r), \qquad 0 < r < \infty \tag{1.4}$$

where the known functions f_1 and f_2 are such that the global equilibrium is satisfied and $0 < \theta_0 \leq 2\pi$. The problem is an ordinary boundary value problem and $r = 0$ is a point of geometric singularity. If in the neighborhood of $r = 0$, f_1 and f_2 are zero, simple application of Mellin transform would indicate that for small values of r the components of the flux vector are of the following form:

$$\frac{1}{r}\frac{\partial u}{\partial \theta} = Kr^{(\pi-\theta_0)/\theta_0}\sin\frac{\pi\theta}{\theta_0} + 0(r^{(2\pi-\theta_0)/\theta_0}),$$
$$\frac{\partial u}{\partial r} = -Kr^{(\pi-\theta_0)/\theta_0}\cos\frac{\pi\theta}{\theta_0} + 0(r^{(2\pi-\theta_0)/\theta_0}), \tag{1.5}$$

where K is a constant. Note that at $r = 0$ the flux becomes unbounded for $\pi < \theta_0 \leq 2\pi$, the corresponding power of singularity being

$$0 > \frac{\pi - \theta_0}{\theta_0} \geq -0.5.$$

If the boundary conditions (1.4) are replaced by

$$u(r, 0) = 0, \qquad \frac{1}{r}\frac{\partial}{\partial \theta}u(r, \theta_0) = f(r), \qquad 0 < r < \infty \tag{1.6}$$

it is seen that the problem is a mixed boundary value problem in which $r = 0$ is a point of both geometric and flux singularity. Hence, in this problem one would expect a stronger flux singularity than in the previous problem. The asymptotic solution of the problem for small r may again be expressed as

$$u(r, \theta) = Kr^{\pi/2\theta_0}\sin\frac{\pi\theta}{2\theta_0} + 0(r^{3\pi/2\theta_0}),$$
$$\frac{1}{r}\frac{\partial u}{\partial \theta} = K\frac{\pi}{2\theta_0}r^{(\pi-2\theta_0)/2\theta_0}\cos\frac{\pi\theta}{2\theta_0} + 0(r^{(3\pi-2\theta_0)/2\theta_0}), \tag{1.7}$$
$$\frac{\partial u}{\partial r} = K\frac{\pi}{2\theta_0}r^{(\pi-2\theta_0)/2\theta_0}\sin\frac{\pi\theta}{2\theta_0} + 0(r^{(3\pi-2\theta_0)/2\theta_0}).$$

From (1.7) it is seen that for $\pi/2 \leq \theta_0 \leq 2\pi$ the corresponding power of the flux singularity is $0 \geq (\pi - 2\theta_0)/2\theta_0 \geq -3/4$, which is stronger than that found in the previous problem for the same angle θ_0.

Let us now consider a special case of this problem in which $\theta_0 = \pi$. Here the boundary is the infinite line; consequently the geometric singularity is removed but $r = 0$ remains to be a point of flux singularity having a power $-1/2$. As will be seen later in this article, $-1/2$ power singularity is quite typical for the points of flux singularity on a smooth boundary. It will also be seen that, however, somewhat contrary to the general expectation, this is not always the case in mixed boundary value problems with boundary conditions containing the potential and the flux.

Finally, consider the following (mixed) boundary conditions

$$u(\mathrm{r},0) = 0, \qquad a < r < b,$$

$$\frac{1}{r}\frac{\partial}{\partial\theta}u(r,0) = g(r), \qquad 0 < r < a, \qquad b < r < \infty, \tag{1.8}$$

$$\frac{1}{r}\frac{\partial}{\partial r}u(r,\theta_0) = f(r), \qquad 0 < r < \infty,$$

where f and g are known functions. It is seen that $r = 0$ is a point of geometric singularity and the points $r = a$ and $r = b$ on the smooth boundary ($\theta = 0$, $0 < r < \infty$) are points of flux singularity. The problem is a mixed boundary value problem. Around the geometric singularity $r = 0$ the solution is expected to behave as in (1.5) and around the points of flux singularity it will have a behavior as in (1.7) with $\theta_0 = \pi$ and necessary coordinate transformations. It should again be emphasized that the standard $-1/2$ power of the singularity at $r = a > 0$ will be a function of θ_0 for $a = 0$ which, depending on the value of θ_0, may be stronger or weaker than $-1/2$. A similar phenomenon will be discussed in connection with a contact problem in elasticity later in this paper.

2 DEFINITIONS: MULTIPLE SERIES EQUATIONS, MULTIPLE INTEGRAL EQUATIONS

In considering the solution of a given mixed boundary value problem perhaps the simplest technique is the direct application of the method of complex potentials provided the problem admits such potentials and the domain and the boundary conditions are suitable for such an application. In this case the problem is reduced to a Riemann–Hilbert problem for a (system of) sectionally holomorphic function(s) which may be solved in a straightforward manner. On the other hand if one applied a more standard technique such as, the separation of variables, integral transforms, or the method of Green's function, the mixed boundary conditions invariably lead to a formulation involving "dual series equations", "dual integral equations" or "singular integral equations". Again, to facilitate the basic understanding of these notions, the definitions will be preceded by the formulation of some simple examples.

2.1 Multiple Series Equations

As a first example consider the following mixed boundary value problem in potential theory for the unit circle:

$$\nabla^2 u(r,\theta) = 0, \qquad 0 \le r < 1, \qquad 0 \le \theta < 2\pi, \tag{2.1}$$

$$\frac{\partial}{\partial r}u(1,\theta) = g(\theta), \qquad \theta \in L_1, \tag{2.2}$$

$$u(1,\theta) = f(\theta), \qquad \theta \in L_2, \tag{2.3}$$

where

$$L_1 = \sum_1^N L_{1i}, \qquad L_{1i} = (r = 1;\, a_i < \theta < b_i), \tag{2.4}$$

and L_2 is the complement of L_1 on the unit circle. Using the technique of the separation of variables the solution of (2.1) may be expressed as

$$u(r,\theta) = A_0 + \sum_1^\infty r^n(A_n \cos n\theta + B_n \sin n\theta). \tag{2.5}$$

Formally, substituting from (2.5) into the boundary conditions one obtains the following system of equations to determine the coefficients A_n and B_n:

$$\begin{aligned} &\sum_1^\infty n(A_n \cos n\theta + B_n \sin n\theta) = g(\theta), \qquad \theta \in L_1, \\ A_0 + &\sum_1^\infty (A_n \cos n\theta + B_n \sin n\theta) = f(\theta), \qquad \theta \in L_2. \end{aligned} \tag{2.6a, b}$$

In the case of Dirichlet ($L_1 \equiv 0$) or Neumann ($L_2 \equiv 0$) problems (2.6b) or (2.6a) give the unknown coefficients directly by expanding $f(\theta)$ or $g(\theta)$ into Fourier series in $(0, 2\pi)$. However, in the present problem the functions $\sin n\theta$ and $\cos n\theta$ ($n = 0, 1, 2, \ldots$) are not orthogonal on L_1 and L_2 and hence, (2.6) is at best equivalent to (or can be transformed into) an infinite system of algebraic equations.

The structure of (2.6) is typical of the mixed boundary value problems defined in a bounded domain $a < x < b$ which may be expressed as

$$\begin{aligned} &\sum_1^\infty A_n k_1(n,x) = f_1(x), \qquad x \in L_1, \\ &\sum_1^\infty A_n k_2(n,x) = f_2(x), \qquad x \in L_2, \end{aligned} \tag{2.7}$$

or, more generally, for J sets of coefficients A_{jn}, ($j = 1, \ldots, J$; $n = 1, 2, \ldots$) one obtains

$$\begin{aligned} &\sum_{j=1}^J \sum_{n=1}^\infty k_{1j}^i(n,x) A_{jn} = f_1^i(x), \qquad x \in L_1, \qquad (i = 1, \ldots, J), \\ &\sum_{j=1}^J \sum_{n=1}^\infty k_{2j}^i(n,x) A_{jn} = f_2^i(x), \qquad x \in L_2, \qquad (i = 1, \ldots, J), \end{aligned} \tag{2.8}$$

where again

$$L_1 = \sum_1^N L_{1k}, \qquad L_{1k} = (a_k, b_k), \qquad L_1 + L_2 = (a, \mathrm{b}),$$
$$a_k < b_k < a_{k+1}, \qquad a \le a_1, \qquad b_N \le b, \tag{2.9}$$

and the input functions f_r or f_r^i and the kernel functions k_r or k_{rj}^i are known. The system of equations such as (2.6), (2.7), or (2.8) are defined as *dual series equations.*

Going back to the problem for the unit circle (2.1), if one assumes that the boundary $L = (0, 2\pi)$ is divided into three parts with the following boundary conditions:†

$$\frac{\partial}{\partial r} u(1, \theta) = f_1(\theta), \qquad \theta \in L_1,$$
$$u(1, \theta) = f_2(\theta), \qquad \theta \in L_2, \tag{2.10a–c}$$
$$h_1 u(1, \theta) + h_2 \frac{\partial}{\partial r} u(1, \theta) = f_3(\theta), \qquad \theta \in L_3,$$

where L_r, $(r = 1, 2, 3)$ is the union of nonintersecting arcs L_{rj}, $(r = 1, 2, 3;$ $j = 1, 2, \ldots, J_r)$ on the unit circle with $L_1 + L_2 + L_3 = L = (0, 2\pi)$. h_1 and h_2 may be functions of θ. Again, formally from (2.1) and (2.10) it follows that

$$\sum_1^\infty n(A_n \cos n\theta + B_n \sin n\theta) = f_1(\theta), \qquad \theta \in L_1,$$
$$A_0 + \sum_1^\infty (A_n \cos n\theta + B_n \sin n\theta) = f_2(\theta), \qquad \theta \in L_2,$$
$$h_1 A_0 + h_1 \sum_1^\infty (A_n \cos n\theta + B_n \sin n\theta) \tag{2.11a–c}$$
$$+ h_2 \sum_1^\infty n(A_n \cos n\theta + B_n \sin n\theta) = f_3(\theta), \qquad \theta \in L_3.$$

In this problem as seen from (2.10) the boundary conditions are defined in terms of three distinct operators on three separate parts of the boundary

† This is a problem, for example, in heat conduction in which, in addition to specifying the heat flux and temperature on parts of the boundary L_1 and L_2, there is free convection taking place along L_3 where h_1 is the coefficient of heat convection, h_2 is the coefficient of heat conduction, and f_3 is related to the environmental temperature u_∞ through $h_1 u_\infty = f_3$.

giving rise to a set of three series equations described by (2.11). Thus, these equations will be called *triple series equations*. In general, then a set of series equations for a system of unknown coefficients $A_1, A_2, \ldots$ of the form

$$\sum_1^\infty A_n k_i(n,x) = f_i(x), \qquad x \in L_i, \qquad \sum_1^M L_i = L, \qquad (i = 1, \ldots, M), \tag{2.12}$$

will be called *multiple series equations* (of multiplicity M).†

In the mixed boundary value problems described by (2.6) and (2.11) the intersections of the boundary segments L_{ij} are points of "flux singularity". The discussion given in the previous section for the wedge would indicate that at least at some of these points the flux will be unbounded, at others the behavior of the solution is not known beforehand. Hence, the technique developed to solve the multiple series equations will not only have to be sufficiently general to apply to a great diversity of problems but will also have to lend itself to the correct treatment of the singular nature of the solution. Aside from the method of complex potentials whenever applicable, it appears that the method of singular integral equations is the only approach which fulfills these requirements. Because of the singularities, since the infinite series giving the components of the flux vector will be divergent at certain points on the boundary, it is clear that any direct method reducing the multiple series equations to an infinite system of algebraic equations in the unknown coefficients (which has no closed form solution) will not be acceptable.

2.2 Multiple Integral Equations

Consider now the equivalent problem in potential theory for the half plane $y > 0$. The problem is stated as follows:

$$\nabla^2 u(x,y) = 0, \qquad (-\infty < x < \infty, 0 < y < \infty), \tag{2.13}$$

$$\frac{\partial}{\partial y} u(x,0) = f_1(x), \qquad x \in L_1, \tag{2.14}$$

$$u(x,0) = f_2(x), \qquad x \in L_2, \tag{2.15}$$

† Note that this definition is different than that found in current literature (e.g. [1]) on "dual series" and "dual integral" equations where the multiplicity of the equations is taken to be the number of independent segments on the boundary rather than the number of independent operators defining the boundary conditions.

$$L_1 = \sum_1^N L_{1i}, \qquad L_{1i} = (a_i, b_i), \qquad a_i < b_i < a_{i+1}, \qquad -\infty \le a_1, \qquad b_N \le \infty, \tag{2.16}$$

where L_2 is the complement of L_1 on $(-\infty < x < \infty)$, f_1 and f_2 are known functions and are such that $u \to 0$ as $x^2 + y^2 \to \infty$ (i.e., any homogeneous solution behaving differently at infinity has been separated). Using Fourier transforms, the solution of (2.13) may be expressed as

$$u(x, y) = \int_{-\infty}^{\infty} A(\alpha)\, e^{-y|\alpha| - i\alpha x}\, d\alpha, \tag{2.17}$$

where $A(\alpha)$ is an unknown function. Substituting from (2.17) into (2.14) and (2.15) formally we find

$$\begin{aligned} -\int_{-\infty}^{\infty} |\alpha|\, A(\alpha)\, e^{-i\alpha x}\, d\alpha &= f_1(x), \qquad x \in L_1, \\ \int_{-\infty}^{\infty} A(\alpha)\, e^{-i\alpha x}\, d\alpha &= f_2(x), \qquad x \in L_2, \end{aligned} \tag{2.18a, b}$$

If L_1 or L_2 is zero, (2.18) may be solved in closed form in terms of inversion integrals. Integral equations of the form (2.18) or more generally the pair of integral equations

$$\begin{aligned} \int_L k_1(x, \alpha) A(\alpha)\, d\alpha &= f_1(x), \qquad x \in L_1, \\ \int_L k_2(x, \alpha) A(\alpha)\, d\alpha &= f_2(x), \qquad x \in L_2, \qquad L_1 + L_2 = L, \end{aligned} \tag{2.19}$$

in which the kernels k_1 and k_2 are different, are called a set of *dual integral equations* for the unknown function $A(\alpha)$. If the problem involves more than one unknown function, the boundary conditions would lead to a *system of dual integral equations* for the unknown functions $A_j(\alpha)$, $(j = 1, \ldots, J)$ of the following form:

$$\int_L \sum_{j=1}^{J} k_{1j}^i(x, \alpha) A_j(\alpha)\, d\alpha = f_1^i(x), \qquad x \in L_1, \qquad i = 1, \ldots, J, \tag{2.20}$$

$$\int_L \sum_1^J k_{2j}^i(x, \alpha) A_j(\alpha)\, d\alpha = f_2^i(x), \qquad x \in L_2, \qquad i = 1, \ldots, J, \qquad L_1 + L_2 = L. \tag{2.20}$$

In this problem too one may consider the following more general boundary conditions:

$$\frac{\partial}{\partial y}u(x,0)=f_1(x), \qquad x\in L_1,$$
$$u(x,0)=f_2(x), \qquad x\in L_2, \tag{2.21a–c}$$
$$h_1u(x,0)+h_2\frac{\partial}{\partial y}u(x,0)=f_3(x), \qquad x\in L_3,$$

where $L_1+L_2+L_3=(-\infty,\infty)$, and L_i, $(i=1,2,3)$ consists of non-intersecting line segments L_{ik} $(k=1,\ldots,K_i)$ on the real line. From (2.17) and (2.21) it follows that

$$-\int_{-\infty}^{\infty}|\alpha|\,A(\alpha)\,e^{-i\alpha x}\,d\alpha=f_1(x), \qquad x\in L_1,$$
$$\int_{-\infty}^{\infty}A(\alpha)\,e^{-i\alpha x}\,d\alpha=f_2(x), \qquad x\in L_2, \tag{2.22a–c}$$
$$\int_{-\infty}^{\infty}(h_1-h_2|\alpha|)A(\alpha)\,e^{-i\alpha x}\,d\alpha=f_3(x), \qquad x\in L_3,$$

Equations (2.22a–c) form a set of *triple integral equations* for the unknown function $A(\alpha)$. More generally

$$\int_L\sum_1^J k_{mj}^n(x,\alpha)A_j(\alpha)\,d\alpha=f_m^n(x), \qquad x\in L_m, \qquad \sum_{m=1}^M L_m=L,$$
$$n=1,\ldots,J, \qquad m=1,\ldots,M, \tag{2.23}$$

is called a *system of multiple integral equations* (of multiplicity M) for the unknown functions $A_1,\ldots,A_J$.

As in multiple series equations, in problems formulated in terms of multiple integral equations the singular nature of the solution is generally not known beforehand. Therefore, in these problems too it is important that the method of solution developed to solve the integral equations be not only sufficiently general and effective but also give the correct behavior of existing singularities. In this respect, particularly in dealing with somewhat unusual mixed boundary value problems, the methods of complex potentials and singular integral equations appear to be far superior to the standard operational techniques. An extensive treatment of the operational techniques for the solution of dual series and dual integral equations may be found in a recent book by Sneddon [1]. References [2–13] are some of the outstanding ones

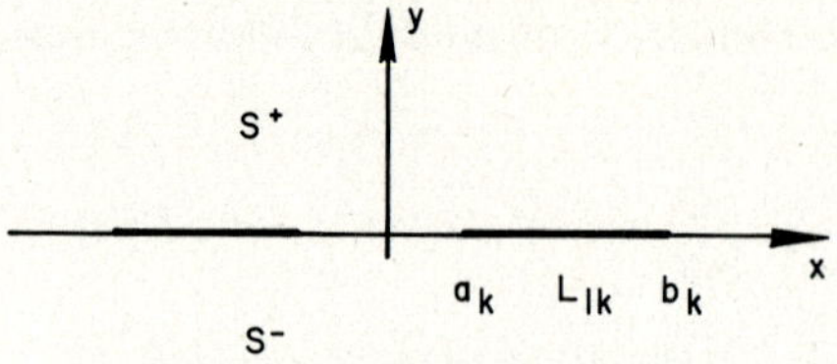

Fig. 1 Plane with collinear cuts.

on the theory and applications of the complex potentials and the singular integral equations. In this article the primary emphasis will be on the recent developments concerning the methods of solution of the singular integral equations and particularly the applications to some mixed boundary value problems with uncommon singularities.

3 APPLICATION OF COMPLEX POTENTIALS

In this section the direct application of complex potentials will be described by considering some relatively simple examples.

3.1 A Problem in Potential Theory

Consider the mixed boundary value problem in potential theory for the half-plane ($-\infty < x < \infty, y > 0$) which is formulated by (2.13)–(2.16) (Fig. 1). Let the harmonic function $u(x, y)$ be the real part of a complex potential $F(z)$, $z = x + iy$. $F(z)$ is holomorphic in the upper half-plane S^+ where the derivatives of u may be expressed as

$$2\frac{\partial u}{\partial x} = F'(z) + \bar{F}'(\bar{z}), \qquad -2i\frac{\partial u}{\partial y} = F'(z) - \bar{F}'(\bar{z}). \tag{3.1}$$

Noting that if $z \in S^+$ and $z \to t + i0$ then $\bar{z} \in S^-$ and $\bar{z} \to t - i0$, taking the boundary values of (3.1), and using the conditions (2.14) and (2.15) we obtain

$$F'^+(t) - \bar{F}'^-(t) = -2if_1(t), \qquad t \in L_1,$$
$$F'^+(t) + \bar{F}'^-(t) = 2f_2'(t), \qquad t \in L_2, \tag{3.2a, b}$$

where (2.15) is used in differentiated form which means that for single-valuedness the solution must satisfy the following conditions:

$$\int_{a_k}^{b_k} \frac{\partial}{\partial x} u(x, 0)\, dx = f_2(b_k) - f_2(a_k), \tag{3.3}$$

where $k = 1, \ldots, N$ if $a_1 > -\infty$, $b_N < \infty$, and $k = 2, \ldots, N-1$ if $a_1 = -\infty$ and $b_N = \infty$.

Since $F'(z)$ is holomorphic in S^+, $\bar{F}'(z)$ will be holomorphic in S^-. If L_1 is finite we define a new sectionally holomorphic function by

$$G(z) = \begin{cases} F'(z), & z \in S^+ \\ -\bar{F}'(z), & z \in S^-. \end{cases} \tag{3.4}$$

From (3.2) and (3.4) it then follows that

$$\begin{aligned} G^+(t) + G^-(t) &= -2if_1(t), \qquad t \in L_1, \\ G^+(t) - G^-(t) &= 2f_2'(t), \qquad t \in L_2. \end{aligned} \tag{3.5a, b}$$

From (3.4) and (3.5) it is seen that $G(z)$ is holomorphic everywhere in the complex plane except on L_1 and on that part of L_2 on which $f_2'(t)$ is not zero. Depending on the behavior of u, G may also have a pole of finite degree at infinity. For example, if there is a uniform "flux" at infinity given by

$$\frac{\partial u}{\partial x} + i\frac{\partial u}{\partial y} = p_1 + ip_2, \qquad (x^2 + y^2 \to \infty), \tag{3.6}$$

we have

$$\lim_{|z| \to \infty} F'(z) = p_1 - ip_2. \tag{3.7}$$

Consider now the related homogeneous Riemann–Hilbert problem given by

$$\begin{aligned} X^+(t) + X^-(t) &= 0, \qquad t \in L_1, \qquad L_1 = \sum_1^N L_{1k} \\ X^+(t) - X^-(t) &= 0, \qquad t \in L_2, \end{aligned} \tag{3.8}$$

where $X(z)$ is the fundamental solution of the original Riemann–Hilbert problem (3.5) which is clearly determinate within an arbitrary multiplicative analytic function. Referring to, for example, [3], the general solution of (3.8) may be expressed as

$$X(z) = P(z) \prod_1^N (z - a_k)^{-(1/2 + \alpha_k)} (z - b_k)^{(1/2 + \beta_k)}, \tag{3.9}$$

where α_k and β_k ($k = 1, \ldots, N$) are arbitrary integers (positive, negative, or zero) and $P(z)$ is an arbitrary polynomial. At this point it should be strongly emphasized that in (3.9) as well as in the applications of the function–

theoretic methods to the singular integral equations elsewhere in this article the arbitrary integers α_k, β_k cannot be determined from purely mathematical considerations. To do this the physics of the problem has to be properly taken into account. In the present problem the points a_k and b_k are the type of singular points at which flux vector has an integrable singularity. Therefore, $\alpha_k = 0$, $\beta_k = -1$, $(k = 1, \ldots, N)$, and ignoring the arbitrary polynomial, the fundamental solution of the problem becomes

$$X(z) = \prod_{1}^{N} (z - a_k)^{-1/2}(z - b_k)^{-1/2}, \tag{3.10}$$

where for the particular branch considered $\lim_{|z|\to\infty} z^n X(z) = 1$.

Dividing both sides of (3.5) by $X^+(t)$ and using (3.8) we find

$$\left(\frac{G(t)}{X(t)}\right)^+ - \left(\frac{G(t)}{X(t)}\right)^- = \begin{cases} -2if_1(t)/X^+(t), & t \in L_1, \\ 2f_2'(t)/X^+(t), & t \in L_2. \end{cases} \tag{3.11}$$

Equation (3.11) is now a simple boundary-value problem the general solution of which having a finite degree at infinity may be written as [3],

$$\frac{G(z)}{X(z)} = -\frac{1}{2\pi i}\int_{L_1} \frac{2if_1(t)}{(t - z)X^+(t)}\,dt + \frac{1}{2\pi i}\int_{L_2} \frac{2f_2'(t)}{(t - z)X^+(t)}\,dt + P_k(z), \tag{3.12}$$

where $P_k(z)$ is an arbitrary polynomial of degree k. From (3.4) and (3.7) it is seen that $G(z)$ has a pole of order zero at infinity. Thus, from (3.10) and (3.12) it follows that the degree of P_k is N, and (3.12) becomes

$$G(z) = -\frac{X(z)}{\pi}\int_{L_1} \frac{f_1(t)\,dt}{(t - z)X^+(t)} + \frac{X(z)}{\pi i}\int_{L_2} \frac{f_2'(t)\,dt}{(t - z)X^+(t)} + X(z)\sum_{0}^{N} c_k z^k \tag{3.13}$$

where $c_0, \ldots, c_N$ are arbitrary constants. From (3.4), (3.7), and (3.13) it is easily seen that

$$c_N = p_1 - ip_2. \tag{3.14}$$

Noting that L_1 is finite, the remaining N constants $c_0, \ldots, c_{N-1}$ are determined from the single-valuedness conditions (3.3). From

$$2\frac{\partial}{\partial x} u(x, +0) = F'^+(x) + \bar{F}'^-(x) = G^+(x) - G^-(x) \tag{3.15}$$

these conditions may be expressed as

$$\frac{1}{2}\int_{a_k}^{b_k} [G^+(t) - G^-(t)]\,dt = f_2(b_k) - f_2(a_k), \qquad (k = 1, \ldots, N), \tag{3.16}$$

giving, with (3.13), a system of N linear algebraic equations in $c_0, \ldots, c_{N-1}$. This completes the solution for finite L_1.

If L_2 instead of L_1 is finite (i.e. $a_1 = -\infty, b_N = \infty$) the procedure to solve the problem is quite similar to that given above with the following main differences: the branch cut should be introduced along L_2 by defining $G(z)$ as

$$G(z) = \begin{cases} F'(z), & z \in S^+ \\ \bar{F}'(z), & z \in S^- \end{cases} \tag{3.17}$$

which would give the fundamental solution as follows:

$$X(z) = \prod_{1}^{N-1} (z - b_k)^{-1/2}(z - a_{k+1})^{-1/2} \tag{3.18}$$

Considering only the solution for which

$$\int_{-\infty}^{\infty} \frac{\partial}{\partial y} u(x, 0)\, dx = Q = \text{finite} \tag{3.19}$$

and noting that $X(z) \to z^{1-N}$ for $|z| \to \infty$, it may be seen that in the expression of $G(z)$ similar to (3.13), the degree of the arbitrary polynomial will be $N-2$, i.e. the solution will contain $N-1$ arbitrary constants $c_0, \ldots, c_{N-2}$. These constants are then determined from (3.19) and (3.3), completing the solution.

3.2 The Case of Periodic Cuts

In the problem considered in the previous section let the cuts (a_k, b_k) be equal in length and be equally spaced. Thus, one may define the end points of $2m + 1$ cuts by

$$\begin{gathered} a_k = 2kb - a, \qquad b_k = 2kb + a, \qquad (k = 0, \mp z1, \ldots, \mp m) \\ L_{1k} = (a_k, b_k), \qquad L_1 = \sum_{-m}^{m} L_{1k}. \end{gathered} \tag{3.20}$$

The fundamental solution (3.10) of the problem then becomes

$$\begin{aligned} X(z) &= \prod_{-m}^{m} (z - a_k)^{-1/2}(z - b_k)^{-1/2} \\ &= A\left[(z + a)\prod_{1}^{m}\left(1 - \left(\frac{z + a}{2kb}\right)^2\right)(z - a)\prod_{1}^{m}\left(1 - \left(\frac{z - a}{2kb}\right)^2\right)\right]^{-1/2}, \end{aligned} \tag{3.21}$$

where A is a constant given by

$$A = (-1)^m (2b)^{-2m} (m!)^{-2}.$$

Because of the homogeneous nature of (3.8), since $X(z)$ is determinate within a multiplicative analytic function, the constant A may be (and will be) ignored. It is seen that if we now let $m \to \infty$, geometrically the problem becomes that of a plane with periodic cuts. In addition to this if the functions f_1 and f_2 are assumed to be periodic then we have a problem for a half-plane with periodic mixed boundary conditions. For simplicity, here it will be assumed that any homogeneous "loading" condition at infinity has been separated through a proper superposition and consequently in the problem under consideration $F(z) \to 0$ as $|z| \to \infty$. First ignoring the constant A and then using the relation [14]

$$\sin \theta = \theta \prod_1^\infty \left(1 - \frac{\theta^2}{k^2 \pi^2}\right), \tag{3.22}$$

in the limit from (3.21) we may obtain $X(z)$ as

$$X(z) = \left(\sin^2 \frac{\pi z}{2b} - \sin^2 \frac{\pi a}{2b}\right)^{-1/2}, \tag{3.23}$$

where the particular branch which is positive for $a < \mathrm{Re}(z) = t < 2b - a$ is to be considered.

With $X(z)$ as given by (3.23) the solution (3.12) is still valid. Since the geometry and the boundary conditions in the problem are periodic in x with a period of $2b$, the pontential $G(z)$ must also be periodic in z with a real period $2b$. On the other hand, from (3.23) it is seen that

$$\begin{aligned} X^+(t) &= -X^-(t) = -X^+(t+2b) = X^+(t+4b), \qquad t \in L_1, \\ X^+(t) &= X^-(t) = -X^+(t+2b) = X^+(t+4b), \qquad t \in L_2, \end{aligned} \tag{3.24}$$

namely, $X(z)$ is periodic with a period of $4b$. Therefore, in order to have a periodic potential with (real) period $2b$, the arbitrary polynomial in (3.12) must be of the following form:

$$P_k(z) = \left(B_1 \sin \frac{\pi z}{2b} + B_2 \cos \frac{\pi z}{2b}\right) Q(z), \tag{3.25}$$

where $Q(z)$ is an arbitrary periodic analytic function with period $2b$ and B_1 and B_2 are arbitrary constants. From

$$\lim_{|z| \to \infty} G(z) = 0 = \lim_{|z| \to \infty} X(z) P_k(z) = (B_1 + iB_2) Q(\infty) \tag{3.26}$$

it follows that $Q(\infty) = 0$, or since Q is analytic in the entire plane, $Q(z) = 0$.
The solution of the problem may then be written as

$$G(z) = -\frac{X(z)}{\pi} \sum_{-\infty}^{\infty} \int_{a_k}^{b_k} \frac{f_1(t)\,dt}{(t-z)X^+(t)} + \frac{X(z)}{\pi i} \sum_{-\infty}^{\infty} \int_{b_k}^{a_{k+1}} \frac{f_2'(t)\,dt}{(t-z)X^+(t)}, \tag{3.27}$$

where $X(z)$ is given by (3.23). Letting

$$\begin{gathered} t = \tau + 2kb, \qquad f_1(t) = f_1(\tau + 2kb) = g_1(\tau), \qquad f_2'(t) = g_2(\tau), \\ X^+(t) = X(\tau), \qquad -b < \tau < b, \qquad k = 0, \mp 1, \ldots, \end{gathered} \tag{3.28}$$

and using (3.24), we may modify the integrals in (3.27) as follows:

$$\begin{aligned} S_1 &= \sum_{-\infty}^{\infty} \int_{a_k}^{b_k} \frac{f_1(t)\,dt}{(t-z)X^+(t)} = \int_{-a}^{a} \frac{g_1(\tau)\,d\tau}{X(\tau)} \sum_{-\infty}^{\infty} \frac{(-1)^k}{(\tau - z) + 2kb} \\ &= \int_{-a}^{a} \frac{g_1(\tau)\,d\tau}{X(\tau)} \left[\frac{1}{\tau - z} + \sum_{1}^{\infty} \frac{(-1)^k 2(\tau - z)}{(\tau - z)^2 - (2kb)^2} \right]. \end{aligned} \tag{3.29}$$

In (3.29) the series may be summed by using [14]

$$\frac{1}{\theta} + \sum_{1}^{\infty} \frac{2\theta}{\theta^2 - n^2} = \pi \cot \pi\theta, \qquad \sum_{1}^{\infty} \frac{2\theta}{\pi^2 \left(\dfrac{2n-1}{2}\right)^2 - \theta^2} = \tan \theta \tag{3.30}$$

which gives

$$\begin{aligned} S_1 &= \frac{\pi}{4b} \int_{-a}^{a} \frac{g_1(\tau)\,d\tau}{X(\tau)} \left[\cot \frac{\pi}{4b}(\tau - z) + \tan \frac{\pi}{4b}(\tau - z) \right] \\ &= \frac{\pi}{2b} \int_{-a}^{a} \frac{g_1(\tau)\,d\tau}{X(\tau) \sin \dfrac{\pi}{2b}(\tau - z)} \end{aligned} \tag{3.31}$$

Similarly

$$\sum_{-\infty}^{\infty} \int_{b_k}^{a_{k+1}} \frac{f_2'(t)\,dt}{(t-z)X^+(t)} = \frac{\pi}{2b} \int_{a}^{2b-a} \frac{g_2(\tau)\,d\tau}{X(\tau) \sin \dfrac{\pi}{2b}(\tau - z)} \tag{3.32}$$

Thus, the solution becomes

$$G(z) = -\frac{X(z)}{2b} \int_{-a}^{a} \frac{g_1(\tau)\,d\tau}{X(\tau) \sin \dfrac{\pi}{2b}(\tau - z)} + \frac{X(z)}{2bi} \int_{a}^{2b-a} \frac{g_2(\tau)\,d\tau}{X(\tau) \sin \dfrac{\pi}{2b}(\tau - z)}, \tag{3.33}$$

where $X(z)$ is given by (3.23) and

$$X(\tau) = X^+(t) = \left(\sin^2\frac{\pi\tau}{2b} - \sin^2\frac{\pi a}{2b}\right)^{-1/2} \tag{3.34}$$

Consider, for example, the simple case of an infinite plane containing uniformly "loaded" periodic cuts for which $f_1(t) = -p_0$ and $f_2(t) = 0$. From (3.33) G may be obtained as

$$G(z) = ip_0\left[1 - X(z)\sin\frac{\pi z}{2b}\right] \tag{3.35}$$

Using now the relations

$$\frac{\partial}{\partial x}u(x, +0) = \tfrac{1}{2}[G^+(x) - G^-(x)],$$
$$\frac{\partial}{\partial y}u(x, +0) = -\frac{1}{2i}[G^+(x) + G^-(x)], \tag{3.36a, b}$$

we express the components of the flux vector on the boundary as follows:

$$\frac{\partial}{\partial x}u(x, +0) = \begin{cases} 0, & a < |x| < b, \\ \dfrac{-p_0\sin(\pi x/2b)}{[\sin^2(\pi a/2b) - \sin^2(\pi x/2b)]^{1/2}}, & 0 \leq |x| < a, \end{cases}$$
$$\frac{\partial}{\partial y}u(x, +0) = \begin{cases} -p_0, & 0 \leq |x| < a \\ \dfrac{p_0\sin(\pi x/2b)}{[\sin^2(\pi x/2b) - \sin^2(\pi a/2b)]^{1/2}} - p_0, & a < |x| < b. \end{cases} \tag{3.37}$$

In limit, for $b \to \infty$, $a =$ finite, (3.35)–(3.37) reduce to the following results for a plane with a single cut, $(-a, a)$ for which $\partial/\partial y\, u(x, 0) = -p_0$, $(-a < x < a)$ is the only external disturbance:

$$G(z) = ip_0[1 - z(z^2 - a^2)^{-1/2}],$$
$$\frac{\partial}{\partial x}u(x, 0) = -\frac{p_0 x}{(a^2 - x^2)^{1/2}}, \quad 0 \leq |x| < a,$$
$$\frac{\partial}{\partial y}u(x, +0) = \frac{p_0 x}{(x^2 - a^2)^{1/2}} - p_0, \quad a < |x| < \infty. \tag{3.38}$$

In the general case the real and imaginary parts of $G(z)$ give the components of the flux vector in the upper half-plane and (3.36) that on the boundary. $G^+(x)$ and $G^-(x)$ may in turn be obtained from (3.33) by using (3.24) and the following Plemelj formulas [3]:

$$\phi(z) = \frac{1}{2\pi i}\int_L \frac{f(t)\,dt}{\sin\dfrac{\pi}{2b}(t-z)},$$

$$\phi^+(x) - \phi^-(x) = \begin{cases} 0, & x \in L', \\ f(x), & x \in L, \end{cases} \tag{3.39a–c}$$

$$\phi^+(x) + \phi^-(x) = \begin{cases} 2\phi(x), & x \in L', \\ \dfrac{1}{\pi i}\displaystyle\int_L \frac{f(t)\,dt}{\sin\dfrac{\pi}{2b}(t-x)}, & x \in L, \end{cases} \tag{3.39a–c}$$

where $f(t)$ is any Hölder-continuous function defined on the open interval L, $L + L' = (-b, b)$.

From (3.13) and (3.33), or more specifically, from (3.37) it is seen that the components of the flux vector $\partial u/\partial x$ and $\partial u/\partial y$ have integrable singularities at the points of intersection a_k, b_k of L_1 and L_2 with a power of $-1/2$ (see also (1.7) for $\theta_0 = \pi$). A close examination of (3.13) and (3.33) would indicate that

$$G(z) = F_1(z) + X(z)F_2(z), \tag{3.40}$$

where F_1 and F_2 are holomorphic in the entire plane. By letting $z - a = r\,e^{i\theta}$, for small values of r, we may express $G(z)$ as

$$G(z) = \frac{\partial u}{\partial x} - i\frac{\partial u}{\partial y} = \frac{1}{\sqrt{r}}\frac{\cos\theta/2 - i\sin\theta/2}{\left[\dfrac{\pi}{2b}\sin\dfrac{\pi a}{b}\right]^{1/2}}F_2(a) + 0(1), \qquad 0 \le \theta < \pi. \tag{3.41}$$

For example, in the case of uniformly "loaded" cuts, from (3.35) it follows that†

$$G(z) = \frac{\partial u}{\partial x} - i\frac{\partial u}{\partial y} = -\left(\frac{b}{\pi}\tan\frac{\pi a}{2b}\right)^{1/2}\frac{p_0}{\sqrt{r}}\left(\sin\frac{\theta}{2} + i\cos\frac{\theta}{2}\right) + 0(1),$$

$$0 \le \theta \le \pi - 0. \tag{3.42}$$

Note that in this simple example $u(x,0) = 0$ on L_1, $\partial/\partial y\; u(x,0) = -p_0$ on L_2. Hence, the boundary conditions around the flux singularity $x = a$ are identical to (1.6). It would then be expected that the resulting asymptotic solutions be the same which may indeed be seen from (1.7) and (3.42).

† In (3.41) and (3.42) the terms 0(1) stands for bounded terms and come from $F_1(z)$ in (3.40). Note that the second term in (3.40) is of the asymptotic form $0(r^{-1/2}) + 0(r^{1/2})$.

3.3 An Elasticity Problem for a Nonhomogeneous Plane

As another example for the direct application of complex potentials to mixed boundary value problems consider the following elasticity problem. Let two linearly elastic isotropic half-planes with material constants‡ μ_i, κ_i ($i = 1$ for $y > 0$, $-\infty < x < \infty$, i.e. S^+; and $i = 2$ for $y < 0$, $-\infty < x < \infty$, i.e. S^-) be bonded along the (nonintersecting) line segments $L_k = (a_k, b_k)$, ($k = 1, \ldots, N$) on the real axis. Let $L = \sum_1^N L_k$ be finite and the x- and y-components of the resultant force acting on the half-planes at infinity be Q and P, respectively. Let L' be the complement of L on $(-\infty < x < \infty$, $y = 0)$ and $-p_1(x)$ and $p_2(x)$ be the normal tractions acting on the half-planes along L'. The problem will have to be solved under the following conditions:†

$$\begin{aligned}
\sigma^+_{1yy}(t) - i\sigma^+_{1xy}(t) &= \sigma^-_{2yy}(t) - i\sigma^-_{2xy}(t), \qquad t \in L + L', \\
\sigma^+_{1yy}(t) - i\sigma^+_{1xy}(t) &= -[p_1(t) + ip_2(t)] = -p(t), \qquad t \in L', \\
[u^+_1(t) + iv^+_1(t)] - [u^-_2(t) + iv^-_2(t)] &= f_1(t) + if_2(t) = f(t), \qquad t \in L, \\
\int_{L+L'} (\sigma^-_{2yy} - i\sigma^-_{2xy})\, dt &= P - iQ,
\end{aligned} \tag{3.43a–d}$$

where the superscripts + and − refer to limits as $y \to +0$ and $y \to -0$, u_k, v_k are the x, y-components of the displacement vector, and σ_{kjl} ($k = 1, 2$; $(j, l) = (x, y)$) is the stress.

The simplest method to solve this problem would be the use of complex potentials known as Kolosov–Muskhelishvili or Goursat functions. In terms of these potentials the stresses and displacements may be expressed as follows [2, 8–10]:

$$\begin{aligned}
\sigma_{kxx} + \sigma_{kyy} &= 2[\Phi_k(z) + \bar{\Phi}_k(\bar{z})], \\
\sigma_{kyy} - \sigma_{kxx} + 2i\sigma_{kxy} &= 2[\bar{z}\Phi'_k(z) - \Psi_k(z)], \\
2\mu_k(u_k + iv_k) &= \kappa_k\phi_k(z) - z\bar{\Phi}_k(\bar{z}) - \bar{\psi}_k(\bar{z}), \\
\Phi_k(z) = \phi'_k(z), \qquad \Psi_k(z) &= \psi'_k(z), \qquad (k = 1, 2)
\end{aligned} \tag{3.44a–e}$$

where $z \in S^+$ for $k = 1$ and $z \in S^-$ for $k = 2$. Noting that Φ_1 and Ψ_1 are defined in S^+ and Φ_2 and Ψ_2 are defined in S^- only, by extending the

‡ Where μ_i is the shear modulus, $\kappa_i = 3 - 4\nu_i$ for plane strain, and $\kappa_i = (3 - \nu_i)/(1 + \nu_i)$ for plane stress, ν_i being the corresponding Poisson's ratio.

† The input function $f(t)$ may be non-zero in, for example, thermal stress and residual stress problems [15].

definition of Φ_1 into S^- and Φ_2 into S^+ in such a way that they are holomorphic on the unloaded parts of the boundary (i.e. the real axis), one could make the following substitution [2]:

$$\Psi_k(z) = -\Phi_k(z) - \bar{\Phi}_k(z) - z\Phi_k'(z), \tag{3.45}$$

where $z \in S^+$ for $k = 1$ and $z \in S^-$ for $k = 2$. From (3.44) and (3.45) it then follows that

$$\sigma_{kyy} - i\sigma_{kxy} = \Phi_k(z) - \Phi_k(\bar{z}) + (z - \bar{z})\bar{\Phi}_k'(\bar{z}), \qquad (k = 1, 2), \tag{3.46}$$

$$2\mu_k \frac{\partial}{\partial x}(u_k + iv_k) = \kappa_k \Phi_k(z) + \Phi_k(\bar{z}) - (z - \bar{z})\Phi_k'(\bar{z}), \qquad (k = 1, 2). \tag{3.47}$$

Substituting from (3.46) into (3.43a) we find

$$\Phi_1^+(t) + \Phi_2^+(t) = \Phi_1^-(t) + \Phi_2^-(t), \qquad t \in L + L', \tag{3.48}$$

meaning that $\Phi_1(z) + \Phi_2(z)$ is holomorphic in the entire plane including the real axis. Noting that the stress state at infinity vanishes and assuming that the rotation at infinity is zero, following [2], for large values of $|z|$ it may be shown that

$$\Phi_1(z) = \frac{Q + iP}{2\pi z} + o(1/z),$$
$$\Phi_2(z) = -\frac{Q + iP}{2\pi z} + o(1/z), \tag{3.49a, b}$$

where, in the usual notation $o(1/z) \leqq c/z$, c being a positive quantity which depends only on $|z|$ and tends to zero as $|z| \to \infty$. Since $\Phi_1 + \Phi_2$ is holomorphic, from (3.49) it is clear that

$$\Phi_1(z) + \Phi_2(z) = 0 \tag{3.50}$$

in the entire plane.

Substituting now from (3.46), (3.47), and (3.50) into the boundary conditions (3.43b) and (3.43c) (after differentiation), we obtain the following Riemann–Hilbert problem for the sectionally holomorphic function $\Phi_2(z)$:

$$\begin{aligned} \Phi_2^+(t) + \omega\Phi_2^-(t) &= h(t), \qquad t \in L, \\ \Phi_2^+(t) - \Phi_2^-(t) &= p(t), \qquad t \in L', \end{aligned} \tag{3.51}$$

where

$$\omega = \frac{\mu_1\kappa_2 + \mu_2}{\mu_2\kappa_1 + \mu_1}, \qquad h(t) = \frac{-2\mu_1\mu_2}{\mu_2\kappa_1 + \mu_1} f'(t). \tag{3.52}$$

Referring to [3], the fundamental solution of (3.51) satisfying the related homogeneous equations

$$\begin{aligned} X^+(t) + \omega X^-(t) &= 0, \qquad t \in L, \\ X^+(t) - X^-(t) &= 0, \qquad t \in L', \\ L = \sum_1^N L_k, \qquad L_k &= (a_k, b_k), \end{aligned} \tag{3.53}$$

may be expressed as

$$\begin{aligned} X(z) &= \prod_1^N (z - b_k)^{\alpha_k}(z - a_k)^{\beta_k}, \\ \alpha_k &= \frac{1}{2\pi i}\log(-\omega) + A_k = \tfrac{1}{2} - i\frac{\log \omega}{2\pi} + A_k, \\ \beta_k &= -\frac{1}{2\pi i}\log(-\omega) + B_k = -\tfrac{1}{2} + i\frac{\log \omega}{2\pi} + B_k, \end{aligned} \tag{3.54}$$

where A_k and B_k $(k = 1, \ldots, N)$ are again arbitrary (positive, zero, or negative) integers.† In the present problem the singular points a_k and b_k correspond to the ends points of interface cracks. Consequently, at these points the stresses and displacement derivatives will have an integrable singularity giving $B_k = 0$, $A_k = -1$, $(k = 1, \ldots, N)$. Here, we will then consider the particular branch of $X(z)$ which is single valued in the plane cut along L and for which

$$\lim_{|z| \to \infty} z^N X(z) = 1. \tag{3.55}$$

Dividing (3.51) by $X^+(t)$ and using (3.53) it is found that

$$\begin{aligned} \left(\frac{\Phi_2(t)}{X(t)}\right)^+ - \left(\frac{\Phi_2(t)}{X(t)}\right)^- &= \frac{h(t)}{X^+(t)}, \qquad t \in L, \\ \left(\frac{\Phi_2(t)}{X(t)}\right)^+ - \left(\frac{\Phi_2(t)}{X(t)}\right)^- &= \frac{p(t)}{X^+(t)}, \qquad t \in L'. \end{aligned} \tag{3.56}$$

Noting that the stress state vanishes at infinity, we observe that the general solution of (3.56) vanishing at infinity becomes

$$\frac{\Phi_2(z)}{X(z)} = \frac{1}{2\pi i}\int_L \frac{h(t)\,dt}{(t - z)X^+(t)} + \frac{1}{2\pi i}\int_{L'} \frac{p(t)\,dt}{(t - z)X^+(t)} + \sum_0^{N-1} c_n z^n, \tag{3.57}$$

† In most physical problems A_k and B_k are such that $-1 < \mathrm{Re}(\alpha_k, \beta_k) < 1$. Even though in literature one finds this as a mathematical condition, clearly A_k and B_k must be determined from the physics of the problem.

where $c_0, \ldots, c_{N-1}$ are arbitrary constants. From (3.49b), (3.55), and (3.57) it may be shown that

$$c_{N-1} = -\frac{Q + iP}{2\pi}. \tag{3.58}$$

The condition of single-valuedness of displacements gives the remaining $N - 1$ constants, $c_0, \ldots, c_{N-2}$. We recall that in deriving (3.51) the continuity condition (3.43c) was used in differentiated form. Thus, this condition requires that

$$\int_{b_k}^{a_{k+1}} \frac{\partial}{\partial x}\left[(u_1 + iv_1)^+ - (u_2 + iv_2)^-\right] dx = f(a_{k+1}) - f(b_k)$$
$$(k = 1, \ldots, N-1) \tag{3.59}$$

or

$$\int_{b_k}^{a_{k+1}} \left[\left(\frac{1+\kappa_1}{2\mu_1} + \frac{1+\kappa_2}{2\mu_2}\right)\Phi_2^+(t) + \left(\frac{1}{2\mu_1} + \frac{\kappa_2}{2\mu_2}\right)p(t)\right] dt = -f(a_{k+1}) + f(b_k)$$
$$(k = 1, \ldots, N-1). \tag{3.60}$$

In (3.60) and elsewhere, to obtain the boundary values of the Cauchy integrals the following general Plemelj formulas may be used:

$$F(z) = \frac{1}{2\pi i}\int_L \frac{g(t)\,dt}{t - z},$$
$$F^+(x) - F^-(x) = \begin{cases} g(x), & x \in L, \\ 0, & x \in L', \end{cases} \tag{3.61a–c}$$
$$F^+(x) + F^-(x) = \begin{cases} \dfrac{1}{\pi i}\displaystyle\int_L \frac{g(t)}{t - x}\,dt, & x \in L, \\ 2F(x), & x \in L', \end{cases}$$

where L' is the complement of L (on an infinite line or on any closed contour in the complex plane). Thus, (3.57) with (3.58), (3.60), (3.50), (3.45) and (3.44) gives the complete solution of the problem. For example, if $L = (-a, a)$ (bonding along a single segment), $p(t) = 0$, and $f(t) = 0$, the solution becomes

$$\Phi_1(z) = -\Phi_2(z) = \frac{Q + iP}{2\pi}\,\frac{1}{(z^2 - a^2)^{1/2}}\left(\frac{z + a}{z - a}\right)^{i(\log\omega)/2\pi}, \tag{3.62}$$

where ω is given by (3.52).

Observing that the general solution of the problem is of the form

$$\Phi_2(z) = F_1(z) + F_2(z)X(z), \tag{3.63}$$

as in the previous example, one may easily investigate the asymptotic behavior of the stresses and displacements around the points of singularity (see [16] for details). If the positive constant ω appearing in (3.51) and defined by (3.52) is not unity, from the behavior of the fundamental function $X(z)$ given by (3.54) it is clear that the stresses and displacement derivatives will have a typical oscillating singularity around the end points of the branch cuts which is of the following form:

$$\sigma_{ij}(r, \theta) = \frac{1}{\sqrt{r}}\left[f_{ij}(\theta)\cos\left(\gamma \log\frac{r}{l}\right) + g_{ij}(\theta)\sin\left(\gamma \log\frac{r}{l}\right)\right] + 0(1),$$

$$\gamma = (\log \omega)/2\pi, \qquad (i, j = x, y), \qquad (r \ll l), \tag{3.64}$$

where r, θ are the polar coordinates around the singular point, f_{ij} and g_{ij} are bounded functions, and the term 0(1) again comes from the analytic function $F_1(z)$ in (3.63).

4 REDUCTION TO SINGULAR INTEGRAL EQUATIONS

As pointed out in the previous section, in order to obtain the correct behavior of the singularities and also in most cases to find a simple closed form solution of a given mixed boundary value problem, whenever possible it is preferable to use the complex potentials and the related complex function theory. However, the technique has its limitations. First, the particular problem may not admit complex potentials. Secondly, the successful application of the technique is severely limited to certain geometries. Finally, there are certain types of boundary conditions which would make the direct use of complex potentials extremely difficult if not impossible (e.g. the boundary conditions described by differential operators containing the unknown function as well as its derivatives).† In such cumbersome cases, the method which is perhaps the most general and the easiest to apply is the reduction of the problem to singular integral equations either by using a Green's function formulation or by formulating the problem first in terms of multiple series or multiple integral equations. Reduction of the boundary

† For this type of mixed boundary conditions even the simplest problems such as that in potential theory for a half-plane do not seem to have closed form solutions (see the following section).

conditions to integral equation is always possible. The main problems here are the selection of the appropriate auxiliary function (i.e. the new unknown function defined on the boundary) and the proper separation of the dominant parts of the kernel for the correct study of the singular behavior of the solution. In this section this important step of reducing the mixed boundary-value problem to a singular integral equation will be discussed by considering some typical examples, and some general remarks will be made regarding the nature of the kernel and the solution.

4.1 Reduction of Dual Series Equations to Singular Integral Equations

Consider the typical simple mixed boundary value problem described by (2.1)–(2.3) and formulated by the dual series equations (2.6). At the generality that the problem is stated, it is not very fruitful to pursue a technique based on the operational methods to solve the problem and, as stated before, because of the importance of existing singularities, a direct numerical solution is nearly useless. To reduce the problem to an integral equation, the first step is the definition or selection of an appropriate auxiliary function. In this problem let this function be

$$\phi(\theta) = \frac{\partial}{\partial\theta} u(1, \theta), \qquad \theta \in L_1 + L_2 = (0, 2\pi). \tag{4.1}$$

Note that

$$\phi(\theta) = \begin{cases} f'(\theta), & \theta \in L_2, \\ \text{unknown}, & \theta \in L_1. \end{cases} \tag{4.2}$$

From (2.5) and (4.1) it follows that

$$\begin{aligned} nA_n &= -\frac{1}{\pi}\int_{L_1+L_2} \phi(t)\sin nt\,dt, \\ nB_n &= \frac{1}{\pi}\int_{L_1+L_2} \phi(t)\cos nt\,dt. \end{aligned} \tag{4.3}$$

Equations (4.3) and (2.5) give the solution once the function $\phi(t)$ and the constant A_0 are determined.

In applying the technique described in this section, in order to circumvent the difficulties arising from the divergent series or integrals giving the kernels, for analytical convenience the boundary condition forming the basis of the integral equation will always be expressed in limit form.† Thus, the boundary condition (2.6a) may be expressed as

$$\lim_{r\to 1-0} \sum_{1}^{\infty} nr^{n-1}(A_n \cos n\theta + B_n \sin n\theta) = g(\theta), \qquad \theta \in L_1. \tag{4.4}$$

Substituting now from (4.3) into (4.4), observing that for $r < 1$ the related infinite series will be uniformly convergent, and hence changing the order of integration and summation, it is found that

$$\lim_{r\to 1-0} \frac{1}{\pi} \int_{L_1+L_2} \phi(t)\, dt \sum_{1}^{\infty} r^{n-1} \sin n(t-\theta) = -g(\theta), \qquad \theta \in L_1. \tag{4.5}$$

First performing the sum as

$$\sum_{1}^{\infty} r^{n-1} \sin nz = \frac{1}{2ir} \sum_{1}^{\infty} [(r\, e^{iz})^n - (r\, e^{-iz})^n] = \frac{\sin z}{1 + r^2 - 2r \cos z} \tag{4.6}$$

and then going to limit, one observes that (4.5) becomes

$$\frac{1}{2\pi} \int_{L_1} \phi(t) \cot \frac{t-\theta}{2}\, dt = -g(\theta) - \frac{1}{2\pi} \int_{L_2} f'(t) \cot \frac{t-\theta}{2}\, dt, \qquad \theta \in L_1, \tag{4.7}$$

giving an integral equation to determine $\phi(\theta)$.

It should be observed that (4.7) is a typical singular integral equation of the following form:

$$\frac{1}{\pi} \int_{L_1} \frac{\phi(t)}{t-\theta}\, dt + \int_{L_1} k(\theta, t)\phi(t)\, dt = g_0(\theta), \qquad \theta \in L_1, \tag{4.8}$$

where the kernel

$$k(\theta, t) = \frac{1}{2\pi} \cot \frac{t-\theta}{2} - \frac{1}{\pi} \frac{1}{t-\theta} \tag{4.9}$$

is bounded everywhere on L_1 (including the end points $a_i, b_i, i = 1, \ldots, N$), and the known function g_0 is the right-hand side of (4.7). Since the fundamental solution (or the behavior of the singularity) depends only on the dominant part of the integral equation, for the purpose of obtaining this solution tentatively expressing (4.8) as

$$\frac{1}{\pi} \int_{L_1} \frac{\phi(t)}{t-\theta}\, dt = g_1(\theta), \qquad \theta \in L_1, \tag{4.10}$$

† The exception here is, of course, the case in which the related series or integrals are uniformly convergent giving bounded kernels. In that case the limit may be put under summation or integration sign before evaluating the kernels.

defining a sectionally holomorphic function by

$$F(z) = \frac{1}{2\pi i} \int_{L_1} \frac{\phi(t)}{t - z} dt, \tag{4.11}$$

and using the Plemelj formulas

$$\begin{aligned} F^+(\theta) - F^-(\theta) &= \phi(\theta), \qquad \theta \in L_1, \\ F^+(\theta) + F^-(\theta) &= \frac{1}{\pi i} \int_{L_1} \frac{\phi(t)}{t - \theta} dt, \qquad \theta \in L, \end{aligned} \tag{4.12a, b}$$

one obtains the following Riemann–Hilbert problem:

$$\begin{aligned} F^+(\theta) - F^-(\theta) &= 0, \qquad \theta \in L_2, \\ F^+(\theta) + F^-(\theta) &= -ig_1(\theta), \qquad \theta \in L_1. \end{aligned} \tag{4.13}$$

Again, following [3] and observing that $L_1 = \sum_1^N L_{1k}$, $L_{1k} = (a_k, b_k)$, the fundamental solution of (4.13) satisfying the related homogeneous Riemann–Hilbert problem is found to be (see (3.5)–(3.10))

$$X(z) = \prod_1^N (z - a_k)^{-1/2 + \alpha_k}(z - b_k)^{1/2 + \beta_k}, \tag{4.14}$$

where the arbitrary integers α_k and β_k $(k = 1, \ldots, N)$ have to be selected in such a way that the solution is consistent with the expected physical behavior. In this problem since at a_k and b_k the "flux" has an integrable singularity, $\alpha_k = 0$ and $\beta_k = -1$, $k = 1, \ldots, N$. Referring, for example, to the solution of (3.5) as given by (3.12) and noting that $X^+(\theta) + X^-(\theta) = 0$, $\theta \in L_1$, from (4.12a) it is seen that $\phi(\theta) \sim X^+(\theta)$, $\theta \in L_1$. We now define the *fundamental function*, $w(\theta)$ of the singular integral equation (4.10) by

$$w(\theta) = \prod_1^N [(\theta - a_k)(b_k - \theta)]^{-1/2} = (-1)^{N/2} X^+(\theta). \tag{4.15}$$

Thus, the solution of (4.10) or (4.7) will be of the following form:

$$\phi(\theta) = w(\theta)p(\theta), \tag{4.16}$$

where p is an unknown bounded function.

Referring again to (3.12) (and [3]), it is seen that the solution of (4.13) vanishing at infinity (and hence $\phi(\theta)$ obtained from (4.12a)) will contain N arbitrary constants. These constants are determined from the single-valuedness conditions of $u(r, \theta)$. We recall that in deriving (4.7) the boundary condition $u(1,\theta) = f(\theta)$, $\theta \in L_2$, was used in differentiated form. Therefore $\phi(\theta)$ must satisfy the following conditions:

$$\int_{a_k}^{b_k} \phi(\theta)\, d\theta = f(b_k) - f(a_k), \qquad (k = 1, \ldots, N). \tag{4.17}$$

To complete the solution of the problem the constant A_0 in (2.5) must be determined. From (2.5), (2.3), and (4.1) it is seen that

$$A_0 = \frac{1}{2\pi}\int_{L_1+L_2} u(1,\theta)\, d\theta = \frac{1}{2\pi}\sum_1^N \int_{b_k}^{a_{k+1}} f(\theta)\, d\theta + \frac{1}{2\pi}\sum_1^N \int_{a_j}^{b_j} d\theta \left[f(a_j) + \int_{a_j}^{\theta} \phi(t)\, dt \right], \qquad (a_{N+1} = a_1). \tag{4.18}$$

Using the Dirichlet transformation

$$\int_a^x dy \int_a^y F(x,y,s) f(s)\, ds = \int_a^x f(s)\, ds \int_s^x F(x,y,s)\, dy \tag{4.19}$$

we reduce Eq. (4.18) to

$$A_0 = \frac{1}{2\pi}\sum_1^N \left[\int_{b_k}^{a_{k+1}} f(\theta)\, d\theta - \int_{a_k}^{b_k} t\phi(t)\, dt + b_k f(b_k) - a_k f(a_k) \right], \qquad (a_{N+1} = a_1). \tag{4.20}$$

The solution of (4.13) is given by (3.12) with $g_1 = 2f_1$, $f_2' = 0$, and $N-1$ as the degree of the polynomial P_k. Thus, using the Plemelj formulas (4.12), we express the solution of (4.7) or (4.8) as

$$\phi(\theta) = 2w(\theta)\sum_0^{N-1} c_n\theta^n - \frac{w(\theta)}{\pi}\int_{L_1} \frac{g_0(t)\, dt}{(t-\theta)w(t)} + \frac{w(\theta)}{\pi}\int_{L_1} \frac{dt}{(t-\theta)w(t)} \int_{L_1} k(t,s)\phi(s)\, ds, \qquad \theta \in L_1. \tag{4.21}$$

Considering now the definition (4.16), it is found that

$$p(\theta) - \frac{1}{\pi}\int_{L_1} n(\theta,s)p(s)\, ds = 2\sum_0^{N-1} c_n\theta^n - \frac{1}{\pi}\int_{L_1} \frac{g_0(t)\, dt}{(t-\theta)w(t)}, \qquad \theta \in L_1, \tag{4.22}$$

where

$$n(\theta,s) = \int_{L_1} \frac{k(t,s)w(s)}{(t-\theta)w(t)}\, dt. \tag{4.23}$$

Equation (4.22) may be treated as a Fredholm integral equation [3, 5] giving the solution of the problem with (4.17) and (4.20).

It should be remarked that any singular integral equation of the form (4.8) may be "regularized" and reduced to a Fredholm-type integral equation by following the foregoing procedure. However, for the solution of (4.8) a considerably simpler numerical technique will be described later in this article. It should also be noted that the particular mixed boundary value problem considered in this section can be solved in closed form. First consider the case of geometric symmetry with respect to the line $\theta = 0$. That is, let us assume that

$$N = 2m, \qquad L_1 = M_1 + \bar{M}_1, \qquad M_1 = \sum_1^m M_{1k}, \qquad M_{1k} = (a_k, b_k),$$
$$\bar{M}_1 = \sum_1^m \bar{M}_{1k}, \qquad \bar{M}_{1k} = (-b_k, -a_k). \tag{4.24}$$

Then the problem may be reduced to a singular integral equation with dominant part (i.e. Cauchy kernel) only. For this the problem is first decomposed into symmetric (i.e. $u(r, \theta) = u(r, -\theta)$) and antisymmetric (i.e. $u(r, \theta) = -u(r, -\theta)$) parts by separating the input functions into even and odd components. In the symmetric problem $g(\theta) = g(-\theta)$, $f(\theta) = f(-\theta)$, $\phi(\theta) = -\phi(-\theta)$, and (4.7) may be expressed as

$$\sum_1^m \frac{1}{2\pi} \int_{a_k}^{b_k} \phi(t) \left[\cot \frac{t-\theta}{2} + \cot \frac{t+\theta}{2} \right] dt = g_0(\theta) = -g(\theta)$$
$$- \frac{1}{2\pi} \int_{M_2} f'(t) \left[\cot \frac{t-\theta}{2} + \cot \frac{t+\theta}{2} \right] dt, \qquad \theta \in M_1, \tag{4.25}$$

where M_2 is the complement of M_1 on $0 \leq \theta \leq \pi$. Or, from

$$\cot(a-b) \mp \cot(a+b) = 2 \begin{Bmatrix} \sin 2b \\ \sin 2a \end{Bmatrix} \Big/ (\cos 2b - \cos 2a) \tag{4.26}$$

it is found that

$$\frac{1}{\pi} \int_{M_1} \frac{\phi(t) \sin t}{\cos \theta - \cos t} dt = g_0(\theta), \qquad \theta \in M_1. \tag{4.27}$$

If we now define

$$\cos t = \alpha, \qquad \phi(t) = \psi(\alpha), \qquad \cos \theta = \beta, \qquad g_0(\theta) = G(\beta),$$
$$t \in M_1 \to \alpha \in \Gamma, \qquad \Gamma = \sum_1^m \Gamma_k \tag{4.28}$$

Eq. (4.27) becomes

$$\frac{1}{\pi}\int_{\Gamma}\frac{\psi(\alpha)}{\alpha-\beta}\,d\alpha = G(\beta), \qquad \beta\in\Gamma. \tag{4.29}$$

Similarly for the antisymmetric problem $g(\theta) = -g(-\theta)$, $f(\theta) = -f(-\theta)$, $\phi(\theta) = \phi(-\theta)$ and (4.7) becomes

$$\frac{1}{2\pi}\int_{M_1}\phi(t)\left[\cot\frac{t-\theta}{2} - \cot\frac{t+\theta}{2}\right]dt = g_0(\theta) = -g(\theta)$$

$$-\frac{1}{2\pi}\int_{M_2} f'(t)\left[\cot\frac{t-\theta}{2} - \cot\frac{t+\theta}{2}\right]dt, \qquad \theta\in M_1, \tag{4.30}$$

with the substitutions (4.28) giving

$$\frac{1}{\pi}\int_{\Gamma}\left(\frac{\psi(\alpha)}{\sqrt{1-\alpha^2}}\right)\frac{d\alpha}{\alpha-\beta} = \left(\frac{G(\beta)}{\sqrt{1-\beta^2}}\right), \qquad \beta\in\Gamma. \tag{4.31}$$

In both cases the solution is given by (4.21) with $k(t,s) = 0$ and appropriate changes in notation. m integration constants arising from the solution are again determined from the single-valuedness conditions (4.17). In the symmetric case A_0 is given by (4.20) (with, again the appropriate changes) and in the antisymmetric case it is zero.

For example, if

$$\begin{gathered} m = 1, \qquad M_1 = (\theta_1, \theta_2), \qquad \alpha_1 = \cos\theta_1, \qquad \alpha_2 = \cos\theta_2, \\ 0 < \theta_1 < \theta_2 < \pi, \qquad g(\theta) = g(-\theta), \qquad f(\theta) = 0, \end{gathered} \tag{4.32}$$

it may be shown that

$$\psi(\alpha) = \begin{cases} \dfrac{1}{[(\alpha-\alpha_2)(\alpha_1-\alpha)]^{1/2}}\left[\dfrac{1}{\pi}\displaystyle\int_{\alpha_2}^{\alpha_1}\frac{G(\beta)(\beta-\alpha_2)^{1/2}(\alpha_1-\beta)^{1/2}}{\beta-\alpha}\,d\beta + c_0\right], \\ \qquad\qquad (\alpha_2 < \alpha < \alpha_1), \\ 0, \qquad (-1 < \alpha < \alpha_2, \qquad \alpha_1 < \alpha < 1), \end{cases} \tag{4.33}$$

where c_0 and A_0 are determined from

$$\int_{\alpha_2}^{\alpha_1}\frac{\psi(\alpha)\,d\alpha}{\sqrt{1-\alpha^2}} = 0, \qquad A_0 = -\frac{1}{\pi}\int_{\theta_1}^{\theta_2} t\phi(t)\,dt. \tag{4.34}$$

Furthermore, if $g(\theta) = a_0 =$ constant, $\psi(\alpha)$ becomes

$$\psi(\alpha) = a_0(\alpha-\alpha_2)^{-1/2}(\alpha_1-\alpha)^{-1/2}\left(\frac{\alpha_1+\alpha_2}{2} - \alpha + b_0\right),$$

$$b_0 = \frac{1+\alpha_2}{K(k)} \prod \left(\frac{\alpha_2 - \alpha_1}{1+\alpha_1}, k \right) - \frac{2+\alpha_1+\alpha_2}{2}, \qquad k^2 = \frac{2(\alpha_1-\alpha_2)}{(1-\alpha_2)(1+\alpha_1)},$$

$$\prod(\sigma, k) = \int_0^1 \frac{dx}{(1+\sigma x^2)(1-x^2)^{1/2}(1-k^2x^2)^{1/2}}, \tag{4.35}$$

where $K(k)$ is the complete elliptic integral of the first kind.

Consider now the general case (4.7), i.e.

$$\frac{1}{2\pi} \int_{L_1} \phi(t) \cot \frac{t-\theta}{2}\, dt = g_0(\theta), \qquad \theta \in L_1, \tag{4.36}$$

where $g_0(\theta)$ is the right-hand side of (4.7) and $L_1 = \sum_1^N L_{1k}$ has no symmetry. Let the origin be selected in such a way that $\theta = \pi$ is not on L_1. Defining

$$\tan\frac{t}{2} = s, \qquad \tan\frac{\theta}{2} = p, \qquad \phi(t) = \psi(s), \qquad g_0(\theta) = G(p),$$
$$t \in L_1 \to s \in \Gamma_1, \qquad \Gamma_1 = \sum_1^N \Gamma_{1k}, \tag{4.37}$$

equation (4.36) may easily be expressed as

$$\frac{1}{\pi} \int_{\Gamma_1} \psi(s) \frac{1+ps}{s-p} \frac{ds}{1+s^2} = G(p), \qquad p \in \Gamma_1, \tag{4.38}$$

or

$$\frac{1}{\pi} \int_{\Gamma_1} \frac{\psi(s)}{s-p}\, ds = G(p) + K, \qquad p \in \Gamma_1, \tag{4.39}$$

$$K = \frac{1}{\pi} \int_{\Gamma_1} \frac{s\psi(s)}{1+s^2}\, ds. \tag{4.40}$$

The singular integral equation (4.39) may be solved in a straightforward manner with (4.40) accounting for the additional constant K. Again, the solution of (4.39) will contain N arbitrary constants which may be determined from (4.17) and (4.20) gives the constant A_0.

4.2 An Example on Triple Series Equations

Consider now the somewhat more general mixed boundary-value problem for the unit circle defined by (2.10) and formulated by the triple series equations (2.11). In this problem let the auxiliary function be†

† See the general remarks and the broad guidelines at the end of Section 4 regarding the selection of the auxiliary (i.e. the new unknown) function.

$$\phi(\theta) = \frac{\partial}{\partial r} u(1, \theta), \qquad \theta \in L_1 + L_2 + L_3 = L = (0, 2\pi). \tag{4.41}$$

From (2.5) it then follows that

$$\begin{aligned} nA_n &= \frac{1}{\pi}\int_L \phi(t) \cos nt \, dt, \\ nB_n &= \frac{1}{\pi}\int_L \phi(t) \sin nt \, dt, \end{aligned} \tag{4.42}$$

where $\phi(t) = f_1(t)$ is known on L_1. After differentiating (2.11b) and, again for analytical expediency expressing them in limiting form, we write the remaining boundary conditions (2.11b and c) as

$$\lim_{r \to 1-0} \sum_1^\infty nr^n(-A_n \sin n\theta + B_n \cos n\theta) = f_2'(\theta), \qquad \theta \in L_2$$

$$\lim_{r \to 1-0} \left\{ h_1 A_0 + h_1 \sum_1^\infty r^n (A_n \cos n\theta + B_n \sin n\theta) \right. \tag{4.43}$$

$$\left. + h_2 \sum_1^\infty nr^{n-1}(A_n \cos n\theta + B_n \sin n\theta) \right\} = f_3(\theta), \qquad \theta \in L_3.$$

Substituting now from (4.42) into (4.43), using (4.6) and [17]

$$\sum_1^\infty \frac{1}{n} \cos nz = -\log\left(2\left|\sin\frac{z}{2}\right|\right), \tag{4.44}$$

we obtain

$$\frac{1}{2\pi}\int_L \phi(t) \cot\frac{t-\theta}{2} dt = f_2'(\theta), \qquad \theta \in L_2,$$

$$-\frac{h_1}{\pi}\int_L \phi(t) \log\left(2\left|\sin\frac{t-\theta}{2}\right|\right) dt + h_2\phi(\theta) = f_3(\theta) - h_1 A_0, \qquad \theta \in L_3. \tag{4.45a, b}$$

Just a superficial observation would indicate that the two integral equations in (4.45) are of entirely different character. Hence, near and at the end points of L_2 and L_3 one would expect the function $\phi(t)$ to behave quite differently. For closer examination let us assume that

$$\phi(t) = \phi_1(t), \qquad t \in L_2, \qquad \phi(t) = \phi_2(t), \qquad t \in L_3. \tag{4.46}$$

Equations (4.45) may then be expressed as

$$\frac{1}{2\pi}\int_{L_2}\phi_1(t)\cot\frac{t-\theta}{2}\,dt+\frac{1}{2\pi}\int_{L_3}\phi_2(t)\cot\frac{t-\theta}{2}\,dt$$
$$=f_2'(\theta)-\frac{1}{2\pi}\int_{L_1}f_1(t)\cot\frac{t-\theta}{2}\,dt=g_1(\theta),\qquad \theta\in L_2,$$

$$-\frac{h_1}{\pi}\int_{L_2}\phi_1(t)\log\left(2\left|\sin\frac{t-\theta}{2}\right|\right)dt-\frac{h_1}{\pi}\int_{L_3}\phi_2(t)\log\left(2\left|\sin\frac{t-\theta}{2}\right|\right)$$
$$+h_2\phi_2(\theta)=f_3(\theta)-h_1A_0+\frac{h_1}{\pi}\int_{L_1}f_1(t)\log\left(2\left|\sin\frac{t-\theta}{2}\right|\right)dt=g_2(\theta),$$
$$0\in L_3.\qquad (4.47\text{a, b})$$

Separating the dominant parts of the kernels it is seen that (4.47) is of the following general form:

$$\frac{1}{\pi}\int_{L_2}\frac{\phi_1(t)}{t-\theta}\,dt=g_1(\theta)-\int_{L_2}k_{11}(\theta,t)\phi_1(t)\,dt-\int_{L_3}k_{12}(\theta,t)\phi_2(t)\,dt,\qquad \theta\in L_2,$$

$$h_2\phi_2(\theta)-\frac{h_1}{\pi}\int_{L_3}\phi_2(t)\log|t-\theta|\,dt=g_2(\theta)-\int_{L_2}k_{21}(\theta,t)\phi_1(t)\,dt$$
$$-\int_{L_3}k_{22}(\theta,t)\phi_2(t)\,dt,\qquad \theta\in L_3,\qquad (4.48\text{a, b})$$

where the kernels $k_{ij}(\theta,t),(i,j=1,2)$ are bounded in their respective domains.

It is clear that (4.48a) is a typical singular integral equation of the general form (4.8), it has a fundamental function of the form (4.15), and if $L_2=\sum_1^N L_{2k}, L_{2k}=(a_k,b_k),\ a_k<b_k<a_{k+1}$, its solution will contain N arbitrary constants. On the other hand, (4.48b) is an integral equation of the second kind with a weakly singular but square integrable kernel. Hence, its solution is bounded everywhere on L_3, including the end points (as they are approached from L_3) and is uniquely determined without any reference to any additional conditions. The coupling of the two integral equations is through Fredholm kernels. Therefore the basic singular behavior of the solution will be unaffected† by coupling. From (2.10) it is clear that if h_1 or h_2 is zero the problem reduces to that considered in the previous section. This can also be seen from (4.47). For $h_1=0$ this is quite clear. For $h_2=0$ differentiating (4.47b) and observing that

$$-\frac{d}{d\theta}\log\left(2\left|\sin\frac{t-\theta}{2}\right|\right)=\tfrac{1}{2}\cot\frac{t-\theta}{2}\qquad (4.49)$$

† As will be pointed out later in this article, if the coupling is through generalized Cauchy kernels, then the singular behavior of the solution will be affected.

we see that (4.47) is reduced to a simple singular integral equation defined on $L_2 + L_3$. It is worthwhile to reemphasize that an integral equation of the second kind with a logarithmic kernel is basically a Fredholm integral equation and has a bounded solution. However, if the integral equation is of the first kind and has a logarithmic kernel, then it is equivalent to a singular integral equation with a simple Cauchy-type kernel. Also, the dominant equation

$$A\phi(x) + B\int_L \phi(t)\log|t-x|\,dt = f(x), \qquad x\in L, \tag{4.50}$$

has apparently no closed-form solution. Even though at the end points of L the solution of (4.50) is bounded, it may easily be shown that its derivative has a logarithmic singularity. We see this by differentiating and modifying (4.50) as follows:

$$\begin{aligned}\phi'(x) &= \frac{B}{A}\int_L \frac{\phi(t)}{t-x}\,dt + \frac{f'(x)}{A}\\ &= \frac{B}{A}\int_L \frac{\phi(t)-\phi(x)}{t-x}\,dt + \frac{B}{A}\phi(x)\int_L \frac{dt}{t-x} + \frac{f'(x)}{A}, \qquad x\in L,\end{aligned} \tag{4.51}$$

which, in the neighborhood of a typical end point $x \leq c$, becomes

$$\phi'(x) = \frac{B}{A}\phi(c)\log|c-x| + G(x), \qquad (x\to c), \tag{4.52}$$

where $G(c)$ is bounded.

If $L_2 = \sum_1^N L_{2k}$, to complete the solution of the problem the arbitrary constants $c_0, c_1, \ldots, c_{N-1}$ arising from the solution of the singular integral equation (4.47a) and A_0 must be determined. Since the boundary condition $u(1,\theta) = f_2(\theta)$, $\theta\in L_2$, was satisfied only in differentiated form and

$$u(1,\theta) = A_0 - \frac{1}{\pi}\int_L \phi(t)\log\left(2\left|\sin\frac{t-\theta}{2}\right|\right)dt, \tag{4.53}$$

we observe that $\phi(t)$ must satisfy the following single-valuedness conditions:

$$A_0 - \frac{1}{\pi}\int_L \phi(t)\log\left(2\left|\sin\frac{t-\theta_k}{2}\right|\right)dt = f_2(\theta_k), \qquad \theta_k\in L_k, \qquad k=1,\ldots,N, \tag{4.54}$$

where θ_k is any convenient point on L_{2k}. Thus, Eqs. (4.54) with the flux equilibrium condition

$$\int_{L_1} f_1(t)\,dt + \int_{L_2} \phi_1(t)\,dt + \int_{L_3} \phi_2(t)\,dt = 0 \qquad (4.55)$$

provide $N + 1$ algebraic equations to determine $c_0, \ldots, c_{N-1}$, and A_0.

4.3 Reduction of Multiple Integral Equations

To demonstrate the technique of reducing a system of multiple integral equations to that of singular integral equations consider the somewhat general mixed boundary value problem in potential theory which is defined by (2.21) for the half-plane $y > 0$. The solution is expressed by (2.17) in terms of the unknown function $A(\alpha)$ which is to be determined from the multiple integral equations (2.22). In this problem let the normal flux

$$\frac{\partial}{\partial y} u(x,0) = \phi(x) = \int_{-\infty}^{\infty} -|\alpha| A(\alpha)\, e^{-i\alpha x}\, d\alpha, \qquad -\infty < x < \infty, \qquad (4.56)$$

be selected as the new unknown function. If $\phi(x)$ is known, the Fourier inversion

$$-|\alpha| A(\alpha) = \frac{1}{2\pi} \int_{-\infty}^{\infty} \phi(t)\, e^{i\alpha t}\, dt \qquad (4.57)$$

with (2.17) would give the complete solution. From (2.21) it is seen that $\phi(x) = f_1(x)$ is known on L_1 and is unknown on L_2 and L_3. Again for reasons of uniform convergence writing the boundary conditions (2.21b) and (2.21c) in limit form, differentiating (2.21b) and using (4.57) we find

$$\lim_{y \to +0} \left[\frac{i}{2\pi} \int_0^{\infty} e^{-\alpha(y+ix)}\, d\alpha \int_{-\infty}^{\infty} \phi(t)\, e^{i\alpha t}\, dt \right.$$
$$\left. - \frac{i}{2\pi} \int_{-\infty}^{0} e^{\alpha(y-ix)}\, d\alpha \int_{-\infty}^{\infty} \phi(t)\, e^{i\alpha t}\, dt \right] = f_2'(x), \qquad x \in L_2,$$

$$(4.58\text{a, b})$$

$$\frac{1}{2\pi} h_1 \lim_{y \to +0} \left[- \int_0^{\infty} e^{-\alpha(y+ix)} \frac{d\alpha}{\alpha} \int_{-\infty}^{\infty} \phi(t)\, e^{i\alpha t}\, dt \right.$$
$$\left. + \int_{-\infty}^{0} e^{\alpha(y-ix)} \frac{d\alpha}{\alpha} \int_{-\infty}^{\infty} \phi(t)\, e^{i\alpha t}\, dt \right] + h_2 \phi(x) = f_3(x), \qquad x \in L_3.$$

Changing the order of integrations and evaluating the inner integrals, we reduce (4.58a) to

$$\frac{i}{2\pi}\lim_{y\to+0}\int_{-\infty}^{\infty}\phi(t)\left[\frac{1}{y-i(t-x)}-\frac{1}{y+i(t-x)}\right]dt=-\frac{1}{\pi}\int_{-\infty}^{\infty}\frac{\phi(t)}{t-x}dt=f_2'(x),$$

$$x\in L_2. \qquad (4.59)$$

Similarly, the first term on the left-hand side of (4.58b) may be expressed as

$$-\frac{h_1}{\pi}\lim_{y\to+0}\int_{-\infty}^{\infty}\phi(t)\,dt\int_0^{\infty}\frac{e^{-\alpha y}}{\alpha}\cos\alpha(t-x)\,d\alpha$$
$$=-\frac{h_1}{\pi}\lim_{y\to+0}\int_{-\infty}^{\infty}\phi(t)\,dt\left\{\int_0^{\infty}\frac{e^{-\alpha y}}{\alpha}d\alpha+\log y\right. \qquad (4.60)$$
$$\left.-\log\left[y^2+(t-x)^2\right]^{1/2}\right\}=\frac{h_1}{\pi}\int_{-\infty}^{\infty}\phi(t)\log|t-x|\,dt,$$

where the following condition of flux equilibrium is used to eliminate the divergent terms:†

$$\int_{-\infty}^{\infty}\phi(t)\,dt=0. \qquad (4.61)$$

Now observing that $L_1+L_2+L_3=(-\infty,\infty)$, $\phi(t)=f_1(t)$, $t\in L_1$, and defining

$$\phi(t)=\phi_1(t),\qquad t\in L_2,\qquad \phi(t)=\phi_2(t),\qquad t\in L_3, \qquad (4.62)$$

we express the integral equations (4.58) as

$$\frac{1}{\pi}\int_{L_2}\frac{\phi_1(t)}{t-x}dt+\frac{1}{\pi}\int_{L_3}\frac{\phi_2(t)}{t-x}dt=-f_2'(x)-\frac{1}{\pi}\int_{L_1}\frac{f_1(t)}{t-x}dt,\qquad x\in L_2,$$

$$h_2\phi_2(x)+\frac{h_1}{\pi}\int_{L_3}\phi_2(t)\log|t-x|\,dt+\frac{h_1}{\pi}\int_{L_2}\phi_1(t)\log|t-x|\,dt$$
$$=f_3(x)-\frac{h_1}{\pi}\int_{L_1}f_1(t)\log|t-x|\,dt,\qquad x\in L_3. \qquad (4.63a,b)$$

In structure the system of Eqs. (4.63) is identical to (4.47) or (4.48). If L_2 is finite with $L_2=\sum_1^N L_{2j}, L_{2j}=(a_j,b_j)$, the solution of (4.63a) will again contain N arbitrary constants which may be obtained from (4.61) and the following $N-1$ single-valuedness conditions:

$$\frac{1}{\pi}\int_{-\infty}^{\infty}\phi(t)\log\left|\frac{t-a_{j+1}}{t-b_j}\right|dt=f_2(a_{j+1})-f_2(b_j),\qquad j=1,\ldots,N-1. \qquad (4.64)$$

† Note that if (4.61) is not satisfied and if u is zero at infinity (as assumed in the present problem) then $u(x,0)$ will not be bounded and will tend to infinity as $\log y$, $y\to 0$.

In the type of problems under consideration the case of non-zero resultant flux may be particularly important. From (4.60) it is clear that if (4.61) is not satisfied the derivation leading to (4.63) is not valid. Let us now assume that L_2 is finite† and

$$\int_{-\infty}^{\infty} \phi(t)\,dt = Q \tag{4.65}$$

(since $u(x,0) = f_2(x)$, $x \in L_2$ is assumed to be bounded) $u(x,y) \sim \log y$ as $y \to \infty$. However, the components of the flux vector may still be expressed in terms of an unknown function $A(\alpha)$ as follows:

$$\frac{\partial}{\partial x}u(x,y) = -\int_{-\infty}^{\infty} i\alpha A(\alpha)\,e^{-y|\alpha| - i\alpha x}\,d\alpha,$$

$$\frac{\partial}{\partial y}u(x,y) = -\int_{-\infty}^{\infty} |\alpha|\,A(\alpha)\,e^{-y|\alpha| - i\alpha x}\,d\alpha. \tag{4.66a, b}$$

Again, defining the auxiliary function $\phi(x)$ by (4.56) and substituting from (4.57) into (4.66a) we obtain

$$\begin{aligned}\frac{\partial}{\partial x}u(x,0) &= \frac{i}{2\pi}\lim_{y\to+0}\left[\int_0^{\infty} e^{-\alpha(y+ix)}\,d\alpha \int_{-\infty}^{\infty}\phi(t)\,e^{i\alpha t}\,dt\right.\\ &\qquad\left. - \int_{-\infty}^{0} e^{\alpha(y-ix)}\,d\alpha \int_{-\infty}^{\infty}\phi(t)\,e^{i\alpha t}\,dt\right]\\ &= \frac{i}{2\pi}\lim_{y\to+0}\int_{-\infty}^{\infty}\phi(t)\frac{2i(t-x)}{y^2+(t-x)^2}\,dt\\ &= -\frac{1}{\pi}\int_{-\infty}^{\infty}\phi(t)\frac{dx}{t-x}, \qquad (-\infty < x < \infty),\end{aligned} \tag{4.67}$$

from which it follows that

$$u(x,0) = f_2(x_0) + \frac{1}{\pi}\int_{-\infty}^{\infty}\phi(t)\log\left|\frac{t-x}{t-x_0}\right|dt, \qquad (-\infty < x < \infty), \tag{4.68}$$

where x_0 is any point on L_2. Thus, the integral equation (4.63a) is still valid and, with the definitions (4.62) and using (4.68), (4.63b) will have to be replaced by

† If L_2 is infinite, by a proper superposition the problem may always be reduced to that in which (4.61) is satisfied.

$$h_2\phi_2(x) + \frac{h_1}{\pi}\int_{L_3} \phi_2(t)\log\left|\frac{t-x}{t-x_0}\right| dt + \frac{h_1}{\pi}\int_{L_2} \phi_1(t)\log\left|\frac{t-x}{t-x_0}\right| dt$$

$$= f_3(x) - h_1 f_2(x_0) - \frac{h_1}{\pi}\int_{L_1} f_1(t)\log\left|\frac{t-x}{t-x_0}\right| dt, \qquad x\in L_3; \qquad (4.69)$$

the N integration constants arising from the solution of (4.63a) are determined from (4.64) (which follows now from (4.68) and from the fact that $u(x, 0) = f_2(x)$, $x\in L_2$) and (4.65). Note that under the stated conditions of the problem (4.63b) and (4.69) are always nonhomogeneous. Also note that if $L_2 = 0$, the problem is reduced to

$$h_2\phi_2(x) + \frac{h_1}{\pi}\int_{L_3} \phi_2(t)\log|t-x|\, dt = f(x), \qquad x\in L_3, \qquad (4.70)$$

where $f(x)$ is a known function. Equation (4.70) has no closed-form solution. In this case it should be noted that $u(x_0, 0) = f_2(x_0)$ which appears in (4.69) is an unknown constant and is determined by the flux equilibrium condition (4.65) where x_0 now is an arbitrary point on the real axis. On the other hand, if $L_3 = 0$ and $L_2 \neq 0$, the problem is formulated by (4.63a) which may easily be solved in closed form.

As pointed out previously the solution of the system of singular integral equations (4.63) will be of the following form:

$$\phi_1(x) = w(x)F_1(x), \qquad w(x) = \prod_1^N [(x-a_k)(b_k-x)]^{-1/2}, \qquad x\in L_2,$$
$$\phi_2(x) = F_2(x), \qquad x\in L_3, \qquad (4.71)$$

where F_1 and F_2 are bounded functions and at the end points of L_3 ϕ_2 is bounded and has a behavior similar to (4.52) whereas at the end points of L_2 ϕ_1 is singular.

In order to have a better understanding of the flux distribution $\phi(x)$ on the boundary it may be worthwhile to relate the foregoing results to a simple physical problem. Consider, for example, the antiplane shear problem for a symmetrically loaded infinite medium shown in Fig. 2. Here $u(x, y)$ is the z-component of the displacement vector, $\phi(x) = \sigma_{yz}(x, 0)/\mu$ represents the traction at $y = 0$, and μQ is the resultant force in the z-direction, μ being the shear modulus of the medium. The part of the x-axis $L_1 = \Sigma L_{1j}$ corresponds to a series of cracks on which the surface traction $\phi(x) = f_1(x) = \sigma_{yz}/\mu$ is specified. L_2 is clearly the uncut portion of the real axis on which (in this problem) $u(x, 0) = f_2(x) = 0$. Physically the part L_3 on which

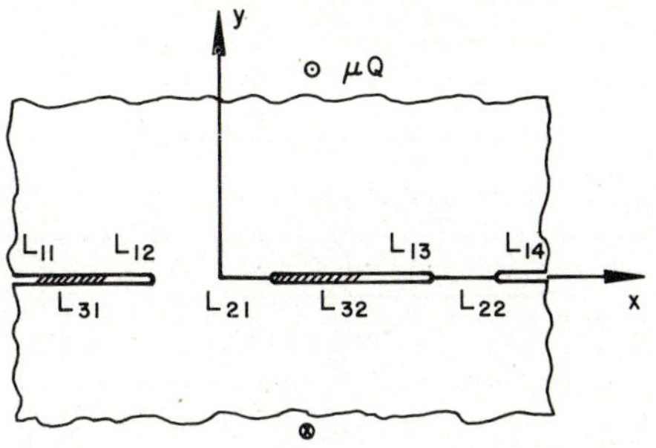

Fig. 2 Plane with collinear cracks under antiplane shear loading.

$$h_1 u(x,0) + h_2 \frac{\partial}{\partial y} u(x,0) = f_3(x) \tag{4.72}$$

corresponds to a series of cracks the surfaces of which are joined through an elastic adhesive layer. If the thickness of the adhesive is sufficiently small it may be represented by a one-dimensional spring and may be modeled by (4.72) with

$$h_1 = \mu_a/h, \qquad h_2 = -\mu, \qquad f_3(x) = 0, \tag{4.73}$$

where μ_a and h are, respectively, the shear modulus and the thickness of the adhesive. Thus, the spring force $\mu_a u/h$ will act as a traction on the crack surface along L_3. Physically, then, the plane is cut along $L_1 + L_3$, which means that the stress $\sigma_{yz}(x,0) = \mu\phi(x)$ must have the expected square root singularity at the end points of L_2 (as x approaches these points from L_2). Note that whether the end point belongs to the intersection of L_2 and L_1 or L_2 and L_3, this singular behavior will remain unchanged. On the other hand, in the adhesive layer (i.e. on L_3) the shear stress will be bounded everywhere, including the end points. Further applications of the technique to the mechanics of bonded joints may be found in [18] and [19].

4.4 Reduction of Multiple Series–Multiple Integral Equations

In some mechanics problems, because of the geometry of the domain, the separation of variables technique may lead to a formulation in which some of the unknown functions are expressed as series of eigenfunctions having a set of undetermined coefficients and some as inversion integrals involving certain undetermined functions. Substitution of these expressions

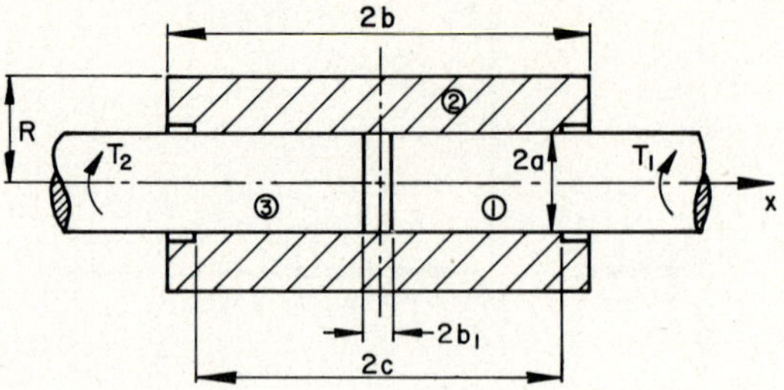

Fig. 3 The torsion problems for two shafts coupled through an elastic sleeve.

into the mixed boundary conditions invariably gives rise to a system of equations involving both multiple series and multiple integral equations. Once the auxiliary functions are properly selected, reduction of this type of problem to a system of singular integral equations is also rather straightforward. In this section we again select a highly representative and a relatively simple example to demonstrate the technique.

Consider the "load transfer" problem shown in Fig. 3. The figure describes either a coupling in which the torque is transmitted from the shaft 1 to the shaft 3 through the sleeve 2, or a gear or pulley in which the external torque acting on the sleeve 2 is not zero. In this problem it is assumed that the shafts are made of the same material with shear modulus μ_1 and the sleeve has the shear modulus μ_2. It is further assumed that (in addition to axial symmetry) $x = 0$ is a plane of symmetry (or antisymmetry) with respect to the external loads and the geometry of the problem. Thus, in both media the circumferential component of the displacement u_i, $(i = 1, 2)$ is the only nonzero displacement which satisfies the following differential equation:

$$\frac{\partial^2 u_i}{\partial r^2} + \frac{1}{r}\frac{\partial u_i}{\partial r} - \frac{u_i}{r} + \frac{\partial^2 u_i}{\partial x^2} = 0, \qquad (r < a : i = 1, \quad a < r < R : i = 2), \tag{4.74}$$

and the nonvanishing stress components are given by

$$\sigma_{ir\theta} = \mu_i\left(\frac{\partial u_i}{\partial r} - \frac{u_i}{r}\right), \qquad \sigma_{i\theta x} = \mu_i \frac{\partial u_i}{\partial x}, \qquad (i = 1, 2). \tag{4.75}$$

For the sake of generality at this point it is assumed that the end clearance $2b_1$ is not zero and $2c < 2b$ (Fig. 3). Because of symmetry it is sufficient to consider one-half $(x > 0)$ of the composite medium only.

In this problem the solution of (4.74) must be found subject to the following conditions:

$$\frac{\partial}{\partial x}[u_2(a+0,x)-u_1(a-0,x)]=f(x), \qquad b_1<x<c, \quad \text{(a)}$$

$$\sigma_{1r\theta}(a,x)=\sigma_{2r\theta}(a,x), \qquad b_1<x<c, \quad \text{(b)}$$

$$\sigma_{1r\theta}(a,x)=0, \qquad x>c, \quad \text{(c)}$$

$$\sigma_{2r\theta}(a,x)=0, \qquad 0\leq x<b_1, \qquad c<x<b, \quad \text{(d)}$$

$$\sigma_{1\theta x}(r,b_1)=0, \qquad 0\leq r<a, \quad \text{(e)}$$

$$\sigma_{2\theta x}(r,b)=0, \qquad a<r<R, \quad \text{(f)}$$

$$\sigma_{2r\theta}(R,x)=\sigma_0(x), \qquad \text{or } u_2(R,x)=0, \qquad 0\leq x<b, \quad \text{(g)}$$

$$\left.\begin{array}{ll}\dfrac{\partial}{\partial x}u_2(r,0)=0 & \text{(symmetric case)}\\ u_2(r,0)=0 & \text{(antisymmetric case)}\end{array}\right\}, \qquad a<r<R, \quad \text{(h)}$$

$$\int_{b_1}^{c}2\pi a^2p(x)\,dx=T, \qquad \text{(i)} \quad (4.76)$$

where the functions $f(x)$, $\sigma_0(x)$ and the torque T are known and $p(x)$ is the contact stress. Referring to Fig. 3, we see that the following symmetry conditions must be satisfied

For the symmetric problem:

$$\begin{aligned}u_i(r,x)&=u_i(r,-x), \qquad \sigma_{ir\theta}(r,x)=\sigma_{ir\theta}(r,-x), \qquad i=1,2,\\ f(x)&=-f(-x), \qquad T_1=T_2, \qquad \sigma_0(x)=\sigma_0(-x);\end{aligned} \qquad (4.77)$$

For the antisymmetric problem:

$$\begin{aligned}u_i(r,x)&=-u_i(r,-x), \qquad \sigma_{ir\theta}(r,x)=-\sigma_{ir\theta}(r,-x), \qquad i=1,2,\\ f(x)&=f(-x), \qquad T_1=-T_2, \qquad \sigma_0(x)=-\sigma_0(-x).\end{aligned} \qquad (4.78)$$

Considering (4.77) and (4.78), we express the solution of (4.74) which satisfies (4.76e, f, and h) as

$$u_1(r,x)=\frac{2}{\pi}\int_0^\infty A(\alpha)I_1(\alpha r)\cos\alpha(x-b_1)\,d\alpha, \qquad x>b_1, \qquad 0\leq r<a,$$

$$u_2(r,x)=\sum_1^\infty[B_nK_1(\alpha_n r)+C_nI_1(\alpha_n r)]\begin{Bmatrix}\cos\alpha_n x\\ \sin\alpha_n x\end{Bmatrix}, \qquad 0\leq x<b, \qquad a<r<R,$$

(4.79a, b)

where, from (4.75) and (4.76f), it is seen that

$$\alpha_n = \pi n/b \quad \text{for symmetric problem, and}$$
$$\alpha_n = (2n-1)\pi/2b \quad \text{for antisymmetric problem.} \tag{4.80a, b}$$

In (4.79b) the upper and lower terms in the series stand for symmetric and antisymmetric problems respectively. Substituting now from (4.79) into the mixed boundary conditions (4.76) (a–d) we find

$$\lim_{r\to a+0} \sum_1^{\infty} \alpha_n[B_n K_1(\alpha_n r) + C_n I_1(\alpha_n r)]\begin{Bmatrix} -\sin \alpha_n x \\ \cos \alpha_n x \end{Bmatrix}$$
$$+ \lim_{r\to a-0} \frac{2}{\pi}\int_0^{\infty} \alpha A(\alpha) I_1(\alpha r) \sin \alpha(x-b_1)\, d\alpha = f(x), \quad b_1 < x < c, \quad \text{(a)}$$

$$\mu_2 \sum_1^{\infty} [-B_n K_2(\alpha_n a) + C_n I_2(\alpha_n a)]\alpha_n \begin{Bmatrix} \cos \alpha_n x \\ \sin \alpha_n x \end{Bmatrix}$$
$$- \frac{2\mu_1}{\pi}\int_0^{\infty} A(\alpha)\alpha I_2(\alpha a) \cos \alpha(x-b_1)\, d\alpha = 0, \quad b_1 < x < c, \quad \text{(b)}$$

$$\frac{2\mu_1}{\pi}\int_0^{\infty} A(\alpha)\alpha I_2(\alpha a) \cos \alpha(x-b_1)\, d\alpha = 0, \quad x > c, \quad \text{(c)}$$

$$\mu_2 \sum_1^{\infty} [-B_n K_2(\alpha_n a) + C_n I_2(\alpha_n a)]\alpha_n \begin{Bmatrix} \cos \alpha_n x \\ \sin \alpha_n x \end{Bmatrix} = 0,$$
$$0 \leq x < b_1, \quad c < x < b. \quad \text{(d)} \tag{4.81}$$

Note that one set of constants B_n or C_n may be eliminated by using the condition (4.76g). For example,†

$$B_n K_1(\alpha_n R) + C_n I_1(\alpha_n R) = 0 \quad \text{for} \quad u_2(R, x) = 0,$$
$$-B_n K_2(\alpha_n R) + C_n I_2(\alpha_n R) = \frac{2}{b\mu_2}\int_0^b \sigma_0(x)\begin{Bmatrix} \cos \alpha_n x \\ \sin \alpha_n x \end{Bmatrix} dx, \tag{4.82a, b}$$
$$\text{for} \quad \sigma_{2r\theta}(R, x) = \sigma_0(x).$$

With (4.82), (4.81) provides a system of dual series–dual integral equations to determine the set of unknown constants B_n (or C_n) and the unknown function $A(\alpha)$.

In this problem, the contact stress

$$\sigma_{1r\theta}(a, x) = \sigma_{2r\theta}(a, x) = p(x) \tag{4.83}$$

† In (4.82b) $\sigma_0(x) = 0$ being the practical case of coupling.

suggests itself as being the most appropriate auxiliary function. Thus, from (4.82) and the expressions

$$\sigma_{1r\theta}(a, x) = \frac{2\mu_1}{\pi}\int_0^\infty A(\alpha)\alpha I_2(\alpha a)\cos\alpha(x - b_1)\,d\alpha = p(x), \qquad (b_1 < x < \infty),$$

$$\sigma_{2r\theta}(a, x) = \mu_2\sum_1^\infty[-B_nK_2(\alpha_na) + C_nI_2(\alpha_na)]\alpha_n\begin{Bmatrix}\cos\alpha_nx\\ \sin\alpha_nx\end{Bmatrix} = p(x),$$

$$(0 \le x < b), \qquad (4.84a, b)$$

we evaluate $A(\alpha)$, B_n, and C_n in terms of $p(x)$ and substitute into (4.81a) to obtain

$$-\lim_{r\to a+0}\frac{1}{b\mu_2}\int_{b_1}^{c} p(t)\,dt\sum_1^\infty L_n(r)[\sin\alpha_n(t - x) \mp \sin\alpha_n(t + x)]$$

$$-\lim_{r\to a-0}\frac{1}{\pi\mu_1}\int_{b_1}^{c} p(t)\,dt\int_0^\infty\frac{I_1(\alpha r)}{I_2(\alpha a)}[\sin\alpha(t - x) - \sin\alpha(t + x - 2b_1)]\,d\alpha = f(x),$$

$$b_1 < x < c, \qquad (4.85)$$

where the upper and lower signs in the series refer to the symmetric and the antisymmetric cases, respectively, and

$$L_n(r) = \frac{K_1(r\alpha_n)I_1(R\alpha_n) - I_1(r\alpha_n)K_1(R\alpha_n)}{K_2(a\alpha_n)I_1(R\alpha_n) + I_2(a\alpha_n)K_1(R\alpha_n)} \quad \text{for} \quad u_2(R, x) = 0,$$

$$L_n(r) = \frac{K_1(r\alpha_n)I_2(R\alpha_n) + I_1(r\alpha_n)K_2(R\alpha_n)}{K_2(a\alpha_n)I_2(R\alpha_n) - I_2(a\alpha_n)K_2(R\alpha_n)} \quad \text{for} \quad \sigma_0(x) = 0. \qquad (4.86a, b)$$

In deriving (4.85) the conditions (4.81b–d) has been used.

We now observe that for $t = x$ the series and the integral giving the kernels in (4.85) are divergent. These divergent parts may be studied and separated by considering the asymptotic behavior of the terms in the series and of the integrand. In (4.86) note that since $R > a$, letting $r = a + \varepsilon$, for large values of α_n we obtain

$$L_n(r) \cong \frac{K_1(a\alpha_n + \varepsilon\alpha_n)}{K_2(a\alpha_n)} \cong e^{-\varepsilon\alpha_n}, \qquad (4.87)$$

where ε is a small positive number and $L_n(r) \to 1$ as $\varepsilon \to 0$ and $\alpha_n \to \infty$. Similarly, note that for large values of α and for a small $\varepsilon = a - r$ we have

$$\frac{I_1(\alpha r)}{I_2(\alpha a)} = \frac{I_1(\alpha a - \alpha\varepsilon)}{I_2(\alpha a)} \cong e^{-\varepsilon\alpha}. \qquad (4.88)$$

Thus making use of the following results:

$$\lim_{\beta\to 0}\int_{b_1}^{c} p(t)\,dt \sum_{1}^{\infty} e^{-\beta n}\sin\lambda n = \lim_{\beta\to 0}\int_{b_1}^{c} p(t)\,dt \frac{\sin\lambda}{2(\cosh\beta-\cos\lambda)}$$
$$= \frac{1}{2}\int_{b_1}^{c} p(t)\cot\frac{\lambda}{2}\,dt,$$

$$\lim_{\varepsilon\to 0}\int_{b_1}^{c} p(t)\,dt\int_{0}^{\infty} e^{-\varepsilon\alpha}\sin\lambda\alpha\,d\alpha = \lim_{\varepsilon\to 0}\int_{b_1}^{c} p(t)\,dt\,\frac{\lambda}{\lambda^2+\varepsilon^2}$$
$$= \int_{b_1}^{c}\frac{p(t)}{\lambda}\,dt, \qquad (4.89\text{a, b})$$

and adding and subtracting the asymptotic values under the summation and integral signs in (4.85) we obtain

$$\frac{1}{\pi}\int_{b_1}^{c} p(t)\,dt\left[\frac{1}{t-x}-\frac{1}{t+x-2b_1}+\frac{\pi\mu_1}{2b\mu_2}\right.$$
$$\left.\times\left\{\begin{matrix}\left[\cot\dfrac{\pi(t-x)}{2b}-\cot\dfrac{\pi(t+x)}{2b}\right]\\[2ex] \left[\operatorname{cosec}\dfrac{\pi(t-x)}{2b}+\operatorname{cosec}\dfrac{\pi(t+x)}{2b}\right]\end{matrix}\right\}\right]$$
$$+\frac{1}{\pi}\int_{b_1}^{c} k(x,t)p(t)\,dt = -\mu_1 f(x), \qquad b_1<x<c, \qquad (4.90)$$

$$k(x,t) = \int_{0}^{\infty}\left(\frac{I_1(\alpha a)}{I_2(\alpha a)}-1\right)[\sin\alpha(t-x)-\sin\alpha(t+x-2b)]\,d\alpha$$
$$+\frac{\mu_1\pi}{b_1\mu_2}\sum^{\infty}[L_n(a)-1][\sin\alpha_n(t-x)\mp\sin\alpha_n(t+x)], \qquad (4.91)$$

where the upper and lower signs and kernels again correspond to symmetric and antisymmetric problems, respectively, the kernel $k(x,t)$ given by (4.91) is bounded and continuous for all x and t in the closed interval $[b_1, c]$, and in (4.91), because of uniform convergence, the limit has been put under the integral and the summation signs. The integral equation (4.90) must be solved under (4.76i), the only condition in (4.76) which has not been satisfied.

At a first glance it may appear that (4.90) is a simple singular integral equation of the type discussed in the previous sections. However, a closer examination would indicate that the term $1/(t + x - 2b_1)$ in the kernel of (4.90) is not bounded in the closed interval $[b_1, c]$ and becomes unbounded (as $1/x$) as x and t go to b_1 together. Hence, this term would be expected to influence the singular behavior of the solution at the end point b_1. Dominant kernels containing, in addition to the Cauchy kernel $1/(t - x)$, terms such as $1/(t + x - 2b_1)$ will be called *generalized Cauchy kernels*. The properties of the solution of singular integral equations with generalized Cauchy kernels will be discussed in Section 6.

4.5 Remarks on the Selection of Auxiliary Functions

In studying multiple series and multiple integral equations if the objective is their reduction to singular integral equations, the selection of the auxiliary function (i.e. the new unknown function) and the procedure followed in the reduction process appear to be quite straightforward. It may be worthwhile to note that in boundary-value problems in mechanics it is always possible to recognize pairs of "complementary functions" on the boundary having basically the same dimension. Such pairs are, for example, the normal and tangential derivatives of the potential, or the potential and the integral of the normal flux along the boundary in potential theory, and the surface tractions and the tangential derivatives of the displacements or the integrals of the tractions and the displacements in solid mechanics. If one considers the structure of the dominant part of a singular integral equation, namely

$$\frac{1}{\pi}\int_L \frac{\phi(t)}{t - x}\,dt = f(x), \qquad x \in L, \tag{4.92}$$

it is clear that (with proper normalizations) the unknown function $\phi(x)$ and the known function $f(x)$ on the boundary have the same physical dimension and in a correct formulation of the problem they invariably are the complementary pair on the boundary. Thus, if one does not pay any attention to the *dimensional consistency* in selecting the auxiliary function $\phi(x)$ at the beginning, the resulting integral equation may have a singular kernel with a singularity either weaker or stronger than the Cauchy singularity, $1/(t - x)$. For example, in the former case the kernel is the integral of $1/(t - x)$, i.e. $\log|t - x|$, and the integral equation may be reduced to the standard form by formally differentiating the both sides with respect to x (i.e. the tangential coordinate), indicating that the particular auxiliary function selected is the complement of $f'(x)$ rather than the input function $f(x)$. Analytically, this selection usually does not create any difficulty, since one may easily recover and isolate the logarithmic kernel by following the procedure described in Sections 4.1 to 4.4. However, if the selection is made in such a way that $\phi(x)$ is the complement of the integral of $f(x)$, then technically the dominant kernel is expected to be the x-derivative of $1/(t-x)$, i.e. $1/(t-x)^2$. For a Hölder-continuous $\phi(t)$ since the integral $\int \phi(t)\,dt/(t-x)^2$ does not exist, this formulation becomes meaningless, and besides it is not possible to recover the strong singularity $1/(t-x)^2$ through a normal procedure outlined in the preceding sections. Clearly, the correct thing to do in such a case is to integrate parts of the multiple series (or integral) equations so that a dimensionally consistent auxiliary function can be defined.

One should again emphasize the importance of writing the boundary conditions in limit form on that part of the boundary which will be the support of the resulting integral equations. Without this, one may not be able to change the order of integrations or integration and summation legitimately to evaluate the kernels. Even if this is done, with the limit under the integral or summation sign, the resulting infinite integrals or series giving the kernels are usually divergent or simply meaningless. Consider, for example, the integral equation (4.5) expressing the boundary condition on L_1. If it is not written in limit form the kernel becomes

$$K(\theta, t) = \sum_{1}^{\infty} \sin n(t-\theta) \tag{4.93}$$

which is not summable. The same thing may be said about the kernels arising from the reduction of multiple integral equations or multiple series–multiple integral equations (see Eqs. (4.58), (4.85), and (4.90)).

5 NUMERICAL SOLUTION OF SINGULAR INTEGRAL EQUATIONS OF THE FIRST KIND

In the previous sections it was shown that, unless the problem has convective boundary conditions,† the mixed boundary value problems in

† Convective boundary conditions generally reduce to integral equations of the second kind with a logarithmic dominant kernel which have bounded solutions and which may be treated as Fredholm-type equations (see Section 4.2, Eqs. (4.48)–(4.49) and Section 4.3, Eqs. (4.63) and (4.70)).

mechanics may invariably be reduced to a system of singular integral equations of the following general form:

$$\sum_1^N a_{ij}\phi_j(x) + \sum_1^N \int_L \left[\frac{1}{\pi}\frac{b_{ij}}{t-x} + k_{ij}^s(x,t) + k_{ij}^f(x,t)\right]\phi_j(t)\,dt = f_i(x),$$

$$i = 1, \ldots, N, \qquad x \in L_i, \qquad (5.1)$$

where $L_i = \sum_1^{m_i} L_{ij}$, the matrices (a_{ij}) and (b_{ij}) are nonsingular, the kernels $k_{ij}^s(x,t)$ consists of terms which become unbounded as x and t approach the end points of L_i and which, with the singular terms $b_{ij}/(t-x)$, constitute the generalized Cauchy kernels, $k_{ij}^f(x,t)$ are bounded Fredholm-type kernels, and $f_i(x)$ are known functions. For the singular equations with generalized Cauchy kernels there does not seem to be any general method of regularization. The singular behavior of the solution of these equations will be studied in Section 6 where a numerical technique for solving the integral equations will also be discussed. Also, the treatment of the singular integral equations of the second kind will be postponed until Section 7. Thus, in this section we will consider only the singular integral equations of the first kind with simple Cauchy-type singularities which represent by far the largest class of mixed boundary-value problems in mechanics. The method will be described for a single equation defined in the normalized interval $(-1, 1)$, namely,

$$\frac{1}{\pi}\int_{-1}^{1} \frac{\phi(t)}{t-x}\,dt + \int_{-1}^{1} k(x,t)\phi(t)\,dt = f(x), \qquad -1 < x < 1. \qquad (5.2)$$

From the development of the method, it will be clear that the extension of the method to a system of singular integral equations (with unknown functions $\phi_1(x), \ldots, \phi_n(x)$) defined in a single interval $a < x < b$ is quite straightforward. It is also easy to show that if each equation in the system is defined on a (different) union of arcs, the technique developed for (5.2) may easily be used for the numerical solution of this general system. For example, consider

$$\frac{1}{\pi}\int_L \frac{\phi(t)}{t-x}\,dt + \int_L k(x,t)\phi(t)\,dt = f(x), \qquad x \in L,$$

$$L = \sum_1^n L_i, \qquad L_i = (a_i, b_i), \qquad b_i < a_{i+1}. \qquad (5.3)$$

Defining the following new variables and functions,

$$t_i = \frac{2t}{b_i - a_i} - \frac{b_i + a_i}{b_i - a_i}, \qquad a_i < t < b_i, \qquad -1 < t_i < 1,$$
$$x_i = \frac{2x}{b_i - a_i} - \frac{b_i + a_i}{b_i - a_i}, \qquad a_i < x < b_i, \qquad -1 < x_i < 1,$$
$$\phi(t) = \phi_i(t_i), \qquad a_i < t < b_i, \qquad -1 < t_i < 1, \tag{5.4}$$
$$f(x) = f_i(x_i), \qquad a_i < x < b_i, \qquad -1 < x_i < 1,$$
$$k(x,t) = k_{ij}(x_i, t_j), \qquad a_i < x < b_i, \qquad a_j < t < b_j,$$
$$-1 < x_i < 1, \qquad -1 < t_j < 1, \qquad i = 1, \ldots, n,$$

and writing (5.3) on each interval $x \in L_i$ separately we obtain

$$\sum_{j=1}^{n} \frac{1}{\pi} \int_{-1}^{1} \frac{[(b_j - a_j)/2]\phi_j(t_j)\, dt_j}{[(b_j - a_j)/2]t_j + [(b_j + a_j)/2] - [(b_i - a_i)/2]x_i - [(b_i + a_i)/2]}$$
$$+ \sum_{j=1}^{n} \int_{-1}^{1} k_{ij}(x_i, t_j)\phi_j(t_j) \frac{b_j - a_j}{2} dt_j = f_i(x_i), \qquad i = 1, \ldots, n, \qquad -1 < x_i < 1. \tag{5.5}$$

It may be noted that all the variables x_i and t_j in (5.5) vary between -1 and 1 hence, the indices i and j in x_i and t_j may be suppressed. Also note that in the first term of (5.5) if $i \neq j$ the kernel is bounded and continuous in the closed interval $[-1, 1]$, i.e. for $-1 \leq (x_i, t_j) \leq 1$. Thus (5.5) is equivalent to the following simple system of singular integral equations defined in the normalized interval $(-1, 1)$:

$$\frac{1}{\pi} \int_{-1}^{1} \frac{\phi_i(t)}{t - x} dt + \sum_{j=1}^{n} \int_{-1}^{1} h_{ij}(x, t)\phi_j(t)\, dt = f_i(x),$$
$$-1 < x < 1, \qquad i = 1, \ldots, n, \tag{5.6}$$

where h_{ij} is the sum of k_{ij} and the corresponding nonsingular kernels in the first term of (5.5).†

Referring to Eqs. (4.10–4.15) and [3], we express the fundamental function of (5.2) as

$$w(x) = (1 + x)^{-1/2 + \alpha_1}(1 - x)^{1/2 + \alpha_2}, \qquad (-1 < x < 1), \tag{5.7}$$

where α_1 and α_2 are (positive, zero, or negative) integers. $-(\alpha_1 + \alpha_2) = \kappa$ is known as the index of the integral equation. The first step in the numerical

† From the analysis and particularly numerical viewpoint another advantage of this procedure is that one now is dealing with a simple fundamental function, for example, of the form $w(t) = (1 - t^2)^{\mp 1/2}$ rather than a complicated function defined by (4.14) or (4.15).

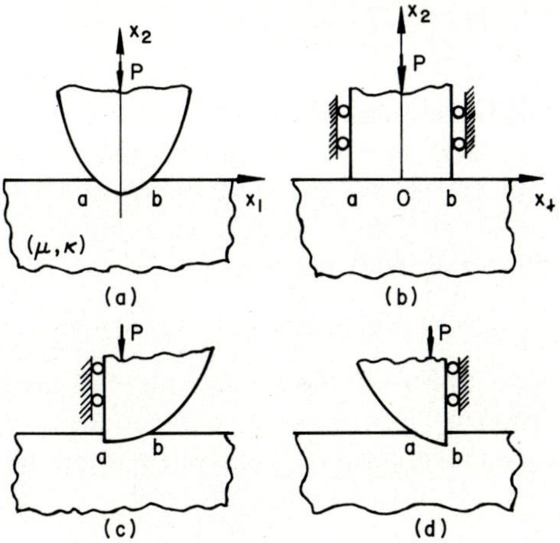

Fig. 4 Contact problem for an elastic half-plane.

procedure which will be described in this section is the determination of the integers α_1 and α_2 or the index of the problem. As pointed out earlier, generally this is not possible without referring to the physics of the problem. For example, consider the plane contact problem shown in Fig. 4 for a rigid stamp with a given profile acting on the elastic for half-plane $x_2 < 0$, $-\infty < x_1 < \infty$. Using the Fourier transforms, or the Green's functions, or the complex potentials [1–4] it can easily be shown that in the absence of friction the mixed boundary-value problem is reduced to the following simple singular integral equation:

$$\frac{1}{\pi}\int_{-1}^{1}\frac{\phi(t)}{t-x}\,dt = f(x), \qquad -1 < x < 1, \tag{5.8}$$

where

$$\begin{aligned} x &= \frac{2x_1}{b-a} - \frac{b+a}{b-a}, \qquad \phi(x) = -\sigma_{yy}(x_1, 0), \\ f(x) &= -\frac{4\mu}{1+\kappa_0}\frac{\partial}{\partial x_1}v(x_1, 0), \qquad a < x_1 < b, \end{aligned} \tag{5.9}$$

with μ and κ_0 being the elastic constants ($\kappa_0 = 3 - 4\nu$ for plane strain, $\kappa_0 = (3-\nu)/(1+\nu)$ for plane stress, and ν is Poisson's ratio). Note that the fundamental function of (5.8) is given by (5.7) and the solution is of the following form [3]:

$$\phi(x) = F(x)w(x), \qquad -1 < x < 1, \tag{5.10}$$

where $F(x)$ is bounded in $-1 \le x \le 1$. Thus, the singular behavior of the solution is completely determined by that of $w(x)$. In Fig. 4a the contact at both ends $x_1 = a$ and $x_1 = b$ is "smooth" and a and b are unknown. Consequently at the end points the contact stress $\phi(x)$ must be bounded (and necessarily zero). Therefore, in (5.7) we have $\alpha_1 = 1$, $\alpha_2 = 0$, and the fundamental function and index become

$$w(x) = (1 - x^2)^{1/2}, \qquad \kappa = -1. \tag{5.11}$$

On the other hand, in Fig. 4b the leading edges of the stamp are "sharp" and a and b are known. Thus, since the contact stress at these end points is known to be unbounded, from (5.7) and (5.10) it follows that

$$\alpha_1 = 0, \qquad \alpha_2 = -1, \qquad \kappa = 1, \qquad w(x) = (1 - x^2)^{-1/2}. \tag{5.12}$$

Similarly for the stamp given in Fig. 4c a is known, b is unknown and

$$\alpha_1 = 0 = \alpha_2, \qquad \kappa = 0, \qquad w(x) = (1 + x)^{-1/2}(1 - x)^{1/2}. \tag{5.13}$$

Also, for the stamp given in Fig. 4d, b is known, a is unknown and

$$\alpha_1 = 1, \quad \alpha_2 = -1, \qquad \kappa = 0, \qquad w(x) = (1 + x)^{1/2}(1 - x)^{-1/2}. \tag{5.14}$$

Referring now to (4.8) and (4.21), we express the general solution of (5.8) as

$$\phi(x) = -\frac{w(x)}{\pi}\int_{-1}^{1}\frac{f(t)\,dt}{(t - x)w(t)} + Cw(x), \qquad -1 < x < 1, \tag{5.15}$$

where C is an unknown constant. Although there are very general rules for determining the unknown constants such as C, a, and b [3], as pointed out earlier in this section, since it is always possible to reduce the general singular integral equations to a system defined only in the normalized interval $(-1, 1)$, for the numerical methods which will be developed in this section it is sufficient to state the rules for a simple singular integral equation such as (5.8) (or (5.2) for which one simply replaces $f(x)$ by $[f(x) - \int_{-1}^{1} k(x,t)\phi(t)\,dt]$). One may also note that the statements made here for the singular integral equations of the first kind are also valid for the equations of the second kind without any modification.

(a) $\kappa = -1$ (Fig. 4a):

In this case the conditions at infinity require that the constant C must be zero and the following consistency condition must be satisfied [3]:

$$\int_{-1}^{1}\left[f(x)-\int_{-1}^{1}k(x,t)\phi(t)\,dt\right]\frac{dx}{w(x)}=0. \tag{5.16}$$

Noting that $f(t)$ contains the constants a and b (see (5.9)), we see that (5.16) provides one equation for the determination of remaining unknowns a and b. The second equation is obtained if we consider the following equilibrium condition:

$$-\int_{a}^{b}\sigma_{yy}(x_1,0)\,dx_1=\frac{b-a}{2}\int_{-1}^{1}\phi(x)\,dx=P. \tag{5.17}$$

(b) $\kappa=1$ (Fig. 4b):

In this case C is the only unknown constant and is determined by substituting from (5.15) into the equilibrium condition (5.17) (which must be satisfied in all cases).†

(c) $\kappa=0$ (Figs. 4c or 4d):

In this case again the conditions at infinity require that the constant C be zero and (5.15) gives the unique solution without any reference to any additional conditions. The problem is solved by assuming that both a and b are known and if, instead of the contact area, the resultant load P is specified, (5.17) is used to relate the two.

The numerical methods used for the solution of singular integral equations may be considered in two separate categories. The first is a rather direct approach which is based on the development Gauss–Jacobi type integration formulas for singular integrals. The second is basically a series solution with Chebyshev or Jacobi polynomials being the related orthogonal polynomials used in the series expansion.

5.1 Solution by Gaussian Integration Formulas

The most common numerical technique to solve a Fredholm-type integral equation of the form

$$\int_{-1}^{1}k(x,t)\phi(t)\,dt=f(x),\qquad -1<x<1 \tag{5.18}$$

† For $\kappa=1$ there is always a physical condition such as (5.17) to be satisfied. For example, in crack problems instead of equilibrium one has the single-valuedness condition (in which P is replaced by zero).

is the use of some kind of integration formula to evaluate the integral in terms of some discrete set of values $\phi(t_1),\ldots,\phi(t_n)$ thereby reducing the integral equation to a system of algebraic equations in $\phi(t_i)$. In particular, if it is appropriate [20, 21] one may prefer an integration formula of Gaussian type for which (5.18) becomes

$$\sum_1^n k(x_i, t_j)\phi(t_j)W_j + R_n(x_i) = f(x_i), \qquad i = 1,\ldots,n, \tag{5.19}$$

where W_j, $(j = 1,\ldots,n)$ is the weighting constant of the related integration formula and R_n is the remainder. If we select n sufficiently large, then R_n can be made as small as necessary for the desired accuracy and, hence, may be neglected. In (5.19) $t_1,\ldots,t_n$ are the roots of the related orthogonal polynomial. This highly appealing simple technique could be used for the solution of singular integral equations if the Gaussian integration formulas for singular integrals were to be available. Some of these formulas will be developed in this and the following sections.

5.1.1 *Gaussian Integration Formula for* $\kappa = 1$. For $\kappa = 1$ the fundamental function of (5.2) is given by (5.12) which is the weight of Chebyshev polynomials (of the first kind) $T_n(x)$. Thus, before deriving the integration formula the following property of the Chebyshev polynomials will be proved: Let

$$T_n(t_k) = 0, \qquad k = 1,\ldots,n;\; U_{n-1}(x_r) = 0, \qquad r = 1,\ldots,n-1. \tag{5.20}$$

Then

$$\sum_{k=1}^n \frac{T_j(t_k)}{n(t_k - x_r)} = \begin{cases} 0, & j = 0, \\ U_{j-1}(x_r), & 0 < j < n, \end{cases} \tag{5.21}$$

where the Chebyshev polynomials are defined by

$$T_n(x) = \cos n\theta, \qquad U_n(x) = \frac{\sin(n+1)\theta}{\sin\theta}, \qquad \cos\theta = x. \tag{5.22}$$

To prove (5.21) note that $T_n(x)$ and $U_n(x)$ are polynomials of degree n and consider the following simple fraction expansion:

$$\frac{U_{n-j-1}(x)}{T_n(x)} = \sum_1^n \frac{a_k}{t_k - x}, \qquad T_n(t_k) = 0,$$

$$a_k = \frac{-U_{n-j-1}(t_k)}{T_n'(t_k)} = \frac{-U_{n-j-1}(t_k)}{nU_{n-1}(t_k)}. \tag{5.23a, b}$$

Using [22]

$$U_{n-j-1}(x) = T_j(x)U_{n-1}(x) - T_n(x)U_{j-1}(x) \tag{5.24}$$

from (5.23) it follows that

$$\sum_{k=1}^{n} \frac{T_j(t_k)}{n(t_k - x)} = -\frac{U_{n-j-1}(x)}{T_n(x)}. \tag{5.25}$$

First part $j = 0$ of (5.21) follows immediately from (5.20) and (5.25). Substituting now from (5.24) into (5.25) we find

$$\sum_{k=1}^{n} \frac{T_j(t_k)}{n(t_k - x)} = U_{j-1}(x) - \frac{T_j(x)U_{n-1}(x)}{T_n(x)} \tag{5.26}$$

which, for $x = x_r$ and $0 < j < n$ is reduced to (5.21) by (5.20), thereby completing the proof.

Let the solution of the singular integral equation now be of the form (5.10) with $w(x)$ as given by (5.12). Let us assume that the unknown bounded function $F(x)$ can be approximated to a sufficient degree of accuracy by

$$F(x) \cong \sum_{0}^{p} A_j T_j(x). \tag{5.27}$$

By using the relation [22]

$$\frac{1}{\pi}\int_{-1}^{1} \frac{T_j(t)\,dt}{(t-x)(1-t^2)^{1/2}} = \begin{cases} 0, & j = 0, \\ U_{j-1}(x), & j > 0, \end{cases} \qquad (-1 < x < 1), \tag{5.28}$$

we express the singular integral in (5.2) or (5.8) as

$$\frac{1}{\pi}\int_{-1}^{1} \frac{\phi(t)}{t-x}\,dt \cong \sum_{j=0}^{p} A_j \frac{1}{\pi}\int_{-1}^{1} \frac{T_j(t)\,dt}{(t-x)(1-t^2)^{1/2}} = \sum_{1}^{p} A_j U_{j-1}(x),$$
$$-1 < x < 1. \tag{5.29}$$

At $x = x_r$, we substitute from (5.21) and (5.27), and reduce (5.29) to

$$\frac{1}{\pi}\int_{-1}^{1} \frac{\phi(t)}{t-x_r}\,dt \cong \sum_{j=1}^{p}\sum_{k=1}^{n} \frac{A_j T_j(t_k)}{n(t_k - x_r)} = \sum_{j=1}^{p}\sum_{k=1}^{n} \frac{A_j T_j(t_k)}{n(t_k - x_r)}$$
$$+ \sum_{k=1}^{n} \frac{A_0 T_0(t_k)}{n(t_k - x_r)} = \sum_{k=1}^{n} \frac{F(t_k)}{n(t_k - x_r)}, \tag{5.30}$$

where

$$t_k = \cos\frac{(2k-1)\pi}{2n}, \qquad k = 1, \ldots, n;\ x_r = \cos\frac{\pi r}{n}, \qquad r = 1, \ldots, n-1. \tag{5.31}$$

Note that if the expansion (5.27) is exact, then there is no approximation in (5.30) for any $n > p$. Also note that (5.30) is identical to the following standard Gauss–Chebyshev integration formula for bounded functions [20, 21]:

$$\frac{1}{\pi}\int_{-1}^{1}\frac{g(x,t)}{(1-t^2)^{1/2}}\,dt \cong \sum_{1}^{n}\frac{1}{n}\,g(x,t_k), \qquad T_n(t_k)=0, \tag{5.32}$$

with the important difference that (5.32) is valid for any x whereas (5.30) holds only for certain discrete set of x_r given by (5.31). Using now (5.10), (5.30), and (5.32) we reduce the integral equation (5.2) to the following system of linear algebraic equations in the unknowns $F(t_1), \ldots, F(t_n)$:

$$\sum_{k=1}^{p}\frac{1}{n}F(t_k)\left[\frac{1}{t_k-x_r}+\pi k(x_r,t_k)\right]=f(x_r), \qquad r=1,\ldots,n-1,$$

$$\sum_{k=1}^{n}\frac{\pi}{n}F(t_k)=A, \tag{5.33a, b}$$

where A is a known constant and (5.33b) comes from the additional condition of the problem (such as (5.17)).

5.1.2 *Gaussian Integration Formula for* $\kappa = -1$. For $\kappa = -1$ the fundamental function of the integral equation is given by (5.11) which is the weight of Chebyshev polynomials of the second kind $U_n(x)$. Thus, we will first prove the following property: Let

$$U_n(t_k)=0, \qquad T_{n+1}(x_r)=0, \tag{5.34}$$

then

$$\sum_{k=1}^{n}\frac{(1-t_k^2)U_j(t_k)}{(n+1)(t_k-x_r)}=-T_{j+1}(x_r), \qquad j<n. \tag{5.35}$$

Proof. Using [22]

$$(1-t^2)U_n'(t)=(n+1)U_{n-1}(t)-ntU_n(t), \tag{5.36}$$

we obtain the following expansion:

$$\frac{U_{n-j-1}(x)}{U_n(x)}=\sum_{1}^{n}\frac{b_k}{t_k-x}, \qquad b_k=-\frac{(1-t_k^2)U_{n-j-1}(t_k)}{(n+1)U_{n-1}(t_k)}. \tag{5.37}$$

Considering the recursion formulas (5.24) and [22]

$$T_n(t)=U_n(t)-tU_{n-1}(t), \quad U_j(t)=T_j(t)+tU_{j-1}(t), \tag{5.38}$$

from (5.37) and (5.34) it follows that

$$\sum_{k=1}^{n} \frac{(1-t_k^2)U_j(t_k)}{(n+1)(t_k-x)} = -\frac{U_{n-j-1}(x)}{U_n(x)} = -T_{j+1}(x) + \frac{U_j(x)T_{n+1}(x)}{U_n(x)}. \tag{5.39}$$

Noting that $T_{n+1}(x_r) = 0$, for $x = x_r$ we reduce (5.39) to (5.35) which completes the proof.

Consider now the singular integral in (5.2) with the solution expressed as in (5.10) and (5.11). Let the following truncated series represent the unknown bounded function $F(t)$ with sufficient accuracy:

$$F(t) \cong \sum_{0}^{p} B_j U_j(t). \tag{5.40}$$

Using (5.35), (5.40), and the relation [22]

$$\frac{1}{\pi}\int_{-1}^{1} \frac{U_j(t)(1-t^2)^{1/2}}{t-x}\,dt = -T_{j+1}, \qquad -1 < x < 1, \tag{5.41}$$

for $x = x_r$ we express the singular integral as

$$\begin{aligned}\frac{1}{\pi}\int_{-1}^{1} \frac{\phi(t)}{t-x_r}\,dt &\cong \sum_{0}^{p} \frac{B_j}{\pi}\int_{-1}^{1} \frac{U_j(t)(1-t^2)^{1/2}}{t-x_r}\,dt = -\sum_{0}^{p} B_j T_{j+1}(x_r)\\ &= \sum_{j=0}^{p}\sum_{k=1}^{n} \frac{(1-t_k^2)B_j U_j(t_k)}{(n+1)(t_k-x_r)} = \sum_{k=1}^{n} \frac{(1-t_k^2)F(t_k)}{(n+1)(t_k-x_r)},\end{aligned} \tag{5.42}$$

where

$$t_k = \cos\frac{k\pi}{n+1}, \qquad k = 1,\ldots,n;\ x_r = \cos\frac{\pi(2r-1)}{2(n+1)}, \qquad r = 1,\ldots,n+1. \tag{5.43}$$

One may again note that (5.42) is identical to the following Gaussian integration formula for a bounded function $g(x, t)$:

$$\frac{1}{\pi}\int_{-1}^{1} g(x,t)(1-t^2)^{1/2}\,dt \cong \sum_{1}^{n} \frac{(1-t_k^2)}{n+1} g(x,t_k), \qquad U_n(t_k) = 0, \tag{5.44}$$

with the important difference that (5.44) is valid for any x whereas (5.42) is valid only for a certain set of $x = x_r$ given by (5.43). Also note that if (5.40) is exact, again there is no approximation in (5.42) for any $n > p$.

Using now (5.10), (5.11), (5.42), and (5.44) we reduce the integral equation (5.2) to

$$\sum_{k=1}^{n} \frac{1-t_k^2}{n+1} F(t_k)\left[\frac{1}{t_k - x_r} + \pi k(x_r, t_k)\right] = f(x_r), \qquad r = 1,\ldots,n+1. \tag{5.45}$$

Referring to (5.10), (5.40), and (5.41) and using the orthogonality condition

$$\frac{1}{\pi}\int_{-1}^{1} T_k(x)T_r(x)\frac{dx}{(1-x^2)^{1/2}} = \begin{cases} 0, & k \neq r, \\ 1, & k = r = 0, \\ \frac{1}{2}, & k = r > 0, \end{cases} \tag{5.46}$$

we express the consistency condition for (5.2) as

$$\begin{aligned} &\int_{-1}^{1}\left[f(x) - \int_{-1}^{1} k(x,t)\phi(t)\,dt\right]\frac{dx}{(1-x^2)^{1/2}} \\ &= \frac{1}{\pi}\int_{-1}^{1}\frac{dx}{(1-x^2)^{1/2}}\int_{-1}^{1}\frac{\phi(t)}{t-x}\,dt \\ &= -\int_{-1}^{1}\sum_{0}^{p} B_j T_{j+1}(x)\frac{dx}{(1-x^2)^{1/2}} = 0. \end{aligned} \tag{5.47}$$

Thus, (5.45) implies that the consistency condition of the integral equation has been satisfied. Note that in (5.45) there are n unknowns $F(t_k)$ and $n+1$ equations. If $\kappa = -1$, usually there are two more unknown constants such as a and b in Fig. 4a and one more condition such as (5.17) giving altogether $n+2$ unknowns and $n+2$ equations.

5.1.3 *Gaussian Integration Formula for* $\kappa = 0$. Let the fundamental function and the index of the integral equation (5.2) be

$$w(x) = (1-t)^{-1/2}(1+t)^{1/2}, \qquad \kappa = 0. \tag{5.48}$$

Expressing again the solution by (5.10) and observing that $w(x)$ is the weight of Jacobi polynomials $P_n^{(-1/2,1/2)}(x)$, it will be assumed that the unknown bounded function $F(t)$ may be approximated to a sufficient degree of accuracy by

$$F(t) \cong \sum_{0}^{p} C_j P_j^{(-1/2,1/2)}(t), \qquad -1 < t < 1. \tag{5.49}$$

From (5.10), (5.49), (5.48), and using the following general property of Jacobi polynomials [23],

$$\frac{1}{\pi}\int_{-1}^{1} P^{(\alpha,\beta)}(t)(1-t)^{\alpha}(1+t)^{\beta}\frac{dt}{t-x} = -\frac{\Gamma(\alpha)\Gamma(1-\alpha)}{\pi}2^{-\kappa}P_{n-\kappa}^{(-\alpha,-\beta)}(x),$$
$$-1 < x < 1, \qquad \kappa = -(\alpha+\beta) = (-1, 0, \text{or } 1), \qquad -1 < (\alpha,\beta) < 1, \tag{5.50}$$

we express the singular integral in (5.2) as

$$\frac{1}{\pi}\int_{-1}^{1}\frac{\phi(t)}{t-x}\,dt = \sum_{0}^{p} C_j P_j^{(1/2,-1/2)}(x), \qquad -1 < x < 1. \tag{5.51}$$

Let

$$P_n^{(-1/2,1/2)}(t_k) = 0, \qquad k = 1,\ldots,n, \tag{5.52}$$

and consider the expansion

$$-\frac{P_n^{(1/2,-1/2)}(x)P_j^{(-1/2,1/2)}(x) - P_n^{(-1/2,1/2)}(x)P_j^{(1/2,-1/2)}(x)}{P_n^{(-1/2,1/2)}(x)} = \sum_{1}^{n}\frac{c_k}{t_k - x}, \qquad j < n. \tag{5.53}$$

Using (5.24) and the relations [21]

$$P_n^{(-1/2,1/2)}(2z^2-1) = \frac{\Gamma(n+1/2)}{n!\pi^{1/2}}\frac{1}{z}T_{2n+1}(z),$$
$$P_n^{(1/2,-1/2)}(2z^2-1) = \frac{\Gamma(n+1/2)}{n!\,\pi^{1/2}}U_{2n}(z), \tag{5.54}$$

we modify equation (5.53) as

$$-\frac{\Gamma(j+1/2)}{j!\,\pi^{1/2}}\frac{U_{2n-2j-1}(y)}{T_{2n+1}(y)} = \sum_{1}^{n}\frac{c_k}{t_k - x}, \qquad x = 2y^2 - 1, \tag{5.55}$$

indicating that the expansion in (5.53) is indeed possible.

Referring now to the results in Subsection 5.1.1 and to (5.54) and using [23]

$$\frac{d}{dt}P_n^{(\alpha,\beta)}(t_k) = \frac{n+\beta}{1+t_k}P_n^{(\alpha+1,\beta-1)}(t_k) = -\frac{n+\alpha}{1-t_k}P_n^{(\alpha-1,\beta+1)}(t_k),$$
$$P_n^{(\alpha,\beta)}(t_k) = 0, \qquad k = 1,\ldots,n, \tag{5.56}$$

we find the coefficients c_k to be

$$c_k = \frac{2(1+t_k)}{2n+1}P_j^{(-1/2,1/2)}(t_k). \tag{5.57}$$

If we let

$$P_n^{(1/2,-1/2)}(x_r) = 0, \qquad r = 1,\ldots,n, \tag{5.58}$$

from (5.53) and (5.57) it follows that

$$\sum_{k=1}^{n} \frac{2(1+t_k)}{2n+1} P_j^{(-1/2,1/2)}(t_k) \frac{1}{t_k - x_r} = P_j^{(1/2,-1/2)}(x_r). \tag{5.59}$$

Finally, from (5.49), (5.51), and (5.59) we find

$$\frac{1}{\pi}\int_{-1}^{1} \frac{\phi(t)}{t-x_r}\,dt = \frac{1}{\pi}\int_{-1}^{1} \frac{F(t)}{t-x_r}\left(\frac{1+t}{1-t}\right)^{1/2} dt \cong \sum_{k=1}^{n} \frac{2(1+t_k)}{2n+1}\frac{F(t_k)}{t_k - x_r}, \tag{5.60}$$

where

$$\begin{aligned} &P_n^{(-1/2,1/2)}(t_k) = 0, \qquad t_k = \cos\left(\frac{2k-1}{2n+1}\pi\right), \qquad k = 1,\ldots,n, \\ &P_n^{(1/2,-1/2)}(x_r) = 0, \qquad x_r = \cos\left(\frac{2r\pi}{2n+1}\right), \qquad r = 1,\ldots,n. \end{aligned} \tag{5.61}$$

Equation (5.60) too is identical to the corresponding Gauss–Jacobi integration formula for a bounded function $g(x,t)$ given by [2]

$$\frac{1}{\pi}\int_{-1}^{1} g(x,t)\left(\frac{1+t}{1-t}\right)^{1/2} dt \cong \sum_{k=1}^{n} \frac{2(t_k+1)}{2n+1} g(x,t_k) \tag{5.62}$$

with again the important difference that (5.62) is valid for any x whereas (5.60) is valid only for $x = x_r$, $r = 1,\ldots,n$.

Using the integration formulas (5.60) and (5.62), we easily reduce the singular integral equation (5.2) to the following system of linear algebraic equations in $F(t_1),\ldots,F(t_n)$:

$$\sum_{k=1}^{n} \frac{2(1+t_k)}{2n+1} F(t_k)\left[\frac{1}{t_k - x_r} + \pi k(x_r, t_k)\right] = f(x_r), \qquad r = 1,\ldots,n, \tag{5.63}$$

where t_k and x_r are given by (5.61).

If the weight function is $w(x) = (1-t)^{1/2}/(1+t)^{-1/2}$ following a similar procedure, it can be shown that

$$\frac{1}{\pi}\int_{-1}^{1} \frac{F(t)}{t-x}\left(\frac{1-t}{1+t}\right)^{1/2} dt \cong \sum_{k=1}^{n} \frac{2(1-t_k)}{2n+1}\frac{F(t_k)}{t_k - x_r}, \tag{5.64}$$

and (5.2) is reduced to

$$\sum_{k=1}^{n} \frac{2(1-t_k)}{2n+1} F(t_k)\left[\frac{1}{t_k - x_r} + \pi k(x_r, t_k)\right] = f(x_r), \qquad r = 1,\ldots,n, \tag{5.65}$$

where

$$P_n^{(1/2,-1/2)}(t_k) = 0, \qquad t_k = \cos\left(\frac{2k\pi}{2n+1}\right), \qquad k = 1,\ldots,n,$$
$$P_n^{(-1/2,1/2)}(x_r) = 0, \qquad x_r = \cos\left(\frac{2r-1}{2n+1}\pi\right), \qquad r = 1,\ldots,n. \tag{5.66}$$

5.2 Solution by Orthogonal Polynomials

In the previous sections it was indicated that the fundamental function of the singular integral equation (5.2) is of the general form

$$w(x) = (1-x)^{\alpha}(1+x)^{\beta}, \qquad \kappa = -(\alpha+\beta) = (0, \mp 1). \tag{5.67}$$

Noting that (5.67) is the weight of Jacobi polynomials $P_n^{(\alpha,\beta)}(x)$, it is natural to look for a solution of (5.2) in the following series form:

$$\phi(t) = w(t)\sum_0^{\infty} C_n P_n^{(\alpha,\beta)}(t), \qquad -1 < t < 1. \tag{5.68}$$

Thus, substituting from (5.68) and (5.50) into (5.2) we obtain

$$\sum_{n=0}^{\infty} C_n\left[-\frac{2^{-\kappa}}{\sin \pi\alpha} P_{n-\kappa}^{(-\alpha,-\beta)}(x) + k_n(x)\right] = f(x), \qquad -1 < x < 1,$$
$$k_n(x) = \int_{-1}^{1} k(x,t)P_n^{(\alpha,\beta)}(t)w(t)\,dt. \tag{5.69}$$

Equation (5.69) can further be reduced to an infinite algebraic system in C_n as follows:

$$-\frac{2^{-\kappa}}{\sin \pi\alpha}\theta_k(-\alpha,-\beta)C_{k+\kappa} + \sum_{j=1}^{\infty} c_{kj}C_j = c_k, \qquad k = 0, 1, \ldots, \tag{5.70}$$

where for $\kappa = -1$, $C_{-1} = 0$ and

$$c_{kj} = \int_{-1}^{1} P_k^{(-\alpha,-\beta)}(x)k_j(x)(1-x)^{-\alpha}(1+x)^{-\beta}\,dx,$$
$$c_k = \int_{-1}^{1} f(x)P_k^{(-\alpha,-\beta)}(x)(1-x)^{-\alpha}(1+x)^{-\beta}\,dx; \tag{5.71}$$

the constants θ_k come from the orthogonality condition

$$\int_{-1}^{1} P_n^{(\alpha,\beta)}(t)P_k^{(\alpha,\beta)}(t)w(t)\,dt = \begin{cases} 0, & n \neq k \\ \theta_k(\alpha,\beta), & n = k, \end{cases} \qquad k = 0,1,2,\ldots,$$

$$\theta_0(\alpha,\beta) = \int_{-1}^{1} w(t)\,dt = \frac{2^{\alpha+\beta+1}\Gamma(\alpha+1)\Gamma(\beta+1)}{\Gamma(\alpha+\beta+2)}, \tag{5.72}$$

$$\theta_k(\alpha,\beta) = \frac{2^{\alpha+\beta+1}\Gamma(k+\alpha+1)\Gamma(k+\beta+1)}{(2k+\alpha+\beta+1)k!\,\Gamma(k+\alpha+\beta+1)}, \qquad k = 1,2,\ldots.$$

The infinite system (5.70) may be solved by the method of reduction [24]. Note that if $\kappa = 1$ in the reduction letting $k = 0,\ldots,n$ in (5.70) involves the unknowns $C_0,\ldots,C_{n+1}$, that is, there is one more unknown than the number of equations. The additional equation is provided by the following physical condition (e.g. (5.17)):

$$\int_{-1}^{1} \phi(x)\,dx = A = C_0\theta_0(\alpha,\beta), \tag{5.73}$$

where A is a known constant. On the other hand, if $\kappa = -1$, from (5.69) and (5.71) it is clear that the first equation in (5.70) (i.e. $k = 0$) is equivalent to the consistency condition (5.16). It should further be noted that the integrations necessary for the evaluation of the constants c_{kj} and c_k in (5.71) are of Gauss–Jacobi type may easily be evaluated if we use [20]

$$\int_{-1}^{1} g(x)(1-x)^{-\alpha}(1+x)^{-\beta}\,dx \cong \sum_{k=1}^{n} W_k g(x_k),$$

$$P_n^{(-\alpha,-\beta)}(x_k) - 0, \qquad k = 1,\ldots,n, \tag{5.74}$$

$$W_k = -\frac{(2n-\alpha-\beta+2)\Gamma(n-\alpha+1)\Gamma(n-\beta+1)}{(n+1)!(n-\alpha-\beta+1)\Gamma(n-\alpha-\beta+1)} \times$$

$$\times \frac{2^{-\alpha-\beta}}{\dfrac{d}{dx}P_n^{(-\alpha,-\beta)}(x_k)P_{n+1}^{(-\alpha,-\beta)}(x_k)}.$$

In the special case of $\kappa = 1$, $w(x) = (1-x^2)^{-1/2}$ the related orthogonal polynomials are $T_n(x)$, $(n = 0,1,2,\ldots)$, and we have

$$\phi(x) = (1-x^2)^{-1/2}\sum_{0}^{\infty} A_n T_n(x), \tag{5.75}$$

$$\frac{1}{\pi}\int_{-1}^{1} U_n(t)U_k(t)(1-t^2)^{1/2}\,dt = \begin{cases} 0, & n \neq k, \\ \pi/2, & n = k, \end{cases} \tag{5.76}$$

$$\frac{\pi^2}{2} A_{k+1} + \sum_{n=0}^{\infty} a_{kn} A_n = a_k, \qquad k = 0, 1, \ldots, \tag{5.77}$$

$$a_k = \int_{-1}^{1} f(x) U_k(x)(1 - x^2)^{1/2}\, dx,$$

$$a_{kn} = \int_{-1}^{1} U_k(x)(1 - x^2)^{1/2}\, dx \int_{-1}^{1} k(x, t) T_n(t)(1 - t^2)^{-1/2}\, dt,$$

$$\int_{-1}^{1} \phi(x)\, dx = A = \pi A_0, \tag{5.78}$$

where (5.28) and (5.46) have been used. Truncating (5.77) at the Nth term and retaining first N equations, from (5.77) and (5.78) we obtain the $N + 1$ unknown constants $A_0, A_1, \ldots, A_n$.

In the other special case $\kappa = -1$, $w(x) = (1 - x^2)^{1/2}$, it may easily be shown that

$$\phi(x) = (1 - x^2)^{1/2} \sum_{0}^{\infty} B_n U_n(x), \tag{5.79}$$

$$-\frac{\pi^2}{2} B_{k-1} + \sum_{n=0}^{\infty} b_{kn} B_n = b_k, \qquad k = 0, 1, \ldots,$$

$$b_k = \int_{-1}^{1} f(x) T_k(x)(1 - x^2)^{-1/2}, \tag{5.80}$$

$$b_{kn} = \int_{-1}^{1} T_k(x)(1 - x^2)^{-1/2}\, dx \int_{-1}^{1} k(x, t) U_n(t)(1 - t^2)^{1/2}\, dt,$$

where (5.41) and (5.46) have been used. Again in (5.80) the first equation ($k = 0$) corresponds to the consistency condition (5.16).

Some typical applications of the numerical methods described in this section may be found in [25–27] which also include extensive references to the solution of mixed boundary value problems in mechanics obtained by applying these methods.

6 INTEGRAL EQUATIONS WITH GENERALIZED CAUCHY KERNELS

As pointed out earlier in this article, in some mixed boundary-value problems formulated in terms of singular integral equations, in addition to Cauchy type singularities the kernel may contain terms which become unbounded as both variables in the kernel approach an end point. For example, in the torsion problem shown in Fig. 3 and formulated by the integral equation (4.90) if $b > c$, separating all the singular terms, we may express (4.90) as

$$\frac{1}{\pi}\int_{b_1}^{c} p(t)\,dt\left[\left(1+\frac{\mu_1}{\mu_2}\right)\frac{1}{t-x}-\frac{1}{t+x-2b_1}\right] + \int_{b_1}^{c} k_{1f}(x,t)p(t)\,dt = -\mu_1 f(x), \qquad b_1 < x < c, \tag{6.1}$$

where $k_{1f}(x,t)$ is a Fredholm type kernel which is bounded in the closed interval $b_1 \leq (x,t) \leq c$ and is the sum of $k(x,t)$ (see (4.91)) and the nonsingular terms in cotangent or cosecant kernels appearing in (4.90). Note that in (6.1) the term $1/(t+x-2b_1)$ becomes unbounded at the end point b_1.

Similarly, if $c = b$ the kernel in (4.90) contains additional "singular" terms which may be separated if we observe that

$$\begin{aligned}
\frac{\pi}{2b}\cot\frac{\pi(t-x)}{2b} &= \frac{1}{t-x} + h_1(x,t),\\
\frac{\pi}{2b}\cot\frac{\pi(t+x)}{2b} &= \frac{1}{t+x-2b} + h_2(x,t),\\
\frac{\pi}{2b}\operatorname{cosec}\frac{\pi(t-x)}{2b} &= \frac{1}{t-x} + h_3(x,t),\\
\frac{\pi}{2b}\operatorname{cosec}\frac{\pi(t+x)}{2b} &= -\frac{1}{t+x-2b} + h_4(x,t),
\end{aligned} \tag{6.2a–d}$$

where $h_1,\ldots,h_4$ are bounded in the closed interval $b_1 \leq (x,t) \leq b$. Thus, substituting from (6.2) into (4.90) we find

$$\frac{1}{b}\int_{b_1}^{b} p(t)\,dt\left[\left(1+\frac{\mu_1}{\mu_2}\right)\frac{1}{t-x}-\frac{1}{t+x-2b_1}-\frac{\mu_1}{\mu_2}\frac{1}{t+x-2b}\right] + \int_{b_1}^{b} k_{2f}(x,t)p(t)\,dt = -\mu_1 f(x), \qquad b_1 < x < b, \tag{6.3}$$

where, again k_{2f} is a bounded kernel. As seen from (6.1) and (6.3), the integral equations for the symmetric and the antisymmetric problems have the same dominant parts but different Fredholm kernels. In (6.3) in addition to the Cauchy singularity, the dominant part of the kernel contains terms which become unbounded at both ends, b_1 and b. The dominant kernels of this type are defined as *generalized Cauchy kernels.*

In the torsion problem under consideration there is one more limiting case which is worth considering. This is the case of $b_1 = 0$ (e.g. a broken shaft). Thus, for $b_1 = 0$, $c = b$, observing that in addition to (6.2a and c), around the end points one may write

$$\frac{\pi}{2b}\cot\frac{\pi(t+x)}{2b} = \frac{1}{t+x} + \frac{1}{t+x-2b} + h_5(x,t),$$
$$\frac{\pi}{2b}\operatorname{cosec}\frac{\pi(t+x)}{2b} = \frac{1}{t+x} - \frac{1}{t+x-2b} + h_6(x,t), \tag{6.4}$$

and the integral equation (4.90) may be expressed as

$$\frac{1}{\pi}\int_0^b p(t)\,dt\left[\left(1+\frac{\mu_1}{\mu_2}\right)-\left(1\pm\frac{\mu_1}{\mu_2}\right)\frac{1}{t+x}-\frac{\mu_1}{\mu_2}\frac{1}{t+x-2b}\right]$$
$$+\int_0^b k_{3f}(x,t)p(t)\,dt = -\mu_1 f(x), \qquad 0 < x < b, \tag{6.5}$$

where h_5, h_6, and k_{3f} are bounded in $0 \le (x, y) \le b$, and the upper and lower signs in (6.5) correspond to the symmetric and the antisymmetric problems, respectively. Structurally, (6.5) is identical to (6.3); however, at $x = 0$ their solutions may have quite different behavior.

6.1 A Plane Elasticity Problem for Nonhomogeneous Media

Perhaps the most typical application of the singular integral equations with a generalized Cauchy kernel arises in the study of crack problems in nonhomogeneous elastic solids. Consider, for example, the three-dimensional problem in which part of periphery of a plane crack extends to the bimaterial interface in a nonhomogeneous medium. Consider the cross-section shown

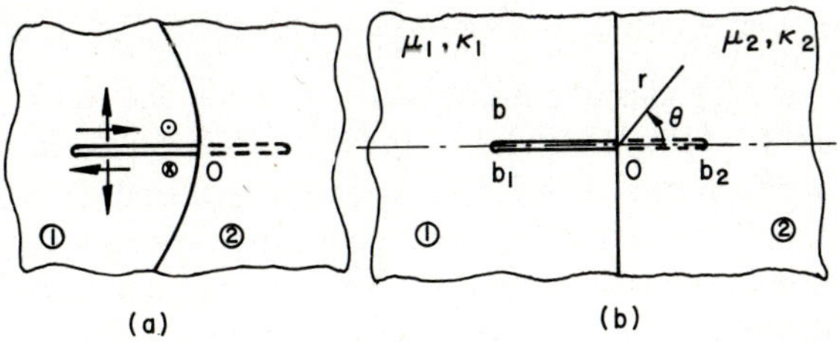

Fig. 5 Bonded elastic materials containing a crack terminating at or going through the interface.

in Fig. 5a. In the neighborhood of the point of interest O let the interface be a smooth surface and, for simplicity, let the crack plane be perpendicular to the interface. Through a proper superposition one can separate the singular or the perturbation part of the problem in which statically self-equilibrating crack surface tractions are the only external loads. From the viewpoint of the stress state around the point O, the perturbation problem in turn may be assumed as having three components, namely, the in-plane extension (mode I), the in-plane shear (mode II), and the antiplane shear (mode III) with the corresponding surface tractions shown in Fig. 5a. Clearly, the behavior of the solution of the antiplane component of this problem around O will be identical to that of the torsion problem shown in Fig. 3 at $x = 0 = b_1$, $r = a$. This behavior should be completely characterized by the first two terms of the generalized Cauchy kernel appearing in (6.5).

Similarly, the singular behavior of the solution under in-plane loading conditions will be equivalent to that of the plane strain problem for two semi-infinite media having a finite crack perpendicular to and ending at the interface shown in Fig. 5b. Formulation of this problem is straightforward. For example, referring to [28] for details, in the case of in-plane tension problem, i.e. for the following symmetric loading conditions:

$$\sigma_{1\theta\theta}(r,\pi) = f(r), \qquad \sigma_{1r\theta}(r,\pi) = 0, \qquad 0 < r < b, \tag{6.6}$$

we express the integral equation of the problem as

$$\frac{1}{\pi}\int_0^b \left[\frac{1}{t-r} + \frac{c_1}{t+r} + \frac{c_2 r}{(t+r)^2} + \frac{c_3 r^2}{(t+r)^3}\right]\phi(t)\,dt = \frac{1+\kappa_1}{2\mu_1} f(r),$$
$$0 < r < b, \tag{6.7}$$

where

$$\phi(r) = -\frac{\partial}{\partial r}[u_{1\theta}(r,\pi+0) - u_1(r,\pi-0)],$$
$$c_1 = \tfrac{1}{2} - m(1+\kappa_1)/[2(m+\kappa_2)] - 3(1-m)/[2(1+m\kappa_1)], \tag{6.8}$$
$$c_2 = 6(1-m)/(1+m\kappa_1), \qquad c_3 = 4(m-1)/(1+m\kappa_1), \qquad m = \mu_2/\mu_1,$$

and μ_i and κ_i are the elastic constants ($i = 1, 2, \kappa_i = 3 - v_i$ for plane strain, $\kappa_i = (3 - v_i)/(1 + v_i)$ for plane stress, and v_i as Poisson's ratio). For the in-plane shear problem one obtains the same (generalized Cauchy) kernel as that shown in (6.7). Thus in the general problem of more complex geometry and loading conditions the system of three integral equations representing the problem will have the dominant kernels given in (6.5) and (6.7) (in uncoupled form) and coupling will be through Fredholm type kernels only.

Referring to Fig. 5 if we now assume that the crack crosses the boundary and extends into the medium 2 (dashed lines), it is again not difficult to argue that the dominant parts of the related system of integral equations will be the same as that of the idealized through crack problem for two elastic half-planes shown in Fig. 5b. The derivation of integral equations is again straightforward. For example, in the symmetric problem selecting the unknown functions

$$\phi_1(r) = -\frac{\partial}{\partial r}[u_{1\theta}(r,\pi+0) - u_{1\theta}(r,\pi-0)], \qquad 0<r<b_1,$$
$$\phi_2(r) = \frac{\partial}{\partial r}[u_{2\theta}(r,+0) - u_{2\theta}(r,-0)], \qquad 0<r<b_2, \tag{6.9}$$

for the following crack surface tractions

$$\sigma_{1\theta\theta}(r,\pi) = f_1(r), \qquad \sigma_{1r\theta}(r,\pi) = 0, \qquad 0<r<b_1,$$
$$\sigma_{2\theta\theta}(r,0) = f_2(r), \qquad \sigma_{2r\theta}(r,0) = 0, \qquad 0<r<b_2, \tag{6.10}$$

we express the integral equations for the perturbation problem as (see [29] for details)

$$\frac{1}{\pi}\int_0^{b_i}\frac{\phi_i(t)}{t-r}dt + \sum_{j=1}^{2}\frac{1}{\pi}\int_0^{b_j}k_{ij}(r,t)\phi_j(t)\,dt = \frac{1+\kappa_i}{2\mu_i}f_i(r),$$
$$i=1,2;\ 0<r<b_i, \tag{6.11}$$

where

$$k_{ii}(r,t) = \sum_{k=1}^{3}\frac{c_{ik}r^{k-1}}{(t+r)^k}, \qquad i=1,2,$$
$$k_{ij}(r,t) = \sum_{k=1}^{2}\frac{d_{ik}r^{k-1}}{(t+r)^k}, \qquad i,j=1,2;\ i\neq j, \tag{6.12a, b}$$

$$c_{11} = 1/2 - m_1(1+\kappa_1)/[2(m_1+\kappa_2)] - 3(1-m_1)/[2(1+m_1\kappa_1)],$$
$$c_{12} = 6(1-m_1)/(1+m_1\kappa_1), \qquad c_{13} = 4(m_1-1)/(1+m_1\kappa_1),$$
$$c_{21} = 1/2 - m_2(1+\kappa_2)/[2(m_2+\kappa_1)] - 3(1-m_2)/[2(1+m_2\kappa_2)],$$
$$c_{22} = 6(1-m_2)/(1+m_2\kappa_2), \qquad c_{23} = 4(m_2-1)/(1+m_2\kappa_2),$$
$$d_{11} = 3(1+\kappa_1)/[2(m_2+\kappa_1)] - (1+\kappa_1)/[2(1+m_2\kappa_2)], \tag{6.13}$$
$$d_{12} = (1+\kappa_1)/(1+m_2\kappa_2) - (1+\kappa_1)/(m_2+\kappa_1),$$
$$d_{21} = 3(1+\kappa_2)/[2(m_1+\kappa_2)] - (1+\kappa_2)/[2(1+m_1\kappa_1)],$$
$$d_{22} = (1+\kappa_2)/(1+m_1\kappa_1) - (1+\kappa_2)/(m_1+\kappa_2),$$
$$m_1 = \mu_2/\mu_1, \qquad m_2 = \mu_1/\mu_2.$$

From (6.11) and (6.12) it is seen that the integral equations contain dominant kernels only which are of generalized Cauchy type.

6.2 The Fundamental Functions

The integral equations (6.5), (6.7), and (6.11) are some special cases of the following system of singular integral equations with generalized Cauchy kernels:

$$\frac{1}{\pi}\sum_{n=1}^{N}\int_{a_n}^{b_n}\left[\frac{A_{mn}}{t-x}+\sum_{k=0}^{K_n}B_{mnk}(x-a_n)^k\frac{d^k}{dx^k}(t-z_{1n})^{-1}\right.$$
$$\left.+\sum_{j=0}^{J_n}C_{mnj}(b_n-x)^j\frac{d^j}{dx^j}(t-z_{2n})^{-1}\right]\phi_n(t)\,dt$$
$$+\sum_{n=1}^{N}\int_{a_n}^{b_n}k_{mn}(x,t)\phi_n(t)\,dt=f_m(x),\qquad m=1,\ldots,N,\qquad a_m<x<b_m, \tag{6.14}$$

where

$$\begin{aligned} z_{1n}&=a_n+(x-a_n)\,e^{i\theta_n}, & 0<\theta_n<2\pi, & \\ z_{2n}&=b_n+(b_n-x)\,e^{i\omega_n}, & -\pi<\omega_n<\pi, & \quad n=1,\ldots,N, \end{aligned} \tag{6.15}$$

A_{mn}, B_{mnk}, C_{mnj} are known constants, $f_1,\ldots,f_N$ are known input functions, $k_{mn}(x,t)$ are Fredholm kernels, and $\phi_1,\ldots,\phi_N$ are the unknown functions. In most practical problems $\theta_n=\pi$ and $\omega_n=0$. However, occasionally one may encounter cases in which $\theta_n=\mp\pi/2$, $\omega_n=\mp\pi/2$ with the related terms in the kernels of the form [28]

$$(t-a_n)/[(t-a_n)^2+(x-a_n)^2],\qquad (b_n-t)/[(b_n-t)^2+(b_n-x)^2]. \tag{6.16}$$

Referring to Fig. 6a and (6.15) we note that as the variable x varies on the line of integration L_n, the variables z_{1n} and z_{2n} vary on lines L_{1n} and L_{2n}, respectively. In some problems the kernels may contain terms of the form (6.16) as well as, for example, (6.12). This means that in Fig. 6a at a given end point there may be more than one auxiliary line L_{1n} or L_{2n}, also meaning that there may be additional terms in (6.14) and (6.15) defined by additional angles θ_n or ω_n.

In this type of problem another point which requires special emphasis is whether the cuts $L_n=(a_n,b_n)$, $(n=1,\ldots,N)$ intersect each other or not. The only type of intersection which is physically meaningful is for two

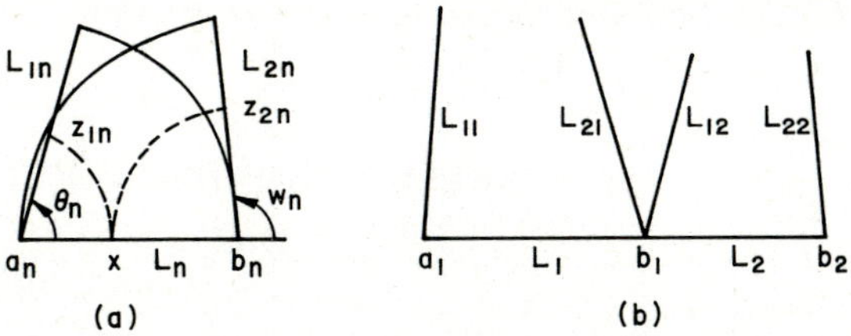

Fig. 6 Cuts for generalized Cauchy kernel.

adjacent cuts to have a common end point (Fig. 6b). A physical example for this would be the through crack problem shown in Fig. 5b. In such problems, aside from the necessary changes in the limits of integrations, the integral equations (6.14) remain unchanged. However, as will be pointed out in this section, one may have to be somewhat more careful in determining the fundamental functions. In the following, first it will be assumed that all end points are distinct, then an example for coinciding ends will be considered.

To find the fundamental functions of (6.14) we let

$$\begin{aligned}\phi_n(t) &= g_n(t)w_n(t) = g_n(t)(b_n - t)^{\alpha_n}(t - a_n)^{\beta_n}\\ &= g_n(t)\,e^{-\pi i\alpha_n}(t - b_n)^{\alpha_n}(t - a_n)^{\beta_n}, \qquad a_n < t < b_n,\\ &-1 < \mathrm{Re}\,(\alpha_n, \beta_n) < 0, \qquad n = 1,\ldots,N, \qquad g_n(a_n) \neq 0, \qquad g_n(b_n) \neq 0,\end{aligned} \tag{6.17}$$

where the function $g_n(t)$, $(n = 1,\ldots,N)$ is Hölder-continuous in $a_n \le t \le b_n$, $(t - b_n)^{\alpha_n}(t - a_n)^{\beta_n}$ is any definite branch which varies continuously in $a_n < t < b_n$, and α_n and β_n are unknown constants which have to be determined. Consider now the following sectionally holomorphic functions:

$$F_n(z) = \frac{1}{\pi}\int_{a_n}^{b_n} \frac{\phi_n(t)}{t - z}\,dt = \frac{e^{-\pi i\alpha_n}}{\pi}\int_{a_n}^{b_n} (t - b_n)^{\alpha_n}(t - a_n)^{\beta_n}\frac{g_n(t)}{t - z}\,dt. \tag{6.18}$$

Examining the singular behavior of $F_n(z)$ around the end points (see, for example, [3], Chapter 4) and separating the principal parts we obtain

$$\begin{aligned}F_n(z) = &-g_n(a_n)(b_n - a_n)^{\alpha_n}\frac{e^{-\pi i\beta_n}}{\sin \pi\beta_n}(z - a_n)^{\beta_n}\\ &+g_n(b_n)(b_n - a_n)^{\beta_n}\frac{1}{\sin \pi\beta_n}(z - b_n)^{\alpha_n} + \mathrm{G}_n(z),\\ &\qquad n = 1,\ldots,N.\end{aligned} \tag{6.19}$$

The function $G_n(z)$ is bounded everywhere except possibly at the ends a_n, b_n near which it may have the following behavior:

$$|G_n(z)| < \frac{A_n}{|z - a_n|^{p_n}}, \qquad p_n < -\mathrm{Re}(\beta_n), \qquad |G_n(z)| < \frac{B_n}{|z - b_n|^{r_n}}, \qquad r_n < -\mathrm{Re}(\alpha_n), \tag{6.20}$$

where A_n, p_n, B_n, and r_n are real constants.

Using the Plemelj formulas (3.61), from (6.19) we find

$$\begin{aligned} \frac{1}{\pi}\int_{a_n}^{b_n} \frac{\phi_n(t)}{t - x}\,dt &= \tfrac{1}{2}[F_n^+(x) + F_n^-(x)] \\ &= -g_n(a_n)(b_n - a_n)^{\alpha_n}\cot(\pi\beta_n)(x - a_n)^{\beta_n} \\ &\quad + g_n(b_n)(b_n - a_n)^{\beta_n}\cot(\pi\alpha_n)(b_n - x)^{\alpha_n} + H_n(x), \\ &\qquad n = 1, \ldots, N, \qquad a_n < x < b_n, \end{aligned} \tag{6.21}$$

where $H_n(x)$ is bounded in $a_n < x < b_n$, and at the end points its behavior is similar to that of $G_n(z)$ which is given by (6.20). From the definition of $F_n(z)$, (6.18), and from Fig. 6a it is clear that for $z = z_{1n} \neq a_n$ and $z = z_{2n} \neq b_n$, i.e. on L_{1n} and L_{2n}, $F_n(z)$ is holomorphic. Therefore

$$\frac{1}{\pi}\int_{a_n}^{b_n} \frac{\phi_n(t)}{t - z_{sn}}\,dt = F_n(z_{sn}), \qquad z_{sn} \in L_{sn}, \qquad s = 1, 2, \qquad n = 1, \ldots, N, \tag{6.22}$$

or, substituting from (6.19), using (6.15) and separating the principal parts, we have

$$\begin{aligned} \frac{1}{\pi}\int_{a_n}^{b_n} \frac{\phi_n(t)}{t - z_{1n}}\,dt &= -g_n(a_n)(b_n - a_n)^{\alpha_n}\frac{e^{i\beta_n(\theta_n - \pi)}}{\sin \pi\beta_n}(x - a_n)^{\beta_n} + h_n(x), \\ &\qquad a_n < x < b_n, \\ \frac{1}{\pi}\int_{a_n}^{b_n} \frac{\phi_n(t)}{t - z_{2n}}\,dt &= g_n(b_n)(b_n - a_n)^{\beta_n}\frac{e^{i\omega_n\alpha_n}}{\sin \pi\alpha_n}(b_n - x)^{\alpha_n} + s_n(x), \\ &\qquad a_n < x < b_n, \end{aligned} \tag{6.23}$$

where $h_n(x)$ is bounded in $a_n \leqq x \leqq b_n$, $s_n(x)$ is bounded in $a_n \leqq x < b_n$, and the behavior of h_n and s_n near the ends a_n and b_n, respectively, is determined by that of $G_n(z)$ as given by (6.20).

Since $F_n(z)$ is holomorphic on L_{1n} and L_{2n} (see (6.18) and Fig. 6a), using (6.23) we may also write

$$\frac{1}{\pi}\int_{a_n}^{b_n} \phi_n(t)(x-a_n)^k \frac{d^k}{dx^k}(t-z_{1n})^{-1}\,dt$$
$$= (x-a_n)^k \frac{d^k}{dx^k} F_n(z_{1n})$$
$$= -g_n(a_n)(b_n-a_n)^{\alpha_n} \frac{e^{i\beta_n(\theta_n-\pi)}}{\sin \pi\beta_n} \beta_n(\beta_n-1)\cdots(\beta_n-k+1)(x-a_n)^{\beta_n}$$
$$+(x-a_n)^k \frac{d^k}{dx^k} h_n(x), \qquad k=1,\ldots,K_n, \qquad a_n<x<b_n,$$
$$\frac{1}{\pi}\int_{a_n}^{b_n} \phi_n(t)(b_n-x)^j \frac{d^j}{dx^j}(t-z_{2n})^{-1}\,dt \tag{6.24}$$
$$= (b_n-x)^j \frac{d^j}{dx^j} F_n(z_{2n})$$
$$= (-1)^j g_n(b_n)(b_n-a_n)^{\beta_n} \frac{e^{i\omega_n\alpha_n}}{\sin \pi\alpha_n} \alpha_n(\alpha_n-1)\cdots(\alpha_n-j+1)(b_n-x)^{\alpha_n}$$
$$+(b_n-x)^j \frac{d^j}{dx^j} s_n(x), \qquad j=1,\ldots,J_n, \qquad a_n<x<b_n, \qquad n=1,\ldots,N.$$

Equations (6.21), (6.23), and (6.24) give all the terms in the dominant part of the integral equations (6.14). Thus, by substituting from (6.21), (6.23), and (6.24) into (6.14), observing that for the mth equation $a_m < x < b_m$, multiplying both sides first by $(x-a_m)^{-\beta_m}$ and letting $x \to a_m$, then by $(b_m - x)^{-\alpha_m}$ and letting $x \to b_m$, and also observing that $g(a_m) \neq 0$, $g(b_m) \neq 0$, $0 \leq r_n < \mathrm{Re}(-\alpha_n)$, $0 \leq p_n < \mathrm{Re}(-\beta_n)$, we obtain

$$A_{mm} \cos \pi\beta_m + e^{i\beta_m(\theta_m-\pi)}\left[B_{mm0} + \sum_{k=1}^{K_m} B_{mmk}\beta_m(\beta_m-1)\cdots(\beta_m-k+1)\right] = 0,$$
$$m=1,\ldots,N,$$
$$A_{mm} \cos \pi\alpha_m + e^{i\omega_m\alpha_m}\left[C_{mm0} + \sum_{j=1}^{J_m} C_{mmj}(-1)^j\alpha_m(\alpha_m-1)\cdots(\alpha_m-k+1)\right] = 0,$$
$$m=1,\ldots,N. \qquad \text{(6.25a, b)}$$

The characteristic Eqs. (6.25) provide two sets of (highly nonlinear) algebraic equations to determine the unknown constants α_m and β_m $(m=1,\ldots,N)$. With α_m and β_m determined, (6.17) gives the fundamental functions $w_m(t)$ $(m=1,\ldots,N)$.

In the foregoing analysis it was assumed that the $2N$ end points a_n, b_n are distinct. In practice this may not always be the case. To give an example in which some of the end points may coincide, consider the case of, for example, $b_1 = a_2$ in (6.14) (Fig. 6b). As seen from Fig. 6b (see also Fig. 5), physically the problem really has $2N - 1$ points of singularity, and mathematically there are only $2N - 1$ irregular points. This implies that in the expressions of the fundamental functions there will be only $2N - 1$ exponents, i.e. $\alpha_1 = \beta_2$. Hence, the fundamental functions of (6.14) given by (6.17) will have to be modified as

$$\begin{aligned} w_1(t) &= (b_1 - t)^{\alpha_1}(t - a_1)^{\beta_1}, & a_1 < t < b_1, \\ w_2(t) &= (b_2 - t)^{\alpha_2}(t - b_1)^{\alpha_1}, & b_1 < t < b_2, \\ w_n(t) &= (b_n - t)^{\alpha_n}(t - a_n)^{\beta_n}, & a_n < t < b_n, \quad n = 3, \ldots, N. \end{aligned} \tag{6.26}$$

The solutions of (6.14) are still of the form $\phi_n(t) = g_n(t)w_n(t)$, and the conditions on α_n, β_n, $g_n(a_n)$, and $g_n(b_n)$, $(n = 1, \ldots, N)$ shown in (6.17) are still valid. Also with $a_2 = b_1$ and $\beta_2 = \alpha_1$ the definition of $F_n(z)$, (6.18) and the asymptotic expressions (6.21), (6.23), and (6.24) are still valid. Thus, proceeding as before, one obtains (6.25a) for $m = 1, 3, 4, \ldots, N$ and (6.25b) for $m = 2, 3, \ldots, N$, giving altogether $2N - 2$ algebraic equations for $2N - 2$ unknowns $\beta_1, \beta_3, \ldots, \beta_N$ and $\alpha_2, \alpha_3, \ldots, \alpha_N$.

Writing (6.14) for $m = 1$ and letting $x \to b_1$ and for $m = 2$ and letting $x \to a_2 = b_1$ one obtains two homogeneous linear algebraic equations for $g_1(b_1)$ and $g_2(b_1)$ of the form

$$\sum_1^2 c_{ij}(\alpha_1)g_j(b_1) = 0, \qquad i = 1, 2. \tag{6.27}$$

Since $g_j(b_1) \neq 0$ (6.27) gives the following characteristic equation to determine α_1:

$$\left| c_{ij}(\alpha_1) \right| = 0. \tag{6.28}$$

Note that $g_1(b_1)$ and $g_2(b_1)$ are not independent and are related through (6.27). This provides the additional condition which was eliminated by letting $a_2 = b_1$ and which is necessary for the unique solution of (6.14).†

As an example for coinciding end points consider the crack problem formulated by (6.11) and (6.12) and shown by Fig. 5a (with dashed lines included). Note that this problem was formulated by using polar coordinates. Hence in the terminology of this section, $a_1 = 0 = a_2$, $\beta_1 = \beta_2 = \beta$, and the fundamental functions and the solution may be expressed as

† For example, in collinear crack problems letting $a_2 = b_1$ essentially eliminates one crack while the number N of integral equations and unknown functions remain the same. Physically, since there are $N - 1$ cuts, there will be only $N - 1$ single-valuedness conditions. However, the general solution of N integral equations will have N arbitrary constants. Thus for a unique solution one more condition is needed which is provided by (6.27) (see [29]).

$$\phi_n(t) = g_n(t)w_n(t) = g_n(t)(b_n - t)^{\alpha_n}t^{\beta} = g_n(t)\, e^{-\pi i\alpha_n}(t - b_n)^{\alpha_n}t^{\beta},$$
$$0 < t < b_n, \qquad -1 < \mathrm{Re}\,(\alpha_n, \beta) < 0, \qquad g_n(b_n) \neq 0, \tag{6.29}$$
$$g_n(0) \neq 0, \qquad n = 1, 2.$$

Consider now the following sectionally holomorphic functions:

$$F_n(z) = \frac{1}{\pi}\int_0^{b_n} \frac{\phi_n(t)}{t - z}\,dt = \frac{e^{-\pi i\alpha_n}}{\pi}\int_0^{b_n} (t - b_n)^{\alpha_n}t^{\beta}\,\frac{g_n(t)}{t - z}\,dt, \qquad n = 1, 2, \tag{6.30}$$

where, noting that (6.11) is derived in polar coordinates, the complex variable $z = r + i\rho$ for $n = 1$ and $n = 2$ is defined in such a way that in each case the cut $(0 < r < b_n)$ lies along the positive real axis. Repeating the analysis given previously, we obtain

$$F_n(z) = -g_n(0)b_n^{\alpha_n}\frac{e^{-\pi i\beta}}{\sin \pi\beta}z^{\beta} + g_n(b_n)b_n^{\beta}\frac{1}{\sin \pi\alpha_n}(z - b_n)^{\alpha_n} + G_n(z),$$

$$\frac{1}{\pi}\int_0^{b_n}\frac{\phi_n(t)}{t - r}\,dt = -g_n(0)b_n^{\alpha_n}\cot \pi\beta r^{\beta} + g_n(b_n)b_n^{\beta}\cot \pi\alpha_n(b_n - r)^{\alpha_n} + H_n(r),$$

$$\frac{1}{\pi}\int_0^{b_n}\frac{\phi_n(t)}{t + r}\,dt = F_n(-r) = -g(0)b_n^{\alpha_n}\frac{1}{\sin \pi\beta}r^{\beta} + u_{0n}(r),$$

$$\frac{1}{\pi}\int_0^{b_n}\frac{r\phi_n(t)}{(t + r)^2}\,dt = -r\frac{d}{dr}F_n(-r) = g_n(0)b_n^{\alpha_n}\frac{\beta}{\sin \pi\beta}r^{\beta} + u_{1n}(r),$$

$$\frac{1}{\pi}\int_0^{b_n}\frac{r^2\phi_n(t)}{(t + r)^3}\,dt = \frac{r^2}{2}\frac{d^2}{dr^2}F_n(-r) = -g_n(0)b_n^{\alpha_n}\frac{\beta(\beta - 1)}{2\sin \pi\beta}r^{\beta} + u_{2n}(r),$$
$$0 < r < b_n, \qquad n = 1, 2, \tag{6.31}$$

where the behavior of $H_n(r)$ and $u_{jn}(r)$, $(n = 1, 2; j = 0, 1, 2)$ around $r = 0$ is similar to that of $G_n(z)$ around $a_n = 0$ as given by (6.20), otherwise they are bounded functions.

If we substitute from (6.31) into (6.11), multiply both sides first by $(b_n - r)^{-\alpha_n}$ and let $r \to b_n$, then by $r^{-\beta}$ and let $r \to 0$ we obtain

$$\cot \pi\alpha_n = 0, \qquad n = 1, 2, \tag{6.32}$$

$$[\cos \pi\beta + c_{11} - \beta c_{12} + \beta(\beta - 1)c_{13}/2][g_1(0)b_1^{\alpha_1}/\sin \pi\beta]$$
$$+ (d_{11} - \beta d_{12})[g_2(0)b_2^{\alpha_2}/\sin \pi\beta] = 0,$$
$$(d_{12} - \beta d_{22})[g_1(0)b_1^{\alpha_1}/\sin \pi\beta] \tag{6.33}$$
$$+ [\cos \pi\beta + c_{21} - \beta_{22} + \beta(\beta - 1)c_{23}/2][g_2(0)b_2^{\alpha_2}/\sin \pi\beta] = 0.$$

Equations (6.32) are the well-known characteristic equations for crack tip singularities which are fully imbedded in a homogeneous medium and give

$\alpha_1 = \alpha_2 = -\frac{1}{2}$. Writing the determinant of the linear system (6.33) one obtains the third characteristic equation giving β as follows:

$$[\cos \pi\beta + c_{11} - \beta c_{12} + \beta(\beta - 1)c_{13}/2][\cos \pi\beta + c_{21} - \beta c_{22} + \beta(\beta - 1)c_{23}/2] - (d_{11} - \beta d_{12})(d_{21} - \beta d_{22}) = 0. \tag{6.34}$$

It should be noted that as expected, (6.34) is identical to the characteristic equation giving the stress singularity at the apex of two bonded quarter planes obtained by using an entirely different method (see, for example, [30]).

Going now back to the other examples discussed in this section, we see that in the torsion problem for $c < b$ the dominant part of the integral equation is given by (6.1) which is a very special case of (6.14). Thus defining the solution and the fundamental function by

$$p(x) = g(x)w(x), \qquad w(x) = (c - x)^{\alpha}(x - b_1)^{\beta}, \qquad b_1 < x < c, \tag{6.35}$$

and applying the procedure described in this section we find

$$\cos \pi\alpha = 0, \qquad \cos \pi\beta = \mu_2/(\mu_1 + \mu_2). \tag{6.36}$$

If $c = b$, there is a contribution to the generalized Cauchy kernel at both ends, the integral equation is given by (6.3), and characteristic equations are found to be†

$$\cos \pi\alpha = \mu_1/(\mu_1 + \mu_2), \qquad \cos \pi\beta = \mu_2/(\mu_1 + \mu_2). \tag{6.37}$$

In the case of the "broken shaft" that is, for $b_1 = 0, c = b$, the integral equation is given by (6.5) from which the characteristic equations are obtained as

$$\cos \pi\alpha = \mu_1/(\mu_1 + \mu_2), \qquad \cos \pi\beta = (\mu_2 \pm \mu_1)/(\mu_1 + \mu_2), \tag{6.38a, b}$$

where the upper sign is for the symmetric and lower sign is for the antisymmetric loading. Note that for symmetric loading (for which $\sigma_{\theta r}(a, x) = p(x) = p(-x)$, Fig. 3) $\beta = 0$, that is the stress state at ($r = a$, $x = 0$) is bounded, whereas for antisymmetric loading the characteristic Eq. (6.38b) is identical to that of a semi-infinite crack perpendicular to a bimaterial interface under antiplane shear loading.

For the plane problem of a crack perpendicular to and terminating at the interface of two bonded half-planes (Fig. 5), the integral equation is given by (6.7). Thus, by defining,

$$\phi(t) = g(t)w(t), \qquad w(t) = (b - t)^{\alpha}t^{\beta}, \tag{6.39}$$

the characteristic equations are found to be

† In this case note that in (6.14) $N = 1$, $K_1 = 0 = J_1$, $\theta_1 = \pi$, $\omega_1 = 0$.

$$\cos \pi\alpha = 0, \qquad 2\gamma_1 \cos \pi(\beta + 1) + \gamma_2(\beta + 1)^2 - \gamma_3 = 0,$$
$$\gamma_1 = (m + \kappa_2)(1 + m\kappa_1), \qquad \gamma_2 = 4(m + \kappa_2)(1 - m), \qquad m = \mu_2/\mu_1,$$
$$\gamma_3 = (1 - m)(m + \kappa_2) + (1 + m\kappa_1)(m + \kappa_2) - m(1 + \kappa_1)(1 + \kappa_1). \tag{6.40a, b}$$

Again (6.40a) is the well-known result giving $\alpha = -\frac{1}{2}$ and (6.40b) is identical to the characteristic equation for a semi-infinite crack perpendicular to and terminating at a bimaterial interface under in-plane loading (see, for example, [28]).

6.3 Numerical Method for Solving the Integral Equations with Generalized Cauchy Kernels

As shown in Section 5 of this chapter, a general system of singular integral equations such as (6.14) can always be expressed by means of a simple system in which both variables x and t vary in the normalized interval $(-1, 1)$. Furthermore, it was also indicated that, once the method of solution is developed for a single equation, it can easily be extended to a system consisting of any number of equations. The numerical method for solving the singular integral equations with generalized Cauchy kernels will therefore be described for a single equation of the following form only:

$$\frac{1}{\pi}\int_{-1}^{1}\left[\frac{1}{t-x} + k_s(x,t) + k_f(x,t)\right]\phi(t)\,dt = f(x), \qquad -1 < x < 1, \tag{6.41}$$

where f is a known function, k_f is a known Fredholm kernel, $k_s(x,t)$ becomes unbounded as x and t approach the end points ∓ 1, is otherwise bounded, and with $1/(t - x)$ form the generalized Cauchy kernel. Generally, (6.41) must be solved under an additional (physical, such as an equilibrium or a single-valuedness) condition of the form

$$\int_{-1}^{1} \phi(t)\,dt = A, \tag{6.42}$$

where A is a known constant. As indicated previously, the unknown and the fundamental functions may be expressed as

$$\phi(t) = g(t)w(t), \qquad w(t) = (1 - t)^{\alpha}(1 + t)^{\beta}, \tag{6.43}$$

where α and β are known constants with $-1 < \mathrm{Re}\,(\alpha, \beta) < 0$ and $g(t)$ is an unknown function which is bounded in $-1 \le t \le 1$. Observing that $w(t)$ is the weight function of Jacobi polynomials $P_n^{(\alpha,\beta)}(t)$, we may solve the integral equation by using a numerical method based on a Gauss–Jacobi integration formula which is similar to the methods described in Section 5. In this case the related integration formula is [20]

$$\int_{-1}^{1} G(x,t)(1-t)^{\alpha}(1+t)^{\beta}\,dt \cong \sum_{1}^{n} W_k G(x,t_k), \qquad -1<\mathrm{Re}\,(\alpha,\beta)<1, \tag{6.44}$$

where t_k are the roots of

$$P_n^{(\alpha,\beta)}(t_k)=0, \qquad k=1,\ldots,n, \tag{6.45}$$

and the weighting constants are given by

$$W_k = -\frac{(2n+\alpha+\beta+2)\Gamma(n+\alpha+1)\Gamma(n+\beta+1)2^{\alpha+\beta}}{(n+1)!(n+\alpha+\beta+1)\Gamma(n+\alpha+\beta+1)P_{n+1}^{(\alpha,\beta)}(t_k)\dfrac{d}{dt}P_n^{(\alpha,\beta)}(t_k)} \tag{6.46}$$

Analogously to the numerical integration methods developed in Section 5, the integral equation (6.41) and the condition (6.42) may now be expressed as

$$\sum_{k=1}^{n} g(t_k)W_k\left[\frac{1}{t_k-x_j}+k_s(x_j,t_k)+k_f(x_j,t_k)\right]\cong \pi f(x_j), \qquad j=1,\ldots,n-1,$$

$$\sum_{1}^{n} g(t_k)W_k = A, \tag{6.47a, b}$$

where

$$-1<\mathrm{Re}\,(\alpha,\beta)<0, \qquad P_n^{(\alpha,\beta)}(t_k)=0, \qquad k=1,\ldots,n,$$
$$P_{n-1}^{(\alpha+1,\beta+1)}(x_j)=0, \qquad j=1,\ldots,n-1, \tag{6.48}$$

and W_k are given by (6.46). Equations (6.47) provide n equations to determine $g(t_1),\ldots,g(t_n)$.

If at one or both ends the solution is required to be bounded (hence, necessarily zero), (6.46) and (6.47a) will still be valid, the points t_k describing the location and the number of the unknowns $g(t_k)$ will still be obtained from (6.45) and the locations x_j giving the number of equations will be obtained from

$$\begin{aligned}
&\text{(a) } 0<\mathrm{Re}\,(\alpha)<1, \quad 0<\mathrm{Re}\,(\beta)<1: P_{n+1}^{(\alpha-1,\beta-1)}(x_j)=0, \quad j=1,\ldots,n+1,\\
&\text{(b) } 0<\mathrm{Re}\,(\alpha)<1, \quad -1<\mathrm{Re}\,(\beta)<0: P_{n}^{(\alpha-1,\beta+1)}(x_j)=0, \quad j=1,\ldots,n,\\
&\text{(c) } -1<\mathrm{Re}\,(\alpha)<0, \quad 0<\mathrm{Re}\,(\beta)<1: P_{n}^{(\alpha+1,\beta-1)}(x_j)=0, \quad j=1,\ldots,n.
\end{aligned} \tag{6.49}$$

In none of the cases given by (6.49) the condition (6.42) or (6.47b) is part of the formulation of the problem. In (6.49a) the additional equation (provided by $x_1,\ldots,x_{n+1}$) is equivalent to the consistency condition of the integral equation and in (6.49b and c) the unique solution is obtained by simply solving (6.47a) with $j = 1, \ldots, n$. As explained in detail in Section 5, the (physical) problem may, however, have additional conditions and unknown constants which may be handled in a straightforward manner.

In the application of the numerical methods described in Section 5 and in this section it is essential that special attention be paid to the convergence of the calculated results. At least one technique regarding the evaluation of the limit value (as $n \to \infty$) of certain calculated results is described in [25] and [27]. In the interest of space no numerical results will be presented in this article. However, there are certain techniques related to the methods described in this article which are quite useful for extracting relevant physical information from the solution. Description of such techniques also falls outside the scope of this article. Among these one may mention the methods used to evaluate such physical quantities as the stress intensity factors, the strain energy release rate, and the crack opening displacement directly in terms of the (calculated) density functions of the integral equations (see, for example, [28], [29], [31]–[34]). The detailed numerical results of the torsion problem discussed in this section may be found in [35] and [36]. In addition to [28], [29], [35], [36], further applications of the singular integral equations with generalized Cauchy kernels may be found in [37–39].

7 SINGULAR INTEGRAL EQUATIONS OF THE SECOND KIND

Some relatively very simple mixed boundary-value problems in mechanics give rise to singular integral equations which are of the second kind. For example, consider the following basic formulas for the elastic half-plane $-\infty < x < \infty$, $y < 0$ relating the surface tractions and the displacement derivatives:

$$\frac{4\mu}{1+\kappa}\frac{\partial}{\partial x}u(x,0) = \gamma\sigma_{yy}(x,0) + \frac{1}{\pi}\int_{-\infty}^{\infty}\sigma_{xy}(t,0)\frac{dt}{t-x}, \qquad -\infty < x < \infty,$$

$$\frac{4\mu}{1+\kappa}\frac{\partial}{\partial x}v(x,0) = -\gamma\sigma_{xy}(x,0) + \frac{1}{\pi}\int_{-\infty}^{\infty}\sigma_{yy}(t,0)\frac{dt}{t-x}, \qquad -\infty < x < \infty,$$

$$\frac{1+\gamma}{2\mu}\sigma_{yy}(x,0) = -\gamma\frac{\partial}{\partial x}u(x,0) - \frac{1}{\pi}\int_{-\infty}^{\infty}\frac{\partial}{\partial t}v(t,0)\frac{dt}{t-x}, \qquad -\infty < x < \infty,$$

$$\frac{1+\gamma}{2\mu}\sigma_{xy}(x,0) = \gamma\frac{\partial}{\partial x}v(x,0) - \frac{1}{\pi}\int_{-\infty}^{\infty}\frac{\partial}{\partial t}u(t,0)\frac{dt}{t-x}, \qquad -\infty < x < \infty,$$

(7.1a–d)

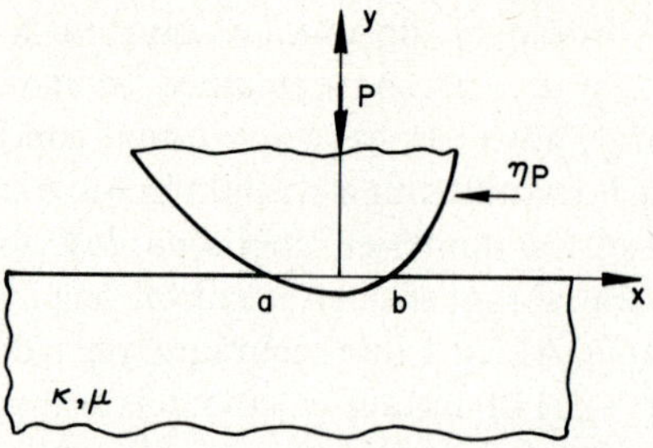

Fig. 7 The contact problem for an elastic half-plane.

where u, v are, respectively, the x, y-components of the displacement vector, σ_{ij}, $(i, j = x, y)$ are the stress components and $\gamma = (\kappa - 1)/(\kappa + 1)$. From (7.1) it is clear that if the displacement vector is specified on part of the boundary $y = 0$ and the traction vector is specified on the remainder, (7.1a and b) or (7.1c and d) give the integral equations of the problem which are coupled and are of the second kind. It is also clear that instead of a half-plane if one is dealing with an axisymmetric or plane problem for a domain bounded by a "smooth" surface the system of integral equations of the related mixed boundary-value problem will have a dominant part which is identical to that given by (7.1).

As a very simple example consider the plane contact problem with friction shown in Fig. 7 in which a rigid stamp of given profile is pressed against an elastic half-plane. Assuming the coefficient of friction η to be constant and defining

$$\sigma_{yy}(t, 0) = -p(t), \qquad \sigma_{xy}(t, 0) = -\eta p(t), \qquad a < t < b,$$

$$\frac{4\mu}{1+\kappa}\frac{\partial}{\partial x}v(x, 0) = f(x), \qquad (a < x < b), \qquad \omega = \eta\gamma \tag{7.2}$$

we see that Eq. (7.1b) gives the integral equation of the problem as follows:

$$\omega p(x) - \frac{1}{\pi}\int_a^b \frac{p(t)}{t - x}dt = f(x), \qquad a < x < b. \tag{7.3}$$

Here the input function $f(x)$ is, aside from a constant multiplier, the derivative of the function describing the stamp profile (i.e. $f(x) = F'(x)$, where $y = F(x)$ is the stamp profile). At the end points if the contact is smooth, the constants a and b defining the contact area are unknown. In this case the solution (or the input function) has to be such that the following consistency condition is satisfied:

$$\int_a^b \left[\omega p(x) - \frac{1}{\pi}\int_a^b \frac{p(t)}{t - x}dt\right]\frac{dx}{w(x)} = \int_a^b \frac{f(x)}{w(x)}dx = 0, \tag{7.4}$$

where $w(x)$ is the fundamental function of the singular integral equation (7.3). Also, if the load P is specified the contact pressure must satisfy the following equilibrium condition:

$$\int_a^b p(t)\,dt = P. \tag{7.5}$$

In this problem the conditions (7.4) and (7.5) account for the unknowns a and b.

For an elastic medium with a more complicated geometry the foregoing formulation basically remains the same with the only difference that the integral equation (7.3) now contains a Fredholm kernel which comes from the change in the geometry. For example, in the symmetric contact problem for an infinite elastic wedge of angle $2\theta_0$ shown in Fig. 8, using the standard Mellin transform technique we obtain the integral equation for the contact pressure as

$$\omega p(r) - \frac{1}{\pi}\int_a^b \left[\frac{1}{t-r} + k(r,t)\right] p(t)\,dt = f(r), \qquad a < r < b, \tag{7.6}$$

where

$$p(r) = -\sigma_{\theta\theta}(r,\theta_0), \qquad \frac{1+\kappa}{4\mu} f(r) = \frac{\partial}{\partial r} u_\theta(r,\theta_0) = F'(r), \tag{7.7}$$

$$\begin{aligned} k(r,t) = {} & \frac{1}{r\log(t/r)} - \frac{1}{t-r} - \frac{\pi\sin^2\theta_0 - \eta\sin\theta_0\cos\theta_0}{r} \frac{}{2\theta_0 + \sin 2\theta_0} \\ & + \int_0^\infty \left(1 - \frac{\sinh 2\theta_0 y}{D(y)}\right)\eta\,\frac{\cos\rho y}{r}\,dy \\ & - \int_0^\infty \left(1 + \frac{\eta\sin 2\theta_0 + \cos 2\theta_0 - \cosh 2\theta_0 y}{D(y)}\right)\frac{\sin\rho y}{r}\,dy, \end{aligned}$$

$$D(y) = \sinh 2\theta_0 y + y\sin 2\theta_0, \qquad \rho = \log(t/r). \tag{7.8}$$

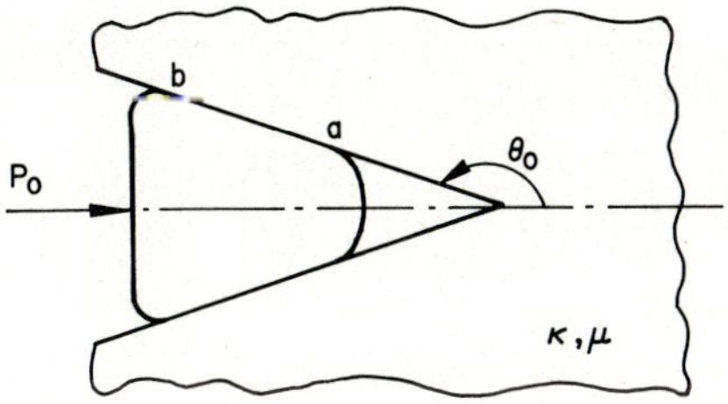

Fig. 8 The contact problem for an elastic wedge in the presence of friction.

The problem is formulated in polar coordinates and $F(r)$ describes the profile of rigid stamp.

7.1 The Fundamental Function

The dominant part of the singular integral equations of the second kind is of the following general form:

$$A\phi(x) + \frac{B}{\pi}\int_a^b \frac{\phi(t)}{t-x}\,dt = f_0(x), \qquad a < x < b, \tag{7.9}$$

where the bounded function f_0 may contain the part of the integral equation with the Fredholm kernel. Defining

$$F(z) = \frac{1}{\pi}\int_a^b \frac{\phi(t)}{t-z}\,dt, \tag{7.10}$$

and using the Plemelj formulas (3.61), we may reduce (7.9) to the following Riemann–Hilbert problem for the sectionally holomorphic function $F(z)$:

$$F^+(x) - \frac{A-iB}{A+iB}F^-(x) = 2if_0(x)/(A+iB). \tag{7.11}$$

Considering the corresponding homogeneous equation

$$X^+(x) - \frac{A-iB}{A+iB}X^-(x) = 0, \tag{7.12}$$

we may obtain the fundamental solution $X(z)$ and the fundamental function $w(x)$ of (7.9) as [3]

$$\begin{gathered} X(z) = (z-b)^\alpha (z-a)^\beta, \qquad w(x) = (b-x)^\alpha (x-a)^\beta, \\ \alpha = \frac{1}{2\pi i}\log\left(\frac{A-iB}{A+iB}\right) + N, \qquad \beta = -\frac{1}{2\pi i}\log\left(\frac{A-iB}{A+iB}\right) + M, \end{gathered} \tag{7.13}$$

where N and M are arbitrary (positive, zero, or negative) integers.† From (7.11) and (7.12) we have

$$\left(\frac{F(x)}{X(x)}\right)^+ - \left(\frac{F(x)}{X(x)}\right)^- = \frac{2if_0(x)}{(A+iB)X^+(x)}, \qquad a < x < b, \tag{7.14}$$

the solution of which is

† Note that if A and B are real then the exponents α and β are also real. For an example of complex exponents see Section 3.3. Also, see [40] for the equivalence of the two formulations.

$$\frac{F(z)}{X(z)} = \frac{1}{\pi}\int_a^b \frac{f_0(t)/(A+iB)}{(t-z)X^+(t)}dt + C, \tag{7.15}$$

where C is an arbitrary constant. From (7.10) and (7.15) the solution of (7.9) may be expressed as

$$\phi(x) = -\frac{B}{A^2+B^2}\frac{w(x)}{\pi}\int_a^b \frac{f_0(t)\,dt}{(t-x)w(t)} + C_0 w(x) + \frac{A}{A^2+B^2}f_0(x),$$
$$a < x < b, \tag{7.16}$$

where (7.13) is used to replace $X^+(x)$ by $w(x)$ and

$$C_0 = -BC\,e^{\pi i\alpha}/(A-iB). \tag{7.17}$$

The index of the integral equation is again defined by

$$\kappa = -(\alpha+\beta) = -(N+M). \tag{7.18}$$

The general remarks made in Section 5 in connection with the contact problem shown in Fig. 4 and the singular integral equation of the first kind regarding the determination of the arbitrary integers N and M or the index of the problem and the constant C_0 which appears in the solution (7.16) are valid for the integral equations of the second kind also and will not be repeated here.

It should be pointed out that the fundamental function of the problem can also be determined directly by applying the method described in Subsection 6.2. For this we let the solution of (7.9) be

$$\phi(x) = g(x)w(x), \qquad w(x) = (b-x)^\alpha(x-a)^\beta. \tag{7.19}$$

From (7.10), (7.19), (6.17), and (6.18) it is clear that (6.21) is still valid. Thus, substituting from (7.19) and (6.21) into (7.9) and multiplying both sides first by $(x-a)^{-\beta}$ and letting $x \to a$ and then by $(b-x)^{-\alpha}$ and letting $x \to b$ we obtain

$$\cot \pi\beta = A/B, \qquad \cot \pi\alpha = -A/B. \tag{7.20}$$

From

$$\cot \pi\theta = \cot \pi(\theta + K), \qquad K = 0, \mp 1, \ldots,$$
$$\frac{1}{2\pi i}\log\left(\frac{A-iB}{A+iB}\right) = -\frac{1}{\pi}\,\mathrm{Arccot}\,(A/B), \tag{7.21}$$

it is seen that (7.13) and (7.20) are identical.

To develop a numerical method for the solution of the singular integral equations of the second kind, it will again be assumed that $\phi(t) = g(t)w(t)$.

At a first glance the exact solution given by (7.16) appears to be not of this form. However, the integral in (7.16) can be evaluated and it can be shown that the expression for $\phi(t)$ indeed has no regular terms. To do this first observe that

$$w(x)\int_a^b \frac{f_0(t)\,dt}{(t-x)w(t)} = X^+(x)\int_a^b \frac{f_0(t)\,dt}{(t-x)X^+(t)},$$

$$\Phi(z) = \int_a^b \frac{f_0(t)\,dt}{(t-z)X^+(t)} = \frac{1}{1-X^+/X^-}\left[\int_a^b \frac{f_0(t)\,dt}{(t-z)X^+(t)} + \int_b^a \frac{f_0(t)\,dt}{(t-z)X^-(t)}\right]$$

$$= \frac{A+iB}{2iB}\int_S \frac{f_0(\tau)\,d\tau}{(\tau-z)X(\tau)}, \tag{7.22}$$

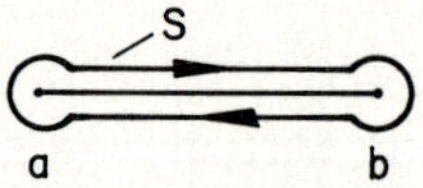

Fig. 9 Contour for evaluating the integral in Eq. (7.22).

where S is the contour shown in Fig. 9 and (7.12) and (7.13) have been used. Now if $f_0(\tau)$ is holomorphic outside S and continuous up to S, the contour integral may be evaluated as follows:

$$\frac{1}{2\pi i}\int_S \frac{f_0(\tau)\,d\tau}{(\tau-z)X(\tau)} = \frac{f_0(z)}{X(z)} - P(z), \tag{7.23}$$

where $P(z)$ is such that for large $|z|$

$$\frac{f_0(z)}{X(z)} = P(z) + 0(1/z). \tag{7.24}$$

On the other hand using the Plemelj formula, from (7.22) it follows that

$$\int_a^b \frac{f_0(t)\,dt}{(t-x)X^+(t)} = \tfrac{1}{2}[\Phi^+(x) + \Phi^-(x)]. \tag{7.25}$$

Thus, substituting from (7.22)–(7.25) into (7.16) and observing that

$$(A+iB)X^+(x) = e^{\pi i\alpha}w(x)(A^2+B^2)^{1/2}\,e^{-\pi i\alpha} \tag{7.26}$$

we obtain

$$\phi(x) = [(A^2+B^2)^{-1/2}P(x) + C_0]w(x). \tag{7.27}$$

7.2 Solution by Orthogonal Polynomials

Consider the following integral equation in the normalized interval $(-1, 1)$:

$$A\phi(x) + \frac{B}{\pi}\int_{-1}^{1} \frac{\phi(t)}{t - x}\, dt + \int_{-1}^{1} k(x, t)\phi(t)\, dt = f(x), \qquad -1 < x < 1. \tag{7.28}$$

Assume that the fundamental function and the index of the problem have been determined and are given by (see Section 7.1)

$$w(t) = (1 - t)^{\alpha}(1 + t)^{\beta}, \qquad \kappa = -(\alpha + \beta) = (0, +1, -1). \tag{7.29}$$

Observing that $w(t)$ is the weight of Jacobi polynomials, we may express the solution of (7.28) as

$$\phi(t) = g(t)w(t), \qquad g(t) = \sum_{0}^{\infty} c_n P_n^{(\alpha,\beta)}(t), \qquad -1 < t < 1, \tag{7.30}$$

where the coefficients c_n are unknown. Consider now the following property of Jacobi polynomials [23, 41, 26]:

$$AP_n^{(\alpha,\beta)}(x)w(x) + \frac{B}{\pi}\int_{-1}^{1} P_n^{(\alpha,\beta)}(t)\frac{w(t)}{t - x}\, dt = -2^{-\kappa}\frac{B}{\sin \pi\alpha} P_{n-\kappa}^{(-\alpha,-\beta)}(x),$$
$$-1 < x < 1, \qquad \operatorname{Re}(\alpha) > -1, \qquad \operatorname{Re}(\beta) > -1, \qquad \operatorname{Re}(\alpha) \neq (0, 1, \ldots,). \tag{7.31}$$

From (7.30), (7.31), and (7.28) it follows that

$$\sum_{n=0}^{\infty} c_n\left[-\frac{2^{-\kappa}B}{\sin \pi\alpha} P_{n-\kappa}^{(-\alpha,-\beta)}(x) + k_n(x)\right] = f(x), \qquad -1 < x < 1,$$
$$k_n(x) = \int_{-1}^{1} k(x, t)P_n^{(\alpha,\beta)}(t)w(t)\, dt \tag{7.32}$$

An effective way of solving for c_n in (7.32) would be to expand both sides into series of Jacobi polynomials $P_j^{(-\alpha,-\beta)}(x)$, and to compare the coefficients. Thus, using the orthogonality relations (5.72) we obtain

$$-\frac{2^{-\kappa}B}{\sin \pi\alpha}\theta_j(-\alpha, -\beta)c_{j+\kappa} + \sum_{n=0}^{\infty} d_{jn}c_n = F_j, \qquad j = 0, 1, \ldots, \tag{7.33}$$

$$d_{jn} = \int_{-1}^{1} P_j^{(-\alpha,-\beta)}(x)k_n(x)\, dx/w(x),$$
$$F_j = \int_{-1}^{1} P_j^{(-\alpha,-\beta)}(x)f(x)\, dx/w(x). \tag{7.34}$$

It is again worthwhile to consider the following three cases separately.

(a) $\kappa = 1$

In this case we solve (7.33) by the method of reduction [24], and if the series is truncated at $n = N$, the system of $N + 1$ linear equation will contain the unknowns $c_0, \ldots, c_{N+1}$. The additional equation to solve the problem is provided by the following condition:

$$\int_{-1}^{1} \phi(t)\, dt = A_0 \tag{7.35}$$

which, substituting from (7.30) and using the orthogonality conditions, may be expressed as†

$$c_0 \theta_0(\alpha, \beta) = A_0. \tag{7.36}$$

(b) $\kappa = 0$

In this case, truncated at $n = N$, (7.33) has $N + 1$ equations and $N + 1$ unknowns and gives the unique solution.

(c) $\kappa = -1$

In this case observing that $P_0^{(\alpha,\beta)}(t) = 1$, we express the first equation ($j = 0$) in (7.33) as‡

$$\begin{aligned} F_0 - \sum_0^\infty d_{0n} c_n &= \int_{-1}^{1} f(x) \frac{dx}{w(x)} - \int_{-1}^{1} \frac{dx}{w(x)} \sum_0^\infty c_n k_n(x) \\ &= \int_{-1}^{1} \frac{dx}{w(x)} \left[f(x) - \int_{-1}^{1} k(x, t) \phi(t)\, dt \right] = 0 \end{aligned} \tag{7.37}$$

which is the consistency condition of the integral equation and is seen to be automatically satisfied if this technique is used. If the boundary value has a symmetry with respect to x in the sense that, for example, in Fig. 7 $a = -b$, then there is only one additional unknown, b and (7.33) and (7.35) provide $N + 2$ equations to determine $c_0, \ldots, c_N$, and b. However, if the problem has no symmetry (which is always the case when the friction is involved), both a and b are unknown. Theoretically, the consistency condition with (7.35) provides the additional equations to determine a and b. In this case an extra equation may be gained by writing (7.33) as

$$-\frac{2B}{\sin \pi\alpha} \theta_j(-\alpha, -\beta) c_{j-1} + \sum_{n=0}^{N-1} d_{jn} c_n = F_j, \qquad j = 0, \ldots, N, \tag{7.38}$$

† Note that for $\kappa = 1$ the arbitrary constant C_0 appearing in the exact solution (7.16) or (7.27) is also determined from (7.35).

‡ Note that (7.32) starts with $c_0 P_1^{(-\alpha,-\beta)}(x)$. Therefore in (7.33) $c_{-1} = 0$.

which, with (7.35) provide $N + 2$ equations to determine $c_0, \dots, c_{N-1}$, a, and b.

It is seen that if $k(x,t)$ is zero and $f(x)$ is a polynomial of finite degree, by expressing it in terms of a series of Jacobi polynomials $P_k^{(-\alpha,-\beta)}(x)$ and using (7.31) we may obtain the solution of the integral equation (7.28) in closed form by simple observation. Such solutions of some simple examples regarding the contact problem shown in Fig. 7 may be found in [26]. The solution of the wedge problem shown in Fig. 8 with and without a crack initiating at the apex will appear elsewhere. The application of the technique described in this section to singular integral equations of the second kind with complex coefficients may be found in [31], [33], [42], and [43]. Extension of the technique to a system of equations in which A, B, and $k(x, y)$ are square matrices is given in [44] and [25].

7.3 Solution by Gauss–Jacobi Integration Formulas

In order to solve the singular integral equation of the second kind (7.28) in a direct way one needs an integration formula for the dominant part

$$K(\phi) = A\phi(x) + \frac{B}{\pi}\int_{-1}^{1}\frac{\phi(t)}{t-x}\,dt. \tag{7.39}$$

Following closely the procedure outlined in Section 5 (see also [26]) we can indeed develop such an integration formula [45]. Analogously to (5.53) first consider the expansion

$$\frac{P_{n-\kappa}^{(-\alpha,-\beta)}(x)P_j^{(\alpha,\beta)}(x) - P_n^{(\alpha,\beta)}(x)P_{j-\kappa}^{(-\alpha,-\beta)}(x)}{P_n^{(\alpha,\beta)}(x)} = -\sum_{k=1}^{n}\frac{b_k}{t_k - x}, \tag{7.40}$$

where

$$P_n^{(\alpha,\beta)}(t_k) = 0, \qquad k = 1, \dots, n, \tag{7.41}$$

$$b_k = [P_{n-\kappa}^{(-\alpha,-\beta)}(t_k)P_j^{(\alpha,\beta)}(t_k)]\Big/\frac{d}{dt}P_n^{(\alpha,\beta)}(t_k). \tag{7.42}$$

If we now select x_r as the roots of

$$P_{n-\kappa}^{(-\alpha,-\beta)}(x_r) = 0, \qquad r = 1, \dots, n-\kappa, \tag{7.43}$$

from (7.40) it follows that

$$P_{j-\kappa}^{(-\alpha,-\beta)}(x_r) = \sum_{k=1}^{n}\frac{P_{n-\kappa}^{(-\alpha,-\beta)}(t_k)P_j^{(\alpha,\beta)}(t_k)}{\dfrac{d}{dt}P_n^{(\alpha,\beta)}(t_k)}\,\frac{1}{t_k - x_r}. \tag{7.44}$$

We next consider (7.30) and assume that $g(t)$ can be approximated to a sufficient degree of accuracy by

$$g(t) \cong \sum_{j=0}^{p} c_j P_j^{(\alpha,\beta)}(t). \tag{7.45}$$

From (7.30), (7.31), (7.39), (7.44), and (7.45) for $n > p$ it then follows that

$$\begin{aligned} K[\phi(x_r)] &\cong -\frac{2^{-\kappa}B}{\sin \pi\alpha} \sum_{j=0}^{p} c_j P_{j-\kappa}^{(-\alpha,-\beta)}(x_r) \\ &= \frac{-2^{-\kappa}B}{\sin \pi\alpha} \sum_{k=1}^{n} \frac{P_{n-\kappa}^{(-\alpha,-\beta)}(t_k)}{\frac{d}{dt} P_n^{(\alpha,\beta)}(t_k)} \frac{g(t_k)}{t_k - x_r}. \end{aligned} \tag{7.46}$$

Again, the only approximation in (7.46) is due to the truncation in (7.45). Using the properties of Jacobi polynomials we can put (7.46) into the standard form [45]

$$K[\phi(x_r)] \cong \sum_{k=1}^{n} W_k \frac{Bg(t_k)}{t_k - x_r}, \tag{7.47}$$

$$\begin{aligned} W_k &= \frac{2^{\alpha+\beta}(2n+\alpha+\beta)\Gamma(n+\alpha)\Gamma(n+\beta)}{\pi n!\Gamma(n+\alpha+\beta+1)P_{n-1}^{(\alpha,\beta)}(t_k)\frac{d}{dt}P_n^{(\alpha,\beta)}(t_k)} \\ &= \frac{2^{\alpha+\beta+1}\Gamma(\alpha+n+1)\Gamma(\beta+n+1)}{\pi n!\Gamma(n+\alpha+\beta+1)(1-t_k^2)\left[\frac{d}{dt}P_n^{(\alpha,\beta)}(t_k)\right]^2} \\ &= -\frac{2^{\alpha+\beta}(2n+\alpha+\beta+2)\Gamma(\alpha+n+1)\Gamma(\beta+n+1)}{\pi(n+1)!\Gamma(n+\alpha+\beta+1)P_{n+1}^{(\alpha,\beta)}(t_k)\frac{d}{dt}P_n^{(\alpha,\beta)}(t_k)}. \end{aligned} \tag{7.48}$$

Using (7.47) we now express the integral equation (7.28) as

$$\sum_{k=1}^{n} W_k \left[\frac{Bg(t_k)}{t_k - x_r} + \pi k(x_r, t_k)\right] = f(x_r), \qquad r = 1, \ldots, n-\kappa, \tag{7.49}$$

where t_k and x_r are given by (7.41) and (7.43). For $\kappa = 1$, (7.49) gives $n - 1$ equations to determine $g(t_1), \ldots, g(t_n)$. The nth equation is obtained from (7.35) which becomes

$$\sum_{k=1}^{n} W_k g(t_k) = A_0/\pi. \tag{7.50}$$

For $\kappa = 0$ (7.49) contains n equations and n unknowns and gives a unique solution. In this case (7.35) may be used to determine the unknown end value (e.g. a or b in Fig. 7).

For $\kappa = -1$ (7.49) and (7.35) give $n+2$ equations to determine $g(t_1), \ldots, g(t_n)$, and the unknown end values a and b. It can again be shown that by using this method the consistency condition of the integral equation

$$\int_{-1}^{1} \left[f(x) - \int_{-1}^{1} k(x,t)\phi(t)\, dt \right] \frac{dx}{w(x)} = \int_{-1}^{1} K[\phi(x)] \frac{dx}{w(x)} = 0 \tag{7.51}$$

is automatically satisfied. This can be seen if we observe that $1/w(x)$ is the weight of $P_n^{(-\alpha,-\beta)}(x)$, $\kappa = -1$, $P_0^{(-\alpha,-\beta)}(x) = 1$,

$$K[\phi(x)] = -\frac{2B}{\sin \pi\alpha} \sum_{j=0}^{p} c_j P_{j+1}^{(-\alpha,-\beta)}(x), \tag{7.52}$$

and by using the orthogonality conditions for the Jacobi polynomials.

As an example consider the wedge problem shown by Fig. 8 and formulated by (7.6–7.8). Assume that the rigid wedge has flat faces with sharp corners at the ends a and b. Thus, (7.6) must be solved for $f(r) = 0$ and under the condition that

$$\int_a^b p(r)\, dr = P = P_0/[2(\sin \theta_0 - \eta \cos \theta_0)] = A_0. \tag{7.53}$$

Let the coefficient of friction be $\eta = 0.5$. From (7.13) it then follows that

$$\begin{aligned} \alpha &= -0.545167, \qquad \beta = -0.454833, \qquad \kappa = 1, \\ p(r) &= g(r)w(r), \qquad w(r) = (b-r)^{\alpha}(r-a)^{\beta}. \end{aligned} \tag{7.54}$$

We obtain the solution in a very straightforward manner by considering (7.6–7.8) with (7.49), (7.50), and (7.54).

Of particular interest in problems of this type is the strength of stress singularity around the singular points a and b. This is known as the stress intensity factor and, in this case, may be defined by

$$\begin{aligned} k(a) &= \lim_{r \to a} \sqrt{2}\,(r-a)^{-\beta} p(r), \\ k(b) &= \lim_{r \to b} \sqrt{2}\,(b-r)^{-\alpha} p(r). \end{aligned} \tag{7.55}$$

Some sample numerical results are shown in Table 1 where $c = (b-a)/2$ and P is defined by (7.53), P_0 being the total wedging force.

Table 1 Stress Intensity Factors for the Contact Problem in an Elastic Wedge (Fig. 8)

$\frac{b+a}{b-a}$	$\theta_0(°)$	$\frac{k(a)}{P/(\pi c^{1+\beta})}$	$\frac{k(b)}{P/(\pi c^{1+\alpha})}$
	150.0	1.012	0.989
10	165.0	1.031	0.971
	172.5	1.037	0.967
	150.0	1.055	0.957
4	165.0	1.108	0.916
	172.5	1.122	0.903
	150.0	1.155	0.906
2	165.0	1.281	0.825
	172.5	1.316	0.803
	150.0	1.346	0.852
4/3	165.0	1.611	0.726
	172.5	1.688	0.689

Acknowledgments. Most of the research which forms the basis of this article was supported by the Engineering Division of the National Science Foundation and by Langley Research Center of the National Aeronautics and Space Administration.

8 REFERENCES

1. Sneddon, I. N., *Mixed Boundary Value Problems in Potential Theory*, North Holland, Amsterdam (1966).
2. Muskhelishvili, N. I., *Some Basic Problems of the Mathematical Theory of Elasticity*, Noordhoff, Groningen, the Netherlands (1953).
3. Muskhelishvili, N. I., *Singular Integral Equations*, Noordhoff, Groningen, the Netherlands (1953).
4. Gakhov, F. D., *Boundary Value Problems*, Pergamon Press, Oxford (1966).
5. Pogorzelski, W., *Integral Equations and their Applications*, Vol. 1, Pergamon Press, Oxford (1966).
6. Mikhlin, S. G., *Integral Equations*, Pergamon Press, Oxford (1964).
7. Galin, L. A., *Contact Problems in the Theory of Elasticity*, North Carolina State College, Raleigh, N.C. (1961).
8. Savin, G. N., *Stress Concentration Around Holes*, Pergamon Press, Oxford (1961).
9. Green, A. E. and Zerna, W., *Theoretical Elasticity*, Oxford University Press (1954).
10. Milne-Thomson, L. M., *Plane Elastic Systems*, Springer-Verlag (1960).
11. Milne-Thomson, L. M., *Antiplane Elastic Systems*, Springer-Verlag (1962).
12. Tricomi, F. G., *Integral Equations*, Interscience Publishers, New York (1963).
13. Vekua, N. P., *Systems of Singular Integral Equations*, Noordhoff, Groningen, the Netherlands (1967).

14. Gradshteyn, I. S. and Ryzhik, I. M., *Tables of Integrals, Series, and Products*, Academic Press, New York (1965).
15. Brown, E. J. and Erdogan, F., "Thermal Stresses in Bonded Materials Containing Cuts on the Interface," *Int. J. Engng Sci.* **6** (1968) 517–529.
16. Erdogan, F., "Stress Distribution in a Nonhomogeneous Elastic Plane with Cracks," *J. Appl. Mech.* **30**, *Trans. ASME* (1963) 232–236.
17. Jolley, L. B. W., *Summation of Series*, Dover, New York (1961).
18. Erdogan, F., "Fracture Problems in Composite Materials," *J. Engng Fract. Mech.* **4** (1972) 811–840.
19. Erdogan, F. and Civelek, M. B., "Contact Problems for an Elastic Reinforcement Bonded to an Elastic Plate," *J. Appl. Mech.* **41**, *Trans. ASME* (1974) 1014–1018.
20. Stroud, A. H. and Secrest, D., *Gaussian Quadrature Formulas*, Prentice-Hall, New York (1966).
21. Abramowitz, M. and Stegun, I. A., *Handbook of Mathematical Functions*, National Bureau of Standards, *Appl. Math.*, Series 55 (1964).
22. Erdelyi, A., *Higher Transcendental Functions*, Vol. 2, McGraw-Hill, New York (1953).
23. Szegö, G., *Orthogonal Polynomials*, Colloquium Publications, 23, American Math. Soc. (1939).
24. Kantorovich, L. V. and Krylov, V. I., *Approximate Methods of Higher Analysis*, Interscience, New York (1958).
25. Erdogan, F., "Complex Function Theory," *Continuum Physics*, Vol. II, A. C. Eringen (ed.), Academic Press (1975) 523–603.
26. Erdogan, F., Gupta, G. D., and Cook, T. S., "Numerical Solution of Singular Integral Equations," *Methods of Analysis and Solutions of Crack Problems*, G. C. Sih (ed.), Noordhoff Int. Publ., Leyden (1973) 368–425.
27. Erdogan, F. and Gupta, G. D., "On the Numerical Solution of Singular Integral Equations," *Q. Appl. Math.* **30** (1972) 525–534.
28. Cook, T. S. and Erdogan, F., "Stresses in Bonded Materials with a Crack Perpendicular to the Interface," *Int. J. Engng Sci.* **10** (1972) 677–697.
29. Erdogan, F. and Biricikoglu, V., "Two Bonded Half Planes with a Crack Going Through the Interface," *Int. J. Engng Sci.* **11** (1973) 745–766.
30. Hein, V. L. and Erdogan, F., "Stress Singularities in a Two-material Wedge," *Int. J. Fract. Mech.* **7** (1971) 317–330.
31. Erdogan, F. and Gupta, G. D., "Layered Composites with an Interface Flaw," *Int. J. Solids Struct.* **7** (1971) 1089–1107.
32. Erdogan, F. and Ratwani, M., "Plasticity and Crack Opening Displacement in Shells," *Int. J. Fract. Mech.* **8** (1972) 413–426.
33. Erdogan, F. and Arin, K., "Penny-shaped Interface Crack Between an Elastic Layer and a Half Space," *Int. J. Engng Sci.* **10** (1972) 115–125.
34. Erdogan, F. and Gupta, G. D., "Stresses Near a Flat Inclusion in Bonded Dissimilar Materials," *Int. J. Solids Struct.* **8** (1972) 533–547.
35. Erdogan, F. and Gupta, G. D., "The Torsion Problems of a Disk Bonded to a Dissimilar Shaft," *Int. J. Solids Struct.* **8** (1972) 93–109.
36. Erdogan, F. and Gupta, G. D., "The Load Transfer Problem in Shafts Coupled Through a Sleeve," *J. Appl. Mech.* **40** (1973) 997–1003.
37. Erdogan, F. and Cook, T. S., "Antiplane Shear Crack Terminating at and Going through a Bimaterial Interface," *Int. J. of Fract.* **10** (1974) 227–240.
38. Erdogan, F. and Gupta, G. D., "The Inclusion Problem with a Crack Crossing the Boundary," *Int. J. of Fract.* **11** (1975) 13–27.
39. Gupta, G. D. and Erdogan, F., "The Problem of Edge Cracks in an Infinite Strip," *J. Appl. Mech.* **41**, *Trans. ASME* (1974) 1001–1006.
40. Erdogan, F., "Simultaneous Dual Integral Equations with Trigonometric and Bessel Kernels," *ZAMM* **48** (1968) 217–225.

41. Tricomi, F. G., "On the Finite Hilbert Transformation," *Q. J. Math. Oxford* **2** (1951) 199.
42. Erdogan, F. and Ozbek, T., "Stresses in Fiber-reinforced Composites with Imperfect Bonding," *J. Appl. Mech.* **36**, *Trans. ASME* (1969) 865–869.
43. Ozbek, T. and Erdogan, F., "Some Elasticity Problems in Fiber-reinforced Composites with Imperfect Bonds," *Int. J. Engng Sci.* **7** (1969) 931–946.
44. Erdogan, F., "Approximate Solution of Systems of Singular Integral Equations," *SIAM J. Appl. Math.* **17** (1969) 1041–1059.
45. Krenk, S., "On Quadrature Formulas for Singular Integral Equations of the First and the Second Kind," *Q. Appl. Math.* **33** (1975) 225–232.

II

On the Problem of Crack Extension in Brittle Solids Under General Loading

K. Palaniswamy† and W. G. Knauss

California Institute of Technology, Pasadena, California

1 INTRODUCTION

By far most analyses of crack propagation involve planar cracks embedded in such geometries and under such loads that the crack tends to extend in its own plane. This special problem of fracture analysis is certainly important since it provides many design-relevant concepts for the prediction of structural failures due to unstable crack propagation. However, there are many situations in which cracks are embedded in stress fields which do not lead to propagation of a crack in its own plane and it appears appropriate that the special circumstances leading to fracture under such conditions be discussed. We shall speak of co-planar and non-planar crack propagation, if an initially planar crack propagates such that its extension coincides with or differs from that plane, respectively.

The following work is devoted to the problem of non-planar crack growth in brittle solids. We restrict ourselves to those problems that allow an *a priori* prescription of the tractions on the crack boundary during crack growth. Thus we exclude those cases for which the crack surfaces are forced together during the crack extension process. For the purpose of this presentation we define a solid as brittle if it exhibits only a vanishingly small plastic zone at the crack tip which may be on the order of a few tens of angstroms. Furthermore, we confine our analysis to those materials which admit the application of the linearized theory of elasticity. We thus rule out from

† Presently at: Northrop Research and Technology Center, Hawthorne, California 90250.

consideration materials which develop what is generally known as large-scale yielding in the presence of cracks. In addition we wish to distinguish between materials which are brittle in the sense just stated and those that are commonly referred to as brittle. For example, uncrosslinked organic glasses such as polystyrene (PS) and polymethylmethacrylate (PMMA) are often called brittle. Yet the domain in which the nonlinear (viscoplastic) material properties dominate at the crack tip is much larger than that which one encounters in the inorganic glasses or in the densely crosslinked rigid polymers such as the epoxies. Materials such as exemplified by polystyrene and polymethylmethacrylate may be called quasi-brittle or mildly ductile for the purpose of the present discussion; a problem germane to them will be discussed in some detail later on.

Although Griffith seemed to believe [1, 2] that an energy balance provides a fundamental criterion for fracture, he could not detach himself from the (historical) idea that the exceedance of a critical tensile stress may be a viable argument for describing fracture processes. We shall be concerned with an extension of the energy-balance argument advanced by Griffith for self-similar crack growth to cases of non-planar crack enlargement. This argument as applied to co-planar crack growth holds that crack propagation becomes possible and will occur when the energy released during crack front advance just equals the energy required to form the new crack surfaces.

In order to compute the amount of energy released during a small amount of crack propagation which is not (necessarily) co-planar, it is required to compare the energy stored in the solid containing the *extended* crack with that stored in the solid prior to crack extension. This comparison presumes that the deformation or stress fields are known for the geometries of both the extended and of the unextended states. Most analyses of the stresses surrounding cracks are made for planar cracks with smooth boundaries. The comparison of the energies for the unextended and the extended states is then easy as long as the extended crack geometry is similar to the unextended one in the sense that the former may be obtained from the latter by merely changing the size scale of the crack. In fact, it can be shown [3–5] that the energy released during co-planar crack propagation can be expressed in terms of the stress intensity factor† corresponding to the stress state *prior* to crack propagation. This result follows essentially from the observation that for co-planar crack propagation the energy is a continuous function of the crack size and that the stress intensity factor changes only infinitesimally for an infinitesimal change in crack size. One cannot claim that the same

† For a definition and significance of this quantity, see for example, Ref. [5].

is true when the crack grows in a non-planar fashion and the requisite analysis, which is not trivial, has not yet been supplied. We hope that the following development will tend to clarify this problem.

Since the stress intensity factor describes the state of stress at the crack front and determines also the energy released per unit of new crack surface generation in *co-planar* crack growth, it is possible to predict the onset of crack growth from conditions that exist only *prior* to crack extension. In other words, for co-planar growth no further statement needs to be made about the *process* of crack extension. The argument is therefore close at hand that the prediction of onset of crack growth under conditions favoring non-planar growth should also be possible in terms of the stress state in the crack front vicinity *prior* to crack extension. While this argument cannot be contradicted out of hand it appears doubtful that prefacture conditions should totally describe the fracture *process*. It appears necessary, therefore, to consider the comparison of the energies appropriate to the (infinitesimally) extended and unextended states allowing for non-planar crack growth in order to arrive at a consistent description of the initial crack propagation process in brittle solids.

To this end, we consider briefly the generalization of Griffith's energy balance to cracks of arbitrary shape in three-dimensional solids subjected to arbitrary loads. The crack extension process is considered to occur in a quasi-static manner such that inertial effects may be neglected. Therefore, when we refer to time (variation) we use it as a parameter that delineates the history of sequence of events such as in loading or crack progression. Next we consider a crack as a discontinuity in the material in the form of a surface† S_1, which is bounded by the crack periphery P_1. We do not insist that S_1 be smooth in the sense that the normal to it vary continuously from point to point, nor that P_1 be similarly smooth. We then consider a surface S_2 obtained from S_1 by extending it by a small amount δS. We assume that δS is smooth but do not require that the periphery of the extension δS be jointed continuously to the periphery P_1.‡ It is thus possible that the extension has only one point in common with the periphery P_1. Let the unfractured solid ($\delta S = 0$) be subject to surface tractions and displacements

† The term surface may refer to either a mathematical surface which defines the geometry of the crack or to the surfaces of the crack which form the boundary of the solid. We believe the meaning of the word to be clear from the context.

‡ Although it is conceivable that the crack enlargement is non-smooth if it starts from a sharp corner in the crack periphery, we disregard this possibility because for our present exposition that complication would only cloud the argument without contributing any fundamental information to the developments we have in mind. For a physically possible counterpart of the postulated crack extension, see Subsection 8.2, in particular Figs. 21 and 22.

on its boundary, and let $\mathscr{L}$ be a parameter that determines the magnitude of the tractions and displacements but not their spatial distribution; thus, as $\mathscr{L}$ increases from, say, zero, all stress components increase proportionately (proportional loading) as long as crack propagation does not occur. During crack extension let $\mathscr{L}$ remain constant. Then we may compute, in principle, the potential energy stored in the solid, $V_1(\mathscr{L}, S_1)$ just prior to crack extension, and the same quantity after the crack has extended by the increment δS, $V_2(\mathscr{L}, S_2)$. We recall for later reference that the extension δS is presumed smooth, that it may be characterized by a periphery δP which may start on one face of the crack and end on the other face, and may possess some average normal direction. Now consider the limit

$$\lim_{S_2 \to S_1} \frac{V_2(\mathscr{L}, S_2) - V_1(\mathscr{L}, S_1)}{S_2 - S_1} = -G(\mathscr{L}, \underset{\approx}{\delta S}). \tag{1.1}$$

We call $G(\mathscr{L}, \delta S)$ the energy release rate which is, in general, a function of the shape and orientation (denoted by $\underset{\approx}{\delta S}$) of the surface extension δS. Of all possible extensions δS there should be one δS^* which makes G an absolute maximum

$$G_{\max} = G(\mathscr{L}, \underset{\approx}{\delta S}^*). \tag{1.2}$$

We now assert that crack propagation in a brittle solid becomes possible when the energy release rate $G_{\max}$ reaches a critical value, say γ_0, which is a material parameter, at a load corresponding to $\mathscr{L}_0$, and that the propagation is initially along the surface increment δS^*.

That the energy release rate should be a maximum is apparent from the following. Let $\mathscr{L}_0$ denote again the load parameters for the proportionately loaded structure at incipient fracture. Furthermore, consider a crack extension $\delta S \neq \delta S^*$ so that $G(\mathscr{L}_0, \underset{\approx}{\delta S}) < G_{\max}$.

In order to cause fracture along the chosen surface increment δS we thus need to increase the load $\mathscr{L}_0$ by $\Delta\mathscr{L}$ in order that $G(\mathscr{L}_0 + \Delta\mathscr{L}, \underset{\approx}{\delta S}) = \gamma_0$. Thus there exists a relation between the magnitude of the load and the orientation of the crack extension at fracture. This relation is such that deviation from δS^* *decreases* the energy release rate, while an increase in the loading $\mathscr{L}$ increases it.† Therefore, if the energy release rate is to be a constant at fracture and equal to γ_0 then *minimal loading* at rupture requires that δS be such as to maximize the energy release rate.

† For elastic solids in general a more precise statement is apparently not readily possible; for linearly elastic solids the energy release rate is proportional to $\mathscr{L}^2$.

So far we have assumed implicitly that the energy γ_0 required to form new surface by fracture is independent of the orientation of the crack extension. In anisotropic solids such is not the case and it is, therefore, appropriate to generalize the just stated hypothesis of maximal energy release rate to materials exhibiting anisotropic fracture behavior.

In fact, one may visualize that in a crystal fracture extension may occur simultaneously on two different crystallographic planes. For reasons of simplicity in this discussion we avoid the complications arising out of that possibility and restrict ourselves to consideration of smooth crack growth surfaces representable by a single plane.

To this end let the (unit) vector **n** denote the normal to a small potential fracture extension surface. Then $\gamma(\mathbf{n})$ represents the energy required to create by fracture a unit of surface in the plane normal to **n**. In order to display the explicit dependence of the energy release rate on **n** we write† $G(\mathscr{L}, \underset{\approx}{\delta S}) = G(\mathscr{L}, \underset{\sim}{\delta S}, \mathbf{n})$. Clearly, for fracture to occur we must have

$$G(\mathscr{L}, \underset{\sim}{\delta S}, \mathbf{n}) = \gamma(\mathbf{n}) \tag{1.3}$$

and this relation must be satisfied for a minimal value of the loading $\mathscr{L}$. If we think of $G(\mathscr{L}, \underset{\sim}{\delta S}, \mathbf{n})$ and $\gamma(\mathbf{n})$ as two (hyper) surfaces in a four-dimensional space, then the requirement of fracture at minimal loading is equivalent to osculation of the two (hyper) surfaces G and γ. That condition requires

$$\frac{\partial G(\mathscr{L}, \underset{\sim}{\delta S}, \mathbf{n})}{\partial n_i} = \frac{\partial \gamma(\mathbf{n})}{\partial n_i}, \qquad i = 1, 2, 3. \tag{1.4}$$

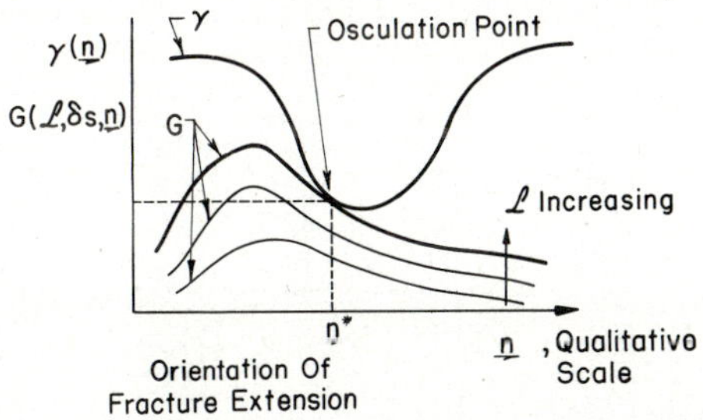

Fig. 1 A symbolic plot of energy release rate and energy required for crack growth.

† Here the second argument δS indicates that the energy release rate G depends on the shape of the extension surface. In reality δS and **n** may not be independent. We treat these quantities as independent here for discussion purposes, although this is not absolutely necessary. G is supposed to be a continuous and smooth function of the vector components of **n**.

Let us assume that the dependence of G on the shape of $\delta \underset{\sim}{S}$ is not very sensitive and that a shape $\delta \underset{\sim}{S}_*$ will allow maximization of G with respect to $\mathscr{L}$ and $\mathbf{n}$. Then (1.3) and (1.4) represent a set of equations for the determination of the four unknowns $\mathscr{L}$, n_1, n_2, and n_3.

We may visualize the solution to these equations in a symbolic two-dimensional graph with $\mathbf{n}$ plotted (symbolically) along the abscissa and γ as well as G along the ordinate as in Fig. 1.

If γ is independent of $\mathbf{n}$, i.e. if the fracture behavior of the solid is isotropic, then $\partial\gamma(\mathbf{n})/\partial n_i = 0$ and $G(\mathscr{L}, \delta \underset{\sim}{S}_* \mathbf{n})$ possesses an extremum for some value of $\mathbf{n}$. For this case we have argued above that G attains a maximum with respect to $\mathbf{n}$.

If γ is not a smooth function of $\mathbf{n}$ such as may be the case in crystals, minerals and other anisotropic solids possessing "weak planes" the derivatives $\partial\gamma(\mathbf{n})/\partial n_i$ do not exist everywhere. In that case the onset of fracture may be determined directly from (1.3) by requiring satisfaction for minimum $\mathscr{L}$ along a weak plane. A symbolic plot for this special case is given in Fig. 2.†

The argument for fracture in strongly anisotropic solids may be extended in an obvious manner to some problems of fracture growth from an interface unbond between two joined brittle solids. The fracture energy of the interface may then be represented by a singularly low value along the interface. If the joined solids are, for example, both isotropic, their fracture properties are represented by (different) constants separated by the singularly low value along the interface. A complete discussion of the unbonding problem under general loads is beyond the scope of this paper and we choose not to discuss the detailed problems arising, for example, out of crack or

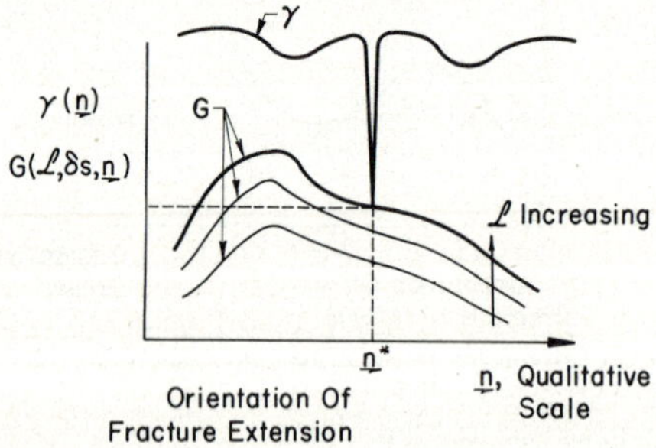

Fig. 2 A symbolic plot of energy release rate and energy required for crack growth in a solid with a "weak" plane.

† We exclude pathological cases such as tension parallel to the crack.

unbond enlargement which occurs simultaneously along the interface and into one of the adherents.

It is clear that the analysis of fracture propagation in arbitrary solids under general loads is not a simple matter. Indeed the stress analysis required for the description of the fracture process just considered is very difficult at best even in linearly elastic and brittle solids containing irregular or even simple plane cracks under arbitrary loads. In one two-dimensional case, however, we can perform the requisite stress analysis. We may thus use those results to compare the criterion of maximal energy release with other criteria based on conditions prevailing before crack growth takes place. Moreover, experiments employing the same two-dimensional geometry may be used to further evaluate the significance of the differences between different crack propagation criteria. If the simple criteria compare favorably with the energy calculations in the two-dimensional case, then one would feel justified in applying the simpler criteria to the more complicated problems resulting from three-dimensional geometries and loads.

In Section 2 we review primarily the problem of non-planar crack growth in two dimensions and comment on solutions to it offered up to this time. In Section 3 we formulate the energy analysis for the non-planar growth of a crack. Because the analysis for the plane problem is not in closed form but involves considerable computational effort we evaluate, primarily for guidance and computational checks, the analogous (three-dimensional) problem under conditions of antiplane shear, a solution to which can be found in closed form if one assumes the geometry to remain two dimensional during crack growth.† These computations are recorded in Subsections 3.2A and 3.2B, while Subsections 3.3 and 3.4 are devoted to the evaluation of the in-plane branched crack problem. A brief Section 4 discusses the numerical procedures, including computational accuracy of determining the crack branching angle and the critical load.

In Section 5 we discuss the results of the two-dimensional analysis in preparation for the comparison with the experimental work. Comparison with related theories is limited to the stress criterion of fracture [6, 8]. In addition to the crack extension by an infinitesimal length we discuss the growth by a finite increment as a possible approximation to predicting fracture in mildly ductile solids. In Section 6 we compare the two-dimensional analysis to fracture experiments on a brittle material conducted in our own laboratory and to data from mildly ductile solids as reported in the literature.

† As we shall see in Section 8, this assumption, made in the present context for purely analytical reasons, is not matched by physical evidence in brittle solids undergoing fracture in this mode.

This is then concluded with an assessment of the ease and practicality (economics) of the application of several proposed fracture criteria which assessment motivates the simple analytical approach adopted for the prediction of fracture in general three-dimensional problems as discussed in the following sections.

In Section 7 we discuss the problem of crack extension for cases when three modes of crack tip deformation occur along a smooth crack front. Drawing on the results of the two-dimensional analysis and the attendant experiments as well as on crack propagation experiments† under antiplane shear discussed in Section 8, we present in Section 10 an approximate analysis of fracture for a general crack tip stress field, which is a generalization of the aforementioned stress criterion to three-dimensional loading. An interspersed Section 9 reviews briefly work related to fracture under mode III loading and examines it in the light of the experimental results in Section 8. A final Section 11 addresses itself to the possible extension of the proposed approximate fracture criterion to mildly ductile solids as a guide for making engineering estimates of fracture strengths.

2 REVIEW OF THE TWO-DIMENSIONAL NON-PLANAR CRACK PROBLEM

In 1963 Erdogan and Sih [6] contributed the first extensive account of non-planar crack growth in two-dimensional geometries.‡ Observing that

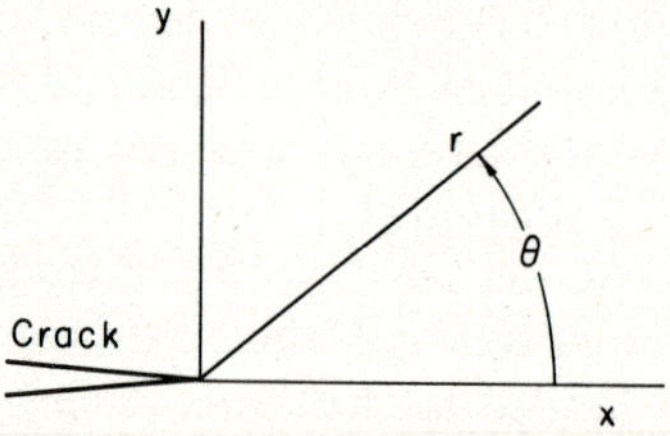

Fig. 3 Definition of polar coordinate system at crack tip.

† These experiments were reported before briefly in Ref. [7]. We include them here in order to discuss them more fully in connection with the present and more general topic of crack propagation in arbitrary geometries.

‡ During the final stages of the manuscript preparation we found that this problem had also been treated analytically and experimentally in Ref. [8]. Ref. [6] discusses also out-of-plane crack growth in cracked plates under transverse loads under the assumption that this loading produced the antiplane (mode III) problem; in reality the tests induced primarily plate-bending stresses.

the stress analysis required for the comparison of the energies for the non-planarly extended crack was difficult they suggested an alternative criterion of fracture. Define the plane polar coordinates as in Fig. 3 and let σ_r, σ_θ and $\tau_{r\theta}$ be the in-plane components of the stress tensor referred to this local coordinate system. In the immediate vicinity of the crack tip the dominant contribution to these stress components derives from the terms

$$\begin{aligned}\sqrt{2\pi r}\,\sigma_r &= K_{\mathrm{I}} \cos\frac{\theta}{2}\left(1+\sin^2\frac{\theta}{2}\right) + K_{\mathrm{II}}\left[\tfrac{3}{2}\sin\theta\cos\frac{\theta}{2} - 2\sin\frac{\theta}{2}\right],\\ \sqrt{2\pi r}\,\sigma_\theta &= K_{\mathrm{I}}\cos^3\frac{\theta}{2} - \tfrac{3}{2}K_{\mathrm{II}}\sin\theta\cos\frac{\theta}{2},\\ \sqrt{2\pi r}\,\tau_{r\theta} &= \tfrac{1}{2}K_{\mathrm{I}}\sin\theta\cos\frac{\theta}{2} + \tfrac{1}{2}K_{\mathrm{II}}\cos\frac{\theta}{2}[3\cos\theta - 1],\end{aligned} \tag{2.1}$$

where K_{I} and K_{II} are the (in-plane) symmetric and asymmetric (mode I and mode II) stress intensity factors, respectively. Erdogan and Sih explore the criterion which demands that:

(a) crack propagation occurs along a ray $\theta = \text{const}$, and in particular along that ray $\theta = \theta_*$ on which $\sqrt{2\pi r}\,\sigma_\theta$ achieves a maximum;† and
(b) fracture is imminent when that maximum of $\sqrt{2\pi r}\,\sigma_\theta$ (for $\theta = \theta_*$) reaches a critical value.

Experiments conducted by Erdogan and Sih corroborated their postulated criterion reasonably well, but some systematic deviation existed which could be due to the material (polymethylmethacrylate) employed in their experiments. Polymethylmethacrylate is an uncrosslinked polymer which behaves macroscopically like a brittle solid but develops nonlinear material behavior at the crack tip such that the Erdogan and Sih analysis involving only singular terms in the stresses may not be adequate. These observations motivated us to reconsider the problem of non-planar crack propagation in two dimensions both from an experimental and analytical viewpoint.‡

A brief account of our re-examination of the non-planar crack propagation problem in terms of the energy release rate for small or infinitesimal extension was summarized in a note [10] and parts of the following develop-

† Because along that particular ray $\tau_{r\theta}$ vanishes σ_θ is a principal stress. Thus the Erdogan and Sih criterion is sometimes referred to as a maximal principal stress criterion. The latter is a misnomer because there exist larger principal stresses but their direction is not normal to a ray, and, by assumption cannot cause crack growth along a ray. It seems more appropriate, therefore, to call the criterion a normal stress criterion, "normal to a ray" being implied, or simply a stress criterion.

‡ The analytical portion was submitted as a Ph.D. thesis to the California Institute of Technology [9] in June 1971.

ments amplify the contents of that note. Since its publication there have appeared a number of papers which relate to the criterion for non-planar fracture growth in two dimensions. J. G. Williams and Ewing [11] pursued further the tensile stress criterion of Erdogan and Sih [6] and pointed out, as Cotterell [12] had done, that the singular terms in the expansions for the crack tip stresses may not be sufficient to discuss the likely orientation of further crack growth. With the inclusion of nonsingular terms in the crack tip stresses Williams and Ewing modify the stress criterion by postulating fracture to occur also along a ray, in particular along that ray on which the circumferential stress σ_θ achieves a maximal value with respect to θ, *but* at some (small) distance ρ from the crack tip. Fracture occurs when the maximal stress σ_θ at the radius ρ reaches a critical value. Finnie and Saith [13] pointed to an oversight in Williams and Ewing's analysis [11], the correction of which substantially improves the agreement of their criterion with experimental data on polymethylmethacrylate as obtained by Erdogan and Sih [6], by Williams and Ewing [11] as well as data on an aluminum alloy reported by Pook [14] (see also Ref. [15]). Additional experiments on non-planar crack growth have been reported for monotonic loading by Shah [16], Liu [17], and Marcus, Ho and Frandsen [18], Ellis [19] and for fatigue loading by Shah [20], Roberts and Kibler [21], and Iida and Kobayashi [22].

Sih recently proposed a strain energy density criterion often referred to as the S-criterion according to which fracture occurs along a ray $\theta = \theta_*$ (cf. Fig. 3) on which the strain energy density reaches a minimum with respect to θ at some small distance from the crack tip [23–27]. This small distance is a material parameter and plays the same role as that introduced by Williams and Ewing [11] in their extension of the tensile stress criterion [6, 8]. Onset of fracture by the minimum energy density criterion occurs when the energy density at $r =$ "a", $\theta = \theta_*$ reaches a critical value. Correlation of this criterion with experiments on two-dimensional geometries is as good as the Williams and Ewing criterion when the latter is corrected as suggested by Finnie and Saith [13]. A more detailed discussion of the application of the S-criterion to two- and three-dimensional stress states at the crack front is deferred until later. A hybrid criterion using the crack-growth orientation as determined from the stress criterion [6, 8] and a failure law based on the stress invariants in the crack tip vicinity has been proposed recently by Lindsay [28].

The criteria mentioned so far all have in common that they base the prediction of the fracture process on conditions that exist prior to the event of crack propagation. The only authors besides the writers who deviate from this practice are, to our knowledge, Hussain, Pu, and Underwood [29].

These authors start with a line crack one end of which has attached a non-planar (angled) small extension of length, say, "a". They next consider a virtual growth of that small angular extension and co-planar with it in terms of a J-integral [30–32] and then maximize the energy release of the virtual growth with respect to the orientation of the small non-planar extension of length "a". The net results of that analysis are close to portions of the work outlined here. There are, however, some difficulties in that investigation for a discussion of which the reader is referred to Ref. [33].†

It seems appropriate to include in this brief review of non-planar crack extension in two-dimensional geometries also phenomena resulting from compressive far-field stresses and from continuous changes of crack geometry during growth (smooth crack trajectories). In a discussion on the paper by Erdogan and Sih, McClintock [35] amplified Griffith's [1, 2] remarks on this subject and pointed out an apparent difference in the non-planar fracture from cracks in tensile and compressive stress fields. It appears that fracture in compression is better predicted by a stress criterion applied to a shallow ellipse than by the co-planar growth of a crack with the sliding friction on the crack surfaces being accounted for [36, 37]. The data collection of Hoek and Bieniawski [38] tends to contradict this claim. However, the stress criterion applied to a shallow ellipse in a tensile field is greatly at variance with the experimental evidence [36]. A plausible explanation for this apparent paradox has been offered by Cotterell [39]. A logical means of resolving the uncertainty connected with this problem is to apply the maximum energy release criterion discussed subsequently, properly accounting for the possible friction on the sliding crack surfaces.

We mention also the problems leading to infinitesimal deviation of the crack extension orientation from that of the original or parent crack such that it gives rise to smooth crack propagation trajectories. This type of non-planar crack propagation makes an undesirable appearance in some fracture-testing procedures, as for example the symmetric or double cantilever beam test, where crack curving destroys the geometric symmetry of the test specimen, which symmetry is a chief asset of the test. Conditions leading to this "geometric motion instability of the crack tip" have been analyzed by Cotterell [40] on the basis of the quasi-static near-tip stress field and under inclusion of terms of higher order than those represented in Eqs. (5).

† Just prior to sending this paper to press it was called to our attention that a closed form solution of the same problem has been attempted in Ref. [34]. The limitations of that work are also discussed in Ref. [33].

Finally we record an observation on smooth, non-planar crack growth that has been made by several investigators; see, for example, Morozov and Fridman [41, 42], Porter and Fairhurst [43]. In experiments involving crack generation and propagation from a boundary of a brittle solid the crack tends to follow a path that is nearly equal to an isostatic line passing through the point of initiation. This observation has been exploited by Morozov and Fridman [44] to formulate an approximate variational scheme for the determination of crack paths; the scheme rests on the minimization of the difference between the surface energy and strain energy of the solid.

3 STATEMENT AND FORMULATION OF THE IN-PLANE PROBLEM

We consider an isotropic linear elastic planar solid with a crack of length l_1 and loaded at infinity by uniform tensile stress σ_∞ oriented at an angle α with respect to the crack axis as shown in Fig. 4. We wish to determine that stress σ_∞ at which the crack will extend and the direction in which it will begin to propagate.

Since the problem remains invariant for a rotation of π radians about an axis normal to the plane of the plate and through the mid-point of the crack, one would expect the crack to extend antisymmetrically from both tips. For reasons of mathematical simplicity in solving the associated elasto-static boundary value problem, we shall assume that the crack extension starts

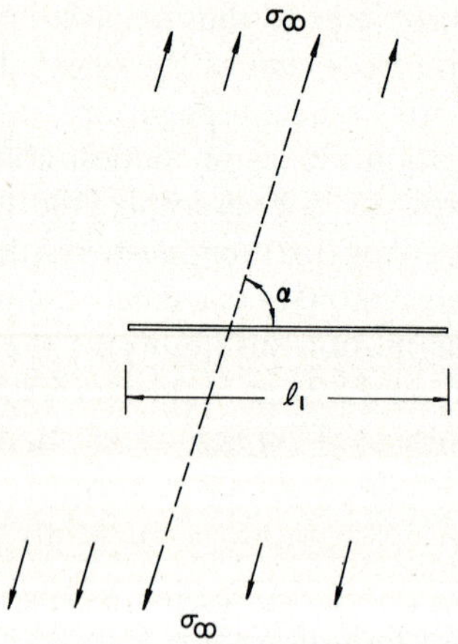

Fig. 4 Geometry and far field loading of crack.

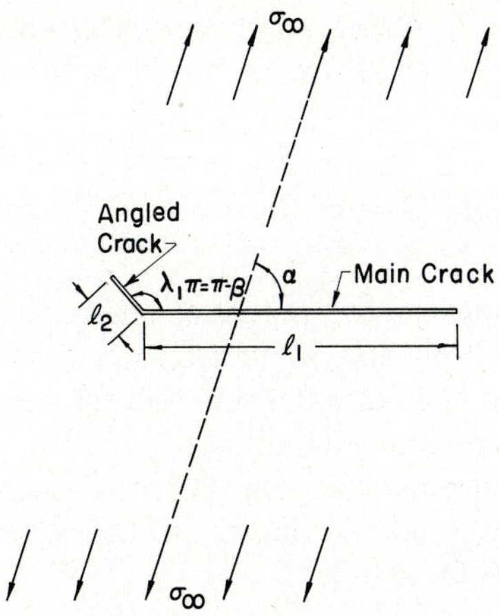

Fig. 5 Angled crack system.

at one tip first—rather than starting from both tips simultaneously—which is also physically reasonable. We shall then speak of the crack as being "branched" or "angled" if it appears as in Fig. 5.

Suppose the elasto-static state for the branched crack is known. Let μ and v be the shear modulus and Poisson's ratio, respectively; let λ_1, l_1 and l_2 be defined as in Fig. 5, and denote by $V(\mu, v, \alpha, \lambda_1, l_1, l_2)$ the change in potential energy per unit (out-of-plane) thickness which is derived from the introduction of the branched crack. In view of Eq. (1.1) the energy release rate for this plane problem becomes, with V_0 denoting the potential energy for the unbranched crack,

$$G(\mu, v; \sigma_\infty, \alpha; l_1; \lambda_1) = -\lim_{l_2 \to 0} \frac{V - V_0}{2l_2}. \tag{3.1}$$

As before let γ_0 be the energy consumed in forming a unit of new surface. For the two-dimensional case Eqs. (1.4) and (1.3), representing the energy criterion, take on the form, respectively,

$$\left.\frac{\partial G}{\partial \lambda_1}\right|_{\lambda_1 = \lambda^*} = 0, \qquad G(\lambda^*) = \gamma_0. \tag{3.2, 3.3}$$

Define the function $g(\alpha, \lambda, l_1)$ so that the energy release rate G can be written for the case of generalized plane stress as

$$G(\mu, v; \sigma_\infty, \alpha; l_1; \lambda) = \frac{\pi}{16} \frac{\sigma_\infty^2}{\mu(1 + v)} g(\alpha, \lambda, l_1). \tag{3.4}$$

Then the maximum of g with respect to λ determines the crack extension angle λ^*; using Eq. (3.4) the relation for the critical load is then obtained as

$$\sigma_\infty \big|_{cr} = \sqrt{\frac{16\mu(1+\nu)}{\pi} \frac{\gamma_0}{g(\alpha, \lambda^*, l_1)}}. \tag{3.5}$$

Hence, once the change in the potential energy of the solid due to the presence of the crack branching in an arbitrary direction is known, we can calculate the direction of crack extension and the value of the loading stress which will cause the crack to extend.

In order to calculate this change in potential energy V which is needed to determine the energy release rate G, it becomes necessary to solve the following boundary-value problem.

3.1 Boundary Value Problem for the Branched Crack

Within the realm of the linearized theory of elasto-statics of the plane possessing homogeneous and isotropic properties, we seek a solution to the pertinent field equations for the angle-cracked (infinite) geometry depicted in Fig. 5 subjected to uniaxial tension σ_∞ far from the crack while the surfaces of the crack (main and angled) remain traction free.

With reference to the cylindrical coordinate system at the branch corner, define the components of the in-plane displacement vector u_r and u_θ and stress tensor σ_r, σ_θ, and $\tau_{r\theta}$; then the boundary conditions on the crack faces correspond to the vanishing of σ_θ and of $\tau_{r\theta}$ on the crack segments, i.e.

$$\begin{matrix} \sigma_\theta = 0 \\ \tau_{r\theta} = 0 \end{matrix} \quad \text{on} \quad L = \begin{cases} \theta = 0, 0 < r < l_1 \\ \theta = \lambda_1 \pi, 0 < r < l_2, \end{cases} \tag{3.6}$$

while for $r \to \infty$

$$\begin{aligned} \sigma_r &= \frac{\sigma_\infty}{2}[1 + \cos 2(\alpha - \theta)] + O\left(\frac{1}{r^2}\right), \\ \sigma_\theta &= \frac{\sigma_\infty}{2}[1 - \cos 2(\alpha - \theta)] + O\left(\frac{1}{r^2}\right), \\ \tau_{r\theta} &= \frac{\sigma_\infty}{2} \sin 2(\alpha - \theta) + O\left(\frac{1}{r^2}\right). \end{aligned} \tag{3.7}$$

We choose to pursue this goal in terms of Kolosov–Muskhelishvili potentials† and in the sequel we adhere to the nomenclature of Ref. [45]. In terms of these complex potentials ϕ and ψ, the components of the stress tensor and of the displacement vector are determined by

$$\begin{aligned} \sigma_r + \sigma_\theta &= 4 \text{ Re } \phi'(z), \\ \sigma_\theta + i\tau_\theta &= 2 \text{ Re } \phi'(z) + \frac{z}{\bar{z}}[\bar{z}\phi''(z) + \psi'(z)], \\ u_r + iu_\theta &= \frac{1}{2\mu} e^{-i\theta}[\kappa\phi(z) - z\overline{\phi'(z)} - \overline{\psi(z)}], \end{aligned} \tag{3.8}$$

where for generalized plane stress $\kappa = (3 - \nu)/(1 + \nu)$.

In terms of the two potential functions the boundary conditions (3.6) become

$$\phi(z) + z\overline{\phi'(z)} + \overline{\psi(z)} = 0 \quad \text{on} \quad L, \tag{3.9}$$

while the conditions at infinity (3.7) reduce to

$$\left.\begin{aligned} \phi(z) &= \Gamma z + \phi_0(z) \\ \psi(z) &= \Gamma' z + \psi_0(z) \end{aligned}\right\}, \quad \text{as} \quad |z| \to \infty, \tag{3.10}$$

where

$$\Gamma = \frac{\sigma_\infty}{4}, \qquad \Gamma' = -\frac{\sigma_\infty}{2} e^{-2i\alpha}, \tag{3.11}$$

and $\phi_0(z)$, $\psi_0(z)$ are holomorphic functions in the plane excluding the crack boundary but including the point at infinity.

We determine the functions $\phi(z)$ and $\psi(z)$ satisfying conditions (3.9) and (3.10) by conformally mapping the region outside the crack boundary L in the z-plane into the region exterior to the unit circle in the ξ-plane through the mapping function [46]

$$z = \omega(\xi) = \frac{A}{\xi}(\xi - e^{i\delta_1})^{\lambda_1}(\xi - e^{i\delta_2})^{\lambda_2}. \tag{3.12}$$

The characteristics of the mapping are documented in Ref. [9], but it is germane to point out for later reference that in the range for $0 < \lambda_1 < 1$ the

† The problem appears to lend itself naturally to attack by a formulation of a set of dual integral equations via Mellin Transforms which can then be reduced to a set of four coupled Fredholm Integral equations. In following this avenue we found some kernels of the integral equations exhibited a complicated singular behavior and believed it easier to exploit the complex variable method instead of tracing the source of the kernel singularity.

mapping is not rational. Also for purposes of later discussion, we record here its derivative

$$\begin{aligned}\omega'(\xi) &= \frac{A}{\xi^2}(\xi - e^{i\delta_1})^{\lambda_1 - 1}(\xi - e^{i\delta_2})^{\lambda_2 - 1}(\xi - e^{i\beta_1})(\xi - e^{i\beta_2}) \\ &= \omega(\xi)\frac{(\xi - e^{i\beta_1})(1 - e^{i\beta_2})}{\xi(\xi - e^{i\delta_1})(\xi - e^{i\delta_2})},\end{aligned} \tag{3.13}$$

where the real parameters A, δ_1, δ_2, β_1, β_2, λ_1, and λ_2 are given in the Appendix. The branches of the functions in (3.12) are chosen such that

$$z = A\xi + O(1) \quad \text{as} \quad |\xi| \to \infty. \tag{3.14}$$

Remembering that we adhere to the notation of Ref. [45], the stress-free boundary condition (3.9) reduces to

$$\phi(\xi) + \frac{\omega(\xi)}{\overline{\omega'(\xi)}}\overline{\phi'(\xi)} + \overline{\psi(\xi)} = 0 \quad \text{on} \quad |\xi| = 1, \tag{3.15}$$

while the conditions (3.10) at infinity become

$$\left.\begin{aligned}\phi(\xi) &= \Gamma\omega(\xi) + \phi_0(\xi) \\ \psi(\xi) &= \Gamma' A\xi + \psi_0(\xi)\end{aligned}\right\}, \quad \text{as} \quad |\xi| \to \infty, \tag{3.16}$$

where $\phi_0(\xi)$ and $\psi_0(\xi)$ are functions holomorphic outside the unit circle $|\xi| = 1$ including the point at infinity. In the right-hand side of the second of Eqs. (3.16) $A\xi$ is used instead of $\omega(\xi)$ for convenience in later use since the regular part of $\omega(\xi)$ can be absorbed into $\psi_0(\xi)$. After drawing upon (3.16) we find from (3.15)

$$\phi_0(\xi) + \frac{\omega(\xi)}{\overline{\omega'\xi}}\overline{\phi_0'(\xi)} + \overline{\psi_0(\xi)} = -2\Gamma\omega(\xi) - \overline{\Gamma' A\xi} \quad \text{on} \quad |\xi| = 1. \tag{3.17}$$

Hence the boundary value problem originally posed in Eqs. (3.8) and (3.10) reduces to finding two functions $\phi_0(\xi)$ and $\psi_0(\xi)$ which are holomorphic (single-valued analytic functions) in the region $|\xi| > 1$ including the point at infinity and satisfying the boundary condition (3.17) on the unit circle. For rational mapping functions, an exact solution (in series representation) for ϕ_0, ψ_0 could be obtained. Here Fourier expansion of the boundary condition is not possible due to the singularity of the factor $\omega/\bar{\omega}'$ at the crack tip image points $\xi = e^{i\beta_1}$ and $\xi = e^{i\beta_2}$†

† Andersson [47] due to an error in selecting the branches of ω and ω' evaluated the expression $\omega/\overline{\omega'}$ incorrectly [10].

3.2 Modification of the Boundary-value Problem and its Effect on the Strain Energy

Due to the difficulties mentioned in Subsection 3.1 the boundary-value problem will be modified in such a way as to obtain a "sufficiently accurate" solution. By "sufficiently accurate" solution we mean one which will give the strain energy† change sufficiently accurately. More concretely let us denote by U_E the strain energy change calculated from the exact problem and by U_M that calculated from the modified problem. It is shown in Section 4 that we need to compare quantities of order (l_2/l_1) as the crack extension l_2 tends to zero. We therefore need to know U_E in our analysis accurate to an order of $(l_2/l_1)^2$ and thus desire a modification which will result in $|U_M - U_E| = O\{(l_2/l_1)^2\}$ or better.

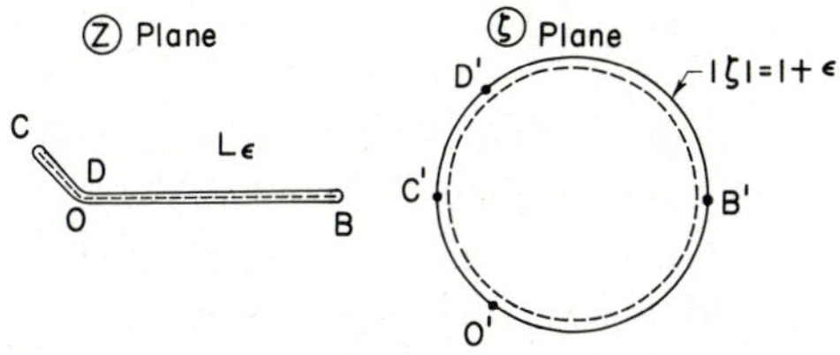

Fig. 6 Mapping of the angled crack into the unit circle.

The modified boundary-value problem is obtained by considering a different "crack boundary," which corresponds to the circle of radius $|\xi| = 1 + \varepsilon$ in the mapped plane (Fig. 6), where $0 < \varepsilon \ll 1$. Because we lack the exact solution, we cannot assess the magnitude of $|U_M - U_E|$ indubitably. However, we shall show in the sequel that in the known cases of the classical "Griffith crack" and the analog of the branched crack problem under antiplane shear load the energy difference is of order ε. Since later in the numerical evaluation of the problem we assign the value $\varepsilon = (l_2/l_1)^4$ it is believed that the modified boundary-value problem provides accurate results for the energy calculations.

Having suggested that moving the boundary of the image crack in the ξ-plane by a distance ε should affect the strain energy change only to order

† In the introduction the fracture problem was formulated in terms of the potential energy. Within the realm of the linearized theory of elasticity, the strain energy for an equilibrium state is equal to the negative of the potential energy.

ε, we document now that this is true for two cases for which exact solutions can be obtained.

We consider first the problem of the centrally cracked sheet loaded at infinity by uniaxial tension oriented arbitrarily with respect to the crack but under the (artificial) assumption that the crack extends along its original axis. Next we deal with the two-dimensional geometry of the problem under consideration but under antiplane shear loading. Sih has presented a partial solution to this problem [48] in which only stress intensity factors at the crack end points are recorded. Since the information contained therein is insufficient for our purpose, we obtained a complete solution by a procedure which duplicates that employed for the in-plane problem treated in this paper.

(a) *Straight Crack.* For the special case of a straight crack of length $l_1 + l_2$, the mapping function reduces to

$$z = \omega(\xi) = A(\xi + 1/\xi)\dagger \tag{3.18}$$

where $A = [(1 + a)/4]l_1$ with $a = l_2/l_1$.

Since this mapping function is rational, the solution for $\phi(\xi)$ and $\psi(\xi)$ satisfying Eqs. (3.15) and (3.16) can be found to be [45]

$$\begin{aligned} \phi(\xi) &= A[\Gamma\xi - (\Gamma + \overline{\Gamma}'p^2)1/\xi], \\ \psi(\xi) &= A\left[\Gamma'\xi + \overline{\Gamma}'p^4\frac{1}{\xi} + \left(\frac{2\Gamma}{p^2} + \overline{\Gamma}'\right)(1 + p^4)\frac{\xi}{1 - \xi^2}\right], \end{aligned} \tag{3.19}$$

where $p = 1$ corresponds to the case when the boundary conditions are applied on the crack, while $p = 1 + \varepsilon$ produces the solution when they are applied on the modified boundary corresponding to the circle $|\xi| = 1 + \varepsilon$. From (3.18) we obtain

$$\left.\begin{aligned} \xi &= \frac{z}{A} - \frac{A}{z} + O\left(\frac{1}{z^3}\right) \\ \frac{1}{\xi} &= \frac{A}{z} + O\left(\frac{1}{z^3}\right) \end{aligned}\right\} \quad \text{as} \quad |z| \to \infty. \tag{3.20}$$

Recall that we denoted the exact strain energy change obtained for the original problem by U_E and by U_M that calculated for the modified problem. Using (3.20) and anticipating (3.50) and (3.51) from Subsection 3.3 we deduce

† We have neglected here a constant term on the R.H.S., which corresponds to a translation of origin, which does not affect the strain energy change.

$$U_E = \frac{\pi(\kappa+1)q^2}{32\mu}\left\{\left(\tfrac{1}{2}+a+\frac{a^2}{2}\right)-\left(\tfrac{1}{2}+a+\frac{a^2}{2}\right)\cos 2\alpha\right\}l_1^2,$$

$$U_M = \frac{\pi(\kappa+1)q^2}{32\mu}\left\{\left[\left(\tfrac{1}{2}+a+\frac{a^2}{2}\right)\cdot(1+\varepsilon)+\varepsilon(1+a)^2\right]\right. \tag{3.21}$$

$$\left.-\left[\tfrac{1}{2}+a+\frac{a^2}{2}\right]\cos 2\alpha\right\}l_1^2+O(\varepsilon^2).$$

It follows now that

$$U_E - U_M = O(\varepsilon). \tag{3.22}$$

It might be pointed out here that, for crack-parallel tension ($\alpha = 0$), $U_E = 0$ while $U_M = O(\varepsilon)$. The exact energy change due to crack extension U_E is zero (and thus smaller than U_M) only if *both* the orientation of the original crack *and* the orientation of its extension align with the tension. This pathological case is clearly of no interest in our later considerations.

(b) *Antiplane Strain Problem.* We consider the same branched crack geometry as in Fig. 5, but loaded at infinity in such a way that only the out-of-plane displacement w does not vanish identically and is a function of the in-plane coordinates x_1 and x_2 only. The only nonvanishing components of the stress-tensor are related to this displacement component through the shear modulus μ by

$$\sigma_{13} = \mu\frac{\partial w}{\partial x_1}$$
$$\sigma_{23} = \mu\frac{\partial w}{\partial x_2} \tag{3.23}$$

so that the equilibrium equations become

$$\nabla^2 w(x_1, x_2) = 0. \tag{3.24}$$

The boundary conditions on the crack boundary L, with n denoting the outward normal and σ_{n3}^* a prescribed function on L reduce to

$$\sigma_{n3} = \mu\frac{\partial w}{\partial n} = \sigma_{n3}^*. \tag{3.25}$$

Let D be the region exterior to the crack boundary. When the loading is a

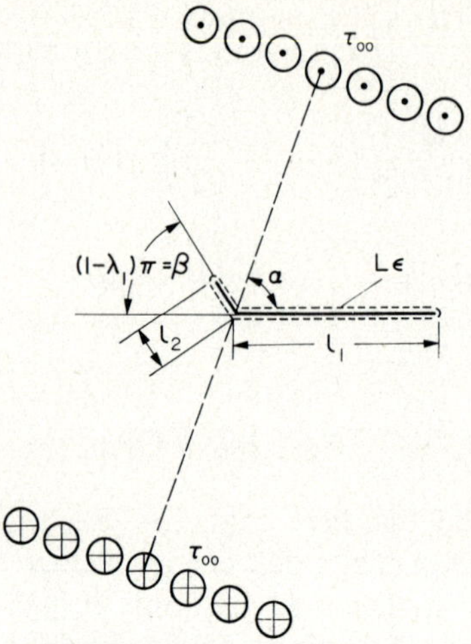

Fig. 7 Angled crack under antiplane shear loading.

uniform shear stress τ_∞ at infinity as shown in Fig. 7, the boundary-value problem reduces to

$$\nabla^2 w = 0 \quad \text{in} \quad D, \tag{3.26}$$

$$\sigma_{n3} = \mu \frac{\partial w}{\partial n} = 0 \quad \text{on} \quad L, \tag{3.27}$$

$$\begin{aligned} \lim_{r\to\infty} \sigma_{13} &= \tau_\infty \cos\alpha \\ \lim_{r\to\infty} \sigma_{23} &= \tau_\infty \sin\alpha \end{aligned} \tag{3.28}$$

In order to employ the complex representation of w, we restate the boundary value problem in Eqs. (3.26) to (3.28). Let w_c be the harmonic conjugate of w and let

$$h(z) = w + iw_c, \qquad z = x_1 + ix_2. \tag{3.29}$$

With s denoting the coordinate along L, condition (3.27) becomes in terms of w_c

$$\frac{\partial w_c}{\partial s} = 0 \quad \text{on} \quad L. \tag{3.30}$$

Without loss of generality this can be written as

$$w_c = 0 \quad \text{on} \quad L. \tag{3.31}$$

From (3.23) and the Cauchy–Riemann conditions we deduce

$$\begin{aligned} \frac{dh}{dz} &= \frac{\partial w}{\partial x_1} - i\frac{\partial w}{\partial x_2} \\ &= \frac{1}{\mu}(\sigma_{13} - i\sigma_{23}). \end{aligned} \tag{3.32}$$

Hence the boundary value problem as posed in Eqs. (3.26) to (3.28) reduces to finding a function $h(z)$, holomorphic-single valued and analytic in the region D excluding the point at infinity. The boundary condition (3.27), making use of (3.29) and (3.31), becomes

$$h(z) - \overline{h(z)} = 0 \quad \text{on} \quad L \tag{3.33}$$

while the conditions at infinity (3.28) reduce to

$$\lim_{|z|\to\infty} \frac{dh}{dz} = \frac{\tau_\infty}{\mu} e^{-i\alpha}. \tag{3.34}$$

The function $\omega(z)$, Eq. (3.12), maps the region D into the region outside the unit circle in the ξ-plane. Let $h_1(\xi) = h(\omega(\xi))$. Then conditions (3.33) and (3.34) lead to

$$\begin{aligned} h_1(\xi) - \overline{h_1(\xi)} &= 0 \quad \text{on} \quad |\xi| = 1 \\ \lim_{|\xi|\to\infty} \frac{h_1'(\xi)}{\omega'(\xi)} &= \frac{\tau_\infty}{\mu} e^{-i\alpha}. \end{aligned} \tag{3.35}$$

Noting from (3.14) that $\lim_{|\xi|\to\infty} \omega'(\xi) = A$, the solution corresponding to (3.35) is

$$h_1(\xi) = \frac{A\tau_\infty}{\mu}\left[e^{-i\alpha}\xi + \frac{e^{i\alpha}}{\xi}\right]. \tag{3.36}$$

Having found the solution, the strain energy can be calculated using a method similar to the one to be described below in Subsection 3.3 for in-plane problems. The result is

$$U_E = \frac{2\pi A^2 \tau_\infty^2}{\mu} \operatorname{Re}\left[1 - \omega_2 e^{-2i\alpha}\right], \tag{3.37}$$

where

$$\omega_2 = \tfrac{1}{2}[\lambda_1(\lambda_1 - 1)\, e^{2i\delta_1} + 2\lambda_1\lambda_2\, e^{i(\delta_1+\delta_2)} + \lambda_2(\lambda_2 - 1)\, e^{2i\delta_2}]. \tag{3.38}$$

The solution to the modified problem for which the boundary corresponds to $|\xi| = 1 + \varepsilon$, can be obtained by replacing the first of the conditions in Eq. (3.35) by

$$h_1(\xi) - \overline{h_1(\xi)} = 0 \quad \text{on} \quad |\xi| = 1 + \varepsilon. \tag{3.39}$$

One obtains in this case

$$h_1(\xi) = \frac{A\tau_\infty}{\mu}\left[e^{-i\alpha}\xi + (1+\varepsilon)^2 \frac{e^{i\alpha}}{\xi}\right] \tag{3.40}$$

and the strain energy change as

$$U_M = \frac{2\pi A^2 \tau_\infty^2}{\mu} \operatorname{Re}\left[(1+\varepsilon)^2 - \omega_2 e^{-2i\alpha}\right]. \tag{3.41}$$

From Eqs. (3.37) and (3.41) we deduce that

$$U_E - U_M = O(\varepsilon).$$

Having shown that in two particular cases the energy change for the modified problem differs only by a term of order ε, from that for the unmodified problem, we return now to the derivation of

3.3 The Strain Energy Change for the In-plane Problem

Let U_R denote the strain energy stored in that part of the solid bounded by the circle of radius R (Fig. 8). Using Clapeyron's theorem [49], it follows that

$$\begin{aligned} U_R &= \frac{1}{2}\int_0^{2\pi} [\sigma_r u_r + \tau_{r\theta} u_\theta]_{r=R} \cdot R\, d\theta \\ &= \tfrac{1}{2} \operatorname{Re} \int_0^{2\pi} [(\sigma_r - i\tau_{r\theta})(u_r + iu_\theta)]_{r=R} R\, d\theta. \end{aligned} \tag{3.42}$$

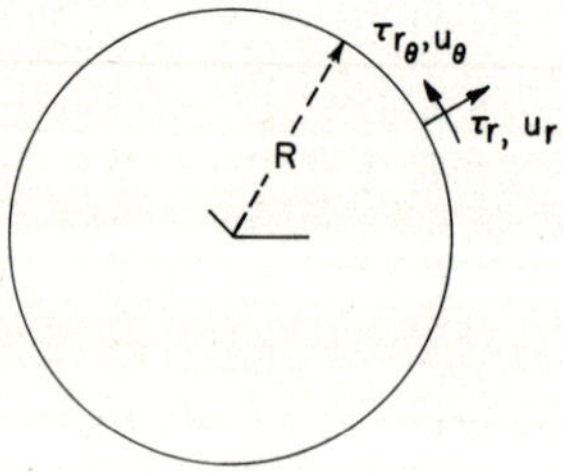

Fig. 8 Definition of radius R surrounding the crack.

In order to make (3.42) more explicit, expand $\phi(z)$ and $\psi(z)$ in power series valid for $|z| > l_1$

$$\phi(z) = \Gamma z + d_0 + \frac{d_1}{z} + \frac{d_2}{z^2} + \cdots$$
$$\psi(z) = \Gamma' z + e_0 + \frac{e_1}{z} + \frac{e_2}{z^2} + \cdots. \tag{3.43}$$

Substituting these series in Eqs. (3.8), and noting that $z = r\,e^{i\theta}$, we obtain for $r = R$, after some rearrangement,

$$\sigma_r - i\tau_{r\theta} = [2\Gamma - \Gamma' e^{2i\theta}] + (e_1 - 3d_1 e^{-2i\theta} - \bar{d}_1 e^{2i\theta})\frac{1}{R^2} + O\left(\frac{1}{R^3}\right),$$
$$2\mu(u_r + iu_\theta) = [(\kappa - 1)\Gamma - \overline{\Gamma'} e^{-2i\theta}]R + \{\kappa d_0 - \bar{e}_0\} e^{-i\theta} + (\kappa d_1 e^{-2i\theta} + \bar{d}_1 e^{2i\theta} - \bar{e}_1)\frac{1}{R} + O\left(\frac{1}{R^2}\right). \tag{3.44}$$

The coefficients d_n, e_n $(n = 0, 1, 2, \ldots)$ are, in general, complex, except e_1 which is real due to the fact that the moment of the applied boundary forces is zero and hence

$$\int_0^{2\pi} (\tau_{r\theta} \cdot r) \cdot r\big|_{r=R}\, d\theta = 0. \tag{3.45}$$

This property of e_1 will be used in deriving Eqs. (3.48) and (3.49) below.

Only the terms in square brackets in Eq. (3.44) are non-zero for a crack-free solid. The terms in curly brackets represent rigid-body displacements. The rest of the terms are generated by the presence of the crack. In order to meaningfully compare the strain energy of the two solids, i.e. one with the crack and one without, and for which stress boundary conditions are prescribed at the outer boundary, one needs to satisfy these conditions on the large circle of radius R to order $1/R^2$. Thus, the stress field must behave like [50, 51]

$$\sigma_r - i\tau_{r\theta} = (2\Gamma - \Gamma' e^{2i\theta}) + O\left(\frac{1}{R^3}\right). \tag{3.46}$$

The alteration† necessary to make the stress field given in Eq. (3.44) conform with that of Eq. (3.46) (elimination of the term of order of $1/R^2$) can be accomplished with the following stress functions $\tilde{\phi}$ and $\tilde{\psi}$:

† We follow here the method outlined in Ref. [51].

$$\tilde{\phi}(z) = \left(-\frac{e_1}{2R^2}\right)z + \left(\frac{\bar{d}_1}{R^4}\right)z^3,$$
$$\tilde{\psi}(z) = \left(\frac{4\bar{d}_1}{R^2}\right)z. \tag{3.47}$$

As a result one incurs additional displacements $\tilde{u}_r$, $\tilde{u}_\theta$

$$2\mu(\tilde{u}_r + i\tilde{u}_\theta) = \left[-(\kappa - 1)\frac{e_1}{2} + \kappa\bar{d}_1\, e^{2i\theta} + d_1\, e^{-2i\theta}\right]\frac{1}{R} \tag{3.48}$$

which when added to those in (3.44) result in

$$2\mu(u_r + iu_\theta) = \{(\kappa - 1)\Gamma + \overline{\Gamma' e^{2i\theta}}\}R + \{\kappa d_0 - \bar{e}_0\}\, e^{-i\theta}$$
$$+(\kappa + 1)\left\{-\frac{e_1}{2} + \bar{d}_1\, e^{2i\theta} + d_1\, e^{-2i\theta}\right\}\frac{1}{R} + \mathrm{O}\left(\frac{1}{R^2}\right). \tag{3.49}$$

Upon combining (3.42), (3.44), and (3.49) and integrating along the circle $r = R$ we derive

$$U_R = \frac{\pi}{2\mu}\{[2(\kappa - 1)\Gamma^2 + \Gamma'\overline{\Gamma'}]R^2 + (\kappa + 1)\,\mathrm{Re}\,(-\Gamma e_1 - \Gamma' d_1)\} + \mathrm{O}\left(\frac{1}{R}\right). \tag{3.50a}$$

Letting $\Gamma = \sigma_\infty/4$ and $\Gamma' = -(\sigma_\infty/2)\, e^{-2i\alpha}$ and rearranging one obtains

$$U_R = \frac{\pi(\kappa + 1)\sigma_\infty^2}{8\mu}R^2 + \frac{\pi(\kappa + 1)\sigma_\infty}{16\mu}\{\mathrm{Re}\,[-2e_1 + 4d_1\, e^{-2i\alpha}]\} + \mathrm{O}\left(\frac{1}{R}\right). \tag{3.50b}$$

The term proportional to R^2 represents the energy stored in the disc of radius R in the absence of a crack. Denote this part by U_R^0. The remaining terms are due to the presence of the crack. Thus taking the limit of $\{U_R - U_R^0\}$ as $R \to \infty$, yields the required expression for the strain energy in an infinite solid due to the crack as

$$U_c = \frac{\pi(\kappa + 1)\sigma_\infty}{16\mu}\mathrm{Re}\,[-2e_1 + 4d_1\, e^{-2i\alpha}]. \tag{3.51}$$

Note that because the energy U contains as the only unknowns the coefficients e_1 and d_1, only these coefficients in the series expansions for ϕ and ψ, Eq. (3.43), need to be determined accurately. From the linear relation between ϕ, ψ and $\Gamma = \sigma_\infty/4$, $\Gamma' = -(\sigma_\infty/2)\, e^{-}2^{i\alpha}$, it follows that e_1 and d_1 are representable in the form

$$\begin{aligned} d_1 &= \sigma_\infty[d_{11} + d_{12}\cos 2\alpha + d_{13}\sin 2\alpha], \\ e_1 &= \sigma_\infty[e_{11} + e_{12}\cos 2\alpha + e_{13}\sin 2\alpha], \end{aligned} \tag{3.52}$$

where the terms with two subscripts are functions of the geometry only as, for example,

$$e_{11} = e_{11}(\lambda_1, l_1, l_2).$$

Substituting (3.52) into (3.51), collecting like terms and grouping combinations of d_{ij} and e_{ij} into coefficients c_i, we obtain

$$U_c = \frac{\pi(\kappa + 1)\sigma_\infty^2}{32\mu}[c_1 + c_2\cos 2\alpha + c_3\sin 2\alpha + c_4\cos 4\alpha + c_5\sin 4\alpha], \tag{3.53}$$

where the c_i are functions of λ_1, l_1 and l_2 only. We recall that U_c represents the energy due to a crack in an infinite sheet; the crack may be either a straight line crack or one having an angular extension as shown in Fig. 5. In order to compute the energy release rate for crack propagation at an angle $\lambda_1\pi$ (cf. Fig. 5) we need to determine the coefficient c_i via ϕ and ψ (d_1 and e_1) from the boundary-value problem for the (modified) branched crack, on the one hand, and for the nonextended crack, on the other. For the latter case one finds easily

$$c_1 = -c_2 = \tfrac{1}{2}l_1^2 \quad \text{and} \quad c_3 = c_4 = c_5 = 0.$$

3.4 Solution of the Modified Boundary-value Problem

In order to reduce the modified boundary to the unit circle, we let $k = 1/(1 + \varepsilon)$ and write

$$z = \omega(\xi) = \frac{A}{k\xi}(\xi - k\,e^{i\delta_1})^{\lambda_1}(\xi - k\,e^{i\delta_2})^{\lambda_2}, \qquad \left(k = \frac{1}{1+\varepsilon}\right). \tag{3.54}$$

Then the original branched crack corresponds to the circle $|\xi| = k$ (cf. Fig. 6). Recall that we need to determine two functions $\phi_0(\xi)$ and $\psi_0(\xi)$, holomorphic in the region $|\xi| > 1$ including the point at infinity and satisfying the following boundary condition on $|\xi| = 1$:

$$\overline{\phi_0(\xi)} + \frac{\overline{\omega(\xi)}}{\overline{\omega'(\xi)}}\phi_0'(\xi) + \psi_0(\xi) = -2\Gamma\overline{\omega(\xi)} - \frac{\Gamma' A}{k}\xi. \tag{3.55}$$

We proceed towards a solution of (3.55) by Fourier series expansion of the functions ϕ_0, ψ_0, $\overline{\omega(\xi)/\omega'(\xi)}$ for the now regularized geometry. Let

$$\left.\begin{aligned}\phi_0(\xi) &= \sum_{m=0}^{\infty} f_m \xi^{-m} \qquad |\xi| \geq 1\\ \psi_0(\xi) &= \sum_{m=0}^{\infty} b_m \xi^{-m} \qquad |\xi| \geq 1\end{aligned}\right\} \tag{3.56}$$

$$\begin{aligned}\frac{\overline{\omega(\xi)}}{\omega'(\xi)} &= \sum_{n=-\infty}^{\infty} h_n \xi^n \qquad |\xi| = 1\\ -2\Gamma\overline{\omega(\xi)} - \frac{\Gamma' A}{k}\xi &= \sum_{n=-\infty}^{\infty} g_n \xi^n \qquad |\xi| = 1.\end{aligned} \tag{3.57}$$

Substituting (3.56) and (3.57) in (3.55)† and equating the coefficients of equal powers of ξ one obtains, first, for the positive powers

$$\bar{f}_n - \sum_{m=1}^{\infty} m h_{n+m+1} f_m = g_n \qquad (n = 1, 2, \ldots). \tag{3.58}$$

For the negative powers we need to record only the equation corresponding to the term ξ^{-1} as the terms b_n for $n > 1$ do not contribute to the energy integral‡

$$b_1 - \sum_{m=1}^{\infty} m h_m f_m = g_{-1}. \tag{3.59}$$

The Eqs. (3.58) are solved for the coefficients f_n by truncating the series for $m > N$ $(n = 1, 2, \ldots, N)$.

Then b_1 is determined by

$$b_1 = g_{-1} + \sum_{m=1}^{N} m h_m f_m. \tag{3.60}$$

4 DETERMINATION OF THE CRACK BRANCHING ANGLE AND OF THE CRITICAL LOAD

Recall that in the introduction we delineated the determination of the crack branching angle as well as the critical load by a maximization of the energy release rate and by a subsequent limit process for "a" $\rightarrow 0$. Since the evaluation of the boundary-value problem is accomplished numerically the further use of these results in the maximization and limit processes just mentioned must also be accomplished numerically. All computations were performed on an IBM 370/185 computing system. The accuracy of solving

† Multiplication of the two series is justified as both series are absolutely convergent.

‡ Since we need only the coefficient of the $1/z$ term in $\psi(z)$ and since $1/\xi = A/kz + O(1/z^2)$ as $|z| \rightarrow \infty$ we need only the coefficient of $1/\xi$.

the truncated set of Eqs. (3.58) was checked by back-substitution and was found to yield at worst three significant digits. In most cases the accuracy was considerably higher.

The numerical procedure may be broken down into several sequential steps. For a chosen value of the load orientation angle α (cf. Fig. 5) we select a sequence of values for the crack extension "a" ($=l_2/l_1$) in the range $0.1 \geq a \geq 0.0025$.

Next, for each value of "a" thus chosen we compute U_c for different values of crack extension angle $\beta = \pi(1 - \lambda_1)$. This is accomplished with the aid of Eq. (3.53), wherein the coefficients c_i are functions of "a" and β, but not of α. The computation of the energy release rate is simple if an expansion of the coefficients c_i exists. Although no expansion of this type was found for the in-plane problem the antiplane shear problem yields

$$\begin{aligned} U_c - U_0 &= \frac{\pi}{32}\frac{\kappa + 1}{\mu}\sigma_\infty^2 C_i(\beta, a) f_i(\alpha) \\ &= \frac{\pi}{32}\frac{\kappa + 1}{\mu}\sigma_\infty^2 \{C_{i1}(\beta)a + C_{i2}(\beta)a^{3/2} + C_{i3}(\beta)a^2 + \cdots\} f_i(\alpha) \end{aligned} \tag{4.1}$$

repeated indices indicating summation.

It follows that the energy release rate G is

$$G \equiv \lim_{a \to 0} \frac{U_c - U_0}{a} = \frac{\pi}{32}\frac{(\kappa + 1)\sigma_\infty^2}{\mu}[C_{i1}(\beta) f_i(\alpha)]. \tag{4.2}$$

Having computed values of $U_c - U_0$ for a matrix of (finite) "a" and β we determine the $C_{ij}(\beta)$, in particular C_{i1}, with the aid of (4.1). The computation was carried out in several ways. For the values of "a" = {0.0025, 0.005, 0.01, 0.025, 0.05, 0.1} the following four *sets* of values for "a" were used to determine as an accuracy check four values for the C_{ij}: [0.0025, 0.005, 0.1}, {0.005, 0.01, 0.025}, {0.01, 0.025, 0.05}, {0.025, 0.05, 0.1}. The maximum variation in the values of C_{i1} (the first and last set) was 1.8%.

For each value of α we now maximize G with respect to β. This is accomplished by parabolic interpolation, the coefficients for which were determined in two ways as a check on accuracy. First, three values around the maximum were used, and then four values in a least squares fit. The difference was less than 1%.† Having determined the angle β^* at which the

† The maxima turned out to be rather shallow. One would therefore expect that the experimental fracture data would be subject to considerable data scatter, a fact borne out by data in the literature as well as those reported here later. This observation was also pointed out to us by Prof. J. Swedlow in a private communication and Ref. [52].

strain energy release rate reaches a maximum, we determine next the load σ_∞ at which this energy release rate attains a critical value. Writing Eq. (4.2) explicitly we compute the critical far-field stress σ_∞ from

$$G = \frac{\pi}{32}\frac{\kappa+1}{\mu}\sigma_\infty^2\{C_{11}(\beta^*) + C_{21}(\beta^*)\cos 2\alpha + C_{31}(\beta^*)\sin 2\alpha + C_{41}(\beta^*)\cos 4\alpha + C_{51}(\beta^*)\sin 4a\}. \qquad (4.3)$$

Before presenting the results of these computations we comment further on the accuracy of the results. We have mentioned earlier in connection with the modified mapping that the boundary modification parameter ε must be "small" compared to the crack extension "a". Otherwise the modified boundary does not represent the extension as a crack.

We are comparing quantities of order "a" in the energy for the advanced and for the original crack. If this comparison is to be accurate to order "a", then the determination of the energy U_c should be at least accurate to order a^2. The error introduced by the boundary modification should be at least as small or smaller, i.e. $\varepsilon \leq a^2$. For added certainty we chose the much stronger constraint $\varepsilon = a^4$.

The determination of the energy U_c requires solution of the infinite set of Eqs. (3.58). The truncation of that system was increased from 25 to 50, and 100 if necessary. The difference between 50 and 100 terms was always less than 2%; if the difference between using 20 and 50 terms was less than 2% the 100 terms were not used.

5 DISCUSSION OF RESULTS FOR THE TWO-DIMENSIONAL PROBLEM

The results of the just outlined computations are summarized in Figs. 9 to 12.

5.1 The Crack Branching Angle

In Fig. 9 we present the dependence of the crack branching angle β on the crack orientation angle α. We also show, for comparison purposes, the results of the simpler stress criterion [6, 8]. First a comment on the value of β for $\alpha = 0$ is in order. We note that the mode I stress intensity factor is proportional to $\sigma_\infty \sin^2 \alpha$ and the mode II stress intensity factor to $\sigma_\infty \sin \alpha \cos \alpha$. If we consider $\sigma_\infty \cdot \sin \alpha$ as constant, then for $\alpha \to 0$, $K_{\rm I} \to 0$ and $K_{\rm II} \neq 0$. This

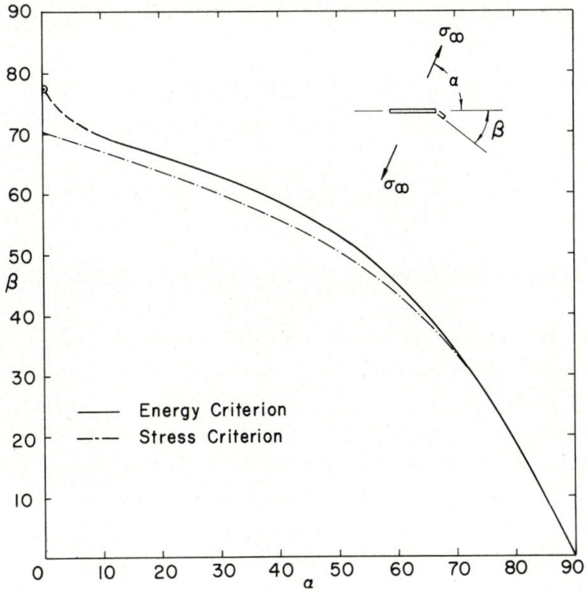

Fig. 9 Crack branching angle β as a function of load orientation angle α for energy criterion and stress criterion [6, 8].

combination corresponds to the pure mode II condition. It was possible to use the computations of the constants C_{ij} to construct the energy release rate for the mode II loading. The result of these computations was thus taken to represent the limit results for $\alpha \to 0$. The angle β thus obtained was 77.4°. The dotted portion of the curve near $\alpha = 0$ represents a guess which is motivated by the observations of the experimental results represented in the subsequent Figs. 17 and 19. The economics of the computations did not allow us to explore that curve segment in more detail.

We note next that the relation between β and α based on the present energy analysis deviates only mildly from that for the stress criterion, virtual agreement prevailing for $\alpha > 60°$.

In order to construct a plausible explanation for the closeness of these results we recall that the stress criterion assumes fracture to occur on a ray (plane) along which the shear stress vanishes before the crack has branched. Consider stress and displacement components before the crack has branched and identify them by a superscript I. Let a superscript II connote corresponding quantities referred to the branched crack with the same coordinate system (cf. Fig. 3). Then, using Betti's reciprocity relation we find that

$$G(\beta) = \lim_{a \to 0} \frac{\Delta U(\beta, \alpha)}{a} \equiv \lim_{a \to 0} \frac{U_c(\beta, \alpha) - U_0}{a}$$
$$= \lim_{a \to 0} \frac{1}{2a} \int_0^a \{\sigma_\theta^{\mathrm{I}}(r, \beta)[u_\theta^{\mathrm{II}}(r, \beta)] + \tau_{r\theta}^{\mathrm{I}}(r, \beta)[u_r^{\mathrm{II}}(r, \beta)]\}\, dr, \tag{5.1}$$

where [ ] denotes the jump in the displacement components across the crack.

Note that the energy criterion requires maximization of $G(\beta)$ under consideration of *both* terms under the integral. In contrast, the stress criterion determines β such that $\tau_{r\theta}^{\mathrm{I}}(r, \beta)$ vanishes, which in turn implies that $\sigma_\theta^{\mathrm{I}}(r, \beta)$ is a maximum. If the jump in u_r^{II} is of the same order of or smaller than that of u_θ^{II} then for a small or vanishing $\tau_{r\theta}^{\mathrm{I}}(r, \beta)$, both the energy and the stress criterion are satisfied (approximately) simultaneously. This situation corresponds to cases when the crack branching angle β is small, which occurs when α is near 90°. In fact, in the limit of $\alpha \to 90°$ and $\beta \to 0$, and only in that limit, the jump $[\mathrm{u}_r^{\mathrm{II}}(\mathrm{r}, b)]$ vanishes identically and the energy criterion is identical to the stress criterion.

In this connection we also comment on the strain energy density theory (S-theory) proposed by Sih [23–26] and alluded to already in the introduction. The S-theory predicts a sensitivity of the orientation and onset of fracture on Poisson's ratio while the computations represented in Figs. 9–12 are independent of that parameter. Thus the equivalent of the single curves in Figs. 9–12 are families of curves with Poisson's ratio as a parameter. We do not reproduce these here but refer the reader to the original papers [23–26]. In connection with the present discussion on the two-dimensional problem it is interesting to note that Finnie and Weiss [53] subjected the sensitivity of the S-theory to Poisson's ratio to a test on cross-rolled beryllium sheet for which Poisson's ratio is close to zero. The limited results agreed significantly better with the stress criterion than with the S-criterion. (See also Sih's discussion [27] on the note by Finnie and Weiss.)

5.2 The Critical Stress

The critical stress is shown in Fig. 10 as a function of the crack orientation angle, together with the results of the stress criterion [6–8]. As we pointed out in Ref. [10] already there is a small minimum in the critical stress

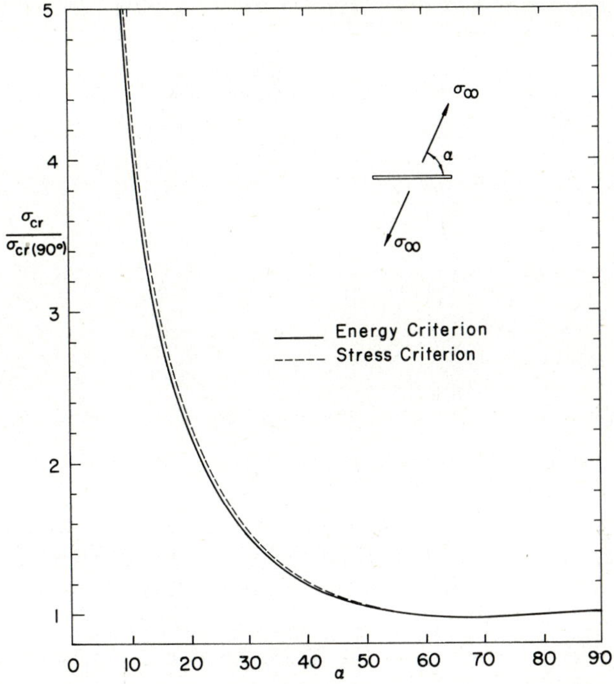

Fig. 10 Critical load σ_∞ as a function of load orientation angle α for energy criterion and stress criterion [6, 8].

around $\alpha = 63°$.† Having determined this minimum for the energy analysis we subsequently checked the stress criterion [6, 8] and found that it, too, yields a minimum which had not yet been reported previously [10].

In practical terms the minimum may not be very significant. If σ_g is the critical stress for the Griffith problem of crack-normal tension and σ_{min} is the minimum value of the critical stress, then $\sigma_{min}/\sigma_g = 0.97$.

5.3 The Combination of Mode I and Mode II Stress Intensity Factors

The stress intensity factors at the onset of fracture are sometimes used to illustrate the effect of a compound stress state at the crack tip. Figure 11

† In Ref. [10] we reported erroneously that this minimum occurs at 72° and has a value of 0.96 of the critical load for $\alpha = 90°$. It is for the stress criterion that the minimum occurs near 72° and its value is about 0.975 of the critical load for $\alpha = 90°$.

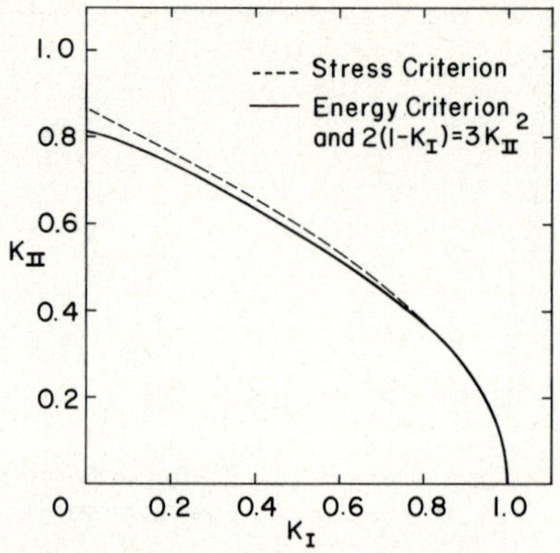

Fig. 11 Mode I and II interaction for the energy criterion and the stress criterion [6, 8].

shows the mode I and mode II interaction for the energy computations in comparison with that resulting from the stress criterion [6, 8]. The energy-based interaction curve intercepts the K_{II} ordinate at $K_{II} = 0.814$ while the stress-derived curve does so at $K_{II} = \sqrt{3/4} = 0.866$.

It is of interest to note that the K_I–K_{II} interaction curve derived from the energy criterion is well represented by the expression

$$2(1 - K_I) = 3K_{II}^2 \tag{5.2}$$

as reference to the comparison in Table 1 will show.

Table 1 Comparison of K_I–K_{II} interaction computed from energy criterion and as represented by Eq. (5.2)

Orientation angle α	K_I	K_{II} computed from energy criterion	K_{II} determined from Eq. (5.2)
0°	0	0.814	0.816
18°	0.235	0.724	0.714
36°	0.444	0.611	0.609
54°	0.658	0.478	0.477
72°	0.879	0.286	0.284
90°	1.000	0	0

In view of this excellent agreement one wonders whether Eq. (5.2) does not represent the exact solution and whether a more clever and closed-form analysis than the one presented here could not be performed to show this.

Finally, it is important to recognize from Fig. 11—as well as 10—that the approximate stress criterion does not predict fracture conservatively.

5.4 Finite Crack Extension

We have emphasized in this presentation the behavior of solids considered brittle in the sense that the extension of the crack must be considered very small or infinitesimal for stability computations. There are many cases involving all degrees of ductile behavior, however, where such a procedure is ill-advised. In fact, in those cases it would be more appropriate to consider advancement of a crack by a finite distance.

Since our numerical procedures yielded the strain energy release rate for finite values of "a", it seemed appropriate to report them for a sequence of such finite values and inquire into their relevance to the fracture of what we like to call mildly ductile solids. This information is compiled in Fig. 12. We shall later compare these relationships with experimental data collected on plexiglass which is known to possess a craze zone at the tip of a crack. If we view this small crazed zone as the smallest dimension admitted by

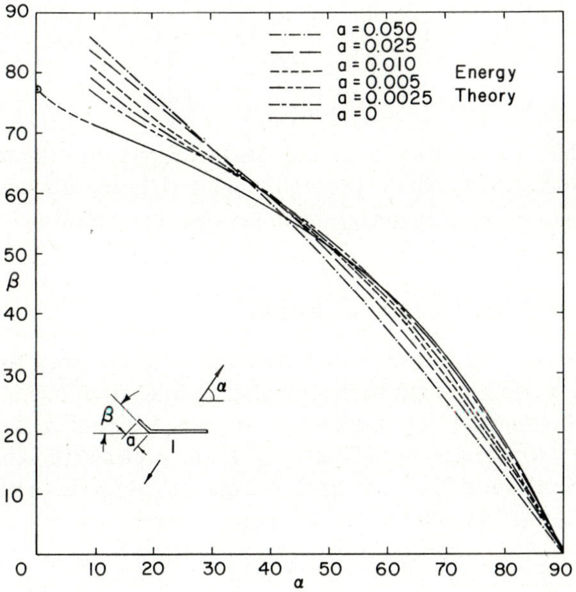

Fig. 12 Maximal energy release rate for finite crack extension.

the material and as giving way virtually simultaneously at the onset of crack propagation, the present energy computations may represent an approximate criterion for fracture of mildly ductile solids of that type.

The approximate nature of this application derives from at least three sources. First, we assume that the fracture process can be described at all by an energy criterion which implies the existence of a fracture energy. Second, the mild plasticity is presumed to be equivalent to a simultaneous crack tip unloading over a small finite length. Third, we assume that the original crack existed in a "virgin state" or in a "virgin material". In reality a crack is produced under crack symmetric loading in a (mildly ductile) material with a small plastic zone symmetric with respect to the crack axis. We presume in the present analysis that such a zone is of secondary importance in the subsequent crack-branching process.

Before concluding this presentation of the effect of a finite crack extension we recall once more that the energy criterion in Figs. 9–11 was deduced from the expression

$$\max_{\text{w.r.t.}\,\theta} \left\{ \lim_{a \to 0} \frac{U_c - U_0}{a} \right\}. \tag{5.3}$$

The data represented in Fig. 12 admit the calculation of the complementary expression

$$\lim_{a \to 0} \left[\max_{\text{w.r.t.}\,\theta} \left\{ \frac{U_c - U_0}{a} \right\} \right]. \tag{5.4}$$

We find within the accuracy of all our computations that the same results obtain regardless of the order of the limit process. If the extrapolation formula (4·1) derived from the antiplane problem is valid, this should, of course, be so.

5.5 Infinitesimal Deviation Angle

There are many situations in which the material at the crack tip is stressed under nearly mode I loading with only a very small component of mode II present. Such situations arise whenever the crack tip stress field changes continuously with the advance of the crack. If one neglects (possibly) material inertia for "fast" moving crack tips† we may encompass in this class of problems all those that lead to curved fracture paths.

† There exists evidence [54] that the path of the crack tip in inorganic glass is only minimally affected, if at all, by inertial effects. The orientation of the crack growth in brittle solids seems to follow as fast as the stress field can be established [55].

From our discussion in Subsection 5.1 it is evident that for these conditions the (infinitesimal) deviation angle is the same as predicted by the normal stress criterion.

We now consider an equation for the crack propagation path which appears to us consistent with the tensile stress criterion. Consider the crack path decomposed into a polygon of infinitesimal line segments, each of length ds. We neglect inertial effects and use time only as a sequence parameter. Let us introduce a local cylindrical coordinate system (r, θ) placed with its origin at the tip of a crack of arbitrary shape such that the ray $\theta = 0$ denotes the line of tangency to the crack path at the current tip.

Consider first the case when the stress field at the crack tip is changed with time due to changing external loads. We assume that the crack tip loading is always sufficiently high such that crack growth does take place. The change in load occasions a change in the stress intensity factors K_{I} and K_{II}. In fact, if at the time $t = \tau$, $K_{\mathrm{II}} = 0$ and at a time dt later we have stress intensity factors

$$K_1 = K_{\mathrm{I}}(\tau) + \frac{dK_{\mathrm{I}}}{ds} ds \quad \text{and} \quad dK_{\mathrm{II}},$$

then the average change in K_{II} is $\frac{1}{2} dK_{\mathrm{II}}$ in the time interval dt. Correspondingly the tensile stress criterion requires that the average orientation for crack propagation in that time interval makes an angle ($dK_{\mathrm{II}} \ll K_{\mathrm{I}}$, always)

$$\frac{d\theta}{ds} = -2 \frac{dK_{\mathrm{II}}/2}{K_{\mathrm{I}}(\tau) + \dfrac{dK_{\mathrm{I}}}{ds} ds} = -\frac{1}{K_{\mathrm{I}}} \frac{dK_{\mathrm{II}}}{ds} \tag{5.5}$$

with the tangent to the crack path at the tip corresponding to $t = \tau$.

As a second case, we consider the changes of the crack tip stress field due to the changing geometry rather than those due to load alterations far from the tip. The uniquely correct way of determining this relation is, of course, to formulate the problem for a crack of arbitrary shape and perform an analysis for the preferred direction of (additional) infinitesimal crack extension from the current tip. Possibly as an approximation we may consider the application of the stress criterion in the following way.

Let us construct at the time $t = \tau$ the local cylindrical coordinate system at the crack tip with θ denoting the ray tangent to the crack path. At the time τ the mode II stress intensity factor is zero by definition. Now allow for crack extension of length ds during the time increment dt along the line of tangency (co-planar growth). At the end of that propagation period there will have built up a mode II stress intensity factor dK_{11} (for some special cases dK_{11} may turn out to be zero). This infinitesimal stress intensity factor causes deviation from the line of tangency through an angle

$$d\theta = -2\frac{dK_{11}}{K_1}. \tag{5.6}$$

Now we assume that the crack follows along the deviated path for a distance *ds* in a further time interval *dt*.† Then at the end of the second time (or distance) interval the crack has advanced with an *average* angle of

$$d\theta = -\frac{dK_{11}}{K_1} \tag{5.7}$$

which is the same relation as encountered in (5.3). This computational procedure corresponds, of course, to assuming stepwise colinear crack growth and use of the relation (5.5) or (5.7) to correct for the effect of the ensuing mode II deformation.

It follows then that the changes in crack path can be computed from Eqs. (5.5) or (5.7), whether the changes in the stress field at the tip are occasioned by changes in the external loading or by those due to the changing geometry of the crack.

6 EXPERIMENTAL WORK

Since we are dealing with the fracture of brittle solids it is appropriate to perform experiments with a solid which exhibits a minimum of plastic deformability. One is therefore tempted immediately to choose an inorganic or organic glass which usually breaks "with a snap". Panasyuk *et al.* [8] chose glass while Erdogan and Sih [6] chose polymethylmethacrylate as test materials because they break in an apparently brittle manner; polymethylmethacrylate has the further advantage of being machined easily into fracture specimens. However, along with other glass-forming polymers polymethylmethacrylate has the ability to form crazes [56–60], a phenomenon barely known nor understood at the time Erdogan and Sih published their results. Such craze material occurs also at the tip of a crack. Craze material is a mechanically transformed phase of bulk polymer which contains a substantial number of voids such that its density is on the order of 50% of the bulk material. Because the craze material is formed at the crack tip

† Whether we consider equal increments in crack length or time is immaterial as long as the crack speed is continuous.

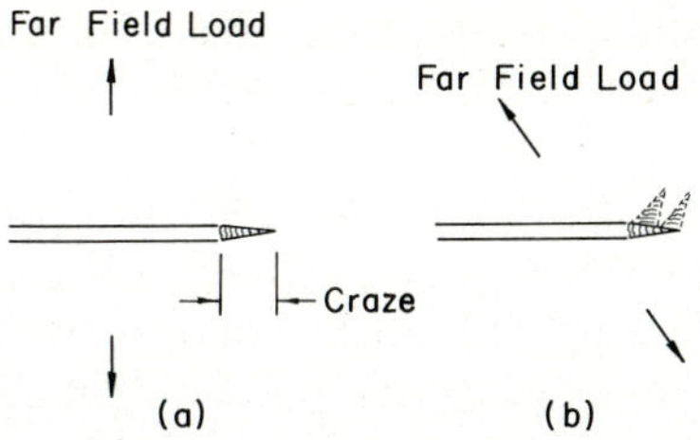

Fig. 13 Crazes at tip of symmetrically grown crack.

under a tensile stress, the craze structure is oriented and consequently possesses anisotropic properties. The craze extension of the crack dominates the stress and deformation fields at the crack tip. For crack propagation along the original direction of the crack axis the existence of a craze extension may not cause a strong deviation from existing theories. However, if a craze-tipped crack attempts to propagate in a direction other than along the original crack axis the craze may exert a recognizable influence on the magnitude of the load at which further crack growth occurs and on the direction which the extension takes. In Fig. 13a we have sketched the (idealized) crack tip craze for self-similar crack propagation (a); if loading on the structure is such that the crack has to propagate away from the crack axis, further crack growth has to occur from somewhere along the length of the craze as indicated by dotted wedges in Fig. 13b for two possible locations. Clearly, if such cohesive craze material is able to constrain the motion of the material at the crack tip, the propagation of angle cracks will be influenced.

To avoid uncertainties due to such cohesive material zones it would be advantageous to reduce the size of the cohesive craze zone to such proportions that it is, practically speaking and in relation to even refined

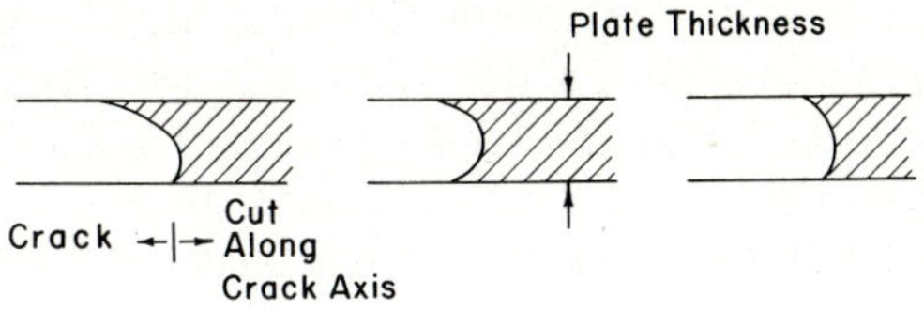

Fig. 14 Crack fronts through glass plates (crack front curvature exaggerated).

measurement techniques, of no consequence in the experiments. The materials which are usually accepted as representing this ideal most closely are the inorganic glasses.

We have experimented, accordingly, with glass plates but found it only possible to produce through-cracks whose fronts took on shapes such as shown in Fig. 14. The curved crack fronts represent, of course, three-dimensional geometries for which the previous two-dimensional analysis is ill-suited; Panasyuk *et al.* [8] do not report on this difficulty and may have overlooked this problem. Their experimental data on the crack-branching angle, though close to the stress criterion, fall below either of the two criteria as does the data of Ref. [11] on polymethylmethacrylate. This suggests that a possibly uneven crack front had an effect similar to that of a small plastic (craze) zone at the crack front. We have considered cracking thick plates and etching away the plate faces along with the uneven portions of the crack front, but decided that the leftover crack front might be damaged and that the dubious result would not warrant the effort.

6.1 Material Choice

In our search for a suitable test material we turned, therefore, to another polymer which distinguishes itself from organic type glasses by chemical crosslinking. Morphologically the polyurethane elastomer must be classified as a rubber, but in terms of its mechanical behavior at room temperature it possesses few of the properties which one associates generally with rubbery solids. This elastomer can sustain only small deformations (stated more specifically later on). Because its glass transition temperature is 18°C below zero its longest relaxation time at 20°C is about 10^{-2} minutes, and it can exhibit time dependences in its fracture behavior. However, this time dependence can be removed almost entirely by swelling the elastomer in a suitable solvent [61]. The remaining time or rate sensitivity is so small that for the small range of time scale in the subsequent measurements it is considered insignificant.

The swollen polyurethane elastomer breaks in an apparently brittle manner much like a glass but under loads that are orders of magnitude smaller than those required for an inorganic glass. The fracture surfaces are as smooth and featureless as those of an inorganic glass. Work on self-similar crack propagation in this material shows that material deterioration at the crack tip occurs—similar to the craze zone in polymethylmethacrylate—and the region in which it occurs is measured in tens of angstroms and is thus below the resolution of optical microscopes.† In this respect the propagation behavior of the crack away from the original crack axis should be similar to that encountered in inorganic glass. The material breaks under low loads in the swollen state so that special care needs to be employed in handling the specimens for testing purposes.

One might argue that a swollen solid is a poor material with which to conduct fracture tests since there exists an interaction between the applied stress and the amount of liquid which a solid can absorb. If a solid is stressed inhomogeneously the absorption of liquid varies from point to point. Thus if a homogeneously swollen polymer—i.e. swollen in a homogeneous state of stress, that is, zero stress—is subsequently subjected to an inhomogeneous stress field the amount of liquid absorbed will begin to vary continuously over the solid. The adjustment of the local liquid absorption requires time to allow for the diffusion of the liquid. The time scale required for this process to occur with the present system of polyurethanetoluene was determined to be on the order of an hour. Typical times from load initiation to rupture was usually below 1 minute, rarely exceeding 2 minutes. We believe therefore that the migration of solvent during the tests did not produce any significantly disturbing effects. Furthermore, one should bear in mind that the loading time for all the tests involving different orientations of the crack with respect to the load was approximately the same so that all test data were affected approximately in the same way.

An additional potential complication in working with swollen polymers is the fact that solvent facilitates the migration of uncrosslinked low molecular weight material out of the solid and into the surrounding solvent. This process can alter the mechanical properties of the swollen material as a function of the time the material is soaked in the liquid; this potential change is large only if a large amount of material is removed and affects the test results (primarily) if different test specimens are soaked initially for different lengths of time. Although for the present polyurethane material the (maximum) amount of low molecular weight material removed‡ is only 1% we attempted to eliminate even this small potential error source by soaking all specimens prior to testing for the same time of 24 hours.

A larger source of error than those associated with the swelling properties were the intrinsic variations in material properties. The material is produced in sheets $12 \times 12 \times \frac{1}{32}$ inches at a time, details of the production process being described in Refs. [61–62]. Here it suffices to report that this poly-

† Strongly nonlinear constitutive behavior comes into play in this small crack tip domain which is approximately of the same order as that in inorganic glasses.

‡ In toluene equilibrium is achieved within a few hours [62].

urethane is produced by mixing equal volumes of two liquids called the prepolymer† and the catalyst.‡ These two liquids are dispensed into a mixer from two 250-ml burettes which are kept at 60°C to provide a constant viscosity of the liquids. Nevertheless, small deviations in mix ratio from the equivoluminal composition are unavoidable. In these experiments eight cracked specimens were cut from each of twelve sheets. These sheets were all produced from the same shipment of prepolymer and catalyst within 1 week.§ No tests were made to assess specifically and quantitatively the sheet-to-sheet variations of mechanical properties. Instead we used one or two specimens from each sheet with the crack oriented normal to the applied load, which all yielded critical loads with little data scatter.

6.2 Specimen Preparation and Test Procedure

The test specimens were prepared in the unswollen state. Let us call that geometry the "reference geometry" for which a centrally located crack in a sheet of dimensions 6 × 3 inches is oriented parallel to the narrow dimension. Assuming that 6 inches is sufficiently long to simulate an infinite strip 3 inches wide we determined the crack length on the basis of Ref. [63] so small that the deviation of the crack tip stresses from those encountered in a doubly infinite sheet (Griffith problem) were less than 2%; the crack length so determined was $\frac{3}{8}$ inch in the unswollen state. We assume that the error introduced by employing a finite sheet instead of an infinite one for cracks oriented at different angles to the tension axis is not larger than that for the reference geometry.

The cracks were grown in a special tension device. After a small initial cut with a razor blade, thin brass rails were joined to the specimen sheet with double-stick adhesive tape parallel and close to the location of the initial cut. These brass rails fitted into the tension device and were then separated slightly to make the cut grow slowly and parallel to the jaws in the unswollen material. When the crack had reached the desired size—normally after 5–8 minutes—the specimen was unloaded. Under a load which was large enough to open the crack very slightly but not large enough to cause further crack growth, the tips of the crack were examined under 160 × magnification and photographed for later reference. A representative photograph is shown in Fig. 15. Although we had hoped to achieve a large number

† A reaction produced of Castor oil and tolylenediisocyanate.

‡ Castor oil.

§ Experience over the years has shown that different batches of the initial liquids yield greater variations in mechanical properties than sheets produced from the same shipment.

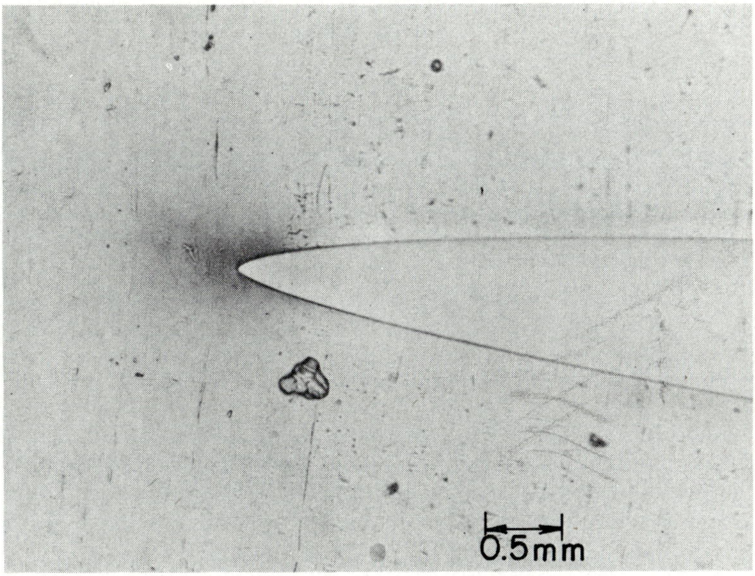

Fig. 15 Appearance of crack tip; deformed in unswollen state.

of specimens for which both crack fronts were sharp and clearly straight through the sheet thickness and normal to the plane, we found that only a few specimens fulfilled this requirement. We therefore accepted also specimens for which only one crack tip was satisfactory; this crack tip was identified for the post fracture examination. Approximately 40% of all specimens were rejected at this point. The actual lengths and orientations of the cracks were measured rather than accepting the nominal values.

The specimens were then placed in toluene. After 24 hours, the specimen was transferred to the test apparatus shown schematically in Fig. 16, where the specimen was loaded and fractured while submersed in toluene. The major problems associated with the test were handling of the fragile specimen (25% loss rate) and the determination of the proper clamp pressure so that the specimen would neither crack at the clamp nor slip out of them. Slippage became problematical at the higher loads associated with crack orientations that made angles smaller than 45° with the tension axis.

From pilot tests the approximate load-carrying capability of specimens was known. Thus loading was accomplished by a set of calibrated weights short of fracture, and further loading by the addition of lead shot to the fracture point. The specimen was oriented by aligning it with lines engraved

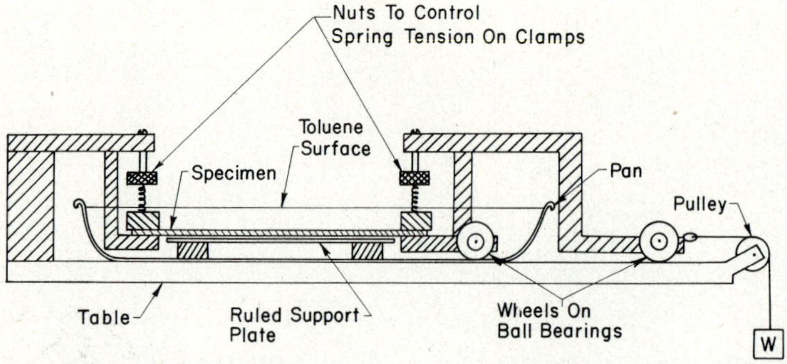

Fig. 16 Schematic of test fixture for testing sheet submerged in toluene (a small vibrator attached to the table virtually eliminated friction in wheels and pulley).

into the orientation plate; the latter served as a specimen rest when no load was applied, but under load the specimen lifted away from this plate by a small amount since it was nearly buoyant in the liquid.

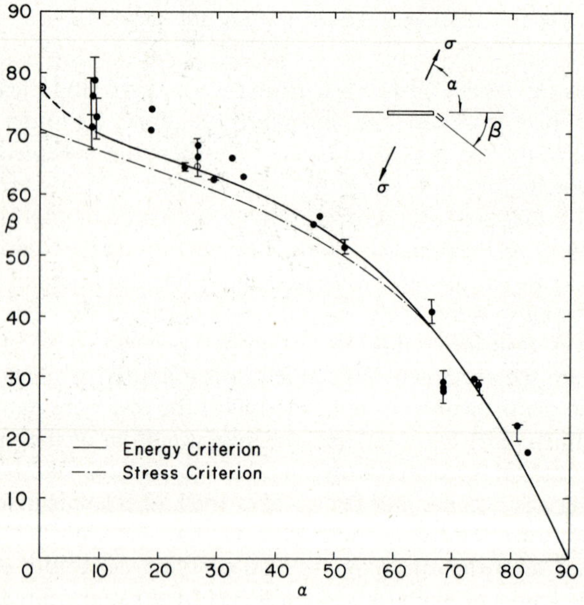

Fig. 17 Experimental results on a toluene swollen polyurethane: crack branching angle as a function of load orientation angle α.

6.3 Data Evaluation

After the specimen was broken it was placed between two glass plates and inserted into a photographic enlarger to produce (approximately) 10× enlarged photographic records. For proper scaling a gage block of 0.500 ± 0.0005 inch in length was included in the view field. Enough solvent was retained around the edges of the specimen that erroneous records due to the miniscus effects were excluded.

The crack lengths prior to rupture were measured and compared with the measurements made in the unswollen state by taking the swelling ratio into account; the agreement was invariably so good that after initial measurements the comparison was abandoned. For the reference geometry an original crack length could not be determined from the photographs nor from inspection of the broken specimen since the continuation of the fracture surfaces was also perfectly smooth and glassy.

It remains to recount the measurement of the crack extension angle. For a specimen having two "good" crack tips we evaluated each crack tip for the two specimen halves separately. Ideally one obtained thus four (not independent) values. Each investigator made these measurements separately; these measurements were then compared and a weighted average was computed. The weight was assigned jointly by both investigators by judging the visual clarity of the fractured corner.

Because the crack extension does not follow a straight line the initial extension orientation must be determined as a tangent (asymptotically) at the original crack tip point. In doing this and following the analytical developments we assumed that the crack extension did break away from the original main crack with a definite angle. It may be argued that the crack could propagate along a line possessing a continuously changing tangent and that the crack path could be highly curved just beyond the initial crack tip. In our experiments we were not able to resolve a radius of crack path curvature smaller than about five-thousandths of an inch; but within the limit of this resolution no crack path curvature emanating from the original crack tip point was detected. Figure 17 shows the crack propagation angle β as a function of the load orientation angle α, and Fig. 18 records the corresponding critical loads normalized by the critical load for the reference geometry.

6.4 Results for Mildly Ductile Solids

We conclude this section on experimental results with a comparison of J. G. Williams and Ewing's ample fracture data on polymethylmethacrylate

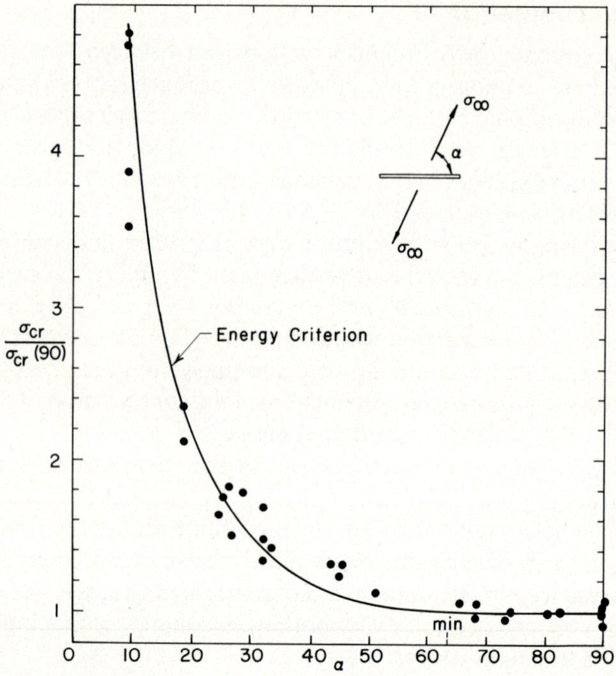

Fig. 18 Experimental results on a toluene swollen polyurethane: critical stress as a function of load orientation angle α.

[11]. Recall that in Subsection 5.4 we suggested that the energy criterion for finite extensions "a" may represent an approximate criterion of fracture for mildly ductile solids. Figure 19 compares selected curves from Fig. 12 with the data from Ref. [11].

7 CRACK GROWTH FROM A CRACK FRONT UNDER A GENERAL, THREE-DIMENSIONAL STATE OF STRESS

We have so far discussed how the propagation of a crack occurs in a two-dimensional stress field.

We turn now to the consideration of crack propagation when one or both of the geometry and the loading are not two-dimensional. In these cases we expect that in addition to the critical load and the orientation of the initial growth the shape of the crack extension enters our consideration. We have

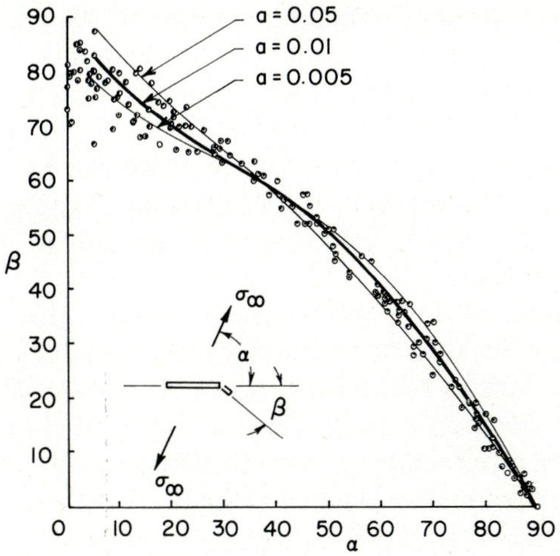

Fig. 19 Comparison of energy-based calculations for finite crack extension with data of Ref. [11].

already discussed in the introduction some of the details of possible crack-propagation geometries. Here we emphasize once more that we wish to discuss only such primary cracks whose fronts, described in general by a spatial curve, are smooth. From a practical viewpoint we may define smooth by the requirement that at any point along the crack front its radius of curvature be large compared to the dimensions of a zone in which the plastic or the nonlinear material behavior dominates. This is probably the case in many naturally grown cracks. Under this smoothness provision we assume that we can approximate the crack geometry in the neighborhood of a small section near its front as planar and as possessing a nearly straight front. Recall now that in Subsection 6.3 we discussed briefly in connection with the experimental two-dimensional results the argument that in a planar geometry the crack might propagate with a tangent to the crack propagation surface that changes but that the curvature of the crack path might be high just past the tip of the original crack. The analogous argument for the growth of a (locally) planar crack from a (locally) straight front would be that under general loading the crack propagates a small distance in its original plane and then develops into a crack surface with very high surface curvature. Within the resolution of the experiments described subsequently such curvature of the crack propagation surface did not appear to be present.

The additional argument that may be brought against this proposition is that if the crack propagates a small amount along its original (local) plane it remains locally self-similar; after the small (infinitesimal) self-similar propagation the crack would be ready to propagate again in a self-similar fashion and so on, unless the external loads change or the crack front stresses change as a result of the geometric change. If we consider the special case of a planar crack with a straight crack front under a general state of stress one would thus tend to conclude that the crack should propagate always along the original plane.

We shall examine below experimental evidence that contradicts this proposition. This contradiction relates to brittle solids for which the minimal structural size parameter at the crack tip ("plastic zone size") is very small. If we deal with a solid possessing a relatively large structural size parameter, say for example as governed by a zone of craze material at the tip of a crack in an uncrosslinked inorganic glass, or by intense void generation and plastic deformation in a metal, then crack propagation is conceivably possible in a path through such a zone of weakened material. While the propagation occurs through the initial zone of decomposing or highly deforming solid, the further development of that zone may not be along the direction of the original one but deviate slightly therefrom. In this way a crack may possibly achieve indeed a growth with a continuously changing tangent plane. If such a propagation process is possible, then the crack extension curvature should be on the order of the size of the structural parameter ("plastic size"). In the limit of such a vanishing microstructural size scale the surface curvature of the crack extension would become very large and crack propagation would tend to become dis-tangential.

To examine how a crack propagates under three-dimensional loading in a brittle solid we may consider as an example the special case of a planar crack with a straight crack front under antiplane shear. Once this special case is presented an avenue exists by which the problem of crack extension under general loading may be formulated.

8 AN EXPERIMENT OF CRACK EXTENSION IN ANTIPLANE DEFORMATION

The object of the following experiment [7] is the visualization of the fracture progression under antiplane deformation. As in the study of the in-plane problem it would be desirable to employ a brittle solid as, for example, an inorganic glass. However, because fracture proceeds very rapidly in these materials the observation of crack propagation details is very difficult even if a high-speed motion camera or ultra-sound fractographic equipment is available. This difficulty is circumvented by a particular choice of the test material.

8.1 Choice of Test Material

We resorted to the same polyurethane material employed in the two-dimensional work, except that it is not swollen in a liquid. It has already been mentioned in Subsection 6.1 that the microstructural size zone of this material is of the same order of magnitude as that for inorganic glasses; moreover, cracks can grow in this material very slowly if the loads are adjusted accordingly. This slow growth can be exploited in stopping the crack-propagation process at any moment and examining the result in a stationary state.

It is natural to ask whether the viscoelastic properties invalidate the arguments based on a linearly elastic analysis. We believe that the material viscosity affects only the rate with which crack propagation occurs but not the direction which a fracture path takes. Our experience with the fracture behavior of visco-elastic solids does not contradict this belief.† Moreover, we shall see that the result of the crack propagation process, namely the appearance of the post-fracture surface, is apparently the same in the polyurethane material as in an inorganic glass [64]. One is thus inclined to conclude that the phenomenon observed with the slowly growing crack in the polymer is—apart from the speed of crack growth—the same and that the "stopped photographs" of the polymer fracture thus represent the fracture in brittle solids.

8.2 Specimen Geometry

In order to induce a pure mode III deformation in the vicinity of the crack front it is sufficient to load a cracked specimen in such a way that all displacement components vanish except the one which is parallel to the crack front. For a finite length specimen the ideal displacement field is approximated away from the ends of the specimen such as shown in Fig. 20 which represents a slab of material into which a long crack has been cut.

One desirable method of loading would be to glue relatively rigid plates

† We assume here that quasi-static processes are involved only. The same assertion could not be readily made if wave phenomena are important.

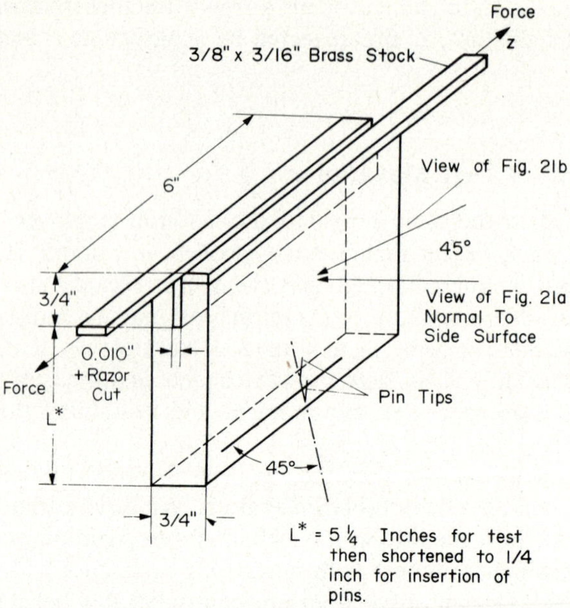

Fig. 20 Geometry of specimen for antiplane fracture test.

to the large slab faces (cf. Fig. 20) and to displace these faces opposite, parallel to each other and to the crack front. However, in order to be able to continuously observe the entire length of the crack front the joined plates as well as the adhesive have to be transparent. It proved difficult to effect a perfect joint over such an extended area without the appearance of view-disturbing bubbles.† We therefore chose to apply the load by displacing the narrow faces of the crack flanks parallel and opposite to each other by means of the glued-on brass rails as shown in Fig. 20. To simulate the infinite solid the slab edges normal to the crack front must be supplied with moments to obtain only deformations parallel to the *z*-axis. If the plate is thus constrained from warping, the central portion of the crack is in a state of antiplane shear. For the crack-growth test the brass rails were displaced a small amount parallel to each other and then held constant so that the crack propagation could occur over a time span of about half an hour.

† The possibility of casting the polymer between two plexiglass plates with a perfect bond did not exist in our laboratory at the time of the tests.

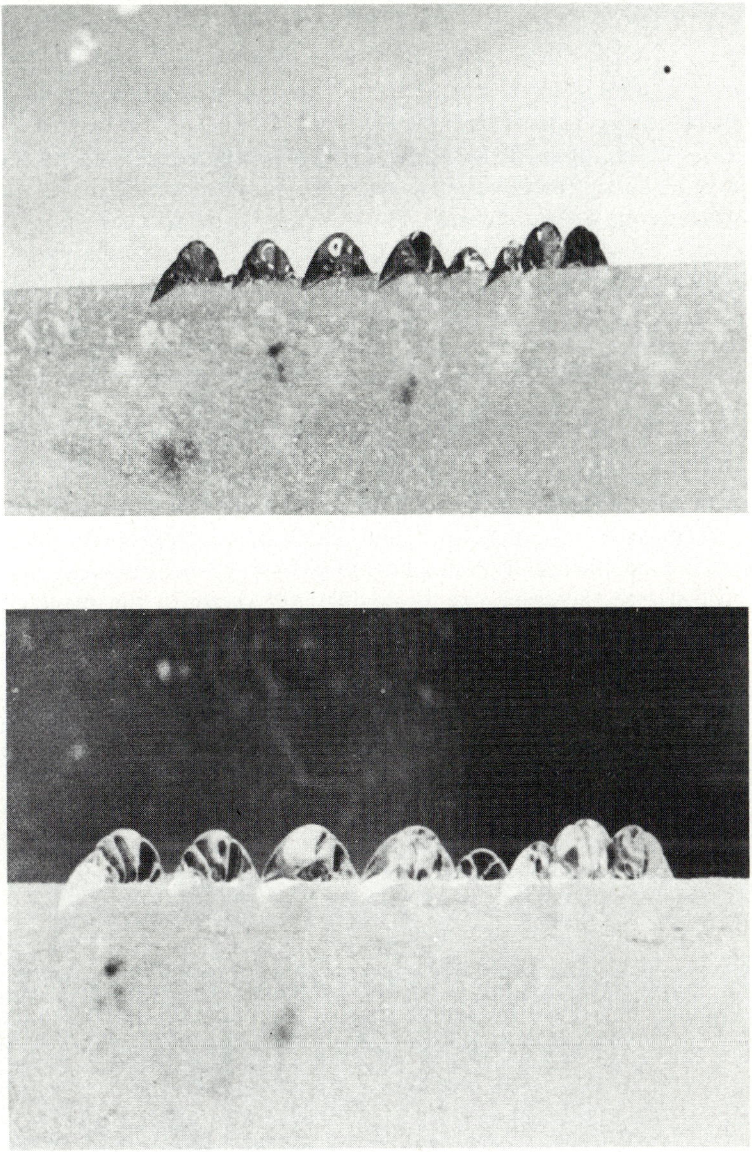

Fig. 21 Parahelical crack extensions. Lower portions of photographs represent original crack. *Top* (a) view at 90° to slab surface; *bottom* (b) 45° to slab surface.

8.3 Crack Propagation Observation

Figure 21 shows that the generation of new fracture surface does not parallel the original crack. Rather, there appear along the crack front small, somewhat lentil-shaped cracks which start on one of the original crack surfaces, spiral through the formerly uncut portion of the solid and terminate on the other surface of the original crack. We shall refer to the shape of these crack extensions as "parahelical". The lines appearing inside the parahelical cracks are caused by partial sticking of the fracture surfaces upon load removal. Figure 22 shows a model of one typical parahelical crack.

The midplane of the parahelical crack extension is inclined to the plane of the original crack by approximately 45°; the refractive properties of the polyurethane makes a direct measurement through the surface of the specimen difficult. Breaking the specimen completely and examining the inclination of the ripples near the initial crack front is possible but not necessarily easy. Immersion in a liquid of the same index of refraction would have overcome this difficulty. But instead of finding such a liquid, possibly the uncured mixture of prepolymer and catalyst, we chose the following approach. Since we are interested here in an average measurement—having established that the crack extension does not occur in the plane of the

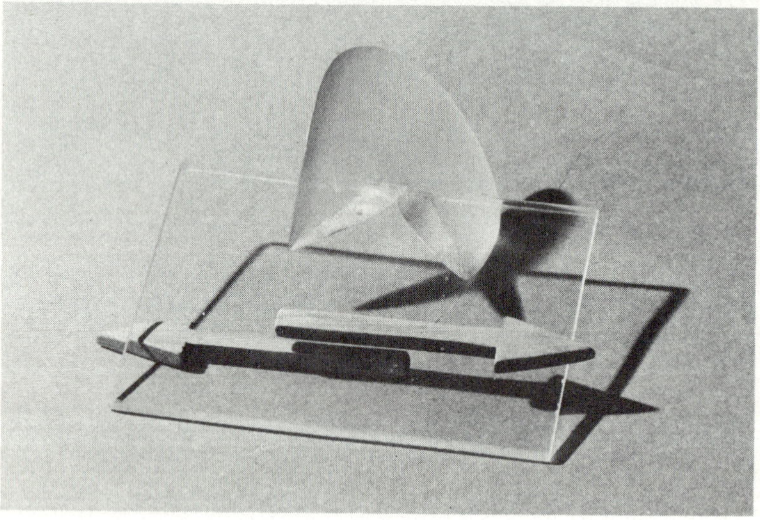

Fig. 22 Model of parahelical crack; plexiglass sheet represents original crack, arrows indicate direction of displacement.

original crack—a simpler demonstration is sufficient. We cut the specimen close to the crack normal to the crack plane and parallel to its front, and insert two needles as shown in Fig. 20 such that they are aligned along a plane making an angle of 45° with the original crack plane. We then examine simultaneously the extension cracks edge-on and the position of the needle points inside the slab relative to each other through the slab faces. The cracks exhibited their smallest (thinnest) dimension when the needles were aligned with each other, thus indicating that their (average) midplane is also oriented 45° with respect to the original crack.

We need to discuss the observation that the individual crack extensions seem to follow, on the average, some regular spacing along the crack front for one might expect that crack extensions should occur simultaneously from all parts of the crack front because the stress state is independent of z. It may be argued that this spacing is induced by irregularities in the crack front which were induced by the crack-cutting operation. Although the razor blade corner was drawn through the saw cut to form the sharp crack front, in a steady, smooth operation, it is not inconceivable that the friction and the elasticity of the material could have induced some stick-slip cutting and consequent unevenness in the crack front from which the crack extension would then begin to propagate. One could argue further that if such an uneven crack front exists, then this unevenness is the primary cause that crack extension occurs in the way observed rather than in the plane of the original crack.

Several points against these arguments can be made. First, the notion that the crack should continuously form a smooth crack propagation front is similar to the one that a continuum solid should fracture "homogeneously" if subjected to a homogeneous state of stress. Materials are never perfect enough to let the latter happen and we conclude that from a practical viewpoint a crack front is similarly imperfect. Second, if imperfections existed in the smoothness of the crack front, they could not be detected with the unaided eye. Third, razor cuts on other specimens of the same material were made and the specimens were then broken in a crack-opening mode to expose the front made by the razor blade; the crack front was smooth. Fourth, we cut the crack front on a circular (torsion) specimen not with the corner of a razor blade but with the straight cutting edge (straight portion of a razor blade). Qualitatively the same result was obtained. Fifth and finally, in Ref. [64] similar experiments were performed on a silica glass, except that only the final rupture surfaces were shown after the specimen had broken completely. In these glass experiments the original crack front was grown naturally, not cut, and the fractures started also along the periphery at

Fig. 23 Post-fracture appearance of externally cracked cylinder (Solithane 113) failed in torsion.

apparently well-spaced locations, each point giving rise to a "lance fracture" in the terminology of that reference.

The spacing of the crack extension is explained most likely in the following way. At some point a parahelical crack extension occurs due to an irregularity in the material or in the crack-front geometry. This fracture relieves the stresses in its vicinity and thus makes the formation of a new fracture likely only some distance away along the crack front. After such further fractures have been established the process repeats itself. Even for a perfectly smooth crack front under a uniform stress field along its length the spacing should be only approximately constant since the stress field for each subsequent fracture depends on the number and spacing of the previously formed fractures. If several fractures occur far apart and independently of each other along a long crack front, then each one can, conceivably, develop a sequence of nearly uniformly spaced cracks.

Once these multiple cracks grow they will begin to interact more strongly, the larger ones overtaking the smaller ones and coalescing with each other. The result is that near the crack front many small features appear in a completely exposed fracture surface and the features become coarser away from the front. This is clearly evident in Fig. 23 which shows the fracture surface of a circularly cracked cylinder under pure torsion.

Furthermore, it appears clear that the coalescence of the small parahelical fractures produces a rough or gross fracture that is oriented on the average along the plane of the original crack, as is apparent, for example, from Fig. 23.

9 RELATED WORK ON FRACTURE INVOLVING MODE III DEFORMATIONS

Before proceeding to exploit this experimental result together with those of the in-plane problem to formulate an approximate criterion for crack instability under combined loading, it seems appropriate to review briefly additional results on mixed mode fracture that is available in the literature. To date energy release rates for mode III deformations have been calculated only under the assumption that crack extension is co-planar [65, 66]. No doubt the prime motivation behind stress analyses of antiplane problems is, although stated only rarely [67, 68], to derive qualitative insight from the mathematically simpler antiplane problem about the in-plane problem. In view of the just recounted parahelical crack growth phenomenon it seems advisable at least to exercise caution in the application of such solutions to the process of crack propagation, under the assumption of co-planar crack growth. Also nonco-planar crack growth which leaves the geometry two-dimensional [69, 70] clearly does not result.

In this context it is appropriate to mention Sih's suggestion that the strain energy density theory of fracture [23, 24] should be applicable to fracture onset under arbitrary, combined modes of crack tip deformation and in particular to the problem of antiplane shear [71]. We note that the stress field for the *anti*plane problem depends only on the *in*-plane coordinates. Thus the strain energy density is also a function only of the in-plane coordinates and hence for this case the *S*-theory can predict, to the best of our understanding, only non-planar crack growth that leaves the geometry of the extended crack two-dimensional. This result [71] leads to the same type of crack growth considered, also from a purely analytical viewpoint, in [70] and [71] for mode III deformations but is at variance with the experimental results reviewed above.†

† For a further examination of the limited applicability of the *S*-theory to fracture in three-dimensional stress fields the reader may consult Ref. [33].

Experimental investigations involving relatively small mode III deformations superposed on mode I and mode II have been reported by Sommer for glass [64], by Pook for an aluminum alloy [14] and by Shah for 4340 steel [72]. In these reports only the fracture surface of the completely broken specimen are recorded. It is interesting to note that in all cases there appear at the front of the original cracks "irregularity" features in the fracture surface that look the same or closely the same as those recorded in Fig. 23. While we do not wish to assert on the basis of these supplementary observations that mode III deformations always induce parahelical or similar crack extensions, it is quite likely that many materials will respond in this fashion, and that such fracture features are (often) observable if one looks for them.

10 AN APPROXIMATE ANALYSIS FOR MULTIMODE FRACTURE IN BRITTLE SOLIDS

Having demonstrated that the classical energy balance analysis agrees well with experiments on crack propagation induced in the two-dimensional problem, it appears reasonable to conclude that the principle upon which it is based applies also to the triple-mode interaction problem. To accomplish that objective and in view of the experiments in purely mode III crack extension one would have to compute first the stress and displacement components for the extended crack geometry having a small discontinuous crack extension of arbitrary (mean) surface shape and possessing an arbitrary crack front periphery. One would next need to compare the energy state for this general extended geometry with that of the unextended geometry, maximize the energy difference with respect to the orientation, curvature and outline of the crack extension and then let the size of the crack extension tend to zero for brittle solids. For mildly ductile solids it might be sufficient to consider small but finite extensions in analogy to the two-dimensional case of Subsection 5.4.

Clearly, these computations are beyond the capability of presently available analytical methods and only numerical discretization methods offer hope for a solution. But even computer calculations tend to be difficult and not necessarily conclusive. We therefore proceed to explore a simple, obviously approximate criterion for multimode fracture in brittle solids.

To this end we recall two previously made observations. First, that in the two-dimensional case the (exact) energy release criterion yielded results which are closely in agreement with the (approximate) stress criterion. Second, for

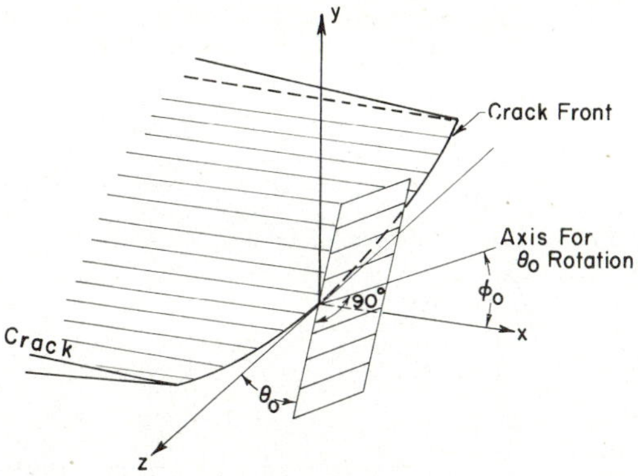

Fig. 24 Local coordinate system at crack front and definition of orientation of crack extension plane.

the two-dimensional problem the fracture occurs in a plane containing the crack front. Third, the mean orientation of the crack extension surface in the mode III fracture exists in a plane oriented normal to the maximal tensile stress.

These observations indicate that a tensile stress criterion may be a reasonable criterion for fracture onset under general loading. We must be aware, however, that we may not simply require that fracture is induced by the maximal principal tensile stress near the crack tip; for in the two-dimensional case the stress criterion assumes that fracture is induced by the *special* principal stress which is oriented normal to a ray originating at the crack tip. Thus, if a stress criterion of fracture is used, it must be formulated such that for the in-plane problem this special two-dimensional result [6, 8] is obtained.

In order to define our approximation scheme we refer to the coordinate system and rotation angles shown in Fig. 24. We assume that in the presence of general crack-front deformations (modes I, II, and III) crack extension occurs locally in a plane obtained by rotating the x–z-plane about the z-axis (locally tangent to the crack front) through an angle ϕ, and the resulting plane about the new x-axis (say, x'-axis) by an angle θ; the magnitude of the angles is determined by the requirement that the stress normal to this plane be maximized.† Fracture is assumed to occur when this stress attains a critical value.

† One could interchange the order of rotation. The resulting angles would, of course, not necessarily be the same but the magnitude of the stress acting on the plane having zero shear would be the same.

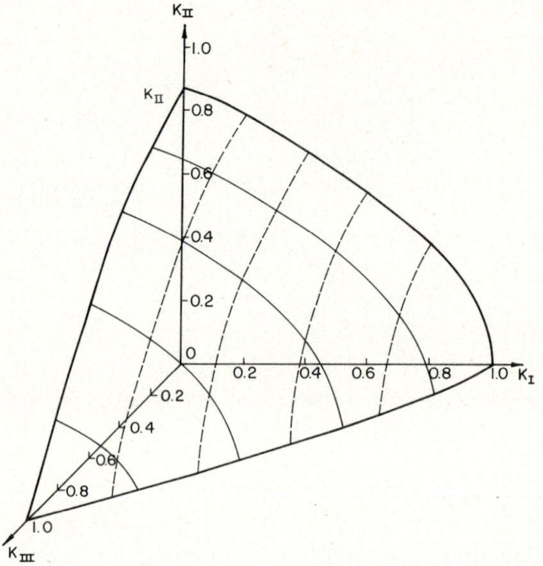

Fig. 25 Mode interaction in general loading of crack front stress criterion.

The application of this criterion to the "stress intensity" $\sqrt{r}\{\tau\}$ where the stress components τ_{ij} are derived from the crack front singularity expansions, clearly yields a normal stress on the doubly rotated plane that depends on the three stress intensity factors. Maximization of the normal stress requires the solution of two transcendental equations for the angles ϕ and θ and results in a relation that connects the three stress intensity factors at the onset of fracture.

In Fig. 25 we show the interaction surface for the three crack tip deformation modes for the case of a unit "stress" normal to the crack surface formed. Any combination of the three modes corresponding to points inside the surface does not lead to fracture while those corresponding to points outside do. The surface is symmetric about the two planes $K_{\mathrm{II}} = 0$ and $K_{\mathrm{III}} = 0$, the intersection of the surface with the plane $K_{\mathrm{III}} = 0$ being the same curve as shown in Fig. 11 for the (normal) stress criterion for in-plane fracture.

We append here a remark on the approximate nature of this criterion which goes beyond the discrepancies documented in its comparison with the two-dimensional computations of the energy release rate. Sommer's experiments on fracture in the presence of modes I and III deformations were conducted on glass cylinders under pressure and torsion in which cracks grew from the center outward and normal to the cylinder axis. The central portion of the naturally grown cracks represented planar crack growth; not until the outwardly growing crack encountered some particular magnitude of the mode III deformation did the planar crack growth branch over into a non-planar one indicative of fracture in the presence of mode III. On the basis of the simple stress criterion one can estimate from data in [64]—without making a complete stress analysis of Sommer's test geometry and loading—that the onset of branching from a planar crack growth into a non-planar one occurs at about $K_{III}/K_I \sim 0.001$. One would expect, therefore, that for larger ratios of K_{III}/K_I non-planar crack growth would occur.

11 EXTENSION TO MILDLY DUCTILE SOLIDS

In connection with the problem of fracture in the plane we have referred to the suggestions of Cottrell [12] and J. G. Williams and Ewing [11] that the stress criterion of fracture may be extended if higher-order terms in the crack-tip stress expansions are included in the analysis. For the two-dimensional problem this suggestion has generally yielded good results [11, 13]. One might inquire therefore whether a similar extension of the triple-mode problem for brittle solids could be extended to the fracture of mildly ductile solids by an analogous procedure.

We shall not explore this suggestion in detail here. The requisite stress analysis is routine and tedious. The results would yield a family of surfaces of the type shown in Fig. 25, each surface corresponding to a different value of the microstructural size parameter employed by J. G. Williams and Ewing [11].

In making this possible suggestion of extending the multi-mode fracture analysis to mildly ductile solids we point out, however, that the formation of an original crack in, say, mode I, is usually associated with the generation of a plastic domain along the crack front. Undoubtedly this original plastic zone influences the direction of crack propagation as well as the failure load upon reloading the crack in a multi-mode fashion. In Section 6 we have already alluded to this probability in connection with failure under in-plane loading. On the other hand, the concepts derived from the experiments on brittle solids may not, in practical terms, be invalidated by the presence of limited ductility (small plastic zone at crack instability); for the fracture surfaces occasioned by multi-mode loading in metal show an appearance that resembles distinctly that obtained in brittle solids when mode III is involved, i.e. lancelike features are present which are the result of the growth of parahelical crack extensions [14, 72]. Clearly the questions raised here can be resolved only by careful experiments which are now under study.

12 Appendix—Supplemental Definitions

Consider the mapping function

$$z = \omega(\xi) = A(\xi - e^{i\delta_1})^{\lambda_1}(\xi - e^{i\delta_1})^{\lambda_2}. \tag{A.1}$$

For real values of A, δ_1, δ_2, λ_1, and λ_2 this function maps the region outside the unit circle in the ξ-plane into the region outside the two radial lines OB and OC in the z-plane (Fig. 6). Let OB be along the x-axis and OC make an angle $\lambda_1\pi$ with the positive direction of the x-axis. Let the line OB be of length l_1 and OC of l_2 and let the points $\xi_1 = e^{i\beta_1}$ and $\xi_2 = e^{i\beta_2}$ correspond to the points B and C respectively. Then the parameters A, δ_1, δ_2, β_1, β_2, and λ_2 are determined in terms of l_1, l_2 and λ_1 by the following relations [47]:

$$\begin{aligned}
\lambda_1 + \lambda_2 &= 2, \\
\lambda_1\delta_1 + \lambda_2\delta_2 &= 2\pi, \\
\lambda_1 \cot\frac{\beta_1 - \delta_1}{2} + \lambda_2 \cot\frac{\beta_1 - \delta_2}{2} &= 0, \\
\lambda_1 \cot\frac{\beta_2 - \delta_1}{2} + \lambda_2 \cot\frac{\beta_2 - \delta_2}{2} &= 0, \\
4A\left|\sin\frac{\delta_1 - \beta_1}{2}\right|^{\lambda_1}\left|\sin\frac{\delta_2 - \beta_1}{2}\right|^{\lambda_2} &= l_1, \\
4A\left|\sin\frac{\delta_1 - \beta_2}{2}\right|^{\lambda_1}\left|\sin\frac{\delta_2 - \beta_2}{2}\right|^{\lambda_2} &= l_2.
\end{aligned} \tag{A.2}$$

When $l_2/l_1 = a \ll 1$ (which case is of main interest to our work), Eqs. (A.2) can be solved in series form for A, δ_1, δ_2, β_1, and β_2 in terms of the small parameter "a". The results obtained are [9]

$$A = \frac{l_1}{4}[1 + p_1^2\lambda_1\lambda_2 a] + O(a^2),$$

$$\begin{aligned}
\delta_1 &= \pi - \lambda_2 p_2 a^{1/2} - \lambda_2 p_3 a^{3/2} + O(a^2), \\
\delta_2 &= \pi + \lambda_1 p_2 a^{1/2} + \lambda_1 p_3 a^{3/2} + O(a^2), \\
\beta_1 &= 2\pi + (\lambda_1 - \lambda_2)p_4 a^{3/2} + O(a^2), \\
\beta_2 &= \pi + (\lambda_1 - \lambda_2)p_2 a^{1/2} + (\lambda_1 - \lambda_2)(p_3 - p_4)a^{3/2} + O(a^2)
\end{aligned} \tag{A.3}$$

with

$$\begin{aligned} p_1^2 &= \lambda_1^{-\lambda_1}\lambda_2^{-\lambda_2}, \\ p_2 &= 2p_1, \\ p_3 &= \frac{p_1^3}{6}(\lambda_1^2 + \lambda_2^2 - 6\lambda_1\lambda_2), \\ p_4 &= \frac{2p_1^3}{3}\lambda_1\lambda_2. \end{aligned} \tag{A.4}$$

Acknowledgment. Portions of this work were supported by the National Aeronautics and Space Administration under Research Grant Nsg-172-60 (Galcit 120).

13 REFERENCES

1. Griffith, A. A., "The Phenomena of Rupture and Flow in Solids," *Phil. Trans. Roy. Soc. (Lond.)*, A, **221** (1920) 163–198.
2. Griffith, A. A., "The Theory of Rupture," in *Proc. of the 1st International Congress for Appl. Mech., Delft*, 1924, pp. 55–63.
3. Barenblatt, G. I., "The Mathematical Theory of Equilibrium Cracks in Brittle Fracture," *Advances in Applied Mechanics*, **7** (1962), Academic Press, pp. 55–129.
4. Irwin, G. R., "Analysis of Stress and Strain near the End of a Crack Traversing a Plate," *J. Appl. Mech.* **24** (1957) 361–364.
5. Rice, J. R., *Mathematical Analysis in the Mechanics of Fracture*, Vol. 2, H. Liebowitz (ed.), (1968), Academic Press, pp. 191–311.
6. Erdogan, F. and Sih, G. E., "On the Crack Extension in Plates under Plane Loading and Transverse Shear," *J. Basic Engng*, **85** (Dec. 1963) 519–527.
7. Knauss, W. G., "An Observation of Crack Propagation in Anti-plane Shear," *Int. J. Fract. Mech.* **6**, No. 2 (June 1970) 183–187.
8. Panasyuk, V. V., Berezhnitskiy, L. T., Kovchik, S. Ye., "Propagation of an Arbitrarily Oriented Rectilinear Crack During Extension of a Plate," NASA Technical Translation F-402 (Dec. 1965).
9. Palaniswamy, K., "Crack Propagation under General In-plane Loading," California Institute of Technology, Ph.D. Thesis, 1972.
10. Palaniswamy, K. and Knauss, W. G., "Propagation of a Crack under General, In-plane Tension," *Int. J. Fract. Mech.* **8** (1972) 114–117.
11. Williams, J. G. and Ewing, P. D., "Fracture under Complex Stress—The Angled Crack Problem," *Int. J. Fract. Mech.* **8**, No. 4 (Dec. 1972) 441–446.
12. Cotterell, B., "Notes on the Paths and Stability of Cracks," *Int. J. Fract. Mech.* **2** (1966) 526–533.
13. Finnie, I. and Saith, A., "A Note on the Angled Crack Problem and the Directional Stability of Cracks," *Int. J. Fract.* **9** (1973) 484–486.
14. Pook, L. P., "The Effect of Crack Angle on Fracture Toughness," *Engng Fract. Mech.* **3** (1971) 205–218.
15. Ewing, P. D. and Williams, J. G., "Further Observations on the Angled Crack Problem," *Int. J. Fract.* **10** (1974) 135.
16. Shah, R. C., "Fracture under Combined Modes in 4340 Steel," presented at the 7th

National Symposium on Fracture Mechanics, University of Maryland, Aug. 27–29, 1973, The Boeing Co., Seattle, Washington.
17. Liu, A. F., "Crack Growth and Failure of Aluminum Plate under In-plane Shear," AIAA 11th Aerospace Sciences Meeting, Washington, D.C., Jan. 10–12, 1973, AIAA Paper No. 73-253.
18. Marcus, H. L., Ho, C. L. and Frandsen, J. D., "Mixed Mode Crack Extension in an Aluminum Alloy," Report of Rockwell International Science Center, Thousand Oaks, Calif. Paper MaB6, *Proc. of 10th Ann. Meeting Soc. of Engng Sci., Raleigh, N.C., Nov. 5–7, 1973* (to be published).
19. Ellis, R., "Residual Strength of Cracked Panels under Shear," Australian Defence Scientific Service, Aeronautical Research Laboratories, Melbourne, May 1973. Structure and Materials Note 393.
20. Shah, R. C., "Effects of Combined Modes Loading on Fracture and Cyclic Flaw Growth in 2219-T87 Aluminum and 6A1-4V Titanium," The Boeing Aerospace Co., Seattle, Washington. AIAA Paper 74-414.
21. Roberts, R. and Kibler, J. J., "Mode II Fatigue Crack Propagation," *J. Basic Engng*, **93** (Dec. 1971) 671–680.
22. Iida, S. and Kobayashi, A. S., "Crack-propagation Rate in 7075-T6 Plates under Cyclic Tensile and Transverse Shear Loadings," Dec. 1969. Paper No. 69-Met-1. University of Washington, Dept. of Mechanical Engineering Report, Seattle, Washington.
23. Sih, G. C., "Some Basic Problems in Fracture Mechanics and New Concepts," *Engng Fract. Mech.* **5** (1973) 365–377.
24. Sih, G. C., Introductory Chapter: "A Special Theory of Crack Propagation," *Mechanics of Fracture, Methods of Analysis and Solutions of Crack Problems*, G. C. Sih (ed.), Noordhoff Int. Publishing Co., Leyden (1973).
25. Sih, G. C. and Kipp, M. E., Discussion on "Fracture under Complex Stress—The Angle Crack Problem," by J. G. Williams and P. D. Ewing. *Int. J. Fract. Mech.* **8** (1972) 441–446.
26. Sih, G. C., "Strain-energy-density Factor Applied to Mixed Mode Crack Problems," *Int. J. Fract.* **10**, No. 3 (Sept. 1974) 305–321.
27. Sih, G. C., Discussion: "Some Observations on Sih's Strain Energy Density Approach for Fracture Prediction," by I. Finnie and H. O. Weiss. *Int. J. Fract.* **10** (1974) 279–283.
28. Lindsey, G. H., "Some Observations on Fracture under Combined Loading," *Progress in Flaw Growth and Fracture Toughness Testing*, ASTM STP 536, American Society for Testing and Materials (1973) 22–31.
29. Hussain, M. A., Pu, S. L. and Underwood, J., "Strain Energy Release Rate for a Crack under Combined Mode I and Mode II." Presented at 7th National Fracture Mechanics Symposium, University of Maryland, August 1972. Benet Weapons Laboratory, Watervliet Arsenal, N.Y.
30. Rice, J. R., "A Path Independent Integral and the Approximate Analysis of Strain Concentration by Notches and Cracks," *J. Appl. Mech.* **35** (June 1968) 378–386.
31. Knowles, J. K. and Sternberg, E., "On a Class of Conservation Laws in Linearized and Finite Elastostatics," *Archive for Rational Mechanics and Analysis* **44**, No. 3 (1972) 187–211.
32. Budiansky, B. and Rice, J. R., "Conservation Laws and Energy-Release Rates," *J. Appl. Mech.* **40** (Mar. 1973) 201–203.
33. Palaniswamy, K. and Knauss, W. G., "On the Problems of Crack Extension in Brittle Solids Under General Loading," Graduate Aeronautical Laboratories, California Institute of Technology Report SM 74-8.
34. Dudukalenko, V. V. and Romalis, N. B., "Direction of Crack Growth under Plane Stress State Conditions," *Izv. AN SSSR Mekhanika Tverdogo Tela.* **8**, No. 2 (1973) 129–136.
35. McClintock, F. A., Discussion of Ref. [6], *J. Basic Engng* (Dec. 1963) 525–527.
36. McClintock, F. A. and Walsh, J. B., "Friction on Griffith Cracks in Rocks under Pressure," *Proc. 4th U.S. Congress Appl. Mech., Berkeley, 1962*, Amer. Soc. Mech. Engng, N.Y. (1963) 1015–1021.

37. Cotterell, B., "Brittle Fracture in Compression," *Int. J. Fract. Mech.* **8**, No. 2 (June 1972) 195–208.
38. Hoek, E. and Bieniawski, Z. T., "Brittle Fracture Propagation in Rock under Compression," *Int. J. Fract. Mech.* **1** (1965) 137–155.
39. Cotterell, B., "The Paradox Between the Theories for Tensile and Compressive Fracture," *Int. J. Fract. Mech.* **5** (1969) 251–252.
40. Cotterell, B., "On Fracture Path Stability in the Compact Tension Test," *Int. J. Fract. Mech.* **6** (1970) 189–192.
41. Morozov, E. M. and Fridman, Ya. B., "Trajectories of Brittle-fracture Cracks as Geodesic Lines on the Surface of a Body," *Dokl. Akad. Nauk SSSR* **139**, No. 1 (July 1961) 87–90; English Language Edition: *Soviet Physics-Doklady* **6**, No. 7 (Jan. 1962) 619–621.
42. Morozov, E. M. and Fridman, Ya. B., "Analysis of Cracks as a Method of Estimating Failure Characteristics," Moscow Engineering-Physical Institute. *Zavodskaya Laboratoriya* **32**, No. 8 (Aug. 1966) 977–984; English Language Edition: *Industrial Laboratory* **32** (1967) 1204–1212.
43. Porter, D. D. and Fairhurst, C., "A Study of Crack Propagation Produced by the Sustained Borehole Pressure in Blasting. In Dynamic Rock Mechanics," *Proc. 12th Symp. on Rock Mechanics, 1971*, Soc. of Mining Engng of AIME, pp. 497–515.
44. Morozov, E. M., Polak, L. S. and Fridman, Ya. B., "Variational Principles in the Development of Cracks in Solids," *Dokl. Akad. Nauk SSSR* **156**, No. 3 (May 1964) 537–540; English Language Edition: *Soviet Physics-Doklady* **9** (1964) 394–397.
45. Muskhelishvili, N. I., *Some Basic Problems of the Mathematical Theory of Elasticity*, P. Noordhoff Ltd, Groningen, Holland (1963).
46. Darwin, C., "Some Conformal Transformations Involving Elliptic Functions," *Phil. Mag.*, Series 7, **41** (1950) 1–11.
47. Andersson, H., "Stress-intensity Factors at the Tips of a Star-Shaped Contour in an Infinite Tensile Sheet," *J. Mech. Phys. Solids* **17**, No. 5 (1969) 405–417.
48. Sih, G. C., "Stress Distribution Near Internal Crack Tips for Longitudinal Shear Problems," *J. Appl. Mech.* **32** (1965) 51–58.
49. Sokolnikoff, I. S., *Mathematical Theory of Elasticity*, McGraw-Hill, New York (1956) 86.
50. Spencer, A. J. M., "On the Energy of the Griffith Crack," *Int. J. Engng Sci.* **3** (1965) 441–449.
51. Sih, G. C. and Liebowitz, H., "On the Griffith Energy Criterion for Brittle Fracture," *Int. J. Solids and Struct.* **3** (1967) 1–22.
52. Swedlow, J. L., "Criteria for Growth of the Angled Crack," Department of Mechanical Engineering, Carnegie Institute of Technology, Pittsburgh, Pennsylvania, Report SM-75-2 (1975).
53. Finnie, I. and Weiss, H. D., "Some Observations on Sih's Strain Energy Density Approach for Fracture Prediction," *Int. J. Fract.* **10** (1974) 136–138.
54. Küppers, H., "The Initial Course of Crack Velocity in Glass Plates," *Int. J. Fract. Mech.* **3**, No. 1 (Mar. 1967) 13–17.
55. Klein, G., *Bestimmung von Spannungsfaktoren bei Gemischten Beanspruckungsarten am Beispiel eines Risses in der Umgebung eines Kreislochs.* (A Method for the Determination of Mixed Mode Stress Intensity Factors and its Application to a Crack in the Neighbourhood of a Circular Hole.) Institut für Festkörpermechanik der Fraunhofer-Gesselschaft e. V., Freiburg (Jan. 1974).
56. Kambour, R. P. and Kopp, R. W., "Cyclic Stress-strain Behavior of the Dry Polycarbonate Craze," *Proc. of the Conf. on Polymer Structure and Mechanical Properties*, U.S. Army Natick Labs., Natick, Mass., April 1967. Also *Journal of Polymer Science*, Part A-2, **7** (1969) 183–200.
57. Kambour, R. P., "Stress-strain Behavior of the Craze," *Polymer Engng Sci.* **8** (1968) 281–289.
58. Kambour, R. P., "Structure and Properties of Crazes in Polycarbonate and Other Glassy Polymers," *Polymer* **5** (1964) 143–155.
59. Kambour, R. P., "Mechanism of Fracture in Glassy Polymers, II. Survey of Crazing

Response during Crack Propagation in Several Polymers," *J. Polymer Sci.*, Part A-2, **4** (1966) 17–24.

60. Kambour, R. P., "Mechanism of Fracture in Glassy Polymers, III. Direct Observation of the Craze Ahead of the Propagating Crack in Poly(methylmethacrylate) and Polystyrene," *J. Polymer Sci.*, Part A-4, **4** (1966) 349–358.
61. Mueller, H. K. and Knauss, W. G., "The Fracture Energy and Some Mechanical Properties of a Polyurethane Elastomer," *Trans. Soc. of Rheol.* **15**:2 (1971) 217–233.
62. Mueller, Hans-Karl, "Stable Crack Propagation in a Viscoelastic Strip," NASA Contractor Report CR-1279 (Mar. 1969). See p. 63.
63. Sneddon, I. N. and Srivastav, R. P., "The Stress Field in the Vicinity of a Griffith Crack in a Strip of Finite Width," *Int. J. Engng Sci.* **9** (May 1971) 479–488.
64. Sommer, E., "Formation of Fracture 'Lances' in Glass," *Engng Fract. Mech.* **1** (Apr. 1969) 539–546.
65. Sih, G. C. and Liebowitz, H., "Mathematical Theories of Brittle Fracture," in *Fracture*, Vol. II, see p. 128.
66. See Ref. 5, p. 230.
67. Westmann, R. A. and Yang, W. H., "Stress Analysis of Cracked Rectangular Beams," *J. Appl. Mech.* **34** (1967) 673–701.
68. Erdogan, F., "Crack Propagation Theories," in *Fracture*, H. Liebowitz (ed.), Vol. II, pp. 497–590. See Section 4, p. 531.
69. Simonson, E. R. and Jones, W. B., "Secondary Crack Trajectory for Longitudinal Shear Problem," *Int. J. Fract. (Mech.)* **6** (Mar. 1970) 65–69.
70. Smith, E., "A Note on Crack-forking in Anti-plane Strain Deformation," *Int. J. Fract.* **9** (June 1973) 181–183.
71. Hartranft, R. J. and Sih, G. C., "Growth Characteristics of a Plane Crack Subjected to Three-dimensional Loading," Lehigh University, IFM Techn. Report NASA-CR-132343 (July 1973).
72. Shah, R. C., "Effects of Proof Loads and Combined Mode Loadings on Fracture and Flaw Growth Characteristics of Aerospace Alloys," Boeing Aerospace Co., Seattle, Wash., NASA CR 134611 (Mar. 1974).

Note added by Editor: This article was received January 1975.

III

Scattering of Elastic Waves

Subhendu K. Datta

University of Colorado, Boulder, Colorado

1 INTRODUCTION

Although the scattering of elastic waves has been the subject of study for over a hundred years, only in the last 20 years has the subject received a good deal of attention from the geophysicists and applied mechanicians. Geophysicists' interest in this problem arises in the use of the scattered wave field to predict the *anomalies*. For this purpose the geophysicists use the gravity and magnetic surveys, electrical conduction and electromagnetic induction methods as well as seismic methods. Of all the physical methods used in geological exploration, the seismic methods are perhaps the most direct and—where applicable—give the least ambiguous results [1]. Seismic methods include the computation of travel times and amplitudes of reflected, refracted, and diffracted seismic signals. Following Rayleigh [2] we shall use the term "scattered wave" in this review to denote the difference between the total wave field observed in the presence of an obstacle and the incident wave. It is well known that a wavefront incident at a surface of discontinuity (of elastic properties) is scattered by it. Mow and Pao [3] give a review of the history of the elastic wave diffraction studies. This excellent monograph discusses in detail the engineering applications of elastic wave diffraction theory. However, they confine their attention mostly to the use of eigen-function expansions† in solving these problems.

In this review we shall outline other methods of analysis that have been found useful. For the sake of completeness and also for its usefulness in seismology and engineering, we shall solve in Section 2 the scattering of elastic waves by fluid sphere and circular cylinder. There we shall concentrate on the short wavelength limit in the relationship with the ray theoretic

† They include a brief discussion and application of a perturbation method developed in Refs. [4, 5].

predictions. These results have been used in predicting the Earth's inner structure. These can also be useful in ultrasonic detection of flaws [6] and studying the effect of caustics of the internally reflected rays on the singular behavior of stresses on the boundary (see, for example, [7–9]).

There are a large number of publications on the internal structure of the Earth as inferred from the travel times of transmitted signals (see Bullen [10]; also, see a lucid exposition by Bolt [11]). Although most of these studies are concerned with analyzing reflected and refracted waves, diffracted waves play an important role in studying the structure at the core–mantle boundary. The analytical technique for obtaining the diffracted field in the shadow of a spherical obstacle is due to Watson [12]. High-frequency diffraction of compressional and shear waves by a liquid core was solved by Duwalo and Jacobs [13] and Scholte [14] using Watson's transformation. Later extensions to lower frequencies were made by Phinney and Alexander [15], who used numerical integration techniques and also considered the effect of spherical-layered transitional region near the core–mantle boundary. Phinney and Cathles [16] made extensive numerical computations of the diffracted wave by a liquid and hollow core. A similar but analytical study was made by Teng and Richards [17] of the diffraction of waves in two dimensions by a circular cavity. Of particular interest in these studies are the broadening of the zone transitional from illuminated to shadow regions with the increasing wavelength and dependence of the shift on the wavelength. Similar studies in acoustics and electromagnetics were made by Rice [18], Rubinow and Keller [19], and Nussenzveig [20]. Of these, perhaps the most detailed and careful analysis is given in [20]. In Section 2 we shall outline the method and some pertinent results of diffraction of P-waves by a liquid sphere or cylinder. It may be noted here that so far all the studies have been concerned with the diffraction by a perfect sphere, but it has been suggested that there may be bumps at the core–mantle boundary in order to explain the precursor waves before the main onset of the core waves. But no detailed analysis of the effect of such bumps on the scattering has been reported. In a recent paper [21] we studied the effect on scattering of deviations from a rigid sphere. However, these results are not directly applicable to the geophysical problem, and one needs to consider the liquid core with boundary perturbations.

The problems of elastic surface wave propagation in layers of nonuniform thickness have been considered as prototypes of seismic wave propagation across continental margins and other regions of variable crustal thickness. Scattering of surface waves by topographic irregularities has also been analyzed. These studies are of importance in understanding site effects on ground response due to an earthquake [22]. A review of the analytical methods applicable to analyze the scattering of surface waves at a corner was done by Knopoff [23]. He also discussed the propagation of surface waves in layers with varying thickness. Since then numerical methods have been used to analyze the Love and Rayleigh wave transmission in non-horizontally layered structures by Drake [24], Lysmer and Drake [25], and others (see [24]). Scattering of surface waves by topographic irregularities has been studied by Hudson and Knopoff [26] and Hudson [27]. The mathematical technique that has been found most useful in solving these problems is an iteration scheme based on integral representations of the displacement field in terms of an appropriate Green's function for the problem (for details the reader may refer to [23] and the references given there). The problem of scattering of waves by cavities in a half space is also of interest in seismology as well as in earthquake engineering. However, this problem has not received much attention. Gregory [28] has recently solved the two-dimensional problem of scattering by an underground circular cylindrical cavity. In Section 3 we discuss the problem of scattering of elastic waves by a cylindrical cavity. There we outline the method of line potentials as developed in [28, 29]. We also show that this problem is solvable by a method of matched asymptotic expansions (MAE). The usefulness of MAE in solving scattering problems has been demonstrated only very recently and we show here its wide applicability in solving nonseparable boundary-value problems.

Wave propagation in a composite medium has obvious geophysical and engineering applications. In the last few years there has been a great interest and a large number of publications in wave propagation in a laminated and fibrous composite medium. However, these studies are limited by the assumption that the structure is periodic or that the distribution is deterministic. On the other hand, there are many geophysical and engineering problems in which the inhomogeneity is random. The scattering of electromagnetic and acoustic waves by a random distribution of scatterers has been considered extensively in the literature (see Twersky [30] and the references given therein). The corresponding literature on elastic wave scattering is rather meager. The case where the elastic properties of the scatterers differ only slightly from those of the surrounding materials and changes continuously were discussed by Karal and Keller [31], Knopoff and Hudson [32]. Hudson [33] later extended the earlier analysis of [32] by allowing for small variations across surfaces of discontinuity. The other problem in which the properties are allowed to vary by finite amounts is rather difficult and only recently has some progress been made for dilute

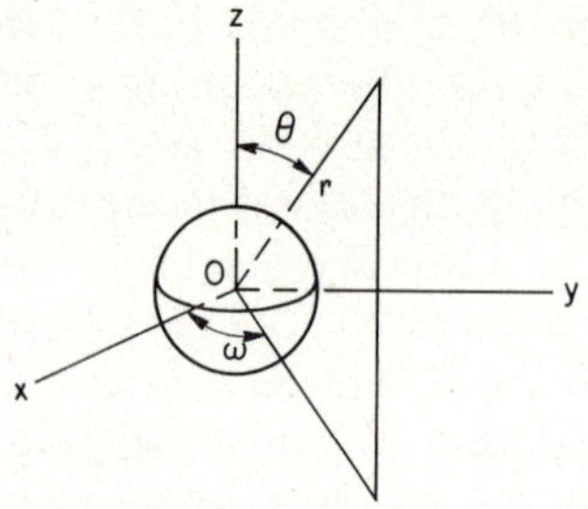

Fig. 1 Spherical polar coordinates systems.

mixtures and wavelengths long compared to the scatterer size. Mal and Knopoff[34] obtained the bulk properties of a composite made up of a dilute random distribution of spheres in an otherwise homogeneous medium. They took the wavelength to be large and also neglected multiple scattering so that the results do not show any dispersion or attenuation. Later Bose and Mal[35, 36, 37], Mal and Bose[38] extended the results to include multiple scattering for a distribution of spheres and circular cylinders. Scattering by a distribution of arbitrarily shaped scatterers is still an unsolved problem. Some limited progress has been made for spheroidal and elliptical shapes. Garbin and Knopoff[39] considered penny-shaped cracks and Datta [40, 41] examined elliptic cylindrical fibers. Multiple scattering effects have been neglected. Furthermore, the analysis of [39] is only suited for the particular geometry considered and does not lend itself to extension to other geometries. In this respect the method used in [40, 41] shows the most promise. There use has been made of MAE, which is believed to have much wider applicability. In Section 4 we analyze the problem of wave propagation in the presence of a distribution of rigid spheroids.

This review would not be complete without some discussion of the integral equation approach. The use of integral equations in solving scattering problems is quite widespread, particularly for mixed boundary value problems involving cracks, planes, etc. There is a vast amount of literature on this subject, and to cover this fully would be another review subject by itself. Some discussion of this approach is given in [3]. Knopoff[23] reviews in some detail the application to diffraction by a half-plane. To bring out some of the essential features of the method we treat in Section 5 a three-dimensional problem of diffraction by a rigid disc. Although we treat a rigid disc, there is no difficulty in treating a crack by the same method. In fact, various authors have treated the problem of diffraction by a crack in two and three dimensions. The reason for treating the rigid disc here is to show that the results verify those obtained in Section 4 taking the limit when the spheroid tends to a disc. The full discussion of the integral equation technique will not be undertaken here.

We have not included here any discussion of the numerical techniques. There are not many papers on scattering that make use of numerical techniques.

2 SCATTERING OF P-WAVES BY A LIQUID SPHERE OR CYLINDER

Diffraction of elastic waves by a spherical or circular cylindrical inclusion has been studied by several authors in recent years and the references to most of these works can be found in [3], [6], and [9]. Some further work in this area has been done by Datta [42], who solves the problem of diffraction of SH-waves in a spherically isotropic medium by a rigid or fluid spherical core. In the following we shall give the method of solution and some results of interest for the diffraction of P-waves by a liquid sphere or a circular cylinder. The results have geophysical applications and are believed to be of engineering interest as well (see [6]). First let us consider the diffraction by a spherical core. For simplicity we shall assume that the incident wave is a plane simple harmonic P-wave. It may be noted that for seismological applications one ought to consider a point source. However, the extension of the present solution to point source is straightforward.

2.1 Diffraction by a Liquid Sphere

Consider spherical polar coordinates (r, θ, ω) with origin at the center of the sphere. These coordinates are related to the cartesian coordinates (x, y, z) (see Fig. 1) by the relations

$$x = r \sin \theta \cos \omega, \qquad y = r \sin \theta \sin \omega, \qquad z = r \cos \theta,$$
$$0 \le \theta \le \pi, \qquad 0 \le \omega \le 2\pi. \tag{2.1}$$

It will be assumed that the material outside the core is elastic, isotropic and homogeneous and that inside the core is a nonviscous homogeneous fluid† that transmits only compressional waves. Thus the displacement field $u(x, y, z, t)$ will satisfy the equation

† The problem of an elastic sphere, although algebraically more complicated, can be solved in a similar manner.

$$(\lambda + 2\mu)\nabla\nabla \cdot \mathbf{u} - \mu\nabla \wedge \nabla \wedge \mathbf{u} = \rho\partial^2\mathbf{u}/\partial t^2 \tag{2.2}$$

in the elastic medium (henceforth referred to as "1"). If **u** is assumed to be simple harmonic in time, then it may be written as $u(x, y, z)\, e^{-i\gamma t}$, where $\gamma/2\pi$ is the frequency. Equation (2.2) may thus be written as

$$c_1^2\nabla\nabla \cdot \mathbf{u} - c_2^2\nabla \wedge \nabla \wedge \mathbf{u} + \gamma^2\mathbf{u} = \mathbf{0}, \tag{2.3}$$

where c_1, c_2 are the compressional (P-) wave and shear (S-) wave speeds, respectively. The solution to (2.3) in spherical polar coordinates can be written as

$$\mathbf{u} = \nabla\Phi + \nabla \wedge (-\partial\Psi/\partial\theta\hat{\omega}) \tag{2.4}$$

if **u** does not depend on ω, which can be assumed to be the case in the present problem with a proper choice of axes. In (2.4) Φ and Ψ satisfy the scalar Helmholtz equation, i.e.

$$\begin{aligned} \nabla^2\Phi + \alpha^2\Phi = 0, \qquad \alpha = \gamma/c_1, \\ \nabla^2\Psi + \beta^2\Psi = 0, \qquad \beta = \gamma/c_2, \end{aligned} \tag{2.5}$$

and $\hat{\omega}$ is the unit vector in the direction of increasing ω. Φ is commonly known as the P-wave potential and Ψ the SV-wave potential. It may be noted that if the medium is infinite in extent with no inhomogeneities, then these two waves propagate independently, the P-wave moving faster than the SV-wave. The particle motions associated with these are in the plane containing the z-axis. Upon encounter with a boundary across which the material property changes, the P- (or SV-) wave is transmitted and reflected into both P- and SV- components. We may note further that another solution to Eq. (2.3) may be written (in spherical coordinates) as

$$\mathbf{u} = \nabla \wedge (\hat{r}r\chi), \tag{2.6}$$

where χ satisfies Eq. $(2.5)_2$. This wave is known as an SH-wave, and is uncoupled from the other two.

Since the fluid core transmits only compressional waves, the displacement $\mathbf{u}'$ there (medium "2") is given in terms of a P-wave potential Φ' that satisfies

$$\nabla^2\Phi' + \alpha'^2\Phi' = 0, \qquad \alpha' = \gamma/c_1'. \tag{2.7}$$

Here prime denotes relevant quantities in medium 2.

Suppose the fluid core is of radius a and a plane P-wave represented by (suppressing the time factor)

$$\Phi^{(i)} = \Phi_0\, e^{i\alpha z} \tag{2.8}$$

is incident upon it. The total field (scattered and incident) outside the core, $r > a$, will be given in terms of $\Phi = \Phi^{(i)} + \Phi^{(s)}$, $\Psi = \Psi^{(s)}$ that solve (2.5) and the field inside, $r < a$, is given by $\Phi' = \Phi^{(\mathrm{tr})}$. The boundary conditions are:

1. The normal component, τ_{rr} of stress and the radial displacement are continuous across $r = a$; (2.9a)

2. The shearing stress, $\tau_{r\theta}$, vanishes on $r = a$; (2.9b)

3. As $r \to \infty$

$$r\left(\frac{\partial \Phi^{(s)}}{\partial r} - i\alpha\Phi\right) \to 0, \quad \text{and} \quad r\left(\frac{\partial \Psi^{(s)}}{\partial r} - i\beta\Psi\right) \to 0. \tag{2.9c}$$

The appropriate solutions $\Phi^{(s)}$, etc., satisfying (2.9c) are given by

$$\Phi^{(s)} = \sum_{n=0}^{\infty} a_n h_n(\alpha r) P_n(\eta), \tag{2.10}$$

$$\Psi^{(s')} = \sum_{n=0}^{\infty} b_n h_n(\beta r) P_n(\eta), \tag{2.11}$$

and

$$\Phi' = \sum_{n=0}^{\infty} a'_n j_n(\alpha' r) P_n(\eta), \tag{2.12}$$

where $\eta = \cos\theta$, $h_n(z)$, and $j_n(z)$ are spherical Bessel functions of the third and first kind, respectively. Note that $\Phi^{(i)}$ can be expanded near $r = 0$ in the form

$$\Phi^{(i)} = \Phi_0 \sum_{n=0}^{\infty} i^n (2n+1) j_n(\alpha r) P_n(\eta). \tag{2.13}$$

Using the boundary conditions (2.9a), (2.9b) and the expressions for stresses τ_{rr}, $\tau_{r\theta}$ and for the displacement u_r from Appendix A, we find the coefficients a_n, b_n, a'_n given by (A.4), (A.5). Thus the solutions for $\Phi^{(s)}$, $\Psi^{(s)}$ and Φ' are formally known. However, these series are very slowly convergent for large values of the wave number α. To obtain a representation that is rapidly convergent for large values of the wave number we first write, following [13],

$$A_n = \frac{1}{2}\left[-1 + \kappa_1 R_{11} + \kappa_1\kappa_4 R_{12} R_{21} \sum_{p=0}^{\infty} (\kappa_4 R_{22})^p\right], \tag{2.14}$$

$$B_n = \frac{1}{2}\left[\kappa_2 R'_{11} + \kappa_2\kappa_4 R_{12} R'_{21} \sum_{p=0}^{\infty} (\kappa_4 R_{22})^p\right], \tag{2.15}$$

and

$$A'_n = \kappa_3 R_{12} \sum_{p=0}^{\infty} (\kappa_4 R_{22})^p, \tag{2.16}$$

where R_{11}, R'_{11}, etc., are given in Appendix A. These series expansions for A_n, B_n, A'_n make evident the physical interpretation of the solution. The solution represented by

$$\bar{\Phi} = \Phi_0 \sum_{n=0}^{\infty} \tfrac{1}{2} i^n (2n+1)[h_n^{(2)}(\alpha r) + \kappa_1 R_{11} h_n^{(1)}(\alpha r)] P_n(\eta) \tag{2.17}$$

corresponds to the incident, reflected and diffracted P-waves in "1", whereas the successive contributions coming from $\kappa_1 \kappa_4 R_{12} R_{21} (\kappa_4 R_{22})^p$ represent the emergent compressional wave in "1" that has been reflected internally p-times at the boundary $r = a$. Similar interpretations can be given to the shear potentials

$$\bar{\Psi} = \Phi_0 \sum_{n=0}^{\infty} \tfrac{1}{2} i^n (2n+1) \kappa_2 R'_{11} h_n^{(1)}(\beta r) P_n(\eta), \tag{2.18}$$

and

$$\Psi^{(p)} = \Phi_0 \sum_{n=0}^{\infty} \tfrac{1}{2} i^n (2n+1) \kappa_2 \kappa_4 R_{12} R'_{21} (\kappa_4 R_{22})^p h_n^{(1)}(\beta r) P_n(\eta). \tag{2.19}$$

On the other hand,

$$\bar{\Phi}' = \tfrac{1}{2} \Phi_0 \sum_{n=0}^{\infty} i^n (2n+1) \kappa_3 R_{12} h_n^{(2)}(\alpha' r) P_n(\eta) \tag{2.20}$$

is the refracted† P-wave in "2" and $\bar{\Phi}'^{(p)}$ represents its pth internal reflection. In the following we shall evaluate $\bar{\Phi}$, $\bar{\Psi}$ and $\Phi^{(0)}$, $\Psi^{(0)}$ for large values of the wave number. The others can be evaluated in a similar manner; see the discussions in [6], [9]. The first step in obtaining an asymptotic solution for large wave numbers is to use Watson's transformation. We shall exemplify this by working with $\bar{\Phi}$. The method of analysis is very similar to that used in [20]. We write

$$\bar{\Phi} = \frac{\Phi_0}{2} \left(\frac{\pi}{2\alpha r} \right)^{1/2} e^{-i\pi/4} \int_{c_1} \frac{\nu\, e^{-i\pi\nu/2}}{\cos \pi\nu} [H_\nu^{(2)}(\alpha r) + \kappa_1 R_{11} H_\nu^{(1)}(\alpha r)] P_{\nu - 1/2}\, d\nu,$$

where contour c_1 is shown in Fig. 2. Then since

$$H_\nu^{(1)}(x) = e^{i\pi\nu} H_\nu^{(1)}(x), \quad H_{-\nu}^{(2)}(x) = e^{-i\pi\nu} H_\nu^{(2)}(x), \quad P_{-\nu-1/2}^{(\eta)} = P_{\nu-1/2}^{(\eta)}, \tag{2.21}$$

contour c_1 can be replaced by the straight line c above the real axis, see Fig. 2, and on c we may use the expansion

† Note that $j_n = \frac{1}{2}(h_n^{(1)} + h_n^{(2)})$ and so the choice depends on the value of θ.

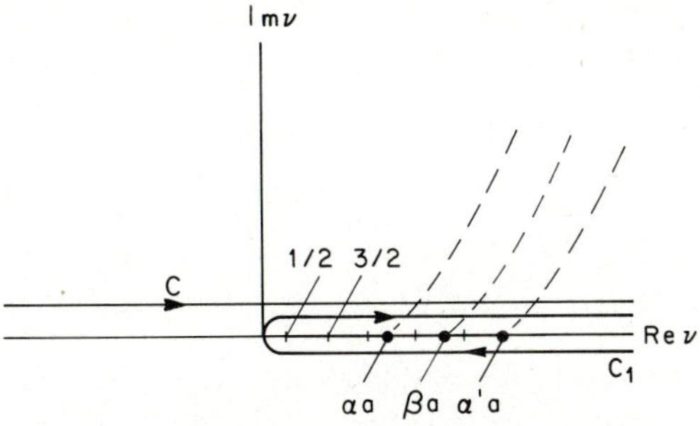

Fig. 2 Contour for Watson's transformation.

$$\frac{e^{-i\pi\nu/2}}{\cos \pi\nu} = 2e^{i\nu\pi/2} \sum_{m=0}^{\infty} (-1)^m e^{2mi\pi\nu} \tag{2.22}$$

so that $\bar{\Phi}$ can be written as

$$\bar{\Phi} = \Phi_0 \sum_{m=0}^{\infty} \bar{\Phi}_m \tag{2.23}$$

with

$$\bar{\Phi}_m = \left(\frac{\pi}{2\alpha r}\right)^{1/2} e^{-i\pi/4} (-1)^m \int_{-\infty}^{\infty} e^{i\nu(2m+1/2)\pi}[\quad] P_{\nu-1/2}\, d\nu. \tag{2.24}$$

This integral is now evaluated by the method of residues. For this purpose we need to know the poles of the integrand. Clearly the poles of the integrand are those of $\kappa_1 R_{11}$, and these are given (see Appendix A) approximately by the zeros of $H_\nu^{(1)}(\alpha a)$, which are symmetrically located about the origin in the first and third quadrants. The ones in the first quadrant in the neighborhood of $\nu = \alpha a$ for large αa are

$$\nu_n = \alpha a + (\alpha a/2)^{1/3} x_n\, e^{i\pi/3}, \tag{2.25}$$

where x_n are the zeros of the Airy function $Ai(-x_n)$. It is shown in Appendix A that these are the only poles. Thus, for $\theta < \pi/2$,

$$\bar{\Phi}_m = \left(\frac{\pi}{2\alpha r}\right)^{1/2} e^{i\pi/4} \sum_n (-1)^m \nu_n\, e^{i\pi\nu_n(2m+1/2)-i\pi/6} \left(\frac{\alpha a}{2}\right)^{1/3} \times H_{\nu_n}^{(1)}(\alpha r) P_{\nu_n-1/2}(\cos\theta)/[Ai'(-x_n)]^2. \tag{2.26}$$

Note that this agrees with the solution for acoustic scattering by an impenetrable sphere obtained in [20]; see Eq. (4.15). For points in the geometric shadow we may express (2.26) as

$$\bar{\Phi}_m = \frac{e^{-i\pi/6}}{(2\pi)^{1/2}}\left(\frac{\alpha a}{2}\right)^{1/3}\left(\frac{a^2}{r^2-a^2}\right)^{1/4}\frac{e^{i\alpha\sqrt{r^2-a^2}}}{(\alpha r\sin\theta)^{1/2}}(-1)^m\sum_n\frac{1}{[Ai'(-x_n)]^2}$$
$$\times[\exp(i\nu_n\gamma_m + i\pi/4) + \exp(i\nu_n\delta_m - i\pi/4)], \qquad (2.27)$$

where we have used the asymptotic expansions of $H_{\nu_n}^{(1)}$ and $P_{\nu_n-1/2}$. Also,

$$\gamma_m = \sin^{-1}\frac{a}{r} - \theta + 2m\pi, \qquad \delta_m = \sin^{-1}\frac{a}{r} + \theta + 2m\pi.$$

The geometrical interpretation of this solution is well known. This represents the diffracted P-wave in the shadow of the core. The boundary of the geometric shadow is given by

$$\theta = \pm\sin^{-1} a/r = \pm\theta_p \qquad (2.28)$$

which represents the grazing rays. These excite diffracted waves that propagate along the sphere with speed c_1 and reach points in the shadow along tangents. The terms corresponding to γ_m represent the rays that have traveled clockwise m times along the sphere, whereas δ_m represents those that have traveled counterclockwise. Since ν_n has a positive imaginary part, these waves decay in amplitude as they travel along the sphere. Note that for small but finite values of the wavelength the shadow boundary depends on the frequency. This frequency dependence has been studied in [17] for a plane wave scattered by a circular cylindrical cavity and in [16] for various earth models. Note that the solution (2.27) is the same as obtained in [20] for acoustic scattering. Thus even in the elastic case the diffracted P-wave is given by the acoustic solution. This interesting fact was first pointed out in [45].

For the diffracted S-wave inside the deep shadow we get, using (A.13) and (2.18),

$$\bar{\Psi} = \Phi_0 \sum_{m=0}^{\infty} \bar{\Psi}_m \qquad (2.29)$$

with

$$\bar{\Psi}_m = \left(\frac{\pi}{2/\alpha r}\right)^{1/2} e^{5i\pi/12}(-1)^m\sum_n\left(\frac{2}{\alpha a}\right)^{1/3}\left(\frac{\alpha a}{\sin\theta}\right)^{1/2}$$
$$\times\frac{2\alpha a(\beta^2/\alpha^2-1)^{1/4}}{(2\alpha^2-\beta^2)a^2(\beta^2r^2/\alpha^2a^2-1)^{1/4}}$$
$$\times\exp[i\beta\{r\sqrt{1-\alpha^2a^2/\beta^2r^2} - a\sqrt{1-\alpha^2/\beta^2}\}] \qquad (2.30)$$
$$\times[e^{i(\nu_n\gamma_m'+\pi/4)} + e^{i(\nu_n\delta_m'-\pi/4)}],$$

where

$$\gamma'_m = 2m\pi + \cos^{-1}\alpha/\beta - \theta + \sin^{-1}\alpha a/\beta r,$$
$$\delta'_m = 2m\pi + \cos^{-1}\alpha/\beta + \theta + \sin^{-1}\alpha a/\beta r. \tag{2.31}$$

The geometric shadow boundary for the S-waves is given by

$$\theta = \pm(\cos^{-1}\alpha/\beta + \sin^{-1}\alpha a/\beta r) = \pm\theta_s. \tag{2.32}$$

Again the geometrical interpretation of (2.30) is quite obvious. The waves moving along the sphere with speed c_1 excite both P- and S-waves in the shadow, the former moves tangentially to the observation point and is given by (2.27) whereas the latter leaves the sphere at an angle $\cos^{-1}\alpha/\beta$ with the tangent and moves with speed c_2. Thus the arrival times of these disturbances can be computed quite easily and can be fruitfully used to calculate the size of the core. It should be stressed here that the foregoing results are useful strictly in the limit of large αa and

$$\theta_p - \theta \gg (\alpha a)^{-1/3}$$

for the P-wave and $\theta_s - \theta \gg (\alpha a)^{-1/3}$ for the S-wave.

In the illuminated region $\theta > \theta_p$, $\gamma_0 < 0$ and so this term in (2.23) becomes exponentially large. Thus in this region we need to obtain a better representation of $\bar{\Phi}$. This we can do by noting that the terms corresponding to $m > 0$ are still exponentially damped and we need only to reexamine $m = 0$ term. This is

$$\bar{\Phi}_0 = \Phi_0\left(\frac{\pi}{2\alpha r}\right)^{1/2} e^{-i\pi/4}\int_{-\infty}^{\infty} e^{i\nu\pi/2}\,\nu[\quad]P_{\nu-1/2}\,d\nu, \tag{2.33}$$

Furthermore, since

$$P_{\nu-1/2} = Q^{(1)}_{\nu-1/2} + Q^{(2)}_{\nu-1/2}, \tag{2.34}$$

and for larger ν

$$Q^{(1,2)}_{\nu-1/2} \sim \frac{\exp\left[\pm i(\nu\theta - \pi/4)\right]}{(2\pi\nu\sin\theta)^{1/2}}, \tag{2.35}$$

the γ_0 term is the contribution of $Q^{(1)}_{\nu-1/2}$. So let us reconsider only the expression

$$\bar{\Phi}_0 = \Phi_0\left(\frac{\pi}{2\alpha r}\right)^{1/2} e^{-i\pi/4}\int_{-\infty}^{\infty} \nu[\quad]\,e^{i\nu\pi/2}\,Q^{(1)}_{\nu-1/2}\,d\nu. \tag{2.36}$$

Examination of the integrand reveals that it has two saddle points on the real ν-axis and the integral can then be evaluated by deforming the contour into steepest descent paths through the saddle points and an arc joining the saddle points. Phinney and Cathles [16] numerically integrated an expression corresponding to (2.36) by choosing a suitable optimum path and presented diffraction amplitudes near the shadow boundary. They took a point source. Here we shall only consider large αa and the region away from the shadow. The method of analysis is again that of [20]. We find

$$\bar{\Phi}_0 = \Phi^{(i)} + \Phi^{(r)}, \tag{2.37}$$

where $\Phi^{(i)}$ is the incident field and $\Phi^{(r)}$ is the reflected field and is given by

$$\Phi^{(r)} = \Phi_0 \left(\frac{a^2 \sin 2\zeta}{4sr \sin \theta} \right)^{1/2} R_{pp} \exp\left[i\alpha(s - \tfrac{3}{2} a \sin \zeta) \right]. \tag{2.38}$$

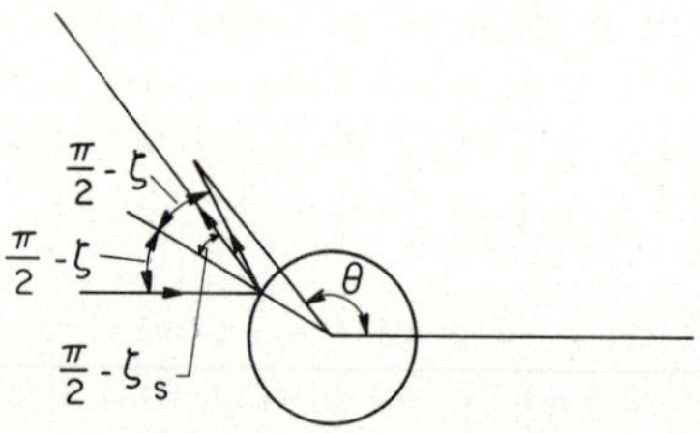

Fig. 3 Incident and reflected rays.

Here

$$s = r \cos \bar{\omega} - a/2 \sin \zeta, \qquad \alpha \cos \zeta = \beta \cos \zeta_s = \alpha' \cos \zeta', \qquad \zeta = \tfrac{1}{2}(\theta - \bar{\omega});$$

see Fig. 3. It is to be noted that the results derived above are consistent with the ray theory prediction. R_{pp} is the reflection coefficient of the PP-reflected ray from a plane boundary separating media "1" and "2". Similar results can be obtained for other types of boundary conditions; see, for example, [21]. Note that R_{pp} for the present case is

$$R_{pp} = \left[\sin \zeta \left(\frac{\lambda'}{2\mu} + 2 \sin \zeta' \cos \zeta' \sin \zeta_s \cos \zeta_s \right.\right.$$
$$\left.\left. + \frac{\alpha}{\alpha'} \left(\frac{\lambda}{2\mu} + \sin^2 \zeta \right) \sin \zeta' \cos 2\zeta_s \right] \middle/ \left[-\sin \zeta \left(\frac{\lambda'}{2\mu} + 2 \sin \zeta' \right.\right.\right. \tag{2.39}$$
$$\left.\left. \times \cos \zeta' \sin \zeta_s \cos \zeta_s \right) + \frac{\alpha}{\alpha'} \left(\frac{\lambda}{2\mu} + \sin^2 \zeta \right) \sin \zeta' \cos 2\zeta_s \right];$$

s is the distance of the observation point from the caustic of the reflected rays (measured along the ray).

In a similar manner we can show that in the illuminated region for the reflected S-waves

$$\Psi^{(r)} \sim i\Phi_0 \frac{4\alpha}{\alpha'\beta a} R_{ps}\left[\frac{\beta a^2 \sin^2 \zeta_s \cos \zeta_s}{r \sin \theta d_1(\beta \sin \zeta_s + \alpha \sin \zeta)}\right]^{1/2} \times e^{i(\beta r \cos \omega - \alpha a \sin \zeta - \beta a \sin \zeta_s)} \tag{2.40}$$

with

$$R_{ps} = \frac{\sin \zeta \sin \zeta' \left(\dfrac{\lambda}{2\mu} + \sin^2 \zeta\right)}{\Delta},$$

$$d_1 = r \cos \omega - \frac{\beta a \sin \zeta \sin \zeta_s}{\alpha \sin \zeta + \beta \sin \zeta_s}.$$

Here Δ is the denominator in (2.39) and d_1 is the distance from the caustic of the reflected S-rays. R_{ps} is the reflection coefficient for P-reflected S-waves. Again the arrival times of these reflected P- and S-waves can be computed and used to get information on the size of the inclusion. To gain knowledge of the property of this one would have to examine the amplitudes as well. But a more direct way is perhaps the use of the arrival times of the transmitted waves. This we now examine in what follows.

The potentials $\Phi^{(0)}$ and $\Psi^{(0)}$ represent the contributions from those that have suffered no internal reflection. As in the case of $\bar{\Phi}$ and $\bar{\Psi}$, we can evaluate $\Phi^{(0)}$ and $\Psi^{(0)}$ using Watson's transformation followed by a residue calculation. The residue series will represent the diffracted waves arising from the grazing rays. Here we shall only give the expressions for the dominant terms in the region illuminated by the rays that are transmitted through medium "2" without suffering internal reflections; see Fig. 4. The calculations for $\Phi^{(p)}$ and $\Psi^{(p)}$ can be performed in similar manner; see, for example, [46].

Consider then

$$\Phi^{(0)} = \left(\frac{\pi}{2\alpha r}\right)^{1/2} e^{i\pi/4} \sum_{m=0}^{\infty} (-1)^m \int_{-\infty}^{\infty} \nu\, e^{i\pi\nu(2m+1/2)} \times \frac{\Phi_{\nu-1/2}^{(1)'} \Phi_{\nu-1/2}^{(2)}}{(\Delta_{\nu-1/2}^{(1)})^2} H_\nu^{(1)}(\alpha r) P_{\nu-1/2}(\cos\theta)\, d\nu. \tag{2.41}$$

In the illuminated region the dominant contribution comes from the $m = 0$ term. In particular, we need to evaluate

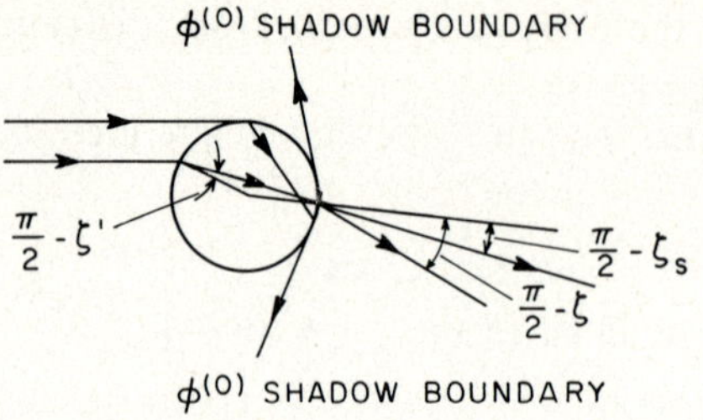

Fig. 4 Incident and transmitted rays.

$$\Phi^{(0)} = \left(\frac{\pi}{2\alpha r}\right)^{1/2} e^{-i\pi/4} \int e^{i\pi\nu/2}\, \nu[\quad] H_\nu^{(1)}(\alpha r) Q_{\nu-1/2}^{(1)}(\cos\theta)\, d\nu \tag{2.42}$$

by the saddle point method. We find

$$\Phi_0^{(0)} \approx -i\left(\frac{a^2 \sin 2\zeta \sin\zeta' \cos\zeta}{4dr\sin\theta k}\right)^{1/2} T_{pp}\, e^{i\alpha r(\cos\omega + 2(a/r)(\alpha'/\alpha \sin\zeta' - \sin\zeta))}, \tag{2.43}$$

where

$$k = \frac{1}{\sin\zeta} - \frac{\alpha}{\alpha'\sin\zeta'}, \qquad d = r\cos\omega - \frac{a}{2k},$$

$$T_{pp} = -2\frac{\lambda'\alpha}{\mu\alpha'}$$

$$\times \frac{\sin\zeta\sin\zeta'\cos 2\zeta_s(\lambda/2\mu + \sin^2\zeta)}{[\alpha/\alpha'(\lambda/2\mu+\sin^2\zeta)\sin\zeta'\cos 2\zeta_s - \sin\zeta(\lambda'/2\mu + 2\sin\zeta'\cos\zeta'\sin\zeta_s\cos\zeta_s)]^2}.$$

For the transmitted S-wave we similarly find

$$\Psi_0^{(0)} \sim \frac{2\beta}{\alpha^2 a}(\sin\zeta\sin\zeta_s)^{1/2}\left(\frac{a^2\cos\zeta_s}{r\sin\theta}\right)^{1/2}\left(\frac{1}{\bar{d}\bar{\kappa}}\right)^{1/2} T_{ps}$$
$$\times\, e^{i\beta(r\cos\omega + a\{2\alpha'/\beta\sin\zeta' - (\alpha/\beta)\sin\zeta - \sin\zeta_s\})} \tag{2.44}$$

with

$$\bar{\kappa} = \frac{\beta}{\alpha\sin\zeta} + \frac{1}{\sin\zeta_s} - \frac{2\beta}{\alpha'\sin\zeta'},$$

$$\bar{d} = r\cos\omega - a/\bar{\kappa},$$

$$T_{ps} = \frac{2\lambda/\mu\cdot\alpha^3/\alpha'\beta^2\sin\zeta\sin\zeta'(\lambda/2\mu + \sin^2\zeta)}{\Delta^2},$$

where Δ is the quantity within square bracket in the expression for T_{pp}. Note that $d, \bar{d}$ are the distances from the caustics.

The solutions (2.27), (2.30), (2.43), and (2.44) represent some of the different waves that arrive in the shadow. These, together with the reflected waves, are then utilized to calculate the size and properties of the core. In the foregone discussion it was assumed that $\alpha < \beta < \alpha'$. The opposite case of $\alpha' < \alpha < \beta$ can also be analyzed in much the same manner. The latter, however, is not the case for the core of the Earth.

2.2 Diffraction by a Liquid Circular Cylinder

The two-dimensional prototype of the problem discussed above is the diffraction of a plane P-wave by a circular fluid cylinder. If reference is again made to Fig. 1 with

$$z = r \cos \theta, \qquad x = r \sin \theta, \tag{2.45}$$

$\Phi^{(i)}$ is still given by (2.8) and $\mathbf{u}$ can be written as

$$\mathbf{u} = \nabla\Phi + \nabla \wedge (-\Psi\mathbf{e}_y). \tag{2.46}$$

All the physical variables are independent of y.

The analysis of this problem is quite similar to the above. The solutions $\Phi^{(s)}$, $\Psi^{(s)}$, Φ' can now be written as

$$\Phi^{(s)} = \sum_{n=-\infty}^{\infty} a_n H_n(\alpha r)\, e^{in\theta}, \tag{2.47}$$

$$\Psi^{(s)} = \sum_{n=-\infty}^{\infty} b_n H_n(\beta r)\, e^{in\theta}, \tag{2.48}$$

and

$$\Phi' = \sum_{n=-\infty}^{\infty} a'_n H_n(\alpha' r)\, e^{in\theta}. \tag{2.49}$$

Note that $\Phi^{(i)}$ has the expansion

$$\Phi^{(i)} = \Phi_0 \sum_{n=-\infty}^{\infty} i^n J_n(\alpha r)\, e^{in\theta}. \tag{2.50}$$

We can determine the constants a_n, b_n, a'_n using the continuity conditions (2.9a)–(2.9b). Then a_n, b_n, a'_n can be expanded as in (2.14)–(2.16), and the different contributions can be identified as before. We refrain from reproducing these calculations. A similar problem in acoustics has been solved by Chen [47]. Instead, let us turn to the problem of scattering by a circular cavity in a half-space.

3 WAVE PROPAGATION IN A HALF-SPACE CONTAINING A CYLINDRICAL CAVITY

Wave propagation in a half-space due to a point or line source has been of interest since the work of Lamb [48]. However, the prototype of the problem considered in the previous section for a half-space has only recently been discussed in the literature. Ben-Menahem and Cisternas [49] solved this for the case of a spherical cavity excited by time-harmonic surface tractions. However, their method leads to an infinite set of equations for the determination of the coefficients of the series expansions used by them. The difficulty arises because of the nonseparable nature of this problem. To get around this difficulty Thiruvenkatachar and Viswanathan [50, 51] presented a method of successive reflections where the wavelength of the disturbance is large compared with the diameter of the cavity. Later Gregory [29] gives a rigorous solution for the case of a cylindrical cavity. The method is based on an expansion theorem proved in his earlier paper [28]. In Sub-section 3.1 we shall give a brief outline of Gregory's solution. In [29] Gregory also gives the solutions for some particular examples where the wavelength is large. Now in the long wavelength case it is possible to obtain an expansion in terms of a small parameter by the method of matched asymptotic expansions (MAE). The use of MAE in solving scattering problems involving nonseparable geometries has been demonstrated in a series of papers [52, 53]. The advantage of this approach is that one need not restrict oneself to circular cavity alone. In Sub-section 3.2 we outline the method and solve two examples, one of which has been discussed in [28]. We show the MAE is ideally suited for such problems to obtain the first few terms. No numerical results are presented. These will be communicated in a separate publication.

3.1 Method of Line Source Potentials

Consider a half-space ($y \geq 0$) of isotropic elastic homogeneous material in which is embedded a circular cylindrical cavity of radius a and center at $y = h (h > a)$. Let the axis of the cylinder be parallel to the z-axis (see Fig. 5).

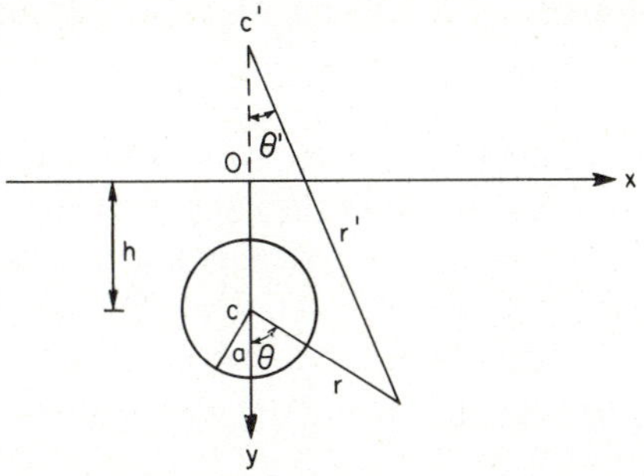

Fig. 5 Circular cavity in a half-space.

We shall also use cylindrical coordinates (r, θ, z) and assume that the problem is planar in the xy-plane. Then the displacement $\mathbf{u}(x, y, t)$ can be expressed in terms of the potentials, see Eq. (2.46), Φ and Ψ as

$$\mathbf{u} = \nabla\Phi + \nabla \wedge (\Psi \mathbf{e}_z), \tag{3.1}$$

where the time factor $e^{-i\omega t}$ is suppressed. Φ and Ψ satisfy the wave equations (2.5). Hence the following boundary value problem is to be solved:

$$(\nabla^2 + \alpha^2)\Phi = 0, \tag{3.2}$$

$$(\nabla^2 + \beta^2)\Psi = 0, \qquad y > 0, \qquad a < r. \tag{3.3}$$

If the boundary $y = 0$ is assumed to be stress free, then

$$2(\Phi_{xy} - \Psi_{xx}) - \beta^2\Psi = 0, \tag{3.4}$$

$$2(\Phi_{xx} + \Psi_{xy}) + \beta^2\Phi = 0. \tag{3.5}$$

On the other hand, at $r = a$ the stresses must satisfy the prescribed tractions

$$\sigma_{rr} = S(\theta), \qquad \sigma_{r\theta} = T(\theta), \qquad 0 \leq \theta \leq 2\pi, \qquad r = a. \tag{3.6}$$

Furthermore, $\mathbf{u}$ must satisfy an appropriate radiation condition. The radiation condition can be put conveniently in an integral form

$$\int_{S_R} \{G_{ki}(\xi, \mathbf{x})\sigma_{ij}(\mathbf{u}) - u_i(\xi)\sigma_{ij}(\mathbf{G}_k)\} n_j \, d\xi = 0, \tag{3.7}$$

where $G_{ki}(\xi, \mathbf{x})$ is the Green's function for a point source in the half-space; see [54]. Here S_R is the semicircle $x^2 + y^2 = R^2$, $y \geq 0$, and σ_{ij} is the stress tensor. An equivalent form of the radiation condition in terms of Green's functions for line sources of the forms $H_0(\alpha|\mathbf{x}-\xi)$ and $H_0(\beta|\mathbf{x}-\xi|)$ is given in [28].

The solution to this boundary-value problem can be constructed by the superposition of the contributions from an infinite number of line sources (compressional and shear) situated at the center of the circle, $y = h$. For this purpose let $\{\Phi_n^p, \Psi_n^p\}$ $(n \geq 0)$ be the field produced in the half-space $y \geq 0$ due to line-compressional source $H_n(\alpha r)\cos n\theta$. Similarly, let $\{\Phi_n^s, \Psi_n^s\}$, $n \geq 1$, be the field due to a shear line source $H_n(\beta r)\sin n\theta$. Then for the symmetric case the expansion theorem proved in [29] asserts that

$$\Phi = \varepsilon^2 a_0 \Phi_0^p + \sum_{m=1}^{\infty} \frac{a_m}{H_m(\varepsilon)} \Phi_m^p + \sum_{m=1}^{\infty} \frac{b_m}{H_m(\tau\varepsilon)} \Phi_m^s, \tag{3.8}$$

$$\Psi = \varepsilon^2 a_0 \Psi_0^p + \sum_{m=1}^{\infty} \frac{a_m}{H_m(\varepsilon)} \Psi_m^p + \sum_{m=1}^{\infty} \frac{b_m}{H_m(\tau\varepsilon)} \Psi_m^s, \tag{3.9}$$

where $\varepsilon = \alpha a$ and $\tau = \beta/\alpha$. The expansion coefficients a_n, $n \geq 0$, and b_n, $n \geq 1$, are to be obtained by the use of the boundary conditions (3.6). The derivation of $\{\Phi_n^p, \Psi_n^p\}$ and $\{\Phi_n^s, \Psi_n^s\}$ are given in Appendix B. Expanding $E_n = \Phi_n^p - H_n(\alpha r)\cos n\theta$, Ψ_n^p, $\Phi_n^{(s)}$ and $F_n = \Psi_n^s - H_n(\beta r)\sin n\theta$ in Fourier–Bessel series near $r = 0$ we obtain

$$\begin{aligned} E_n &= \sum_{m=0}^{\infty} P_{mn} J_m(\alpha r)\cos m\theta, \\ \Psi_n^p &= \sum_{m=1}^{\infty} Q_{mn} J_m(\beta r)\sin m\theta, \qquad n > 0, \end{aligned} \tag{3.10}$$

$$\begin{aligned} \Phi_n^s &= \sum_{m=1}^{\infty} R_{mn} J_m(\alpha r)\cos m\theta, \\ F_n &= \sum_{m=1}^{\infty} S_{mn} J_m(\beta r)\sin m\theta, \qquad n > 1. \end{aligned} \tag{3.11}$$

The determination of the coefficients P_{mn}, Q_{mn}, R_{mn} and S_{mn} is outlined in Appendix B. Combining (3.8) to (3.11) we find

$$\Phi = \sum_{m=0}^{\infty} \cos m\theta \left[a_m \frac{H_m(\alpha r)}{H_m^*(\varepsilon)} + \sum_{n=0}^{\infty} a_n \frac{P_{mn} J_m(\alpha r)}{H_n^*(\varepsilon)} + \sum_{n=1}^{\infty} b_n \frac{R_{mn} J_m(\alpha r)}{H_n(\tau\varepsilon)} \right], \tag{3.12}$$

and

$$\psi = \sum_{m=1}^{\infty} \sin m\theta \left[b_m \frac{H_m(\beta r)}{H_m(\tau\varepsilon)} + \sum_{n=0}^{\infty} a_n \frac{Q_{mn} J_m(\beta r)}{H_n^*(\varepsilon)} + \sum_{n=1}^{\infty} b_n \frac{S_{mn} J_m(\beta r)}{H_n(\tau\varepsilon)} \right], \tag{3.13}$$

where

$$H_n^*(\varepsilon) = \varepsilon^2, \qquad n = 0,$$
$$= H_n(\varepsilon), \qquad n > 0.$$

The above expressions for Φ and Ψ can now be used to calculate the stresses σ_{rr}, $\sigma_{r\theta}$ on $r = a$ and then using Eq. (3.6) we can solve for the unknown constants. This process leads to an infinite set of algebraic equations which may be written in the form

$$[a] = [T][a] + [U][b] + [d], \tag{3.14}$$

$$[b] = [V][a] + [W][b] + [e], \tag{3.15}$$

where $[a]$, $[b]$ are column matrices of the coefficients a_m, b_m, respectively. $[T]$, $[U]$, $[V]$, and $[W]$ are infinite square matrices that depend on the material properties and the geometry of the medium. $[d]$, $[e]$ are column matrices that represent the forcing term. It is shown in [28] that the system of Eqs. (3.14), (3.15) has a unique solution which may be approximated by successive truncation for $0 < \alpha a < \alpha h$. Furthermore, when αa is small, the solution can be obtained by iteration. For arbitrary ε, however, the successive truncation approach can involve formidable algebra and has not yet been used. The iteration approach is much simpler. In the following we consider an example where the surface $r = a$ is subject to uniform pulsating radial pressure. The solution to this problem will be obtained for small ε by iteration.

Example A. *Pulsating radial pressure on* $r = a$

Let

$$S(\theta) = -p, \qquad T(\theta) = 0, \qquad 0 \leq \theta \leq 2\pi. \tag{3.16}$$

In this case it is found that

$$\begin{aligned} a_0 &= d_0 + T_{00}d_0 + 0(\varepsilon^4), \\ a_1 &- T_{10}d_0 + 0(\varepsilon^5), \\ a_2 &= T_{20}d_0 + 0(\varepsilon^4), \\ b_1 &= V_{10}d_0 + 0(\varepsilon^5), \\ b_2 &= V_{20}d_0 + 0(\varepsilon^4), \\ a_{m'} b_m &= 0(m^2\varepsilon^m), \quad m \geq 3, \end{aligned} \tag{3.17}$$

where

$$d_0 = -(p/\mu\alpha^2)\,\frac{1}{F_0(H,\varepsilon)},$$

$$T_{0n} = P_{0n}\,\frac{-F_0(J,\varepsilon)}{\varepsilon^2 F_0(H,\varepsilon)H_n^*(\varepsilon)}, \qquad n \geq 0,$$

$$T_{mn} = P_{mn}\,\frac{G_m(H,\tau\varepsilon)G_m(J,\varepsilon) - F_m(H,\tau\varepsilon)F_m(J,\varepsilon)}{D_m H_m(\tau\varepsilon)H_n^*(\varepsilon)}$$

$$+ Q_{mn}\,\frac{G_m(H,\tau\varepsilon)F_m(J,\tau\varepsilon) - F_m(H,\tau\varepsilon)G_m(J,\tau\varepsilon)}{D_m H_m(\tau\varepsilon)H_n^*(\varepsilon)} \qquad m \geq 1,\ n \geq 0,$$

$$V_{mn} = P_{mn}\,\frac{G_m(H,\varepsilon)F_m(J,\varepsilon) - F_m(H,\tau\varepsilon)G_m(J,\tau\varepsilon)}{D_m H_m(\varepsilon)H_n^*(\varepsilon)}$$

$$+ Q_{mn}\,\frac{G_m(H,\varepsilon)G_m(J,\tau\varepsilon) - F_m(H,\varepsilon)F_m(J,\tau\varepsilon)}{D_m H_m(\varepsilon)H_n^*(\varepsilon)}, \quad m \geq 1, \quad n \geq 0,$$

$$F_m(H,Z) = \{2m(m+1) - (\tau\varepsilon)^2\}H_m(Z) - 2ZH_{m-1}(Z),$$

$$G_m(H,Z) = 2m(m+1)H_m(Z) - 2mZH_{m-1}(Z),$$

$$D_m = \frac{1}{H_m(\varepsilon)H_m(\tau\varepsilon)}\begin{vmatrix} F_m(H,\varepsilon) & G_m(H,\tau\varepsilon) \\ G_m(H,\varepsilon) & F_m(H,\tau\varepsilon) \end{vmatrix}.$$

To obtain an expansion for Φ, Ψ in powers of ε one now has to expand T_{mn}, V_{mn}, etc., for small ε. This is a rather lengthy and tedius algebra. Instead in the next section we present an alternative method that gives the successive terms in such an expansion in a systematic and, what appears to be, a much simpler manner.

3.2 Method of MAE

The example considered above is ideally suited for the use of matched asymptotic expansion. The advantages of this approach are threefold: (a) the successive terms of the expansion are obtained in a systematic manner without the need to expand the complicated expressions like T_{mn}, V_{mn}, etc., (b) the solution to the problem is reduced to solving some related problems that are simpler than the original problem, and (c) the major advantage is in its applicability to nonseparable boundary-value problems. Application of MAE to solve scattering problems has been demonstrated only very recently. In [53, 54] this has been used to analyze elastic wave scattering by spheroids and ellipses. Sangster [55] uses this approach to solve scattering by a rigid spheroid of P- and S-waves for arbitrary angle of incidence. This problem cannot be solved by the usual techniques of separation of variables. This solution together with the effect of a large number of inclusions will be discussed in Section 4. MAE has also been used to solve acoustic scattering problems; see [56].

The example considered in Subsection 3.1 is nonseparable and has been formulated in a way ($\varepsilon \ll 1, \alpha h$ finite) that is encountered in many applications of MAE. The method will be outlined in its application to Example A considered above. We shall also apply this to solve the scattering by a circular cavity in a half-space. To our knowledge this has not been solved previously. We shall give detailed solutions for distances both near and far from the cavity. These can be extracted for Example A from [28, 50], but with considerable algebra.

Example A. *Pulsating radial pressure on* $r = a$

It will be assumed that $\varepsilon \ll 1$. Then the region $r > a,\ y > 0$ can be thought of as composed of two regions: one for which $r \gg a,\ \varepsilon r = 0(1)$ or larger and the other for which $r/a = 0(1)$. The former will be denoted as the far-field and the latter the near-field. We shall develop two asymptotic expansions in these two regions and then these will be matched in an intermediate region; see Cole [57] for a detailed discussion. In the near-field we shall nondimensionalize all distances with respect to a:

$$\bar{x}_i = x_i/a \tag{3.18}$$

and assume an expansion for $\mathbf{u}$ in the form

$$\begin{gathered}\mathbf{u} = \mathbf{u}_0 + \mu_1(\varepsilon)\mathbf{u}_1 + \mu_2(\varepsilon)\mathbf{u}_2 + \cdots,\\ \lim_{\varepsilon \to 0} \mu_{n+1}/\mu_n = 0.\end{gathered} \tag{3.19}$$

In the far-field we shall use the variables

$$x_i' = \varepsilon \bar{x}_i, \tag{3.20}$$

$$\begin{gathered}\mathbf{u}' = \nu_0(\varepsilon)\mathbf{u}_0' + \nu_1(\varepsilon)\mathbf{u}_1' + \cdots,\\ \lim_{\varepsilon \to 0} \nu_{n+1}/\nu_n = 0.\end{gathered} \tag{3.21}$$

$\mathbf{u}$ and $\mathbf{u}'$ satisfy the following equations and boundary conditions:

$$\bar{\nabla}\bar{\nabla}\cdot\mathbf{u} - \tau^{-2}\bar{\nabla} \wedge \bar{\nabla} \wedge \mathbf{u} = -\varepsilon^2\mathbf{u}, \tag{3.22}$$

$$\sigma_{rr}(\mathbf{u}) = S(\theta), \qquad \sigma_{r\theta}(\mathbf{u}) = T(\theta) \quad \text{at} \quad \bar{r} = 1, \tag{3.23}$$

and

$$\nabla'\nabla'\cdot\mathbf{u}' - \tau^{-2}\nabla' \wedge \nabla' \wedge \mathbf{u}' = -\mathbf{u}'. \tag{3.24}$$

$\mathbf{u}'$ satisfies the radiation condition (3.7). Note that in the present case $S(\theta) = -p$, $T(\theta) = 0$.

Note also that we have taken $\mu_0(\varepsilon) = 1$ motivated by the reasoning that $\mathbf{u}_0$ represents the static solution of a pressurized hole in an infinite space. Thus

$$u_{0r} = pa/2\mu\bar{r}, \qquad u_{0\theta} = 0; \tag{3.25}$$

see [58].

Introduce the intermediate variables

$$r_\delta = \bar{r}\delta = r'\,\delta/\varepsilon, \qquad \varepsilon \ll \delta \ll 1; \tag{3.26}$$

Note that $\delta \to 0$ and $\delta/\varepsilon \to \infty$ as $\varepsilon \to 0$. Then

$$u_{0r} = pa\delta/2\mu r_\delta, \qquad u_{0\theta} = 0. \tag{3.27}$$

To obtain the first term in the far-field expansions of $\mathbf{u}$ we first use the representation

$$u'_r = \frac{\partial\Phi}{\partial r'} + \frac{1}{r'}\frac{\partial\Psi}{\partial\theta}, \qquad u'_\theta = \frac{1}{r'}\frac{\partial\Phi}{\partial\theta} - \frac{\partial\Psi}{\partial r'}. \tag{3.28}$$

Then

$$\begin{aligned} \Phi &= \Sigma\, a_n\Phi_n^p + \Sigma\, b_n\Phi_n^s, \\ \Psi &= \Sigma\, a_n\Psi_n^p + \Sigma\, b_n\Psi_n^s, \end{aligned} \tag{3.29}$$

with $\Phi_n^p = H_n(r')\cos n\theta + E_n$, $\Psi_n^s = H_n(\tau r')\sin n\theta + F_n$.

Furthermore, E_n, Ψ_n^p, F_n, Φ_n^s can be expanded in the forms (3.10), (3.11) for small r'. Since $\mathbf{u}'_n$ satisfies (3.24), it can also be formally expressed in the form (3.28) with (3.29). However, as will be shown, the expansions (3.29) contain finite number of terms at any stage.

Consider $\mathbf{u}'_0$. To determine this we note

$$\begin{aligned} H_0(\varepsilon r_\delta/\delta) = {} & \frac{2i}{\pi}\ln(\varepsilon r_\delta/\delta) + \frac{2i}{\pi}(\gamma - \ln 2 - i\pi/2) \\ & + (\varepsilon r_{\delta/\delta})^2\left\{-\frac{i}{2\pi}\ln\frac{\varepsilon r_\delta}{\delta} - \left(\gamma - \ln 2 - 1 - \frac{i\pi}{2}\right)\frac{i}{2\pi}\right\} \\ & + 0(\varepsilon^4 r_\delta^4/\delta^4 \ln(\varepsilon r_{\delta/\delta})), \end{aligned} \tag{3.30}$$

$$\begin{aligned} H_1(\varepsilon r_{\delta/\delta}) = {} & -\frac{2i}{\pi}\,\delta\varepsilon/r_\delta + \frac{i}{\pi}\frac{\varepsilon r_\delta}{\delta}\ln\frac{\varepsilon r_\delta}{\delta} + \frac{er_\delta}{\delta}\{i/\pi(\gamma - \ln 2 - \tfrac{1}{2}) + \tfrac{1}{2}) \\ & + 0(\varepsilon^3 r_\delta^3/\delta^3 \ln(\varepsilon r_\delta/\delta))\} \end{aligned} \tag{3.31}$$

$$H_2(\varepsilon r_\delta/\delta) = -\frac{4i}{\pi}\frac{\delta^2}{\varepsilon^2 r_\delta^2} - \frac{i}{\pi} + (\varepsilon r_\delta/\delta)^2\left\{\frac{i}{4\pi}(\gamma - \ln 2 + \ln(\varepsilon r_\delta/\delta) - \tfrac{3}{4}) + \tfrac{1}{8}\right\}$$
$$+ 0(\varepsilon^4 r_\delta^4/\delta^4 \ln(\varepsilon r_\delta/\delta)), \tag{3.32}$$
$$H_n(er_\delta/\delta) = 0(\delta^n/\varepsilon^n r_\delta^n).$$

Furthermore,

$$E_n = \sum_{m=0}^{\infty} P_{mn} J_m(\varepsilon r_\delta/\delta)\cos m\theta$$
$$= P_{0n}(1 - \tfrac{1}{4}\varepsilon^2 r_\delta^2/\delta^2) + P_{1n}\frac{\varepsilon r_\delta}{2\delta}\cos\theta + P_{2n}\frac{\varepsilon^2 r_\delta^2}{8r_\delta^2}\cos 2\theta + 0(\varepsilon^3 r_\delta^3/\delta^3) \tag{3.33}$$

with similar expansions for Ψ_n^p, F_n, Φ_n^s.

Examination of (3.30)–(3.33) reveals that v_0 must be chosen as ε and then $\mathbf{u}_0'$ is given by (3.28) with Φ, Ψ chosen as

$$\Phi_0 = a_0\Phi_0^p, \qquad \Psi_0 = a_0\Psi_0^p, \tag{3.34}$$

in order to match with $\mathbf{u}_0$ given by (3.27). Then $a_0 = -i\pi pa/4\mu$. Thus, as is to be expected, in the far-field the first term represents the solution due to a dilatational line source in a half-space; see Lapwood [59]. Using (3.30), (3.33), etc., in (3.34) and the resulting expressions in (3.28) we get

$$\varepsilon u_{0r}' = a_0\Bigg[\frac{2i\delta}{\pi r_\delta} + \tfrac{1}{2}\varepsilon(P_{10} + \tau Q_{10})\cos\theta + \varepsilon^2\ln\varepsilon\,\frac{-i}{\pi}\frac{r_\delta}{\delta}$$
$$+ \varepsilon^2\{-\tfrac{1}{2}(1 + P_{00} - \tfrac{1}{2}(P_{20} + \tau^2 Q_{20})\cos 2\theta) \tag{3.35}$$
$$-\frac{i}{\pi}(\gamma - \ln 2 - \tfrac{1}{2} + \ln r_\delta/\delta)\}r_\delta/\delta + 0(\varepsilon^4\ln\varepsilon)\Bigg],$$

$$\varepsilon u_{0\theta}' = -a_0[\tfrac{1}{2}\varepsilon(P_{10} + \tau Q_{10})\sin\theta + \varepsilon^2\tfrac{1}{4}(P_{20} + \tau^2 Q_{20})r_\delta/\delta\sin 2\theta]. \tag{3.36}$$

Examination of (3.35) and (3.36) reveals that $\mu_1(\varepsilon) = \varepsilon$ and

$$u_{1r} = \tfrac{1}{2}a_0(P_{10} + \tau Q_{10})\cos\theta, \qquad u_{10} = -\tfrac{1}{2}a_0(P_{10} + \tau Q_{10})\sin\theta, \tag{3.37}$$

which define a rigid-body translation which does not affect the stress field. The third term in the expansion (3.19) is similarly obtained by setting $\mu_2(\varepsilon) = \varepsilon^2\ln\varepsilon$ and

$$u_{2r} = -\frac{ia_0}{\pi}\bar{r} + \frac{A_1}{\bar{r}}, \qquad u_{2\theta} = 0, \tag{3.38}$$

where A_1 is to be chosen so that (3.23) is satisfied. In the present case this implies that A_1 is found by using

$$\sigma_{rr}^{(2)} = \sigma_{r\theta}^{(2)} = 0 \quad \text{on} \quad \bar{r} = 1. \tag{3.39}$$

Thus

$$A_1 = -\frac{ia_0}{\pi}(\tau^2 - 1). \tag{3.40}$$

Now we are in a position to find the second term in the expansion (3.21). To do so we note

$$u_2 = -\frac{ia_0}{\pi}\frac{r_\delta}{\delta} + A_1\frac{\delta}{r_\delta}, \qquad u_{2\theta} = 0. \tag{3.41}$$

So we must take $\nu_2(\varepsilon) = \varepsilon^3 \ln \varepsilon$ with

$$\Phi_1 = a_0^{(1)}\Phi_0^P, \qquad \Psi_1 = a_0^{(1)}\Psi_0^P. \tag{3.42}$$

The matching then gives

$$a_0^{(1)} = -\frac{i\pi}{2}A_1 = -\frac{a_0}{2}(\tau^2 - 1). \tag{3.43}$$

Proceeding in this manner we can find the successive terms in the expansions (3.21), (3.19). Carrying out the expansion to $0(\varepsilon^2)$ gives

$$\mu_3(\varepsilon) = \varepsilon^2,$$

$$\begin{aligned} u_{3r} = a_0\Bigg[&-\frac{i}{\pi}\bar{r}\ln\bar{r} + \bar{r}\left\{-\frac{i}{\pi}\left(\gamma - \ln 2 - \tfrac{1}{2} - i\frac{\pi}{2}P_{00}\right) - \tfrac{1}{2}\right. \\ &\left.+ \tfrac{1}{4}(P_{20} + \tau^2 Q_{20})\cos 2\theta\right\}\Bigg] + U_r, \end{aligned} \tag{3.44}$$

$$u_{3\theta} = -\tfrac{1}{4}a_0(P_{20} + \tau^2 Q_{20})\bar{r}\sin 2\theta + U_\theta. \tag{3.45}$$

Where $\mathbf{U}$ must satisfy (3.22) with the right-hand side set equal to zero and the stresses $\sigma_{rr}(\mathbf{u}_3)$, $\sigma_{r\theta}(\mathbf{u}_3)$ vanish on $\bar{r} = 1$. We find

$$U_r = -\frac{1}{2\mu}\left[\frac{1}{\bar{r}}(F - 4D(1 - \sigma)\cos 2\theta) - \frac{2c}{\bar{r}^3}\cos 2\theta\right], \tag{3.46}$$

$$U_\theta = \frac{1}{2\mu}\left[-\frac{1}{\bar{r}}2D(1 - 2\sigma) + \frac{2c}{\bar{r}^3}\right]\sin 2\theta,$$

$$F = \frac{ia_0}{\pi}\left[(\lambda + 2\mu) + 2(\lambda + \mu)\left(\gamma - \ln 2 - \tfrac{1}{2} - \frac{i\pi}{2}P_{00} - \frac{i\pi}{2}\right)\right], \tag{3.47}$$

$$D = \tfrac{1}{2}\mu a_0(P_{20} + \tau^2 Q_{20}),$$

$$C = -\tfrac{1}{4}\mu a_0(P_{20} + \tau^2 Q_{20}).$$

The near-field expansion is thus

$$u = \mathbf{u}_0 + \varepsilon \mathbf{u}_1 + \varepsilon^2 \ln \varepsilon \mathbf{u}_2 + \varepsilon^2 \mathbf{u}_3 + 0(\varepsilon^3 \ln \varepsilon), \tag{3.48}$$

and the far-field expansion is

$$\mathbf{u}' = \varepsilon \mathbf{u}'_0 + \varepsilon^3 \ln \varepsilon \mathbf{u}'_1 + \varepsilon^3 \mathbf{u}'_2 + 0(\varepsilon^4 (\ln \varepsilon)^2) \tag{3.49}$$

with

$$\Phi_2 = a_0^{(2)} \Phi_0^P + a_2^{(2)} \Phi_2^P + b_2^{(2)} \Phi_2^s, \tag{3.50}$$

$$\begin{gathered} \Psi_2 = b_2^{(2)} \Psi_2^s + a_0^{(2)} \Psi_0^P + a_2^{(2)} \Psi_2^P, \\ a_0^{(2)} = \frac{i\pi F}{4\mu}, \qquad a_2^{(2)} = b_2^{(2)}/\tau^2, \qquad a_2^{(2)} = \frac{i\pi D(1-2\sigma)}{2\mu}. \end{gathered} \tag{3.51}$$

The above expression for **u** can now be used to calculate the effect of the free boundary on the dynamic hoop stress on the circular cavity. Note that this effect is not felt to $0(\varepsilon^2 \ln \varepsilon)$. To this order the expansion is simply that of the solution in an infinite medium.

$\mathbf{u}'$ given by (3.49) can be used to calculate the displacement at the free surface $y = 0$. The first term is simply due to a point pressure of strength $pa^2/4\mu$. The successive terms are the effects of the cavity on this displacement.

The method outlined above together with conformal mapping can be used to study the effect of an elliptical cavity. These will be published elsewhere. In the next example we consider the scattering by the cavity of a plane compressional wave incident normal to the free surface. We shall assume normal incidence for the sake of algebraic simplicity. The case of general incidence can similarly be dealt with. Again we shall restrict our attention to small ε. The solution for arbitrary ε would have to be done formally as outlined in Subsection 3.1.

Example B. *Scattering by a Circular Cavity in a Half-space*

Consider a plane compressional wave propagating in the direction of the negative y-axis and is represented by the potential

$$\Phi^{(i)} = \frac{iu_0}{\alpha} e^{-i\varepsilon \bar{y} - i\omega t}. \tag{3.52}$$

In the absence of the cavity the total field $\bar{\mathbf{u}}$ will be given by $\nabla \bar{\Phi}$ with

$$\bar{\Phi} = \Phi^{(i)} + \Phi^{(r)} = \frac{iu_0}{\alpha} [e^{-i\varepsilon \bar{y}} - e^{i\varepsilon \bar{y}}], \tag{3.53}$$

where the factor $e^{-i\omega t}$ is suppressed. To study the effect of the cavity we now proceed in the same manner as before. In this case (3.19) and (3.20) take the forms

$$\frac{1}{u_0}\mathbf{u} = 2\cos\alpha h\mathbf{e}_y + \varepsilon\mathbf{u}_1 + 0(\varepsilon^2), \tag{3.54}$$

$$\frac{1}{u_0}\mathbf{u}' = \frac{1}{u_0}\bar{\mathbf{u}} + \varepsilon^2\mathbf{u}_2' + 0(\varepsilon^2). \tag{3.55}$$

Here

$$u_{1r} = \frac{A}{2(1-2\sigma)}\left[(1-2\sigma)\bar{r} + \frac{1}{\bar{r}} + (1-2\sigma)\left\{\bar{r} - \frac{1}{\bar{r}^3} + 4(1-\sigma)\frac{1}{\bar{r}}\right\}\cos 2\theta\right], \tag{3.56}$$

$$u_{1\theta} = -\tfrac{1}{2}A\sin 2\theta\left[\bar{r} + \frac{1}{\bar{r}^3} + 2(1-2\sigma)\frac{1}{\bar{r}}\right],$$
$$A = -2\sin\alpha h, \tag{3.57}$$

$$\Phi_2 = a_0\Phi_0^P + a_2\Phi_2^P + b_2\Phi_2^s, \qquad \Psi_2 = b_2\Psi_2^s + a_0\Psi_0^P + a_2\Psi_2^P,$$

$$a_0 = -\frac{i\pi A}{4(1-2\sigma)}, \qquad a_2 = b_2/\tau^2, \qquad a_2 = \frac{i\pi A}{2}(1-2\sigma). \tag{3.58}$$

The successive terms of the expansion can be obtained in a manner similar to that used in the previous example.

The approach taken here is applicable to noncircular boundaries as well if use is made of conformal mapping technique to get the inner solution. The outer solution will still have the same structure. This will be explored elsewhere. In the next section we use the MAE approach to solve scattering by a rigid spheroid embedded in an infinite medium.

4 SCATTERING OF ELASTIC WAVES BY RIGID SPHEROIDS

The object of this section is to study the scattering of a plane P- or S-wave by a rigid spheroid and then use the single scattering results to obtain the effect of several such inclusions. For arbitrary angle of incidence this problem cannot be solved by eigenfunction expansions. However, it will be shown that application of MAE leads to a solution in powers of ε when the wavelength is large compared to the linear dimensions of the spheroid.

4.1 Scattering by a Single Rigid Spheroid

We shall introduce oblate spheroidal coordinates (ξ, η, ω) by the relations

$$\begin{gathered} x = \rho \cos \omega, \qquad y = \rho \sin \omega, \qquad z = c\xi\eta, \\ \rho = c[(\xi^2 + 1)(1 - \eta^2)]^{1/2}, \qquad \xi \geq \xi_0 > 0, \qquad -1 \leq \eta \leq 1, \end{gathered} \tag{4.1}$$

where $\xi = \xi_0$ defines the surface of the oblate spheroidal inclusion. The incident wave will be assumed to be either

$$\mathbf{u}^{(i)} = u_0 \, e^{i\varepsilon(\bar{z} \cos \zeta + \bar{y} \sin \zeta)} (\cos \zeta \mathbf{e}_z + \sin \zeta \mathbf{e}_y) \quad \text{for a P-wave,} \tag{4.2}$$

or

$$\mathbf{u}^{(i)} = u_0 \, e^{i\tau\varepsilon(\bar{z} \cos \zeta + \bar{y} \sin \zeta)} (\sin \zeta \mathbf{e}_z - \cos \zeta \mathbf{e}_y) \quad \text{for an SV-wave.} \tag{4.3}$$

Here $\bar{y} = y/c$, $\bar{z} = z/c$.

It is seen from above that the solution to the SV-case could be easily obtained from the P-wave case by some simple transformations; see [55]. Here we shall give the solution for the case of incident P-waves. For an incident SH-wave the solution is somewhat different. This will be discussed in a separate communication.

The boundary condition on the rigid inclusion is that

$$\mathbf{u} = \mathbf{u}^{(s)} + \mathbf{u}^{(i)} = \mathbf{U}, \quad \text{on} \quad \xi = \xi_0, \tag{4.4}$$

where $\mathbf{U}$ is the displacement of the surface of the inclusion. Of course, $\mathbf{U}$ is an unknown and has to be determined by use of the equations of motion of the inclusion. We can write $\mathbf{U}$ as

$$\mathbf{U} = \mathbf{U}_c + \boldsymbol{\Theta}_c \wedge \mathbf{r}_0, \tag{4.5}$$

where $\mathbf{U}_c$ is the translational motion of the center and $\boldsymbol{\Theta}_c$ is the rotation about the center. Clearly,

$$\mathbf{U}_c = U_y \mathbf{e}_y + U_z \mathbf{e}_z, \tag{4.6}$$

and

$$\boldsymbol{\Theta}_c = \Theta_0 \mathbf{e}_x. \tag{4.7}$$

For simplicity of analysis we can write

$$\mathbf{u}^{(s)} = u_0 \mathbf{u}^{(d)} + U_y \mathbf{u}^{(t)}_{(y)} + U_z \mathbf{u}^{(t)}_{(z)} + c\Theta_0 \mathbf{u}^{(\mathrm{rot})}_{(x)} \tag{4.8}$$

such that on $\xi = \xi_0$

$$\mathbf{u}^{(d)} + \mathbf{u}^{(i)}/u_0 = 0, \tag{4.9a}$$

$$\mathbf{u}_{(y)}^{(t)} - \mathbf{e}_y = 0, \tag{4.9b}$$

$$\mathbf{u}_{(z)}^{(t)} - \mathbf{e}_z = 0, \tag{4.9c}$$

$$\mathbf{u}_{(x)}^{(\text{rot})} - \mathbf{e}_x \wedge \mathbf{r}_0/c = 0. \tag{4.9d}$$

Then the equations of motion of the spheroid give

$$U_j = -u_0 \iint_S (\boldsymbol{\tau}^{(i)} + \boldsymbol{\tau}^{(d)}) \cdot \mathbf{e}_j \, ds \Big/ \left[M\gamma^2 + \iint_S \boldsymbol{\tau}_{(j)}^{(t)} \cdot \mathbf{e}_j \, ds \right], \qquad j = y, z, \tag{4.10}$$

and

$$\Theta_0 = -u_0 \mathbf{e}_x \cdot \iint_S \mathbf{r}_0 \wedge (\boldsymbol{\tau}^{(d)} + \boldsymbol{\tau}^{(i)}) \, ds \Big/ \left[I_x \gamma^2 + c\mathbf{e}_x \cdot \iint_S \mathbf{r}_0 \wedge \boldsymbol{\tau}_{(x)}^{(\text{rot})} \, ds \right]. \tag{4.11}$$

Here M is the mass of the spheroid, I_x is its moment of inertia about the x-axis, and $\boldsymbol{\tau}^{(d)}$, $\boldsymbol{\tau}^{(i)}$, etc., are the stress vectors acting on S arising out of the displacement fields $\mathbf{u}^{(d)}$, $\mathbf{u}^{(i)}/u_0$, etc.

The boundary-value problem for arbitrary ζ cannot be solved by eigenfunction expansions. If $\zeta = 0$ then $U_y = \Theta_0 = 0$ and the solution can formally be obtained by eigenfunction expansions. However, even in this case, the boundary conditions (4.9a)–(4.9b) lead to an infinite set of algebraic equations for the determination of the unknown expansion coefficients. This system was solved by Oien and Pao [60] by truncation. They found that the rate of convergence is very slow if the shape of the spheroid departs very much from a sphere. On the other hand, the MAE approach was found [53] to give excellent results in this case. In what follows we shall present the solution as obtained by this method for arbitrary ζ. The details of the matching process will be omitted and can be found in [55].

Assume $\varepsilon \ll 1$. The inner variables will be defined as

$$\bar{x} = x/c, \qquad \bar{\mathbf{u}} = \mathbf{u}/\mathbf{u}_0, \tag{4.12}$$

and the outer variables as

$$x' = \varepsilon \bar{x}, \qquad \mathbf{u}' = \mathbf{u}/u_0, \qquad \xi' = \varepsilon \xi. \tag{4.13}$$

The inner and outer expansions are then assumed to be

$$\bar{\mathbf{u}} = \bar{\mathbf{u}}_0 + \varepsilon \mathbf{u}_1' + \cdots, \tag{4.14}$$

$$\mathbf{u}' = \mathbf{u}_0' + \varepsilon \mathbf{u}_1' + \cdots. \tag{4.15}$$

Consider $\mathbf{u}^{(d)}$ and denote $\mathbf{u} = \mathbf{u}^{(d)} + \mathbf{u}^{(i)}/u_0$. Then it is easily shown that $\mathbf{u}_0' = \mathbf{u}^{(i)}/u_0$, because the dominant contribution in the far-field comes from the incident field in order to satisfy the radiation condition on $\mathbf{u}^{(d)}$. In terms of the intermediate variables (see Section 3) $\mathbf{u}_0'$ can be written as

$$\mathbf{d}\, e^{i\varepsilon/\delta(z_\delta \cos\zeta + y_\delta \sin\zeta)} = [1 + i(\varepsilon/\delta)(z_\delta \cos\zeta + y_\delta \sin\zeta) + 0(\varepsilon^2/\delta^2)]\mathbf{d}, \qquad \mathbf{d} = \cos\zeta \mathbf{e}_z + \sin\zeta \mathbf{e}_y. \tag{4.16}$$

Thus to match with the inner solution to 0(1), we assume

$$\bar{\mathbf{u}}_0 = \mathbf{d} + \mathbf{v}_0, \tag{4.17}$$

where

$$\bar{\nabla}\bar{\nabla} \cdot \mathbf{v}_0 - \tau^{-2}\bar{\nabla} \wedge \bar{\nabla} \wedge \mathbf{v}_0 = 0, \tag{4.18}$$

$$\mathbf{v}_0 + \mathbf{d} = 0 \quad \text{on} \quad \xi = \xi_0. \tag{4.19}$$

The solution to (4.18)–(4.19), that is of $0(1/r)$ for large r, is found to be (see Appendix C)

$$\mathbf{v}_0 = \bar{\nabla}F_0 + \bar{y}\bar{\nabla}Y_0 + \bar{z}\bar{\nabla}Z_0 - (3 - 4\sigma)(Y_0\mathbf{e}_y + Z_0\mathbf{e}_z) \tag{4.20}$$

with F_0, Y_0, Z_0 given in Eq. (C.2). Now expressing $\bar{\mathbf{u}}_0$ in terms of the intermediate variables and expanding in powers of δ we get

$$\begin{aligned}\bar{\mathbf{u}}_0 = \mathbf{d} + [&\mathbf{e}_\xi i4(1 - \sigma)(C_{00}P_1^0 \cos\zeta + B_{00}\sin\zeta P_1^1 \sin\omega)\\ &+ \mathbf{e}_\eta i(3 - 4\sigma)(C_{00}P_1^0 \cos\zeta - B_{00}P_1^1 \sin\zeta \sin\omega)\\ &+ \mathbf{e}_\omega i(3 - 4\sigma)B_{00}\sin\zeta \cos\omega]\, \delta/\xi_\delta + 0(\delta^3/\xi_\delta^3).\end{aligned} \tag{4.21}$$

We now turn to the determination of $\mathbf{u}_1'$. Note first that $\mathbf{u}_n'$ must satisfy

$$\nabla'\nabla' \cdot \mathbf{u}_n' - \tau^{-2}\nabla' \wedge \nabla' \wedge \mathbf{u}_n' = -\mathbf{u}_n' \tag{4.22}$$

for all n. The general solution to this equation satisfying the radiation condition as $\bar{r} \to \infty$ is given by Eqs. (C.3)–(C.4). Now $\mathbf{u}_1'$ has to be chosen so that when expressed in intermediate variables $\varepsilon\mathbf{u}_1'$ matches to $0(\delta)$ with the bracketed term in (4.21). We thus find

$$\mathbf{u}_1' = \nabla'\Phi_1 + \tau^2(r'\Psi_1)\mathbf{e}_r + \nabla'\left(\frac{\partial}{\partial r'}(r'\Psi_1)\right), \tag{4.23}$$

where

$$\Phi_1 = a_{01}Y_{01}^e(\theta,\omega)h_1(r')\cos\zeta + a_{11}Y_{11}^0(\theta,\omega)h_1(r')\sin\zeta,$$
$$\Psi_1 = b_{01}Y_{01}^e(\theta,\omega)h_1(\tau r')\cos\zeta + b_{11}Y_{11}^0(\theta,\omega)h_1(\tau r')\sin\zeta.$$

Using the expansions for $h_1(\varepsilon r_\delta/\delta)$ as $\varepsilon \to 0$ in (4.23) and matching with (4.21), the coefficients a_{01}, etc., are found to be

$$\begin{aligned} a_{01} &= b_{01}/\tau^2, \qquad a_{11} = b_{11}/\tau^2, \\ b_{01} &= C_{00}(4\sigma - 4), \qquad b_{11} = B_{00}(4\sigma - 4). \end{aligned} \tag{4.24}$$

Proceeding in this manner we can show that the expansions (4.14)–(4.15) are simple power series in ε. This is to be contrasted with the expansions obtained in Section 3, where terms of the type $\varepsilon^p(\ln \varepsilon)^q$ appear. The expansions (4.14), (4.15) have been carried out to $0(\varepsilon^2)$ and $0(\varepsilon^3)$, respectively, and can be found in [55]. Let us now turn to the determination of $\mathbf{u}^{(t)}$ and $\mathbf{u}^{(\mathrm{rot})}$.

As in the determination of $\mathbf{u}^{(d)}$ we denote

$$\mathbf{u}_{(y)} = \mathbf{u}^{(t)}_{(y)} - \mathbf{e}_y, \qquad \mathbf{u}_{(z)} = \mathbf{u}^{(t)}_{(z)} - \mathbf{e}_z. \tag{4.25}$$

Clearly then $\mathbf{u}_{(y)}$, $\mathbf{u}_{(z)}$ will have expansions very similar to (4.14)–(4.15). In fact

$$\begin{aligned} \bar{\mathbf{u}}_{(y)} &= \mathbf{u}_{0(y)} + \varepsilon \mathbf{u}_{1(y)} + \cdots, \\ \bar{\mathbf{u}}_{(z)} &= \mathbf{u}_{0(z)} + \varepsilon \mathbf{u}_{1(z)} + \cdots. \end{aligned} \tag{4.26}$$

The far-field expansions will be

$$\begin{aligned} \mathbf{u}^{(t)}_{(y)} &= \varepsilon \mathbf{u}'_{1(y)} + \cdots, \\ \mathbf{u}^{(t)}_{(z)} &= \varepsilon \mathbf{u}'_{1(z)} + \cdots, \end{aligned}$$

where we obtain $\mathbf{u}'_{1(y)}$, $\mathbf{u}'_{1(z)}$ from $-\mathbf{u}'_1$ by setting $\zeta = \pi/2$ and 0, respectively. On the other hand, $\mathbf{u}_{0(y)}$, $\mathbf{u}_{0(z)}$ are obtained from $-\mathbf{v}_0$ if we take $\zeta = \pi/2$ and 0, respectively. The higher order terms are obtained by somewhat similar transformations of $\bar{\mathbf{u}}$ and $\mathbf{u}'$.

The determination of $\mathbf{u}^{(\mathrm{rot})}$ follows the same steps. However, there is no simple transformation of $\mathbf{u}$ to yield this.

Assume the following expansions for $\mathbf{u}^{(\mathrm{rot})}_{(x)}$:

$$\mathbf{u}^{(\mathrm{rot})}_{(x)} = \mathbf{u}^{(\mathrm{rot})}_0 + \varepsilon \mathbf{u}^{(\mathrm{rot})}_1 + \ldots \text{ in the near-field,} \tag{4.27}$$

and

$$\mathbf{u}^{(\mathrm{rot})}_{(x)} = \varepsilon^2 \mathbf{u}'^{(\mathrm{rot})}_1 + \ldots \text{ in the far-field.} \tag{4.28}$$

The solution $\mathbf{u}^{(\mathrm{rot})}_0$ is given by (C.1) with

$$\begin{aligned} F &= f_0 = \mathscr{A}_{12} Q^1_2 P^1_2 \sin \omega, \\ Y &= y_0 = \mathscr{B}_{01} Q^0_1 P^0_1, \qquad Z = z_0 = \mathscr{C}_{11} Q^1_1 P^1_1 \sin \omega. \end{aligned} \tag{4.29}$$

The coefficients $\mathscr{A}_{12}$, etc., are obtained by using the boundary condition (4.9d) and are given by Eq. (C.5). $\mathbf{u}'^{(\mathrm{rot})}_1$ is given by (C.3) with

$$\begin{aligned}\Phi &= a_{12} Y_{12}^{0} h_2(r'),\\ \Psi &= b_{12} Y_{12}^{0} h_2(\tau r'),\\ \chi &= c_{11} Y_{11}^{e} h_1(\tau r'),\end{aligned} \tag{4.30}$$

$$a_{12} = 2b_{12}/\tau^3 = -\frac{2i}{9}(1-2\sigma)(2\mathscr{C}_{11} - \mathscr{B}_{01}),$$

$$c_{11} = i\frac{\tau^3}{3}(\mathscr{B}_{01} + 2\mathscr{C}_{11})/(1-\tau^2).$$

Once $\mathbf{u}^{(d)}$, $\mathbf{u}^{(t)}$ and $\mathbf{u}^{(\mathrm{rot})}$ are known they can be used to calculate $\boldsymbol{\tau}^{(d)}$, $\boldsymbol{\tau}^{(t)}$ and $\boldsymbol{\tau}^{(\mathrm{rot})}$ on S. Then (4.10)–(4.11) are used to find the amplitudes U_j and Θ_0 to be

$$U_j/u_0 = -cF_j/[M\gamma^2 + cF_j^{(t)}], \qquad j = y, z, \tag{4.31}$$

$$c\Theta_0/u_0 = -cT_x/[I_x\gamma^2 + cT_x^{(\mathrm{rot})}]. \tag{4.32}$$

The expansions for F_j, etc., are given by (C.6)–(C.8). Numerical results for the amplitudes $|U_j/u_0|$ for different values of ξ_0 and for different angles of incidence have been presented in [55].

From the results derived above one can obtain the solution for a prolate spheroid by changing $\xi_0 \to -i\xi_0$ and $c \to ic$.

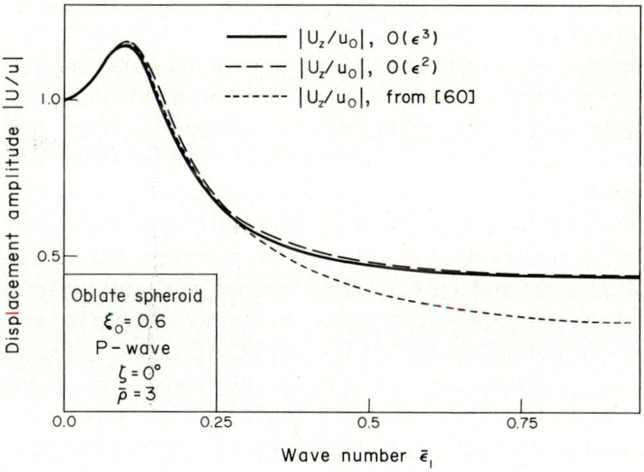

Fig. 6 Plot of $|U_z/u_0|$ vs. $\bar{\varepsilon}_1$ for an oblate spheroid ($\xi_0 = 0.6$) in comparison with that of Ref. [60].

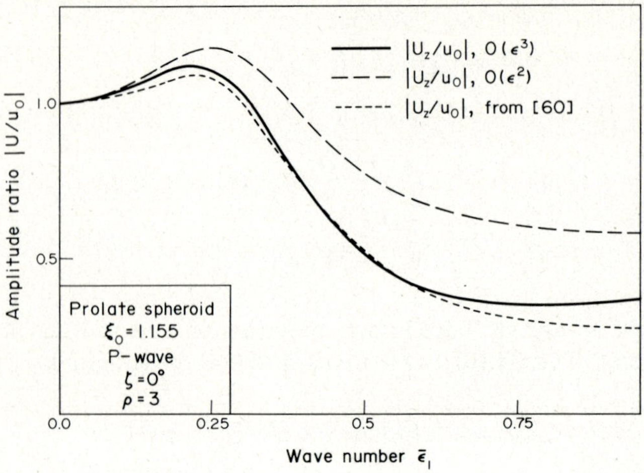

Fig. 7 Plot of $|U_z/u_0|$ vs. $\bar{\varepsilon}_1$ for a prolate spheroid ($\xi_0 = 1.155$) in comparison with that of Ref. [60].

It is to be noted that $\Theta_0 = 0(\varepsilon)$ since $T_x = \int\int \bar{\mathbf{r}}_0 \wedge (\boldsymbol{\tau}^{(d)} + \boldsymbol{\tau}^{(i)})\, ds$ is $0(\varepsilon)$ whereas $T_x^{(\text{rot})}$ is 0(1). It is also easily shown that

$$c\Theta_0/u_0 = \Gamma\varepsilon + 0(\varepsilon^3), \tag{4.33}$$

where Γ is a constant depending on ξ_0 and the angle of incidence. It is zero when $\zeta = 0$ and $\pi/2$.

For purposes of comparison with the solution obtained for the axisymmetric problem by the eigenfunction method we show in Figs. 6 and 7 the amplitude ratio $|U_z/u_0|$ (where $\zeta = 0°$) obtained to $0(\varepsilon^2)$ and $0(\varepsilon^3)$ by the author and by Oien and Pao [60]. It is to be noted that for the prolate spheroid with $\xi_0 = 1.155$ the third-order approximation shows a marked improvement over the second-order solution and is very close to that reported in [60]. For the oblate spheroid with $\xi_0 = 0.6$, however, both approximations are almost the same and are reasonably close to that obtained in [60]. The two choices of the ξ_0 values correspond to the cases where the length of the minor axis is about one-half of the length of the major axis. Figures 8 and 9 show the ratio $|U_z/u_0|_0$ (when $\zeta = 0°$) for a sphere and a disc in comparison with that obtained in [60] and [62]. For the disc $\bar{\rho}$ represents the ratio of the mass of the disc and ρc^3. It is found that the third-order approximation agrees quite well with the numerical integration results reported in [62]. For the sphere, however, the agreement is reasonably good for ε up to about 0.5

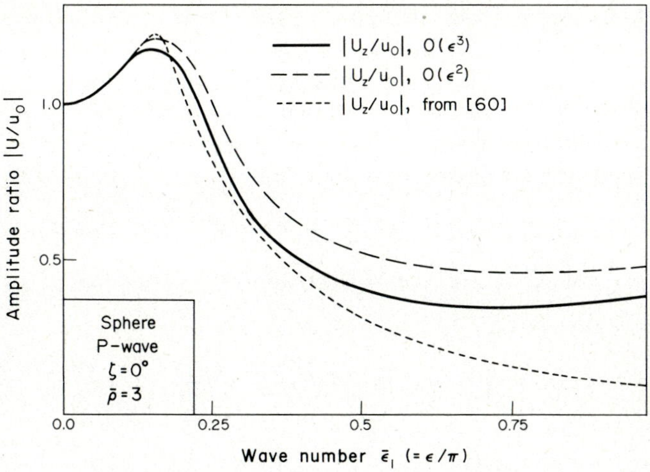

Fig. 8 Plot of $|U_z/u_0|$ for a sphere in comparison with the exact solution.

except for the fact that the peak value is underestimated by about 5%. Also, the approximate solution does not show the sharp rise and fall as predicted by the exact solution.

As limiting cases we can derive the solutions for scattering by a sphere and a rigid disc. The former is derived by taking the limit $\xi_0 \to \infty$, $c \to 0$

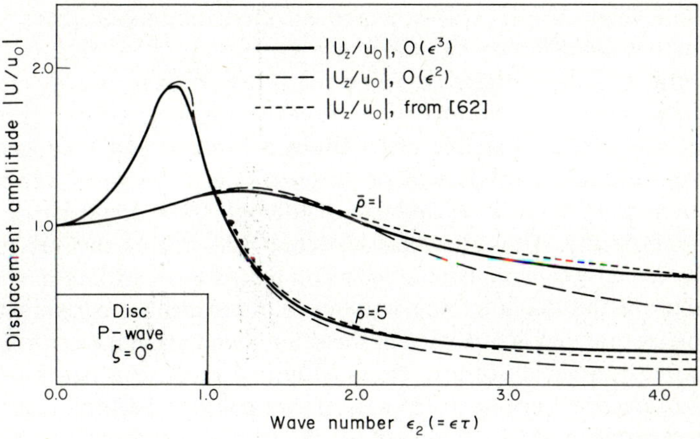

Fig. 9 Plot of $|U_z/u_0|$ for a disc in comparison with the numerical integration results.

such that $c\xi_0 \to a$, a being the radius of the sphere. This limiting solution has been discussed in [55]. The disc limit is obtained by letting $\xi_0 \to 0$. c then becomes the radius of the disc. We give here the expressions for U_z, U_y and Θ_0 correct to $0(\varepsilon^2)$. These will be shown to agree with those obtained by the method of integral equations outlined in Section 5.

For a rigid circular disc of mass m we find

$$
\begin{aligned}
U_z/u_0 &= -c\cos\zeta[F_z - \tfrac{1}{6}\varepsilon^2\sin^2\zeta]/[m\gamma^2(1+\tau^{-2})/16\mu - cF_z],\\
U_y/u_0 &= -c\sin\zeta[F_y - \tfrac{1}{6}\varepsilon^2\sin^2\zeta]/[m\gamma^2(3+\tau^{-2})/32\mu - cF_y],\\
c\Theta_0/u_0 &= i\varepsilon\sin\zeta\cos\zeta, \qquad (4.34)\\
F_z &= 1 - t_1\varepsilon_1 + \varepsilon_1^2(t_1^2 + (3+\tau^{-4})/6(1+\tau^{-2})), \quad t_1 = \frac{4i}{3\pi}\frac{2+\tau^{-3}}{1+\tau^{-2}},\\
F_y &= 1 - t_2\varepsilon_1 + \varepsilon_1^2(t_2^2 + (5+3\tau^{-4})/3(3+\tau^{-2})), \quad t_2 = \frac{8i}{3\pi}\frac{2+\tau^{-3}}{3+\tau^{-2}},\\
\varepsilon_1 &= \varepsilon\cdot\tau.
\end{aligned}
$$

Having obtained the solution for the single scattering problem we now turn to the analysis of the effect of a distribution of inclusions.

4.2 Wave Propagation in the Presence of a Random Distribution of Rigid Spheroids

The following analysis will be based on the assumptions that: (1) the wavelength is much larger than the linear dimensions of the spheroids, and (2) the concentration is dilute. Our objective is to derive the average material properties of the composite. Over and above the two assumptions made here we shall also assume that the distribution is random and homogeneous. Furthermore, the spheroids will be taken to be of identical properties. Multiple interactions will be neglected, although these could be included with more algebra. Thus the total scattered field will be the sum of the fields scattered by each individual inclusion placed in a medium having the average properties that are to be determined. If the inclusions have the axes of symmetry oriented in a particular direction then the composite will show anisotropy; transverse isotropy. So longitudinal waves can be propagated only in a direction parallel to the axis of symmetry or perpendicular to it. Our objective here will be to determine the average propagation speeds of a P-wave in these directions.

Now the scattered field due to the jth inclusion with center at ξ_j is given by

$$\Delta\mathbf{u}_j = \varepsilon^3\left[\nabla'\Phi + \tau^2 r_j'\Psi\mathbf{e}_{r_j} + \nabla'\left(\frac{\partial}{\partial r_j'}(r_j'\Psi)\right)\right] + 0(\varepsilon^4), \tag{4.35}$$

where

$$\begin{aligned}\Phi &= \frac{iV_j}{4\pi c^3}[(\rho'/\rho - 1)h_1(r_j')P_1(\cos\theta_j) + a_0h_0(r_j')\\ &\quad + a_2h_2(r_j')P_2(\cos\theta_j)]u_0\,e^{i\varepsilon\zeta_{3j}},\\ \Psi &= \frac{iV_j}{4\pi c^3}\Big[\tau^2(\rho'/\rho - 1)h_1(\tau r_j')P_1(\cos\theta_j)\\ &\quad + \frac{\tau^3}{2}a_2h_2(\tau r_j')P_2(\cos\theta_j)\Big]u_0\,e^{i\varepsilon\zeta_{3j}},\end{aligned} \tag{4.36}$$

for $\zeta = 0$, and

$$\begin{aligned}\Phi &= \frac{iV_j}{4\pi c^3}[(\rho'/\rho - 1)h_1(r_j')P_1^1(\cos\theta_j)\sin\omega + \bar{a}_0h_0(r_j')\\ &\quad + \bar{a}_2h_2(r_j')P_2(\cos\theta_j) + \bar{a}_4h_2(r_j')P_2^2(\cos\theta_j)\cos 2\omega]u_0\,e^{i\varepsilon\eta_{3j}},\\ \Psi &= \frac{iV_j}{4\pi c^3}\Big[(\rho'/\rho - 1)\tau^2h_1(\tau r_j')P_1^1(\cos\theta_j)\sin\omega + \frac{\tau^3}{2}\bar{a}_2h_2(\tau r_j')P_2(\cos\theta_j)\\ &\quad + \frac{\tau^3}{2}\bar{a}_4h_2(\tau r_j')P_2^2(\cos\theta_j)\cos 2\omega\Big]u_0\,e^{i\varepsilon\eta_{3j}},\end{aligned} \tag{4.37}$$

for $\zeta = \pi/2$. Here ζ_{3j} and η_{3j} are the z- and y-coordinates of the center of the jth spheroid, and V_j is its volume; r_j, θ_j, ω_j are the spherical coordinates of the observation point $\mathbf{x}$ referred to the center of the jth inclusion. The coefficients a_0, a_2, $\bar{a}_0$, $\bar{a}_2$, $\bar{a}_4$ are given in Appendix C by (C.9)–(C.13). Note that $\Delta\mathbf{u}_j$ given by (4.35) and (4.36) has the same expression (with different values of a_0, a_2) as for an elastic sphere; see [38]. Thus it would appear that the scattered field $\Delta\mathbf{u}_j$ for any symmetric inclusion will be, correct to $0(\varepsilon^3)$, of the form (4.36) for axial incidence of a P-wave.

The intermediate field expansion to $0(\varepsilon\delta^2)$ of (4.35) matches with that of the near-field given by

$$\frac{1}{u_0}\Delta\mathbf{u}_j = \varepsilon\mathbf{v}_{1j} + 0(\varepsilon^2), \tag{4.38}$$

where, for $\zeta = 0$,

$$\mathbf{v}_{1j} = \bar{\nabla} F_1 + \bar{z}\bar{\nabla} Z_1 - (3 - 4\sigma) Z_1 \mathbf{e}_z,$$

$$F_1 = \frac{i}{3}(\xi_0^2 + 1)\xi_0 \left[a_0 - \frac{5 - 4\sigma}{4(2\sigma - 1)} a_2 \right] Q_0 P_0$$

$$+ \frac{\xi_0^2}{Q_2(i\xi_0)} \left[\frac{(\xi_0^2 + 1) Q_1(i\xi_0)}{2} a_2 - \frac{i}{3} \right] Q_2 P_2, \tag{4.39}$$

$$Z_1 = - \frac{3\xi_0(\xi_0^2 + 1)}{4(2\sigma - 1)} a_2 Q_1 P_1.$$

Now (4.38) represents the perturbed static displacement field due to a rigid spheroid when the stress state at infinity is

$$\tau_{zz} = u_0 \frac{i\varepsilon(\lambda + 2\mu)}{c} = \tau_0, \qquad \tau_{yy} = \tau_{xx} = \frac{\lambda}{\lambda + 2\mu} \tau_0. \tag{4.40}$$

Thus the average elastic property derived from Eq. (4.35) for small concentrations will be the corresponding average static modulus. This was found to be the case in [40]; see also [38]. This average modulus can be found from the use of (4.38) in a manner similar to that of [63]. The assumptions under which such an approximation is valid have been listed at the beginning of this section. It may, however, be mentioned here that in deriving (4.35) it was assumed that ρ' is not very much larger than ρ, that is, the inclusions are not very much heavier than the surrounding medium.

The average modulus $\langle \lambda + 2\mu \rangle$ in the axial direction can be found in the following manner. Consider a large volume V containing a large number of the inclusions which are uniformly and homogeneously distributed with the number of inclusions per unit volume being n. Let Σ be the surface of V and assume that Σ is subjected to a displacement field given by

$$u_z^{(0)} = e_0 z, \qquad u_x^{(0)} = u_y^{(0)} = 0. \tag{4.41}$$

Then for dilute concentration we have

$$\langle \lambda + 2\mu \rangle = (\lambda + 2\mu) - \frac{n}{e_0^2} \int_\Sigma [T_i^{(0)} \Delta u_{ij} - \Delta T_{ij} u_i^{(0)}] \, d\Sigma. \tag{4.42}$$

The integral can be evaluated easily if Σ is taken to be a sphere of large radius. We find

$$C_{33} = \langle \lambda + 2\mu \rangle = (\lambda + 2\mu) - \text{in}\, v\mu \frac{2(1 - \sigma)}{(1 - 2\sigma)} (a_0 - a_2). \tag{4.43}$$

Here $v = V_j$ is the volume of each spheroid. The average density to this order of approximation follows from (4.36) and is given by

$$\langle \rho \rangle = \rho[1 + nv(\rho'/\rho - 1)]. \tag{4.44}$$

The average modulus $\langle \lambda + 2\mu \rangle$ in the x- (or y-) direction can similarly be found. For this we take $\zeta = \pi/2$. Then $\Delta \mathbf{u}_j$ is given by (4.35) and (4.37). We find

$$C_{22} = C_{11} = \langle \lambda + 2\mu \rangle = \lambda + 2\mu - \text{in } v\mu \frac{1 - \sigma}{1 - 2\sigma}(2\bar{a}_0 + \bar{a}_2 + 6\bar{a}_4). \tag{4.45}$$

In the limit when the spheroid tends to a sphere we obtain

$$C_{11} = C_{22} = C_{33} = (\lambda + 2\mu)\left[1 + nv\frac{3(3 - 5\sigma)}{4 - 5\sigma}\right],$$

which agrees with that derived in [63].

When spheroid tends to a circular disc of radius c we find $C_{33} = (\lambda + 2\mu)$ and $C_{11} = C_{22}$ with

$$C_{11} = (\lambda + 2\mu)\left[1 + 16\pi c^3 n \frac{(1 - 2\sigma)(13 - 16\sigma)}{(3 - 4\sigma)(7 - 8\sigma)}\right]. \tag{4.46}$$

Equation (4.35) can also be used to obtain the attenuation factors Q of the composite. These and numerical results based on (4.43)–(4.45) can be found in [64].

5 SCATTERING BY A RIGID CIRCULAR DISC

In the previous section the problem of the scattering of a plane P-wave by a rigid spheroid was solved by the methods of MAE and the solution for the disc was obtained as a limiting case. The latter problem can also be solved by integral equation techniques and this approach has been taken in [62] and [65]. The analysis presented in this section is that reported in [65], so some of the details will be omitted.

As before the displacement $\mathbf{u}(\mathbf{r}, t)$ will be assumed of the form $\mathbf{u}(\mathbf{r})\, e^{-i\gamma t}$ and the time factor will be suppressed. We shall use cylindrical polar coordinates (r, ω, z) with the z-axis being the axis of the disc. The displacement components u_r, u_ω, u_z will be expressed in terms of scalar wave functions Φ, Ψ, x as

$$u_r = \frac{\partial \Phi}{\partial r} + \frac{\partial^2 \Psi}{\partial r\, \partial z} + \frac{1}{r}\frac{\partial x}{\partial \omega},$$
$$u_\omega = \frac{1}{r}\frac{\partial \Phi}{\partial \omega} + \frac{1}{r}\frac{\partial \Psi}{\partial \omega\, \partial z} - \frac{\partial x}{\partial r}, \tag{5.1}$$
$$u_z = \frac{\partial \Phi}{\partial z} + \frac{\partial^2 \Psi}{\partial z^2} + \beta^2 \Psi.$$

The incident field $\mathbf{u}^{(i)}$ is given by (4.2), c being the radius of the disc. The boundary condition (4.4) becomes

$$\mathbf{u} = \mathbf{u}^{(i)} + \mathbf{u}^{(s)} = \mathbf{U} \quad \text{on} \quad z = 0. \tag{5.2}$$

Writing $\mathbf{U}$ in the form (4.5) with (4.6)–(4.7) we get, at $z = 0$,

$$\begin{aligned} u_r^{(i)} + u_r^{(s)} &= U_y \sin \omega, \qquad 0 \le r < c,\\ u_\omega^{(i)} + u_\omega^{(s)} &= U_y \cos \omega, \qquad 0 \le r < c,\\ u_z^{(i)} + u_z^{(s)} &= U_z + r\Theta_0 \sin \omega, \qquad 0 \le r < c. \end{aligned} \tag{5.3}$$

Besides, the stresses and the displacement must be continuous across the plane $z = 0$ for $r > c$. We obtain the unknown displacement components U_y, U_z, and the rotation Θ_0 using (4.10)–(4.11).

To conform with the notation used in [65] we shall replace ω by $\pi/2 + \omega$. Then on $z = 0$ we can write $\mathbf{u}^{(i)}$ as

$$\mathbf{u}^{(i)} = u_0\, e^{i\varepsilon\bar{r} \sin \zeta \cos \omega} (\cos \zeta \mathbf{e}_z + \sin \zeta \mathbf{e}_y). \tag{5.4}$$

Using the relation

$$e^{i\varepsilon\bar{r} \sin \zeta \cos \omega} = \sum_{n=0}^{\infty} \varepsilon_n i^n J_n(\varepsilon\bar{r} \sin \zeta) \cos n\omega, \tag{5.5}$$

when ε_n equals 1 or 2 depending on whether $n = 0$ or $n > 0$, we can express the incident displacement components as

$$u_\omega^{(i)} = -u_0 \sin \zeta \sum_{n=1}^{\infty} \{i^{n-1} J_{n-1}(\varepsilon\bar{r} \sin \zeta) - i^{n+1} J_{n+1}(\varepsilon\bar{r} \sin \zeta)\} \sin n\omega,$$
$$u_r^{(i)} = u_0 \sin \zeta \left[iJ_1(\varepsilon\bar{r} \sin \zeta) + \sum_{n=1}^{\infty} \{i^{n+1} J_{n+1} + i^{n-1} J_{n-1}\} \cos n\omega \right], \tag{5.6}$$
$$u_z^{(i)} = u_0 \cos \zeta \sum_{n=0}^{\infty} \varepsilon_n i^n J_n \cos n\omega.$$

For the scattered field we now choose the potentials $\Phi^{(s)}$, $\Psi^{(s)}$, $x^{(s)}$ as

$$\Phi^{(s)} = u_0 \sum_{n=0}^{\infty} \varepsilon_n \Phi_n \cos n\omega,$$

$$\Psi^{(s)} = u_0 \sum_{n=0}^{\infty} \varepsilon_n \Psi_n \cos n\omega, \tag{5.7}$$

$$x^{(s)} = u_0 \sum_{1}^{\infty} \varepsilon_n x_n \sin n\omega.$$

The expansion coefficients Φ_n, etc., are functions of r and z, and will be assumed of the forms

$$\Phi_n = \frac{1}{\gamma^2} \int_0^{\infty} \left[\frac{k^2 A_n}{\nu_1} \mp kP_n\right] J_n(kr)\, e^{-\nu_1|z|}\, dk,$$

$$\Psi_n = \frac{1}{\gamma^2} \int_0^{\infty} \left[\mp C_n + \frac{k}{\nu_2} Q_n\right] J_n(kr)\, e^{-\nu_2|z|}\, dk, \tag{5.8}$$

$$x_n = \frac{1}{\gamma^2} \int_0^{\infty} \frac{\beta^2}{\nu_2} B_n J_n(kr)\, e^{-\nu_2|z|}\, dk,$$

where $\nu_1 = (k^2 - \alpha^2)^{1/2}$, $\nu_2 = (k^2 - \beta^2)^{1/2}$. The square roots are chosen so that $\mathrm{Re}\,(\nu_1)$, $\mathrm{Re}\,(\nu_2) > 0$.

Using the boundary conditions (5.3) for $0 \le r < c$ and the continuity conditions for $r > c$ we derive a set of dual integral equations for the determination of the unknown constants A_n, P_n, etc.

5.1 Equations for P_n, Q_n.

$$P_n = -Q_n, \qquad n \ge 0,$$

$$\int_0^{\infty} k\left(\frac{k^2}{\nu_2} - \nu_1\right) P_0 J_0(kr)\, dk = -\gamma^2 (U_z/u_0 - \cos\zeta J_0(\varepsilon\bar{r}\sin\zeta)),$$

$$0 \le r < c, \tag{5.9}$$

$$\int_0^{\infty} kP_0 J_0(kr)\, dk = 0, \qquad r > c,$$

$$\int_0^\infty k\left[\frac{k^2}{\nu_2} - \nu_1\right] P_1 J_1(kr)\, dk = -\gamma^2(\tfrac{1}{2}r\Theta_0/u_0 - i\cos\zeta J_1(\varepsilon\bar{r}\sin\zeta)),$$

$$0 \le r < c, \qquad (5.10)$$

$$\int_0^\infty kP_1 J_1(kr)\, dk = 0, \qquad r > c,$$

$$\int_0^\infty k\left[\frac{k^2}{\nu_2} - \nu_1\right] P_n J_n(kr)\, dk = i^n \cos\zeta\gamma^2 J_n(\varepsilon\bar{r}\sin\zeta), \qquad 0 \le r < c,$$

$$\int_0^\infty kP_n J_n(kr)\, dk = 0, \qquad r > c. \qquad (5.11)$$

We find then that P_0 depends on the vertical displacement of the disc as well as on incident field contribution, whereas P_1 is related to the angle of rotation Θ_0 and an incident field contribution. Equations (5.9)–(5.11) are all of the same form and in the following we shall outline a method of solving for P_n. This is due to Noble [66].

Assume that

$$P_n(k) = \left(\frac{2k}{\pi}\right)^{1/2} \int_0^c \xi^{1/2}\lambda_n(\xi) J_{n-1/2}(k\xi)\, d\xi. \qquad (5.12)$$

Then $(5.11)_2$ is automatically satisfied and $(5.11)_1$ can be written as

$$\int_0^\infty k^{3/2}\left(\frac{k^2}{\nu_2} - \nu_1\right) \int_0^c \xi^{1/2}\lambda_n(\xi) J_{n-1/2}(k\xi) J_n(kr)\, dk$$

$$= \left(\frac{\pi}{2}\right)^{1/2} i^n\gamma^2 \cos\zeta J_n(\varepsilon\bar{r}\sin\zeta), \qquad 0 \le r < c; \qquad (5.13)$$

(5.13) is a Fredholm integral equation of the first kind to solve for $\lambda_n(\xi)$. It has been shown in Appendix D that (5.13) can be rewritten as a Fredholm equation of the second kind that is suitable for iteration at small ε. Using the results from Appendix D we can easily show that (see also [65])

$$\begin{aligned}
\lambda_0(\xi) &= \lambda_0^{(0)}(\xi) + \varepsilon\lambda_0^{(1)}(\xi) + \varepsilon^2\lambda_0^{(2)}(\xi) + \cdots,\\
\lambda_1(\xi) &= \lambda_1^{(0)}(\xi) + \varepsilon\lambda_1^{(1)}(\xi) + \varepsilon^2\lambda_1^{(2)}(\xi) + \cdots,\\
\lambda_2(\xi) &= \varepsilon^2\lambda_2^{(2)}(\xi) + \cdots,\\
\lambda_n(\xi) &= 0(\varepsilon^n), \qquad n > 2.
\end{aligned} \qquad (5.14)$$

The convergence of the series has been discussed in [62].

The different terms in these expansions are

$$\lambda_0^{(0)}(\xi) = -\frac{2c^2\gamma^2}{\varepsilon^2(1+\tau^2)}(U_z/u_0 - \cos\zeta),$$

$$\lambda_0^{(1)}(\xi) = -\lambda_0^{(0)}(\xi)t_1, \qquad t_1 = \frac{4i}{3\pi}\frac{1+2\tau^3}{1+\tau^2},$$

$$\lambda_0^{(2)}(\xi) = -\frac{c^2\gamma^2}{\varepsilon^2(1+\tau^2)}\xi^2\sin^2\zeta\cos\zeta + \lambda_0^{(0)}\left[t_1^2 + \frac{1+3\tau^4}{8(1+\tau^2)}(1+\xi^2)\right], \tag{5.15}$$

$$\lambda_1^{(0)}(\xi) = -\frac{2c^2\gamma^2}{\varepsilon^2(1+\tau^2)}\cdot\frac{c\Theta_0}{u_0}\cdot\xi,$$

$$\lambda_1^{(1)}(\xi) = \frac{2c^2\gamma^2}{\varepsilon^2(1+\tau^2)}\cdot i\sin\zeta\cos\zeta\cdot\xi, \tag{5.16}$$

$$\lambda_1^{(2)}(\xi) = \frac{c^2\gamma^2(1+3t^4)}{4\varepsilon^2(1+\tau^2)^2}\cdot\frac{c\Theta_0}{u_0}\xi(1-\tfrac{1}{3}\xi^2),$$

$$\lambda_2^{(2)}(\xi) = -\frac{2c^2\gamma^2}{3\varepsilon^2(1+\tau^2)}\sin^2\zeta\cos\zeta\cdot\xi^2. \tag{5.17}$$

To compute U_z/u_0 we now use the equation of the z-component of the motion of the disc, which is

$$\begin{aligned}-m\gamma^2U_z/u_0 &= \int_0^c\int_0^{2\pi}[\tau_{zz}|_+ - \tau_{zz}|_-]r\,dr\,d\omega\\ &= \frac{4\pi\mu}{c_2^2}\int_0^1\int_0^\infty kP_0J_0(k\bar{r})\bar{r}\,dk\,d\bar{r}\\ &= \frac{8\mu c}{c_2^2}\int_0^1\lambda_0(\xi)\,d\xi,\end{aligned} \tag{5.18}$$

where (5.12) has been employed. Using the expansion of $\lambda_0(\xi)$ given by (5.14) and (5.15) we obtain U_z/u_0 to be given by (4.34).

The angle of rotation is similarly obtained from the equation

$$\begin{aligned}-I\gamma^2\Theta_0/u_0 &= \int_0^c\int_0^{2\pi}[\tau_{zz+}| - \tau_{zz-}|]r^2\cos\omega\,dr\,d\omega\\ &= \frac{16\mu c^2}{c_2^2}\int_0^1\xi\lambda_1(\xi)\,d\xi.\end{aligned} \tag{5.19}$$

Expressions for U_z/u_0 and $c\Theta_0/u_0$ derived here agree with those of [65] after some corrections are made in Eqs. (19), (70) and (71). The sign in front

of k_2^2 in the expression for $\mathscr{F}$ should be a "+". In (70) and (71) the factors 8 and 16 should be replaced by 16 and 32, respectively.
Thus

$$c\Theta_0/u_0 = i \sin\zeta \cos\zeta \Big/ \left[1 - \varepsilon^2 \frac{1 + 3\tau^4}{10(1 + \tau^2)} - \frac{3I(1 + \tau^2)}{32\rho c^5}\varepsilon^2\right]. \tag{5.20}$$

Here I is the moment of inertia of the disc about a diameter. This completes the solutions for λ_n correct to $0(\varepsilon^2)$. These together with (5.12) and (5.8) will determine Φ_n, Ψ_n partly.

In subsection 5.3 we shall determine the scattered far-field.

5.2 Equations for A_n, B_n and C_n

$$A_n = -C_n, \qquad n \geq 0, \tag{5.21}$$

$$\int_0^\infty k\left(\frac{k^2}{\nu_1} - \nu_2\right)A_0 J_1(kr)\,dk = i\gamma^2 \sin\zeta J_1(\varepsilon\bar{r}\sin\zeta), \qquad 0 \leq r < c,$$

$$\int_0^\infty kA_0 J_1(kr)\,dk = 0, \qquad r > c, \tag{5.22}$$

$$\int_0^\infty k\left[\left(\frac{k^2}{\nu_1} - \nu_2\right)A_1 + \frac{\beta^2}{\nu_2}B_1\right]J_0(kr)\,dk = \gamma^2[U_y/u_0 - \sin\zeta J_0(\varepsilon\bar{r}\sin\zeta)],$$
$$0 \leq r < c,$$

$$\int_0^\infty k\left[\frac{k^2}{\nu_1} - \nu_2\right)A_1 - \frac{\beta^2}{\nu_2}B_1\right]J_2(kr)\,dk = -\gamma^2 \sin\zeta J_2(\varepsilon\bar{r}\sin\zeta),$$
$$0 \leq r < c,$$

$$\int_0^\infty k(A_1 + B_1)J_0(kr)\,dk = 0, \qquad r > c,$$

$$\int_0^\infty k(A_1 - B_1)J_2(kr)\,dk = 0, \qquad r > c. \tag{5.23}$$

The equation for A_n, B_n $(n > 1)$ are obtained from (5.23) if J_0, J_2 are replaced by J_{n-1} and J_{n+1}, respectively, and then by dropping the term U_y/u_0 and multiplying the right-hand sides of $(5.23)_1$, $(5.23)_2$ by i^{n-1} and $-i^{n+1}$, respectively.

A_0 can now be solved in a similar manner as before. Determination of A_n and B_n is complicated by the fact that they are governed by four coupled integral equations. To derive appropriate Fredholm equations we assume

$$A_n + B_n = p_n, \qquad A_n - B_n = q_n, \tag{5.24}$$

$$\begin{aligned} p_n &= \left(\frac{2k}{\pi}\right)^{1/2} c \int_0^1 \xi^{1/2}\theta_n(\xi) J_{n-3/2}(k\xi)\, d\xi, \\ q_n &= \left(\frac{2k}{\pi}\right)^{1/2} c \int_0^1 \xi^{1/2}\mu_n(\xi) J_{n+1/2}(k\xi)\, d\xi. \end{aligned} \tag{5.25}$$

This leads to a pair of coupled Fredholm integral equations of the second kind in θ_n and μ_n. It is found that

$$\theta_n(\xi) + \frac{1}{\pi}\int_0^1 \theta_n(\eta) M_n(\xi, \eta)\, d\eta = \xi^{1-n} F_n(\xi), \tag{5.26}$$

$$\mu_n(\xi) + \frac{1}{\pi}\int_0^1 \mu_n(\eta) M_{n+2}(\xi, \eta)\, d\eta = \xi^{-1-n} G_n(\xi), \tag{5.27}$$

where

$$M_n(\xi, \eta) = \pi \int_0^\infty (\xi\eta)^{1/2} \gamma k M(k) J_{n-3/2}(\xi t) J_{n-3/2}(\eta t)\, dt,$$

$$1 + \gamma M(k) = \frac{2kc^2}{\varepsilon^2(1 + 3\tau^2)}\left[\frac{k^2}{v_1} - v_2 + \frac{\beta^2}{v_2}\right],$$

$$F_n(\xi) = \frac{d}{d\xi}\int_0^\xi \left[\frac{4c^2\gamma^2}{\varepsilon^2(1 + 3\tau^2)}\{U_y/u_0\, \delta_{1n} - i^{n-1} \sin \zeta J_{n-1}(\varepsilon p \sin \zeta)\} - f_n(\rho)\right] \frac{\rho^n}{(\xi^2 - \rho^2)^{1/2}}\, d\rho,$$

$$G_n(\xi) = \frac{d}{d\xi}\int_0^\xi \left[\frac{4c^2\gamma^2}{\varepsilon^2(1 + 3\tau^2)} i^{n+1} \sin \zeta J_{n+1}(\varepsilon \rho \sin \zeta) - g_n(\rho)\right] \frac{\rho^{n+2}}{(\xi^2 - \rho^2)^{1/2}}\, d\rho,$$

$$f_n(\rho) = \frac{1 - \tau^2}{1 + 3\tau^2}\int_0^\infty (1 + \gamma N(k)) q_n J_{n-1}(k\rho)\, dk,$$

$$g_n(\rho) = \frac{1 - \tau^2}{1 + 3\tau^2}\int_0^\infty (1 + \gamma N(k)) p_n J_{n+1}(k\rho)\, dk,$$

$$1 + \gamma N(k) = \frac{2kc^2}{\varepsilon^2(1 - \tau^2)}\left(\frac{k^2}{v_1} - v_2 - \frac{\beta^2}{v_2}\right).$$

Equations (5.26)–(5.27) can now be solved iteratively for small ε in a manner similar to that outlined in the previous subsection. The lateral displacement U_y of the disc is determined by the equation

$$m\gamma^2 U_y/u_0 = \frac{8\mu c}{c_2^2}\int_0^1 \theta_1(\xi)\, d\xi. \tag{5.28}$$

The amplitude U_y/u_0 can be shown to agree with (4.34) derived in Section 4; see also [65], Eq. (105).†

Here we have presented a way of solving the dual integral equations that arise in problems of elastic wave scattering by rigid circular discs. The problem of a penny-shaped crack can similarly be solved. Of course, there are alternative ways of reducing the dual equations to Fredholm equations of the second kind. As indicated before, a review of these would necessarily be long, and we hope to do this at a later date. Interested readers are referred to [67–71].

5.3 Far-field Scattering Amplitudes

In this subsection we present the scattered field far away from the disc when $\zeta = 0$. For arbitrary angle of incidence this can be obtained in a similar manner.

For $\zeta = 0$ the far-field amplitudes can be calculated by evaluating Φ_n and Ψ_n for large values of R. To do so note that when $\zeta = 0$ we need only to consider (5.7) with $n = 0$ and then

$$\Phi_n = -\frac{1}{\gamma^2}\int_0^\infty kP_n J_n(kr)\, e^{-\nu_1 z}\, dk, \qquad z > 0, \qquad n = 0, \tag{5.29}$$

and

$$\Psi_n = -\frac{1}{\gamma^2}\int_0^\infty \frac{k}{\nu_2} P_n J_n(kr)\, e^{-\nu_2 z}\, dk, \qquad z > 0, \qquad n = 0; \tag{5.30}$$

(5.19) and (5.20) can be asymptotically evaluated for large R by the method of steepest descent. For this purpose we write

$$J_n(kr) = -\frac{i}{\pi} e^{-in\pi/2}\int_0^\infty [e^{ikr\cosh t} - (-1)^n e^{-ikr\cosh t}] \cosh nt\, dt. \tag{5.31}$$

† There is a mistake in Eq. (105). The factor 16 should be replaced by 8 and the expression for $\mathscr{G}$ should read

$$\mathscr{G} = 1 - \frac{1}{\pi} M_1^{(1)} + k_2^2\left[\frac{8m_1^{(2)}}{3\pi(3+\gamma^2)} + (M_1^{(1)})^2/\pi^2 k_2^2\right].$$

Then (5.19) can be written as

$$\Phi_n = \frac{i}{\gamma^2 \pi} e^{-in\pi/2} \int_0^\infty \cosh nt \, dt \int_{-\infty}^{\infty} kP_n(k)\, e^{ikr\cosh t - \nu_1 z}\, dk, \qquad z > 0. \quad (5.32)$$

The integral $\int_{-\infty}^{\infty}$ is now evaluated by the method of steepest descent to be

$$\alpha^{3/2} \sin x \cos x P_n(\alpha \cos x) \sqrt{\frac{2\pi}{R_1}}\, e^{i\alpha R_1 - i\pi/4},$$

keeping only the dominant term. Here

$$R_1 = R\sqrt{\cos^2\theta + \sin^2\theta \cosh^2 t}, \qquad \tan x = \cot\theta \operatorname{sech} t.$$

Substitution of these in (5.22) gives

$$\Phi_n \sim i\sqrt{\frac{2}{\pi}}\, e^{in\pi/2}\, \alpha^{3/2} \int_0^\infty \sin x \cos x P_n(\alpha \cos x) R_1^{-1/2}\, e^{i\alpha R_1 - i\pi/4} \cosh nt \, dt. \quad (5.33)$$

For large R (5.23) gives

$$\Phi_n \sim \frac{i\alpha}{\gamma^2 R} e^{-in\pi/2} \cos\theta P_n(\alpha \sin\theta)\, e^{i\alpha R}. \quad (5.34)$$

Similarly

$$\Psi_n \sim -\frac{1}{\gamma^2 R} e^{-in\pi/2} \cdot P_n(\beta \sin\theta)\, e^{i\beta R}. \quad (5.35)$$

Using (5.24) and (5.25) in (5.7) for $n = 0$ we obtain the potentials $\Phi^{(s)}$ and $\Psi^{(s)}$. From (5.1) we then obtain

$$u_r \sim -\frac{1}{c_1^2 R} e^{i\alpha R} \sin\theta \cos\theta P_0(\alpha \sin\theta) + \frac{1}{c_2^2 R} e^{i\beta R} \sin\theta \cos\theta P_0(\beta \sin\theta), \quad (5.36)$$

and

$$u_z \sim -\frac{1}{c_1^2 R} e^{i\alpha R} \cos^2\theta P_0(\alpha \sin\theta) - \frac{1}{c_2^2 R} e^{i\beta R} \sin^2\theta P_0(\beta \sin\theta). \quad (5.37)$$

The first terms in these expressions represent a P-wave and the second a S-wave. It can be seen that these agree with (4.35)–(4.36) in the limiting case when the spheroid tends to a disc; $\xi_0 \to 0$. For in this limit a_0 and a_2 tend to zero, $V_j\rho' \to m_j$, $V_j \to 0$ and for large r', $h(r'_j) \sim -(1/r'_j)\, e^{ir'_j}$. Furthermore,

$$P_0(\alpha \sin \theta) \approx P_0(\beta \sin \theta) \approx \frac{2}{\pi} \lambda_0^{(0)} + 0(\varepsilon). \tag{5.38}$$

Acknowledgment. I express my thanks to Professors Y. H. Pao and A. K. Mal for their helpful comments. I am also grateful to Dr. R. D. Gregory for giving permission to reproduce some results from his papers for the material in Section 3. The partial support from the Solid Mechanics Program of the Engineering Division of the National Science Foundation under grant ENG 73-03525-A01 is gratefully acknowledged.

6 REFERENCES

1. Grant, F. S. and West, G. F., *Interpretation Theory in Applied Geophysics,* McGraw-Hill Book Company, New York (1965).
2. Lord Rayleigh, "Investigation of the Disturbance Produced by a Spherical Obstacle on the Waves of Sound," *Proc. London Math. Soc.* **4** (1872) 253–283.
3. Pao, Y. H. and Mow, C. C., *The Diffraction of Elastic Waves and Dynamic Stress Concentrations*, Crane & Russak, New York (1973).
4. Thau, S. A. and Pao, Y-H., "A Perturbation Method for Boundary Value Problems in Dynamic Elasticity," *Q. Appl. Math.* **25** (1967) 243–260.
5. Pao, Y-H. and Thau, S. A., "A Perturbation Method for Boundary Value Problems in Dynamic Elasticity, Part II," *Q. Appl. Math.* **28** (1970) 191–204.
6. Pao, Y-H. and Sachse, W., *Interpretation of Time Records and Power Spectra of Scattered Ultrasonic Pulses in Solids,* Report, Department of Theoretical and Applied Mechanics, Cornell University, Ithaca, N.Y. (1974).
7. Achenbach, J. D., Hermann, G. and Ziegler, F., "Separation at the Interface of a Circular Inclusion and the Surrounding Medium Under an Incident Compressive Wave," *J. Appl. Mech.* **37** (1970) 298–304.
8. Ko, W. L., "Scattering of Stress Waves by a Circular Elastic Cylinder Embedded in an Elastic Medium," *J. Appl. Mech.* **37** (1970) 345–355.
9. Griffin, J. H. and Miklowitz, J. H., "Wave Front Analysis of a Plane Compressional Pulse Scattered by a Circular Cylindrical Inclusion," *Int. J. Solids and Str.* **10** (1974) 1333–1356.
10. Bullen, K. E., *An Introduction to the Theory of Seismology*, 3rd Edition, Cambridge University Press (1963).
11. Bolt, B. A., "The Fine Structure of the Earth's Interior," *Scientific American* **228** (1973) 24–33.
12. Watson, G. N., "The Diffraction of Electric Waves by the Earth," *Proc. Roy. Soc. London* **A95** (1918) 83–99.
13. Duwalo, G. and Jacobs, J. A., "Effects of Liquid Core on the Propagation of Seismic Waves," *Can. J. Phys.* **37** (1959) 109–127.
14. Scholte, J. G. J., "On Seismic Waves in a Spherical Earth," *Koninkl. Ned. Meteorol. Inst. Publ.* **65** (1956) 1–55.
15. Phinney, R. A. and Alexander, S., "P Wave Diffraction Theory and the Structure of the Core-Mantle Boundary," *J. Geophys. Res.* **71** (1966) 5959–5975.
16. Phinney, R. A. and Cathles, L. M., "Diffraction of P by the Core: A Study of Long-

Period Amplitudes Near the Edge of the Shadow," *J. Geophys. Res.* **74** (1969) 1556–1574.
17. Teng, T. L. and Richards, P. G., "Diffracted P, SV, and SH Waves and Their Shadow Boundary Shifts," *J. Geophys. Res.* **74** (1969) 1537–1555.
18. Rice, S. O., "Diffraction of Plane Radio Waves by a Parabolic Cylinder," *Bell System Tech. J.* **33** (1954) 417–504.
19. Rubinow, S. I. and Keller, J. B., "Shift of the Shadow Boundary and Scattering Cross Section of an Opaque Object," *J. Appl. Phys.* **32** (1961) 814–820.
20. Nussenzveig, H. M., "High-frequency Scattering by an Impenetrable Sphere," *Ann. Phys.* **34** (1965) 23–95.
21. Datta, Subhendu, K., "Interaction of a Plane Compressional Elastic Wave with a Rigid Spheroidal Inclusion," *Q. Appl. Math.* **31** (1973) 217–235.
22. Mal, A. K., "Application of Continuum Mechanics," in *Applied Mechanics in Earthquake Engineering*, W. D. Iwan (ed.), The American Society of Mechanical Engineers, N.Y. (1974).
23. Knopoff, L., "Elastic Wave Propagation in a Wedge," in *Wave Propagation in Solids*, J. Miklowitz (ed.), The American Society of Mechanical Engineers, N.Y. (1969).
24. Drake, L. A., "Love and Rayleigh Waves in Nonhorizontally Layered Media," *Bull. Seismol. Soc. Amer.* **62** (1972) 1241–1257.
25. Lysmer, J. and Drake, L. A., "The Propagation of Love Waves Across Nonhorizontally Layered Structures," *Bull. Seismol. Soc. Amer.* **61** (1971) 1233–1251.
26. Hudson, J. A. and Knopoff, L., "Statistical Properties of Rayleigh Waves Due to Scattering by Topography," *Bull. Seismol. Soc. Amer.* **57** (1967) 83–90.
27. Hudson, J. A., "The Attenuation of Surface Waves by Scattering," *Proc. Camb. Philos. Soc.* **67** (1970) 215–223.
28. Gregory, R. D., "The Propagation of Waves in an Elastic Half Space Containing a Circular Cylindrical Cavity," *Proc. Camb. Philos. Soc.* **67** (1970) 689–710.
29. Gregory, R. D., "An Expansion Theorem Applicable to Problems of Wave Propagation in an Elastic Half-space Containing a Cavity," *Proc. Camb. Philos. Soc.* **63** (1967) 1341–1367.
30. Twersky, V., "Acoustic Bulk Parameters of Random Volume Distributions of Small Scatterers," *J. Acoust. Soc. Amer.* **36** (1964) 1314–1326.
31. Karal, F. C. and Keller, J. B., "Elastic, Electromagnetic, and Other Waves in a Random Medium," *J. Math. Phys.* **5** (1964) 537–547.
32. Knopoff, L. and Hudson, J. A., "Scattering of Elastic Waves by Small Inhomogeneities," *J. Acoust. Soc. Amer.* **36** (1964) 338–343.
33. Hudson, J. A., "The Scattering of Elastic Waves by Granular Media," *Quart. J. Mech. Appl. Math.* **21** (1968) 487–502.
34. Mal, A. K. and Knopoff, L., "Elastic Wave Velocities in Two-component Systems," *J. Inst. Math. Appl.* **3** (1967) 376–387.
35. Bose, S. K. and Mal, A. K., "Axial Shear Waves in a Medium with Randomly Distributed Cylinders," *J. Acoust. Soc. Amer.* **55** (1974) 519–523.
36. Bose, S. K. and Mal, A. K., "Longitudinal Shear Waves in Fiber-Reinforced Composite," *Int. J. Solids and Str.* **9** (1973) 1075–1085.
37. Bose, S. K. and Mal, A. K., "Elastic Waves in a Fiber-reinforced Composite," *J. Mech. Phys. Solids* **23** (1974) 217–229.
38. Mal, A. K. and Bose, S. K., "Dynamic Elastic Moduli of a Suspension of Imperfectly Bonded Spheres," *Proc. Camb. Philos. Soc.* **76** (1974) 587–800.
39. Garbin, H. D. and Knopoff, L., "The Compressional Modulus of a Material Permeated by a Random Distribution of Circular Cracks," *Q. Appl. Math.* **30** (1973) 453–464.
40. Datta, S. K., "Propagation of SH-waves Through a Fiber-reinforced Composite—Elliptic Cylindrical Fibers," *J. Appl. Mech.* **42** (1975) 165–170.
41. Datta, S. K., "Visco-elastic Properties of a Medium Permeated by Viscous Fluid-filled Cavities," *Proc. IUTAM Symposium on Visco-Elastic Media and Bodies* (1975).
42. Datta, S. K., "Diffraction of SH-waves in a Spherically Isotropic Medium by a Rigid or Fluid Spherical Core," *Geophys. J. Royal Astron. Soc.* **21** (1970) 33–46.

43. Nagase, M., "On the Zeros of Certain Transcendental Functions Related to Hankel Functions. Parts I and II," *J. Phys. Soc. Japan* **9** (1954) 826–853.
44. Keller, J. B., Rubinow, S. I. and Goldstein, M., "Zeros of Hankel Functions and Poles of Scattering Amplitude," *J. Math. Phys.* **4** (1963) 829–832.
45. Knopoff, L. and Gilbert, F., "Diffraction of Elastic Waves by the Core of the Earth," *Bull. Seismol. Soc. Amer.* **51** (1961) 35–49.
46. Rubinow, S. I., "Scattering from a Penetrable Sphere at Short Wavelength," *Ann. Phys.* **14** (1961) 305–332.
47. Chen, Y. M., "Diffraction by a Smooth Transparent Object," *J. Math. Phys.* **5** (1964) 820–832.
48. Lamb, H., "On the Propagation of Tremors Over the Surface of an Elastic Solid," *Phil. Trans. Roy. Soc. London* **A203** (1904) 1–42.
49. Ben-Menahem, A. and Cisternas, A., "The Dynamic Response of an Elastic Half-space to an Explosion in a Spherical Cavity," *J. Math. Phys.* **42** (1963) 122–125.
50. Thiruvenkatachar, V. R. and Viswanathan, K., "Dynamic Response of an Elastic Half-space with Cylindrical Cavity to Time Dependent Surface Tractions Over the Boundary of the Cavity," *J. Math. Mech.* **14** (1965) 541–571.
51. Thiruvenkatachar, V. R. and Viswanathan, K., "Dynamic Response of an Elastic Half-space to Time-dependent Surface Tractions Over an Embedded Spherical Cavity," *Proc. Roy. Soc.* A **287** (1965) 549–567.
52. Datta, S. K., "Diffraction of SH-waves by an Elliptic Elastic Cylinder," *Int. J. Solids and Str.* **10** (1974) 123–133.
53. Datta, S. K. and Sangster, J. D., "Response of a Rigid Spheroidal Inclusion to an Incident Plane Compressional Elastic Wave," *SIAM J. Appl. Math.* **26** (1974) 350–369.
54. Herrera, I., "Contributions to the Linearized Theory of Surface Wave Transmission," *J. Geophys. Res.* **69** (1964) 4791–4800.
55. Sangster, J. D., *The Scattering of Plane Elastic Waves by Rigid Spheroidal Inclusion*, Ph.D. Thesis, Department of Mechanical Engineering, University of Colorado, Boulder (1974).
56. Lesser, M. B. and Crighton, D. G., "Physical Acoustics and the Method of Matched Asymptotic Expansions," *Physical Acoustics*, W. P. Mason and R. N. Thurson (eds.), Academic Press (1975).
57. Cole, J. D., *Perturbation Methods in Applied Mathematics*, Ginn/Blaisdell, Mass (1968).
58. Sokolnikoff, I. S., *Mathematical Theory of Elasticity*, McGraw-Hill, New York (1956).
59. Lapwood, E. R., "The Disturbance Due to a Line Source in a Semi-Infinite Elastic Medium," *Phil. Trans. Roy. Soc. London* A **242** (1949) 63–100.
60. Oien, M. A. and Pao, Y. H., "Scattering of Compressional Waves by a Rigid Spheroidal Inclusion," *J. Appl. Mech.* **40** (1973) 1073–1077.
61. Stratton, J. A., *Electromagnetic Theory*, McGraw-Hill, New York (1941).
62. Mal, A. K., "Motion of a Rigid Disc in an Elastic Solid," *Bull. Seismol. Soc. Amer.* **61** (1971) 1717–1729.
63. Hashin, Z., "The Elastic Moduli of Heterogeneous Materials," *J. Appl. Mech.* **29** (1962) 143–150.
64. Datta, S. K., "Scattering of Elastic Waves by a Distribution of Inclusions," Presented at the *Symposium on Continuum Model of Discrete Systems*, Poland, June 22–28, 1975. *Arch. Mech.* **28** (1976) 317–324.
65. Datta, S. K., "The Diffraction of a Plane Compressional Elastic Wave by a Rigid Circular Disc," *Q. Appl. Math.* **28** (1970) 1–14.
66. Noble, B., "The Solution of Bessel Function Dual Integral Equation by Multiplying Factor Method," *Proc. Camb. Philos. Soc.* **59** (1963) 351–362.
67. Jain, D. L. and Kanwal, R. P., "An Integral Equation Method for Solving Mixed Boundary Value Problems," *SIAM J. Appl. Math.* **20** (1971) 642–657.
68. Mal, A. K., "Dynamic Stress Intensity Factors for a Non-axisymmetric Loading of the Penny-shaped Crack," *Int. J. Engng Sci.* **6** (1968) 725–733.

69. Sneddon, I. N., *Fourier Transforms*, McGraw-Hill, New York (1956).
70. Williams, W. E., "The Reduction of Boundary Value Problems to Fredholm Equations of the Second Kind," *ZAMP* **13** (1962) 133–152.
71. Noble, B., *Electromagnetic Waves*, R. E. Langer (ed.), Univ. of Wisconsin Press, Madison (1962).

APPENDIX A

The displacement components u_r, u_θ in terms of the potentials Φ and Ψ are given by

$$u_r = \frac{\partial \Phi}{\partial r} - \frac{1}{r}\frac{\partial}{\partial \eta}\left[(1-\eta^2)\frac{\partial \Psi}{\partial \eta}\right],$$
$$u_\theta = \frac{1}{r}\frac{\partial \Phi}{\partial \theta} + \frac{1}{r}\frac{\partial}{\partial r}\left(r\frac{\partial \Psi}{\partial \theta}\right), \tag{A.1}$$

and the stress components τ_{rr}, $\tau_{r\theta}$ are expressed as

$$\frac{\tau_{rr}}{2\mu} = -\frac{\alpha^2 \sigma}{1-2\sigma}\Phi + \frac{\partial^2 \Phi}{\partial r^2} - \frac{\partial}{\partial \eta}\left[(1-\eta^2)\frac{\partial^2}{\partial r\,\partial \eta}\left(\frac{1}{r}\Psi\right)\right],$$
$$\frac{\tau_{r\theta}}{\mu} = 2\left[\frac{1}{r}\frac{\partial^2 \Phi}{\partial r\,\partial \theta} - \frac{1}{r^2}\frac{\partial \Phi}{\partial \theta}\right] - \frac{2}{r^2}\frac{\partial \Psi}{\partial \theta}$$
$$-\frac{1}{r^2}\frac{\partial}{\partial \theta}\left[\frac{1}{\sin\theta}\frac{\partial}{\partial \theta}\left(\sin\theta\frac{\partial \Psi}{\partial \theta}\right)\right] + \frac{\partial^3 \Psi}{\partial r^2\,\partial \theta}. \tag{A.2}$$

The boundary conditions at $r = a$ are

$$u_r = u_r', \qquad \tau_{rr} = \tau_{rr}', \qquad \tau_{r\theta} = 0. \tag{A.3}$$

Using (A.3) we determine the constants a_n, b_n, a_n' in the following manner. We write

$$a_n = i^n(2n+1)\Phi_0 A_n, \qquad b_n = i^n(2n+1)\Phi_0 B_n, \qquad a_n' = i^n(2n+1)\Phi_0 A_n'. \tag{A.4}$$

Thus, bearing in mind the ray theory description of the different waves, we express A_n, B_n, A_n' in the forms (2.14)–(2.16) with

$$\kappa_1 R_{11} = \frac{\Phi_n^{(1)}}{\Delta_n^{(1)}}, \qquad \kappa_2 R_{11}' = \frac{\Psi_n^{(1)}}{\Delta_n^{(1)}}, \qquad \kappa_3 R_{12} = \frac{\Phi_n^{(1)\prime}}{\Delta_n^{(1)}},$$
$$\kappa_4 R_{22} = -\frac{\Delta_n^{(2)}}{\Delta_n^{(1)}}, \qquad \frac{\kappa_1 \kappa_4}{\kappa_3} R_{21} = \frac{\Phi_n^{(2)}}{\Delta_n^{(1)}}, \tag{A.5}$$
$$\frac{\kappa_2 \kappa_4}{\kappa_3} R_{21}' = \frac{\Psi_n^{(2)}}{\Delta_n^{(1)}},$$

$$
\Delta_n^{(1)} = \begin{vmatrix} h_n^{(1)'}(\alpha a)/\alpha & \dfrac{n(n+1)}{\beta^2 a} h_n^{(1)}(\beta a) & \dfrac{-h_n^{(2)'}(\alpha' a)}{\alpha'} \\ h_n^{(1)''}(\alpha a) - \dfrac{\sigma}{1-2\sigma} h_n^{(1)}(\alpha a) & n(n+1)\left(\dfrac{1}{\beta a} h_n^{(1)}(\beta a)\right)' & \dfrac{\lambda'}{2\mu} h_n^{(2)}(\alpha' a) \\ 2\left(\dfrac{1}{\alpha a} h_n^{(1)}(\alpha a)\right)' & h_n^{(1)''}(\beta a) + \dfrac{(n+2)(n-1)}{\beta^2 a^2} h_n^{(1)}(\beta a) & 0 \end{vmatrix}
$$

$$
\Phi_n^{(1)''} = -\begin{vmatrix} h_n^{(2)'}(\alpha a)/\alpha & \dfrac{n(n+1)}{\beta^2 a} h_n^{(1)}(\beta a) & \dfrac{-h_n^{(2)'}(\alpha' a)}{\alpha'} \\ h_n^{(2)''}(\alpha a) - \dfrac{\sigma}{1-2\sigma} h_n^{(2)}(\alpha a) & n(n+1)\left(\dfrac{1}{\beta a} h_n^{(1)}(\beta a)\right)' & \dfrac{\lambda'}{2\mu} h_n^{(2)}(\alpha' a) \\ 2\left(\dfrac{1}{\alpha a} h_n^{(2)}(\alpha a)\right)' & h_n^{(1)''}(\beta a) + \dfrac{(n+2)(n-1)}{\beta^2 a^2} h_n^{(1)}(\beta a) & 0 \end{vmatrix}
$$

$$
\Phi_n^{(2)} = \begin{vmatrix} h_n^{(1)'}(\alpha' a)/\alpha' & \dfrac{n(n+1)}{\beta^2 a} h_n^{(1)}(\beta a) & \dfrac{-h_n^{(2)'}(\alpha' a)}{\alpha'} \\ -\dfrac{\lambda'}{2\mu} h_n^{(1)}(\alpha' a) & n(n+1)\left(\dfrac{1}{\beta a} h_n^{(1)}(\beta a)\right)' & \dfrac{\lambda'}{2\mu} h_n^{(2)}(\alpha' a) \\ 0 & h_n^{(1)''}(\beta a) + \dfrac{(n+2)(n-1)}{\beta^2 a^2} h_n^{(1)}(\beta a) & 0 \end{vmatrix} \dfrac{\alpha'^2}{\alpha^2},
$$

$$
\Phi_n^{(1)'} = -\begin{vmatrix} h_n^{(1)'}(\alpha a)/\alpha & \dfrac{n(n+1)}{\beta^2 a} h_n^{(1)}(\beta a) & h_n^{(2)'}(\alpha a)/\alpha \\ h_n^{(1)''}(\alpha a) - \dfrac{\sigma}{1-2\sigma} h_n^{(1)}(\alpha a) & n(n+1)\left(\dfrac{1}{\beta a} h_n^{(1)}(\beta a)\right)' & h_n^{(2)''}(\alpha a) - \dfrac{\sigma}{1-2\sigma} h_n^{(2)}(\alpha a) \\ 2\left(\dfrac{1}{\alpha a} h_n^{(1)}(\alpha a)\right)' & h_n^{(1)''}(\beta a) + \dfrac{(n+2)(n-1)}{\beta^2 a^2} h_n^{(1)}(\beta a) & 2\left(\dfrac{1}{\alpha a} h_n^{(2)}(\alpha a)\right)' \end{vmatrix} \dfrac{\alpha^2}{\alpha'^2},
$$

$$
\Psi_n^{(1)} = -\begin{vmatrix} h_n^{(1)'}(\alpha a)/\alpha & h_n^{(2)'}(\alpha a)/\alpha & -h_n^{(2)'}(\alpha' a)/\alpha \\ h_n^{(1)''}(\alpha a) - \dfrac{\sigma}{1-2\sigma} h_n^{(1)}(\alpha a) & h_n^{(2)''}(\alpha a) - \dfrac{\sigma}{1-2\sigma} h_n^{(2)}(\alpha a) & \dfrac{\lambda'}{2\mu} h_n^{(2)}(\alpha' a) \\ 2\left(\dfrac{1}{\alpha a} h_n^{(1)}(\alpha a)\right)' & 2\left(\dfrac{1}{\alpha a} h_n^{(2)}(\alpha a)\right)' & 0 \end{vmatrix} \dfrac{\alpha^2}{\beta^2},
$$

$$
\Psi_n^{(2)} = \begin{vmatrix} h_n^{(1)'}(\alpha a)/\alpha & h_n^{(1)'}(\alpha' a)/\alpha' & -h_n^{(2)'}(\alpha' a)/\alpha \\ h_n^{(1)''}(\alpha a) - \dfrac{\sigma}{1-2\sigma} h_n^{(1)}(\alpha a) & -\dfrac{\lambda'}{2\mu} h_n^{(1)}(\alpha' a) & \dfrac{\lambda'}{2\mu} h_n^{(2)}(\alpha' a) \\ 2\left(\dfrac{1}{\alpha a} h_n^{(1)}(\alpha a)\right)' & 0 & 0 \end{vmatrix} \dfrac{\alpha'^2}{\beta^2}; \tag{A.6}
$$

$\Delta_n^{(2)}$ is obtained from $\Delta_n^{(1)}$ on replacing $h_n^{(2)}(\alpha' a)$ appearing in its last column by $h_n^{(1)}(\alpha' a)$. Note that

$$
\begin{aligned} \kappa_1 &= h_n^{(2)}(\alpha a)/h_n^{(1)}(\alpha a), \qquad & \kappa_2 &= h_n^{(2)}(\alpha a)/h_n^{(1)}(\beta a), \\ \kappa_3 &= h_n^{(2)}(\alpha a)/h_n^{(2)}(\alpha' a), \qquad & \kappa_4 &= h_n^{(1)}(\alpha' a)/h_n^{(2)}(\alpha' a). \end{aligned} \tag{A.7}
$$

In order to write (2.24) in the form of a residue series it is necessary to find the zeros of $\Delta^{(2)}_{\nu-1/2}$ qua-function of ν. Of greatest interest are the zeros with positive imaginary parts in the neighborhood of the points $\nu = \alpha a$, βa, or $\alpha' a$. These can be found in a manner similar to that of [43, 44]. For the zeros in the neighborhood of $\nu = \alpha a$ (it will be assumed that αa is large) we shall use the Debye expansions of the Bessel functions $H^{(1)}_\nu(\beta a)$, $H^{(2)}_\nu(\alpha' a)$ as

$$H^{(1)}_\nu(\beta a) \sim (2/\pi)^{1/2}(\beta^2 a^2 - \nu^2)^{-1/4} \times \exp\left[i\left\{(\beta^2 a^2 - \nu^2)^{1/2} - \nu \cos^{-1}\frac{\nu}{\beta a} - \pi/4\right\}\right],$$

$$H^{(2)}_\nu(\alpha' a) \sim (2/\pi)^{1/2}(\alpha'^2 a^2 - \nu^2)^{-1/4} \times \exp\left[-i\left\{(\alpha'^2 a^2 - \nu^2)^{1/2} - \nu \cos^{-1}\frac{\nu}{\alpha' a} - \pi/4\right\}\right]. \tag{A.8}$$

These expansions fail for $H^{(1)}_\nu(\alpha a)$, $H^{(2)}_\nu(\alpha a)$ in the neighborhood of $\nu = \alpha a$. In this region we have

$$H^{(1,2)}_\nu[\nu - (\nu/2)^{1/3} e^{i\pi/3} x] \sim 2e^{\mp i\pi/3}\left(\frac{2}{\nu}\right)^{1/3}\{Ai(-x), Ai(x\, e^{-i\pi/3})\}. \tag{A.9}$$

Substitution of these in the expression for $\Delta_{\nu-1/2}$ gives, after some algebra,

$$\Delta^{(1)}_{\nu-1/2} = (\pi/2)^{1/2}(2/\nu)^{1/3}\frac{2}{\alpha}e^{-2i\pi/3} Ai'(-x)(\alpha'^2 a^2 - \nu^2)^{-1/4}(\beta^2 a^2 - \nu^2)^{-1/4}$$
$$\times\left\{\frac{Ai(-x)}{Ai'(-x)}e^{-i\pi/3}\frac{-i\alpha}{\alpha'}\frac{\sigma}{1-2\sigma}\frac{(2\nu^2 - \beta^2 a^2)(\alpha'^2 a^2 - \nu^2)^{1/2}}{\alpha'\beta^2 a^3}\right.$$
$$\left.+\left(\frac{2}{\nu}\right)^{1/3}\left[\frac{\lambda'}{2\mu} + \frac{2\nu^2(\alpha'^2 a^2 - \nu^2)^{1/2}(\beta^2 a^2 - \nu^2)^{1/2}}{\alpha'^2\beta^2 a^4}\right]\right\}(\alpha\alpha'\beta a^3)^{-1/2}$$
$$\times\exp\left[-i\left\{(\alpha'^2 a^2 - \nu^2)^{1/2} - (\beta^2 a^2 - \nu^2)^{1/2} - \nu\cos^{-1}\frac{\nu}{\alpha' a} + \nu\cos^{-1}\frac{\nu}{\beta a}\right\}\right].$$

Thus the zeros are given by (2.25) with x_n being a solution of the equation

$$\frac{Ai(-x)}{Ai'(-x)} = e^{-i\pi/6}\left(\frac{2}{\nu}\right)^{1/3}\frac{\lambda'/2\mu + 2\nu^2(\alpha'^2 a^2 - \nu^2)^{1/2}(\beta^2 a^2 - \nu^2)^{1/2}/\alpha'^2\beta^2 a^4}{\dfrac{\sigma}{1-2\sigma}\dfrac{\alpha}{\alpha'}(2\nu^2 - \beta^2 a^2)(\alpha'^2 a^2 - \nu^2)^{1/2}/\alpha'\beta^2 a^3}. \tag{A.10}$$

For large v, x_n is approximately a zero of $Ai(-x)$ with an error of $0(v^{-1/3})$. In a similar manner we can show that the zeros in the neighborhood of $v = \beta a$ and $\alpha' a$ are

$$\begin{aligned} v_n &= \beta a + (\beta a/2)^{1/3} x_n e^{i\pi/3}, \\ v_n &= \alpha' a + (\alpha' a/2)^{1/3} x_n e^{i\pi/3}, \end{aligned} \tag{A.11}$$

respectively. However, it is easily shown that these last roots are not the poles of $\kappa_1 R_{11}$. The residues at the poles (2.25) are

$$\Phi^{(1)}_{v-1/2}/\partial[\Delta^{(1)}_{v-1/2}]/\partial v\,|_{v=v_n} = \left(\frac{v_n}{2}\right)^{1/3} \frac{e^{-i\pi/6}}{2\pi[Ai'(-x_n)]^2}. \tag{A.12}$$

In a similar manner we can show that the poles of $\kappa_2 R'_{11}$ are also given by (2.25) and the residues are

$$\begin{aligned} \Psi^{(1)}_{v-1/2}/\partial[\Delta^{(1)}_{v-1/2}]/\partial v\,|_{v=v_n} &= \left(\frac{\pi}{2}\right)^{1/2} (2/v_n)^{1/3} e^{i\pi/3} \frac{2\alpha a(\beta^2 a^2 - v_n^2)^{1/4}}{2v_n^2 - \beta^2 a^2} \\ &\times \frac{e^{-i\pi/6}}{2\pi Ai'(-x_n)} \exp\left[-i\left\{(\beta^2 a^2 - v^2)^{1/2} - v\cos^{-1}\frac{v}{\beta a} - \pi/4\right\}\right]. \end{aligned} \tag{A.13}$$

APPENDIX B

To find $\{\Phi^p_n, \Psi^p_n\}$ we assume

$$\Phi^p_n = H_n(\alpha r)\cos n\theta - H_n(\alpha r')\cos n\theta' + \int_{-\infty}^{\infty+\pi i} A(t)\, e^{y\sinh t}\cos(\alpha x\cosh t)\, dt. \tag{B.1}$$

The contour is shown in Fig. 10. (r, θ), (r', θ') are defined in Fig. 5. Now $H_n(\alpha r)$ may be written as

$$= \frac{1}{\pi i}\int_{-\infty}^{\infty+\pi i} \exp[\alpha r\sinh(t + i\theta) - n(t + i\theta)]\, dt, \qquad -\pi/2 < \theta < \pi/2. \tag{B.2}$$

Thus

$$H_n(\alpha r)\cos n\theta = \frac{1}{\pi i}\int_{-\infty}^{\infty+\pi i} e^{\alpha(h-y)\sinh t - nt}\cos(\alpha x\cosh t)\, dt. \tag{B.3}$$

Similarly,

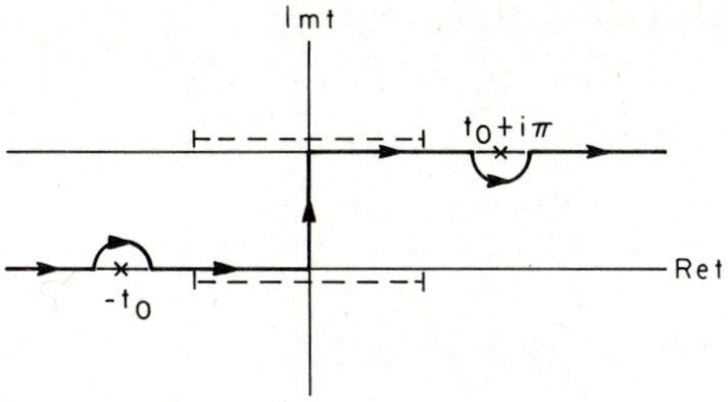

Fig. 10 Contour for the integrals in (B.1) and (B.5).

$$H_n(\alpha r') \cos n\theta' = \frac{1}{\pi i} \int_{-\infty}^{\infty + \pi i} e^{\alpha(h+y)\sinh t - nt} \cos(\alpha x \cosh t)\, dt. \qquad \text{(B.4)}$$

We also assume

$$\Psi_n^p = \int_{-\infty}^{\infty + \pi i} B(t)\, e^{-\Delta(t)y} \sin(\beta x \cosh t)\, dt, \qquad \text{(B.5)}$$

where $\Delta(t) = (\alpha^2 \cosh^2 t - \beta^2)^{1/2}$, and where the square root is chosen so that $\text{Re}\, \Delta(t) \geq 0$ and the cuts are shown in Fig. 10. Using (B.1), (B.5) together with (B.3), (B.4) in the boundary conditions (3.4)–(3.5) we obtain $A(t)$, $B(t)$ as

$$A(t) = -\frac{8i}{\pi} \frac{\alpha^3 \Delta \cosh^2 t \sinh t \exp[\alpha h \sinh t - nt]}{f(\alpha \cosh t)}, \qquad \text{(B.6)}$$

and

$$B(t) = \frac{4i}{\pi} \frac{\beta^2 \cosh t \sinh t (2\alpha^2 \cosh^2 t - \beta^2) \exp[\alpha h \sinh t - nt]}{f(\alpha \cosh t)} \qquad \text{(B.7)}$$

with

$$f(x) = (2x^2 - \beta^2)^2 - 4x^2(x^2 - \alpha^2)^{1/2}(x^2 - \beta^2)^{1/2}$$

being the Rayleigh Determinant. The square roots are taken so that $\text{Re}\,(x^2 - \alpha^2)^{1/2} \geq 0$, $\text{Re}\,(x^2 - \beta^2)^{1/2} \geq 0$. The zeros of $f(x)$ are $\pm x_0$, where $x_0 > \beta$. So the expressions for Φ_n^p, Ψ_n^p are given by (B.1), (B.5) together with (B.6), (B.7). Similarly one obtains

$$\Phi^s = \int_{c_2} -\frac{4i}{\pi} \frac{\beta^2 \cosh s \sinh s (2\beta^2 \cosh^2 s - \beta^2) \exp(\beta h \sinh s - ns)}{f(\beta \cosh s)} \times e^{-\Gamma(s)y} \cos(\beta x \cosh s)\, ds, \qquad \text{(B.8)}$$

$$\Psi_n^s = H_n(\beta r)\sin n\theta + H_n(\beta r')\sin n\theta' + \int_{c_2} \frac{8i}{\pi} \frac{\beta^2 \cosh s \sinh s\Gamma(s) \exp(\beta h \sinh s - ns)}{f(\beta \cosh s)} \times e^{\beta y \sinh s} \sin(\beta x \cosh s)\, ds, \tag{B.9}$$

with

$$\Gamma(s) = (\beta^2 \cosh^2 s - \alpha^2)^{1/2},$$

where c_2 is shown in Fig. 11.

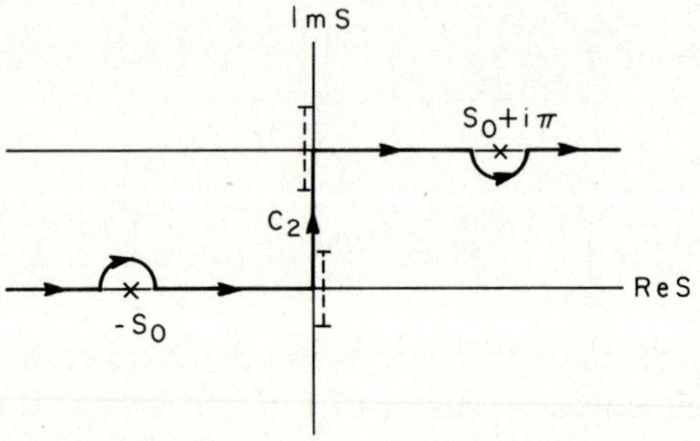

Fig. 11 Contour for the integral in (B.8).

We shall now expand E_n, Ψ_n^p, etc., near $\bar{r} = 1$. To do so we note

$$H_n(\alpha r')\, e^{-in\theta'} = \sum_{m=-\infty}^{\infty} H_{n+m}(2\alpha h) J_m(\alpha r)\, e^{im\theta}, \qquad r < 2h, \tag{B.10}$$

$$e^{\alpha(y \sinh t - ix \cosh t)} = e^{\alpha h \sinh t} \sum_{m=-\infty}^{\infty} e^{-mt} J_m(\alpha r)\, e^{im\theta}. \tag{B.11}$$

Using these in (B.1) we get

$$P_{mn} = -\varepsilon_m (H_{n+m}(2\alpha h) + H_{n-m}(2\alpha h)(-1)^m) + \varepsilon_m \int_{c_1} \frac{(-8i/\pi)\alpha^3 \Delta \cosh^2 t \sinh t \exp(2\alpha h \sinh t)}{f(\alpha \cosh t)} \times e^{-nt}\left((-1)^m e^{mt} + e^{-mt}\right) dt, \tag{B.12}$$

where

$$\varepsilon_m = \begin{cases} \frac{1}{2}, & m = 0. \\ 1, & m \neq 0. \end{cases}$$

Similar expressions can be obtained for Q_{mn}, etc.

APPENDIX C

The general solution to (4.18) can be written as (see [58])

$$\mathbf{v} = \overline{\nabla} F + y\overline{\nabla} Y + z\overline{\nabla} Z - (3 - 4\sigma)(Y\mathbf{e}_y + Z\mathbf{e}_z), \tag{C.1}$$

where F, Y, Z are the solutions of Laplaces' equation. The use of the boundary conditions determines the choice of these solutions. In view of (4.19) and the matching condition we find

$$\begin{aligned} F = F_0 &= A_{01}Q_1^0(i\xi)P_1^0(\eta)\cos\zeta + A_{11}Q_1^1P_1^1\sin\omega\sin\zeta, \\ Z = Z_0 &= C_{00}Q_0^0\cos\zeta, \qquad Y = Y_0 = B_{00}Q_0^0\sin\zeta, \end{aligned} \tag{C.2}$$

with

$$\begin{aligned} A_{01} &= -(\xi_0 - (3 - 4\sigma)C_{00}\xi_0 Q_0^0(i\xi_0))/Q_1^0, \\ A_{11} &= -(\xi_0^2 + 1)^{1/2}(1 - (3 - 4\sigma)B_{00}Q_0^0)/Q_1^1, \\ B_{00} &= [(\xi_0^2 + 1)Q_1^{1\prime}/Q_1^1 - \xi_0]/[(\xi_0^2 + 1)(Q_0^{0\prime} + (3 - 4\sigma)Q_0^0Q_1^{1\prime}/Q_1^1) \\ &\quad - (3 - 4\sigma)\xi_0 Q_0^0], \\ C_{00} &= (\xi_0 Q_1^{0\prime}/Q_1^0 - 1)/[\xi_0(Q_0^{0\prime} + (3 - 4\sigma)Q_0^0Q_1^{0\prime}/Q_1^0) - (3 - 4\sigma)Q_0^0]. \end{aligned}$$

Here $Q_n^{m\prime} = \partial Q_n^m/\partial\xi\,|_{\xi=\xi_0}$, and $Q_n^m(i\xi)$, $P_n^m(\eta)$ are Legendre functions of the second and first kind, respectively.

The solution to (4.22) can be written as (see [61])

$$\mathbf{u}' = \nabla'\Phi + \tau^2 r'\Psi\mathbf{e}_r + \nabla'\left(\frac{\partial}{\partial r'}(r'\Psi)\right) + \tau\nabla' \wedge (r'x\mathbf{e}_r), \tag{C.3}$$

where Φ, Ψ, x satisfy the equations

$$(\nabla'^2 + 1)\Phi = 0, \qquad (\nabla'^2 + \tau^2)\left[{\Psi \atop \psi}\right] = 0,$$

and hence they can be written as

$$\begin{aligned} \Phi &= \sum_{n=0}^{\infty}\sum_{m=-n}^{n}[a_{mn}^1 Y_{mn}^e(\theta,\omega) + a_{mn}^2 Y_{mn}^0(\theta,\omega)]h_n(r'), \\ \Psi &= \sum_{n=0}^{\infty}\sum_{m=-n}^{n}[b_{mn}^1 Y_{mn}^e + b_{mn}^2 Y_{mn}^0]h_n(\tau r'), \end{aligned} \tag{C.4}$$

where χ has a similar expansion as Ψ. Here

$$Y^e_{mn} = P^m_n(\cos\theta)\cos m\omega, \qquad Y^0_{mn} = P^m_n(\cos\theta)\sin m\omega.$$

The coefficients $\mathscr{A}_{12}$, $\mathscr{B}_{01}$, and $\mathscr{C}_{11}$ appearing in (4.29) are given by

$$\mathscr{A}_{12} = \frac{2(2\sigma-1)\xi_0}{3Q_2^1}\left[\frac{\sqrt{\xi_0^2+1}}{4(\sigma-1)} - \mathscr{C}_{11}\right],$$

$$\mathscr{B}_{01} = \frac{\xi_0}{Q_1^0}\left[\frac{Q_1^1}{\sqrt{\xi_0^2+1}}\mathscr{C}_{11} - \frac{1}{2(\sigma-1)}\right], \tag{C.5}$$

$$\begin{aligned}\mathscr{C}_{11} = \Bigg[&(\xi_0^2+1)^{-1/2} - \frac{2\sigma-1}{2(\sigma-1)}\xi_0(\xi_0^2+1)^{1/2}Q_2^1/Q_2^1 + \frac{\xi_0}{2(\sigma-1)Q_1^0} \\ &\times\left\{(\xi_0^2+1)^{1/2}Q_1^{0\prime} + (4\sigma-3)\xi_0 Q_1^0/\sqrt{\xi_0^2+1}\right\}\Bigg] \\ &\div [\xi_0 Q_1^{1\prime} - (3-4\sigma)Q_1^1 + \xi_0 Q_1^1 Q_1^{0\prime}/Q_1^0 - (3-4\sigma)\xi_0^2 Q_1^1 \\ &\div (\xi_0^2+1) + 2(1-2\sigma)\xi_0 Q_1^1 Q_2^{1\prime}/Q_2^1].\end{aligned}$$

The expressions for F_y, F_z, etc., are

$$F_y = -16\pi\mu ic(1-\sigma)\left[B_{00} + \varepsilon\hat{B}_{00} + \varepsilon^2\left\{\hat{\hat{B}}_{00} + {}^2B_{00}^1 + \frac{\xi_0^2+3}{6}\tau^2 B_{00}\right\}\right] \times \sin\zeta,$$

$$\begin{aligned}F_y^{(t)} = 16\pi\mu ic(1-\sigma)\Bigg[&B_{00} + \varepsilon\hat{B}_{00} + \varepsilon^2\Bigg\{\hat{\hat{B}}_{00} - {}^2\overline{B}_{00}^1 + \frac{\xi_0^2+3}{6}\tau^2 B_{00} \\ &+ \frac{i\xi_0(\xi_0^2+1)\tau^2}{12(1-\sigma)}\Bigg\}\Bigg],\end{aligned} \tag{C.6}$$

$$F_z = -16\pi\mu ic(1-\sigma)\left[C_{00} + \varepsilon\hat{C}_{00} + \varepsilon^2\left\{\hat{\hat{C}}_{00} + {}^2\overline{C}_{00}^1 + \frac{\xi_0^2-1}{6}\tau^2 C_{00}\right\}\right]\cos\zeta,$$

$$\begin{aligned}F_z^{(t)} = 16\pi\mu ic(1-\sigma)\Bigg[&C_{00} + \varepsilon\hat{C}_{00} + \varepsilon^2\Bigg\{\hat{\hat{C}}_{00} - {}^2\overline{C}_{00}^1 + \frac{\xi_0^2-1}{6}\tau^2 C_{00} \\ &+ i(\xi_0+1)\xi_0\tau^2/12(1-\sigma)\Bigg\}\Bigg],\end{aligned} \tag{C.7}$$

$$\begin{aligned}T_x &= -16\pi\mu c\varepsilon(1-\sigma)(B_{01} + 2C_{11})/3, \\ T_x^{(\mathrm{rot})} &= -16\pi\mu c(1-\sigma)(\mathscr{C}_{01} + 2\mathscr{C}_{11})/3.\end{aligned} \tag{C.8}$$

Here

$$\hat{B}_{00} = \frac{4(1+2\tau^3)}{3\tau^2}(1-\sigma)B_{00}, \qquad \hat{C}_{00} = \frac{4(1+2\tau^3)}{3\tau^2}(1-\sigma)C_{00},$$

$$\hat{\hat{B}}_{00} = \frac{16(1+2\tau^3)^2}{9\tau^4}(1-\sigma)^2 B_{00}, \hat{\hat{C}}_{00} = \frac{16(1+2\tau^3)^2}{9\tau^4}(1-\sigma)^2 C_{00}.$$

The other constants ${}^2B_{00}^1$, $2\overline{B}_{00}^1$, ${}^2C_{00}^1$, ${}^2\overline{C}_{00}^1$, B_{01}, and C_{11} are rather lengthy in expression and can be found in [55].

The coefficients a_0, a_2 are given by

$$a_0 = \frac{i}{\Delta}\left[\frac{4\sigma-5}{3}(2Q_2^0 - \xi_0 Q_2^{0\prime})/(\xi_0^2+1) - \xi_0 Q_2^{0\prime}(\xi_0 Q_1^{0\prime} - Q_1^0)\right], \quad \text{(C.9)}$$

$$a_2 = \frac{i}{\Delta}\tfrac{4}{3}(1-2\sigma)(2Q_2^0 - \xi_0 Q_2^{0\prime})/(\xi_0^2+1), \quad \text{(C.10)}$$

with

$$\Delta = 2[(2\sigma-1)\xi_0 Q_1^0 Q_2^{0\prime} - \xi_0 Q_1^{0\prime} Q_2^0 + (3-4\sigma)Q_1^0 Q_2^0];$$

$\bar{a}_0$, $\bar{a}_2$, and $\bar{a}_4$ are given by

$$\bar{a}_0 = \frac{i}{\xi_0}\left[B_1\left\{\sqrt{\xi_0^2+1}\,Q_1^{1\prime} - (3-4\sigma)\frac{\xi_0}{\sqrt{\xi_0^2+1}}Q_1^1 - \frac{2}{3}\frac{4\sigma-5}{\sqrt{\xi_0^2+1}}\right\}\right.$$
$$\left. + C_1\left\{\xi_0 Q_1^{0\prime} - (3-4)Q_1^0 + \frac{1}{3\sqrt{\xi_0^2+1}}\right\} + \frac{\xi_0}{\sqrt{\xi_0^2+1}}\right], \quad \text{(C.11)}$$

$$\bar{a}_2 = \frac{i}{\xi_0}\frac{4(1-2\sigma)}{3\sqrt{\xi_0^2+1}}(B_1 + C_1), \quad \text{(C.12)}$$

$$\bar{a}_4 = \frac{i}{\xi_0}\frac{2(1-2\sigma)}{3\sqrt{\xi_0^2+1}}B_1, \quad \text{(C.13)}$$

where

$$B_1 = \frac{2Q_2^2\xi_0 - (\xi_0^2+1)Q_2^{2\prime}}{\Delta_1},$$

$$\Delta_1 = (4\sigma-2)(\xi_0^2+1)Q_1^1 Q_2^{2\prime} - 2(\xi_0^2+1)Q_1^{1\prime}Q_2^2 + 2(3-4\sigma)\xi_0 Q_1^1 Q_2^2,$$

$$C_1 = -\frac{i}{2\Delta}[B_1\{2(\xi_0^2+1)Q_1^{1\prime}Q_2^0 - 2(3-4\sigma)\xi_0 Q_1^1 Q_2^0$$
$$-2(2\sigma-1)Q_1^1 Q_2^{0\prime}(\xi_0^2+1)\} + 2\xi_0 Q_2^0 - (\xi_0^2+1)Q_2^{0\prime}].$$

APPENDIX D

To reduce (5.13) into an equation of the second kind we write it as

$$\int_0^c \xi^{1/2}\lambda_n(\xi)\,d\xi \int_0^\infty k^{1/2}(\gamma H(k)+1)J_{n-1/2}(k\xi)J_n(kr)\,dk$$

$$= \frac{2i^n\gamma^2}{\alpha^2+\beta^2}\left(\frac{2}{\pi}\right)^{1/2}\cos\zeta J_n(\varepsilon\bar{r}\sin\zeta), \qquad \text{(D.1)}$$

where

$$1+\gamma H(k) = \frac{2k}{\alpha^2+\beta^2}\left(\frac{k^2}{v_2} - v_1\right).$$

Since

$$\int_0^\infty k^{1/2}J_{n-1/2}(k\xi)J_n(kr)\,dk = \begin{cases} 0, & 0<r<\xi, \\ \left(\dfrac{2}{\pi}\right)^{1/2}\dfrac{\xi^{n-1/2}}{r^n(r^2-\xi^2)^{1/2}}, & 0<\xi<r, \end{cases} \qquad \text{(D.2)}$$

(D.1) can be written as

$$\int_0^r \frac{\lambda_n(\xi)\xi^n\,d\xi}{r^n(r^2-\xi^2)^{1/2}} + \left(\frac{\pi}{2}\right)^{1/2}\int_0^c \xi^{1/2}\lambda_n(\xi)\,d\xi \int_0^\infty \gamma k^{1/2}H(k)J_{n-1/2}(k\xi)J_n(kr)\,dk$$

$$= \left(\frac{\pi}{2}\right)\frac{2i^n\gamma^2}{\alpha^2+\beta^2}\cos\zeta J_n(\varepsilon\bar{r}\sin\zeta). \qquad \text{(D.3)}$$

Since the solution of the equation

$$\int_0^r \frac{g(\eta)\,dn}{(r^2-\eta^2)^{1/2}} = f(r), \qquad 0<r<1,$$

is

$$g(\eta) = \frac{2}{\pi}\frac{d}{d\eta}\int_0^\eta \frac{rf(r)\,dr}{\sqrt{\eta^2-r^2}},$$

we get from (D.3), replacing ξ by $c\xi$ and suppressing c in the argument of λ_n,

$$\lambda_n(\xi) + \frac{1}{\pi}\int_0^1 \lambda_n(\eta)T_n(\xi,\eta)\,d\eta = \xi^{-n}H_n(\xi), \qquad 0<\xi<1, \qquad \text{(D.4)}$$

with

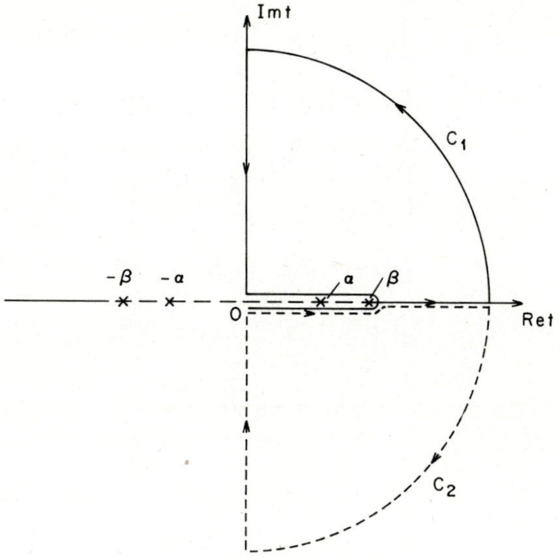

Fig. 12 Contour of integration for evaluating $T_n(\xi, \eta)$ (Eq. (D.5)).

$$T_n(\xi, \eta) = \pi \int_0^{\infty} (\xi\eta)^{1/2} \gamma t H(t) J_{n-1/2}(\xi t) J_{n-1/2}(\eta t)\, dt, \tag{D.5}$$

and

$$H_n(\xi) = \frac{2i^n c^2 \gamma^2}{\varepsilon^2(1+\tau^2)} \cos\zeta \frac{d}{d\xi} \int_0^{\xi} \frac{\rho^{n+1} J_n(\varepsilon\rho \sin\zeta)}{(\xi^2 - \rho^2)^{1/2}}\, d\rho. \tag{D.6}$$

In writing (D.4) we have also used the result

$$\int_0^{\eta} \frac{r^{n+1} J_n(kr)\, dr}{(\eta^2 - r^2)^{1/2}} = \left(\frac{\pi}{2k}\right)^{1/2} \eta^{n+1/2} J_{n+1/2}(k\eta). \tag{D.7}$$

Equation (D.4) is in a form suitable for iteration at small ε.

To do so we express $T_n(\xi, \eta)$ in a modified form. Let $\xi > \eta$. Write

$$J_{n-1/2}(\xi t) = \tfrac{1}{2}[H^{(1)}_{n-1/2}(\xi t) + H^{(2)}_{n-1/2}(\xi t)],$$

and use the contour C_1 (Fig. 12) to integrate

$$f(t) = \gamma t H(t) H^{(1)}_{n-1/2}(\xi t) J_{n-1/2}(\eta t).$$

This gives

$$\varepsilon^2(1+\tau^2)\int_0^\infty f(t)\,dt = -\int_0^\infty t\left\{2t\left(\sqrt{t^2+\alpha^2}-\frac{t^2}{\sqrt{t^2+\beta^2}}\right)\right.$$

$$\left.-\,\varepsilon^2(1+\tau^2)\right\}H^{(1)}_{n-1/2}(i\xi t)J_{n-1/2}(i\eta t)\,dt$$

$$+\,4i\int_0^{\tau\varepsilon}\frac{t^4}{\sqrt{\tau^2\varepsilon^2-t^2}}H^{(1)}_{n-1/2}(\xi t)J_{n-1/2}(\eta t)\,dt$$

$$+\,4i\int_0^{\tau} t^2\sqrt{\varepsilon^2-t^2}\,H^{(1)}_{n-1/2}(\xi t)J_{n-1/2}(\eta t)\,dt.$$

Similarly, using the contour C_2 to integrate

$$g(t) = \gamma t H(t) H^{(2)}_{n-1/2}(\xi t)J_{n-1/2}(\eta t)$$

one obtains

$$\varepsilon^2(1+\tau^2)\int_0^\infty g(t)\,dt = \int_0^\infty t\{\quad\}H^{(2)}_{n-1/2}(-i\xi t)J_{n-1/2}(-i\eta t)\,dt.$$

Since

$$H^{(2)}_{n-1/2}(-i\xi t)J_{n-1/2}(-i\eta t) = -H^{(1)}_{n-1/2}(i\xi t)J_{n-1/2}(i\eta t),$$

we get

$$T_n(\xi,\eta) = \frac{2\pi i(\xi\eta)^{1/2}\varepsilon^2}{1+\tau^2}\left[\left(\int_0^\tau\frac{t^4}{\sqrt{\tau^2-t^2}}+\int_0^1 t^2\sqrt{1-t^2}\right)\right.$$

$$\left. H^{(1)}_{n-1/2}(\varepsilon\xi t)J_{n-1/2}(\varepsilon\eta t)\,dt\right],\qquad \xi>\eta. \tag{D.8}$$

For $\xi<\eta$, we have to interchange ξ and η. $T_n(\xi,\eta)$ can now be expanded in powers of ε and (D.4) can be used to obtain expansions for λ_n given by (5.14).

IV

Electromagnetic Forces in Deformable Continua

Yih-Hsing Pao

Cornell University, Ithaca, New York

1 INTRODUCTION

In the beginning, the theory of electricity and magnetism was founded on principles of mechanics and experiments with electric charges and bar magnets in air. Thus the subject of forces exerted by the electric field on a charge, or magnetic field on a magnetic pole, was well understood because the static electric, or magnetic field was defined in terms of forces acting on a charge, or a pole. These forces are governed by Coulomb's laws.

With the discovery of current and magnetic induction, the problems became somewhat more complex. However, on the basis of Ampère's law, the force exerted by a magnetic field on a current loop in air still remained a well-defined quantity.

Additional complications arose when matter (ponderable bodies) were introduced in electromagnetic fields. In *A Treatise on Electricity and Magnetism*, James Clerk Maxwell (1873) treated the forces on ponderable bodies in electric and magnetic fields with two different approaches. In Chapter V of Part I, he discussed the mechanical action between two systems of electric charges and applied Coulomb's law to calculate the total force of one system of charges on the other. If the concept of action through contact as opposed to action at a distance were followed, this mechanical action could be expressed in terms of stresses. In Chapter XI of Part IV, forces acting on an element of a body placed in the electromagnetic field were calculated from the change of magnetostatic energy and Ampère's law of force on current circuits. These forces could be explained by the hypothesis of *stresses* in a medium. They are now known as the *Maxwell stresses*.

Already, fields in matter are no longer definable in terms of forces between charges, poles, or current circuits. Instead, they are expressed by some abstract quantities which satisfy the Maxwell equations and the forces and stresses are determined by these abstract quantities, and derived from other principles or postulations.

For stationary and undeformable (rigid) matter, Maxwell's treatment, with some additions and modifications, seemed sufficient. However, for deformable bodies in motion, not only the original concept of the Maxwell stresses was unclear, but also the applicability of the Maxwell equations to the fields inside the moving bodies was questionable. Many theories were advanced at the end of the last and the beginning of this century. The theories are generally called *electrodynamics of moving media* or simply *electrodynamics*. The most well-known theories were proposed by H. A. Lorentz (1892) and by H. Minkowski (1908). Historical development of these and other theories are described by E. Whittaker (1951).

Minkowski's theory of electrodynamics was based on the *special principle of relativity*. The theory is simple and elegant, and the governing field equations for moving media have the same form as the Maxwell equations.

Lorentz attributed all electric and magnetic phenomena to the presence or motion of electrical charges embedded in material particles. By averaging the microscopic fields associated with these charged particles, called *electrons*, Lorentz deduced a set of electrodynamical equations for moving media. One equation in the set is different from the corresponding Maxwell equation. He also postulated the force acting on the moving electrons by the field, which is now known as the *Lorentz force*.

In 1960 another form of the electrodynamic equations was proposed by L. J. Chu. In addition to the moving charges, the theory included magnetic poles and dipoles as sources for electromagnetic fields. The final form of the field equations differs from all previous formulations.

In the meantime, considerable interest has been revived in the subject of electrodynamics of moving and deformable media. This is witnessed by the appearance of many monographs and research articles on the subject. The monographs by W. F. Brown, Jr. (1966), P. Penfield, Jr. and H. A. Haus (1967), C. Møller (1972), S. R. de Groot and L. G. Suttorp (1972), and the volume in the *Handbuch der Physik* by C. Truesdell and R. A. Toupin (1960) are of particular interest as each book centers on one principle or theory, and concentrates on one formulation of electrodynamics. It is the purpose of this article to review and to compare the principles and theories, and also the equations, formulas, and results of various formulations. Our main objective is to determine the interaction of the stress–strain field with the electromagnetic field in deformable bodies, which is a part of the general subject of matter–field interaction.

The interaction of electromagnetic fields with matter is indeed very complicated. First, the action of the electromagnetic field may induce changes in the state of matter. This is the subject studied in solid-state physics and materials science. Macroscopically, the induced state of the matter is described in terms of charges, currents, electric moments (electric polarizations), magnetic moments (magnetizations), and various constitutive laws. Next, matter can act as a source and cause changes in electromagnetic fields both internally and externally. This is investigated in the theory of electromagnetism which is embodied in the Maxwell equations. Finally, the electromagnetic fields exert forces and torques on the induced sources in the matter and cause the latter to move and deform; simultaneously, the motion and deformation of the matter cause further change of these sources.

In this article, we mainly address the last category of interactions for materials which can be polarized and magnetized, and which are thermally and electrically conductive, and are mechanically deformable. Nonrelativistic equations of mechanics are used and the results are valid when the particle speeds are much less than the speed of light.

That there are so many coexisting theories and results for a subject so fundamental in nature may sound very surprising to experimentalists, for theories can usually be sorted out, or proven to be fallacious by carefully designed experiments. The difficulty here is that the electromagnetic fields inside matter are expressed in terms of field variables which cannot be directly measured in laboratories. Existing experiments may lend support to one theory or another but they do not exclude the rest. Besides, most comparisons are made on the basis of incomplete theories of electrodynamics. Hence, at this stage, we can only judge various theories, if we must make a judgement, on the mathematical rigor, soundness of assumptions, and applicability of the principles on which the theory is based.

To trace the origins of many principles and theories, some background material which can be found in standard texts of electromagnetism are included in this article. The system of International Units (S.I.), known also as MKSA units, is used throughout this article. The unification of units is no small feat by itself as no less than four systems of units are used in the five monographs cited above. They are: Gaussian, Heaviside–Lorentz (rationalized Gaussian), generalized Gaussian (Gaussian MKS system), and Giorgi (rationalized MKS, or MKSA as in this article) systems. To students in rational mechanics where units are only complicated by nationalities, the units in electromagnetics are most irrational.

In the following two sections, we summarize the basic principles, theories, and equations in mechanics and electromagnetism. Section 4 outlines the four formulations of electrodynamics which form the basis for later discussions of field–matter interactions. In Section 5, two theories of interactions for stationary media are discussed, these yield consistent results. Starting in Section 6, we present five different theories of the interaction of field with moving matter, one in each section. Each theory is based on different concepts, principles, and postulations. A final comparison of these theories is made in Section 11, in which we also summarize two sets of equations and formulas. Both sets are supposedly applicable to deformable bodies moving in electromagnetic fields. In Section 12 we discuss briefly the corresponding constitutive equations for solids and boundary conditions, thus complete the presentation of two theories of interaction of electromagnetic and elastic fields in solids.

Except for a few subsections (e.g. Subsections 4.5, 5.4, 10.2, and 12), the substance of this article is based upon the five aforementioned monographs and related texts or reference books. Each monograph contains a long list of research articles that are pertinent to our discussion. We shall make reference to the monographs and books, instead of citing the original articles, whenever possible.

2 BALANCE EQUATIONS OF CONTINUUM MECHANICS

In continuum physics, many basic laws may be stated in the following form of the equation of balance:

$$\frac{d}{dt}\int_V \Psi\, dV = \oint_S \mathbf{A}\cdot\mathbf{n}\, dS + \int_V \Phi\, dV, \tag{2.1}$$

where Ψ is the density of the physical *quantity* in balance, $\mathbf{A}$ its *flux* vector relative to the material, and Φ is its *supply*. The integral is over a material volume V bounded by a closed surface S with an outward normal unit vector $\mathbf{n}$. Since V contains the same set of material particles at each instant t, both V and S move in space and the region of integration changes with time (see Ref. [18] for contents in this section).

For deformable media the general motion is described by a transformation. The motion of all material particles are referred to a *common frame* in which a rectangular Cartesian coordinate system is embedded. Let $\mathbf{X}$ be the coordinates of a given particle at an arbitrary initial time $t = t_0$ in the *reference* (*undeformed*) *configuration* of a body, and $\mathbf{x}$ be the coordinates of

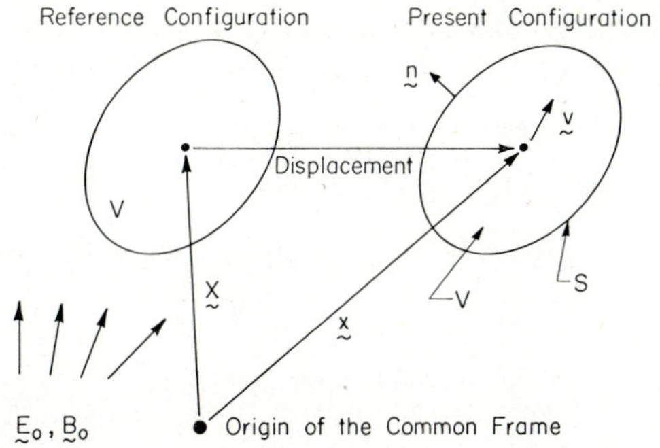

Fig. 1 Geometry of a deformable body in applied electromagnetic fields $\mathbf{E}_0$, $\mathbf{B}_0$, or $\mathbf{H}_0$.

the point occupied by the same particle at time $t > t_0$ in the *present* (*deformed*) *configuration.* The transformation

$$\mathbf{x} = \hat{\mathbf{x}}(\mathbf{X}, t) \tag{2.2}$$

with t as the parameter describes the motion of the body (Fig. 1).

The rate of change of position for a given particle is the *velocity* $\mathbf{v}$,

$$\mathbf{v} = \dot{\mathbf{x}} = \frac{d\mathbf{x}}{dt} = \frac{\partial \hat{\mathbf{x}}(\mathbf{X}, t)}{\partial t}.$$

The motion is completely specified once the function $\hat{\mathbf{x}}$, or $\mathbf{v}$ is known.

The global (integral) form of the balance laws may be transformed into local (differential) form by applying the divergence theorem and the Stokes theorem,

$$\begin{aligned} \oint_S \mathbf{n} \cdot \mathbf{A}\, dS &= \int_V \nabla \cdot \mathbf{A}\, dV, \\ \int_S \mathbf{n} \cdot \nabla \times \mathbf{A}\, dS &= \oint_C \mathbf{A} \cdot d\mathbf{C}, \end{aligned} \tag{2.3}$$

and the transport theorems (Sections 79–81 of [18])

$$\begin{aligned} \frac{d}{dt}\int_V \Psi\, dV &= \int_V \left[\frac{\partial \Psi}{\partial t} + \nabla \cdot (\mathbf{v}\Psi)\right] dV, \\ \frac{d}{dt}\int_S \mathbf{A} \cdot \mathbf{n}\, dS &= \int_S \overset{*}{\mathbf{A}} \cdot \mathbf{n}\, dS, \end{aligned} \tag{2.4}$$

where

$$\overset{*}{\mathbf{A}} = \partial \mathbf{A}/\partial t + \mathbf{v}\nabla \cdot \mathbf{A} + \nabla \times (\mathbf{A} \times \mathbf{v}).$$

In Eq. (2.3), S is bounded by a closed curve C and $d\mathbf{C}$ is an element of the curve in the direction tangent to the curve; in Eq. (2.4), $\mathbf{v}$ is the velocity of the moving particle contained in V or S. Thus the local form of Eq. (2.1) is

$$\frac{\partial \Psi}{\partial t} + \nabla \cdot (\mathbf{v}\Psi) = \nabla \cdot \mathbf{A} + \Phi. \tag{2.5}$$

The four basic balance laws of continuum mechanics are (Sections 156, 200, 241 of [18]):

Mass
$$\frac{d}{dt}\int_V \rho \, dV = 0; \tag{2.6}$$

Linear momentum
$$\frac{d}{dt}\int_V \rho \mathbf{v} \, dV = \oint_S \mathbf{t}^{(n)} \, dS + \int_V \mathbf{F} \, dV; \tag{2.7}$$

Angular momentum
$$\frac{d}{dt}\int_V (\mathbf{x} \times \rho \mathbf{v}) \, dV = \oint_S (\mathbf{x} \times \mathbf{t}^{(n)}) \, dS + \int_V (\mathbf{L} + \mathbf{x} \times \mathbf{F}) \, dV; \tag{2.8}$$

Energy
$$\frac{d}{dt}\int_V (\tfrac{1}{2}\rho \mathbf{v} \cdot \mathbf{v} + \rho U) \, dV = \oint_S (\mathbf{t}^{(n)} \cdot \mathbf{v} - \mathbf{Q} \cdot \mathbf{n}) \, dS + \int_V (\mathbf{F} \cdot \mathbf{v} + \phi_Q + \phi) \, dV. \tag{2.9}$$

In the above equations, ρ is the *mass density*, $\mathbf{t}^{(n)}$ the *stress vector* (*traction*) at the surface with unit normal $\mathbf{n}$, $\mathbf{F}$ the *body force* per unit volume, $\mathbf{L}$ the *body couple* per unit volume, U the *energy density*, $\mathbf{Q}$ the *heat flux* vector, ϕ_Q the *energy supply* due to heat source, and ϕ the *energy supply* due to sources other than heat. In the absence of externally applied electromagnetic fields, $\mathbf{L}$ and ϕ vanish for nonpolar media, and the body force $\mathbf{F}$ is usually a prescribed quantity. The quantities $\mathbf{Q}$ and ϕ_Q are possibly affected by the presence of electromagnetic fields only through constitutive laws which are discussed at the end of this article.

However, when a material body which carries electric charges and currents is placed in the electromagnetic field, $\mathbf{L}$ and ϕ may not vanish and $\mathbf{F}$ and $\mathbf{t}^{(n)}$ are augmented by the so-called electromagnetic forces. Since body forces of mechanical origin like the gravitational force and centrifugal force are well known, they are excluded from this discussion, and in this article $\mathbf{F}$ is solely due to the electromagnetic field.

By applying Eqs. (2.3) and (2.4) to Eqs. (2.6)–(2.9), and introducing the *stress tensor* $\underset{\sim}{\tau}$ with

$$\mathbf{t}^{(n)} = \mathbf{n} \cdot \underset{\sim}{\tau}, \tag{2.10}$$

we obtain the balance equations in the local form:

Mass	$\partial\rho/\partial t + \nabla \cdot (\rho \mathbf{v}) = 0;$	(2.11)
Linear momentum	$\rho\, d\mathbf{v}/dt = \nabla \cdot \underset{\sim}{\tau} + \mathbf{F};$	(2.12)
Angular momentum	$(\underset{\sim}{\tau} - \underset{\sim}{\tau}^T)/2 = -\underset{\sim}{\boldsymbol{\eta}};$	(2.13)
Energy	$\rho\, dU/dt = \underset{\sim}{\tau} : \nabla \mathbf{v} - \nabla \cdot \mathbf{Q} + \phi_Q + \phi,$	(2.14)

where $\underset{\sim}{\tau}^T$ is the transpose of the stress tensor $\underset{\sim}{\tau}$ and $\underset{\sim}{\boldsymbol{\eta}}$ is the dual of the couple $\mathbf{L}$. In Cartesian tensor notation, $L_i = \varepsilon_{ijk}\eta_{jk}$ where ε_{ijk} is the permutation symbol with $\varepsilon_{ijk} = 1$ or -1 according to whether the indices are in a cyclic or an anticyclic order, respectively, and $\varepsilon_{ijk} = 0$ otherwise. Thus Eq. (2.13) may be written as

$$(\tau_{ij} - \tau_{ji})/2 = -\eta_{ij}, \quad \text{or} \quad \varepsilon_{ijk}\tau_{jk} = -L_i. \tag{2.13A}$$

The double dot product of two dyadics in (2.14) is a scalar with $\underset{\sim}{\tau} : \nabla \mathbf{v} = \tau_{ij}\, \partial_i v_j$ where $\partial_i \equiv \partial/\partial x_i$.

The following sections discuss how the quantities $\underset{\sim}{\tau}$, $\mathbf{F}$, $\mathbf{L}$, and ϕ in the mechanical balance equations are determined in terms of the electromagnetic field variables that satisfy the Maxwell equations. We begin with a summary of the Maxwell equations for stationary (Section 3) and moving media (Section 4).

3 MAXWELL EQUATIONS FOR MEDIA AT REST

The Maxwell equations as proposed in the original memoir by James Clerk Maxwell (1864) were based on experiments for isolated electric charges and electric current circuits in the air. The equations contain several basic laws of electromagnetism. The Gauss–Coulomb law is based on Charles Augustin Coulomb's experiment (1785) on forces between two electric charges. The Faraday law of induction is based on Michael Faraday's discovery and experiments on electric currents induced by a changing magnetic flux (1831). The Ampère–Maxwell law is based on Hans Christian Oersted's discovery (1820) that magnetic fields were generated by electric current in a thin wire, and André–Marie Ampère's experiments on forces between two circuits conducting electric currents (1825).

In the vacuum, all electromagnetic phenomena can be described by two field variables, one for electric and one for magnetic fields, and two source variables, the electric charge and the electric current. The electric field is represented by the *electric field intensity* **E**. The magnetic part is traditionally represented by the *magnetic field intensity* **H** when it is generated by permanent magnets, or by the *magnetic induction* **B** when the field is generated by currents. Another electric field variable, the *electric displacement* **D**, is used in connection with the field between capacitors. In the vacuum

$$\mathbf{B} = \mu_0 \mathbf{H}, \qquad \mathbf{D} = \varepsilon_0 \mathbf{E}, \tag{3.1}$$

where μ_0 and ε_0 are two universal constants. They are related by the equation $\varepsilon_0 \mu_0 = c^{-2}$ where c is the light speed *in vacuo*.

3.1 The Maxwell Equations

For material bodies at rest (stationary matter), two field variables are not sufficient to account for all electromagnetic phenomena because of the wide variance of material properties. In the original Treatise by Maxwell (Chapter IX, Part IV of [12]), all four field variables mentioned above are used and the governing equations are:

$$\text{(I)}\dagger \left\{ \begin{array}{ll} \nabla \cdot \mathbf{B} = 0, & \text{(3.2A)} \\ \nabla \times \mathbf{E} + \partial \mathbf{B}/\partial t = \mathbf{0}, & \text{(3.2B)} \\ \nabla \cdot \mathbf{D} = \sigma, & \text{(3.2C)} \\ \nabla \times \mathbf{H} = \partial \mathbf{D}/\partial t + \mathbf{J}. & \text{(3.2D)} \end{array} \right.$$

In the above, σ is the *free charge density* and **J** is the *free current density*. **D**, **J**, and **H** are related to **E** and **B** respectively by the following constitutive equations for linear isotropic materials:

$$\mathbf{D} = \varepsilon \mathbf{E}, \qquad \mathbf{B} = \mu \mathbf{H}, \qquad \mathbf{J} = \nu \mathbf{E}. \tag{3.3}$$

ε is called the *dielectric constant*, μ the *magnetic permeability*, and ν the *electric conductivity*. In the vacuum, $\varepsilon = \varepsilon_0$ and $\mu = \mu_0$.

The constitutive equations can be generalized for nonlinear rigid materials as

$$\mathbf{D} = \hat{\mathbf{D}}(\mathbf{E}), \qquad \mathbf{B} = \hat{\mathbf{B}}(\mathbf{H}), \qquad \mathbf{J} = \hat{\mathbf{J}}(\mathbf{E}), \tag{3.4}$$

where $\hat{\mathbf{D}}$, $\hat{\mathbf{J}}$, and $\hat{\mathbf{B}}$ are general vector functions.

The Maxwell equations, Eq. (3.2), are not independent of each other. From Eqs. (3.2B), (3.2C), and (3.2D), we can derive the equation of conservation of charge

† Various forms of the Maxwell equations are also designated by roman numerals for quick identification in this article.

$$\nabla \cdot \mathbf{J} + \partial \sigma / \partial t = 0. \tag{3.2E}$$

Often, Eqs. (3.2B), (3.2D), and (3.2E) are taken as the set of independent equations for the unknown variables **E**, **B**, and σ.

Modern texts in physics (e.g. [6], [11]) tend to regard **E** and **B** fields as the basic variables for electric and magnetic fields in the vacuum. For material media two more variables, the *polarization density* **P** (electric moment per unit volume) and the *magnetization density* **M** (magnetic moment per unit volume), are introduced. These two are related to the other four variables by†

$$\mathbf{B} \equiv \mu_0(\mathbf{H} + \mathbf{M}), \qquad \mathbf{D} \equiv \varepsilon_0 \mathbf{E} + \mathbf{P}. \tag{3.5}$$

In terms of **E**, **B**, **P**, and **M**, Eqs. (3.2) may be expressed as

$$\text{(III)} \begin{cases} \nabla \cdot \mathbf{B} = 0, & \text{(3.6A)} \\ \nabla \times \mathbf{E} + \partial \mathbf{B}/\partial t = 0, & \text{(3.6B)} \\ \varepsilon_0 \nabla \cdot \mathbf{E} = \sigma - \nabla \cdot \mathbf{P}, & \text{(3.6C)} \\ \mu_0^{-1} \nabla \times \mathbf{B} - \varepsilon_0 \partial \mathbf{E}/\partial t = \partial \mathbf{P}/\partial t + \nabla \times \mathbf{M} + \mathbf{J}. & \text{(3.6D)} \end{cases}$$

The terms on the right-hand side of Eq. (3.6) may all be regarded as sources for electricity and magnetism. The $-\nabla \cdot \mathbf{P}$ is the *polarization charge*, $\partial \mathbf{P}/\partial t$ the *polarization current*, and $\nabla \times \mathbf{M}$ the *magnetization current*. The constitutive equations (3.4) are replaced by

$$\mathbf{P} = \hat{\mathbf{P}}(\mathbf{E}), \qquad \mathbf{M} = \hat{\mathbf{M}}(\mathbf{B}), \qquad \mathbf{J} = \hat{\mathbf{J}}(\mathbf{E}). \tag{3.7}$$

Although none of the variables, **E**, **B**, **D**, **H**, **P**, and **M**, can be measured directly inside a material body, the Maxwell equations in the form of either Eq. (3.2) or Eq. (3.6) are accepted as the true law for electromagnetism in stationary media or rigid bodies at rest. So far all results derived from these equations, that can be measured in laboratories have been confirmed experimentally.

3.2 Forces on Free Charges and Free Currents

Let a point charge q be placed at $\mathbf{x}$, and a second point charge q' at $\mathbf{x}'$. According to Coulomb's law, the force acting on q by the field generated by the second charge is (Fig. 2)

$$\mathbf{f}(\mathbf{x}) = q\mathbf{E}(\mathbf{x}), \qquad \mathbf{E}(x) = \frac{q'}{4\pi\varepsilon_0} \frac{\mathbf{x} - \mathbf{x}'}{|\mathbf{x} - \mathbf{x}'|^3}. \tag{3.8}$$

† The relation $\mathbf{B} = \mu_0 \mathbf{H} + \mathbf{M}$ has also been used (e.g. [1] and [6]).

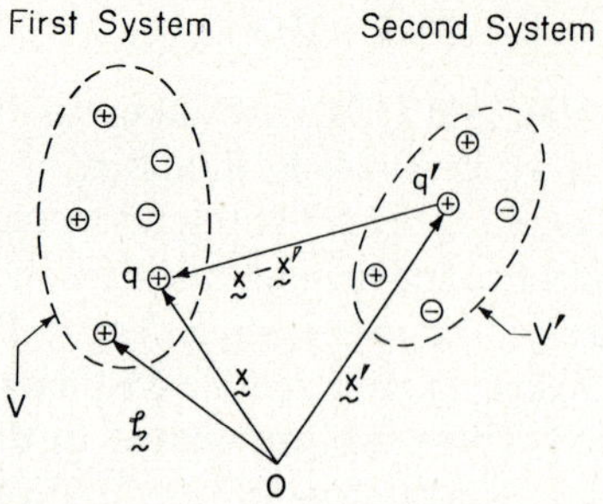

Fig. 2 Interaction of two distributions of point charges.

The second of Eq. (3.8) is the definition of **E** in electrostatics. For two systems of charges, let $\sigma(\mathbf{x})$ be the distribution function of charges within the volume V of the first system, and $\sigma'(\mathbf{x}')$ be that within V' of the second system. The resultant force on the first system is

$$\mathbf{F}_r = \int_V \sigma(\mathbf{x}) E(\mathbf{x})\, d^3x, \tag{3.9}$$

$$\mathbf{E}(\mathbf{x}) = \int_{V'} \frac{\sigma'(\mathbf{x}')(\mathbf{x} - \mathbf{x}')}{4\pi\varepsilon_0 |\mathbf{x} - \mathbf{x}'|^3}\, d^3x'. \tag{3.10}$$

In the above, d^3x and d^3x' denote volume elements at **x** and **x**′ respectively.

In magnetostatics, a similar inverse square law of forces between two magnetic poles, p at **x** and p' at **x**′, was established by John Michell (1750) and Coulomb (1777). The force on p is

$$\mathbf{f}(\mathbf{x}) = p\mathbf{H}(\mathbf{x}), \qquad \mathbf{H}(\mathbf{x}) = \frac{p'}{4\pi\mu_0} \frac{\mathbf{x} - \mathbf{x}'}{|\mathbf{x} - \mathbf{x}'|^3}. \tag{3.11}$$

A formula similar to Eq. (3.9) can be written for a distribution of poles.

For two current conducting circuits in vacuum, the force between the circuits is determined from Ampère's law. Let the current flowing in the first circuit be i, and that in the second be i', and consider two arc elements, $d\mathbf{C}$ of the first circuit at **x**, and $d\mathbf{C}'$ of the second at **x**′ (Fig. 3). The force on the element $d\mathbf{C}$ is (Chapter 5 of [7])

$$d\mathbf{f}(\mathbf{x}) = i\, d\mathbf{C} \times \mathbf{B}(\mathbf{x}), \qquad \mathbf{B}(\mathbf{x}) = \frac{\mu_0}{4\pi} \oint_{C'} \frac{i'\, d\mathbf{C}' \times (\mathbf{x} - \mathbf{x}')}{|\mathbf{x} - \mathbf{x}'|^3}. \tag{3.12}$$

For a distribution of currents, the $i\, d\mathbf{C}$ is replaced by $\mathbf{J}\, d^3x$ where **J** is the

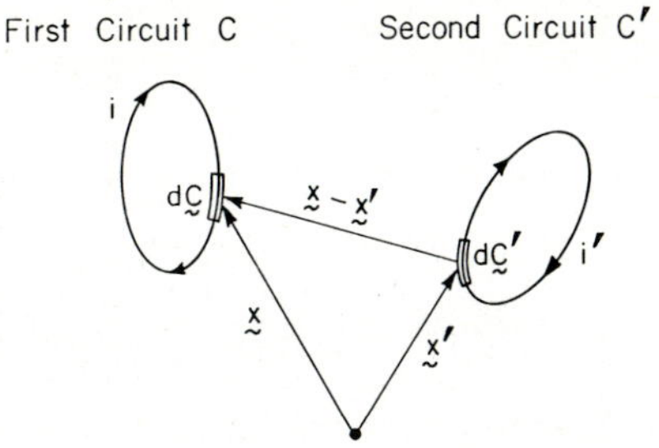

Fig. 3 Interaction of two current-conducting circuits.

current density, and $i'\,d\mathbf{C}'$ is replaced by $\mathbf{J}'\,d^3x'$. The resultant force on the first system is

$$\mathbf{F}_r(\mathbf{x}) = \int_V \mathbf{J}(\mathbf{x}) \times \mathbf{B}(\mathbf{x})\,d^3x, \tag{3.13}$$

$$\mathbf{B}(\mathbf{x}) = \frac{\mu_0}{4\pi}\int_{V'} \frac{\mathbf{J}'(\mathbf{x}') \times (\mathbf{x} - \mathbf{x}')}{|\mathbf{x} - \mathbf{x}'|^3}\,d^3x'. \tag{3.14}$$

The volume V contains the first system of current circuits, and V' contains the second.

Note that Eqs. (3.9) and (3.13) give rise to the resultant force on a system of charges and currents respectively. $\mathbf{E}(x)$ and $\mathbf{B}(x)$ are the fields generated by a second system of charges and current circuits respectively; they are from sources external to the first system. So far as the calculation of $\mathbf{F}_r$ is concerned, $\mathbf{E}$ or $\mathbf{B}$ could be fields generated by any other external sources like batteries or bar magnets. We shall call them the *external fields* and denote them by a subscript 0 when a distinction is to be made. If the first system contains both charges and current circuits, the resultant force acting on it by external fields is

$$\mathbf{F}_r = \int_V \sigma \mathbf{E}_0(\mathbf{x})\,d^3x + \int_V \mathbf{J} \times \mathbf{B}_0(\mathbf{x})\,d^3x. \tag{3.15}$$

In addition to the force due to external fields, there are Coulomb forces acting between charges within the first system. Since the action of any point charge P on another Q is equal and opposite to that of Q on P, the resultant force on the first system arising from all charges in the system itself vanishes. However, when the force on one volume element d^3x inside the first system is to be determined, we must consider the interacting forces between the charges within d^3x and those without, all belonging to the first system. The force per unit volume at location **x** is

$$\mathbf{F}(\mathbf{x}) = \sigma(x)\mathbf{E}(\mathbf{x}), \tag{3.16}$$

where

$$\mathbf{E}(x) = \mathbf{E}_0(\mathbf{x}) + \int_V \frac{\sigma(\zeta)(\mathbf{x}-\zeta)}{4\pi\varepsilon_0|\mathbf{x}-\zeta|^3}\, d^3\zeta \tag{3.17}$$

is the sum of the external field and the internal field. A similar expression can be obtained for the magnetic component.

In view of Eq. (3.2C), the **E** field within the volume V of the first system should satisfy the equation

$$\nabla \cdot \mathbf{E} = \sigma/\varepsilon_0; \tag{3.18}$$

it is seen that Eq. (3.17) is indeed the general solution of Eq. (3.18) because $\nabla \cdot \mathbf{E}_0 = 0$ within V, and a particular integral of Eq. (3.18) is

$$\mathbf{E}(\mathbf{x}) = -\nabla_x \int \frac{\sigma(\zeta)}{4\pi\varepsilon_0|\mathbf{x}-\zeta|}\, d^3\zeta = \int \frac{\sigma(\zeta)(\mathbf{x}-\zeta)}{4\pi\varepsilon_0|\mathbf{x}-\zeta|^3}\, d^3\zeta. \tag{3.19}$$

In the above, ∇_x is the vector differential operator with respect to **x** coordinates.

With this understanding, we express the force density within the first system as

$$\mathbf{F}(\mathbf{x}) = \sigma\mathbf{E} + \mathbf{J} \times \mathbf{B}, \tag{3.20}$$

where **E** and **B** satisfy the Maxwell equations within the volume V. Often, **F** in Eq. (3.20) is called the *Lorentz force* although originally Lorentz considered the force on a single charge moving in external fields $\mathbf{E}_0$ and $\mathbf{B}_0$ (see Eq. (7.1)).

The formula (3.20) is valid only for a distribution of charges and current circuits in the vacuum. Nevertheless, it has been extended to charge and current sources inside material media. This is discussed in Section 8. In Section 9 we shall discuss the resultant force, like $\mathbf{F}_r$ in Eq. (3.15), on a material body.

4 MAXWELL EQUATIONS FOR MOVING MEDIA

When a material body moves relative to a fixed reference frame, the description of electromagnetic phenomena in the body becomes very complicated. In the literature there appeared many formulations of the Maxwell equations for moving media, none of which is universally accepted as the "true" law (see a review in Chapter 8 of [15]). We shall only discuss four formulations in this article.

According to the special theory of relativity, the equations of electrodynamics are invariant under the Lorentz transformation. The special principle of relativity postulates that if physical laws hold good in relation to a system of coordinates K, the same laws also hold good in relation to any other system of coordinates K', moving in uniform translation relative to K [26]. The constancy of light speed in both systems is maintained when the spatial and temporal coordinates are transformed from one system to the other according to the Lorentz transformation.

To investigate the electrodynamics of moving bodies, an *inertial frame* is first chosen as the reference frame and it is called the *laboratory frame.* When a rigid body moves with constant velocity $\mathbf{v}$ relative to the laboratory frame, we may attach a reference frame to the moving body. It is called the *rest frame* because to an observer situated at the rest frame, every material particle in the rigid body appears to be at rest. Hence if the K system of coordinates is embedded in the laboratory frame, and the K' in the rest frame, the equations of electrodynamics in these two systems should be invariant.

For a rigid body in rotation or for a body undergoing deformation, the velocity of material particles not only changes in time, but also varies from point to point in space. Thus it is not possible to define a single rest frame for the entire body. In such a case, it is customary to attach one reference frame to each and every particle at each instant. This is called the *pointwise rest frame* as each particle appears to be at rest in its own frame. Physical laws are then supposed to be invariant in the K coordinates of the laboratory frame, and the K' coordinates which are embedded to the pointwise rest frame (p. 119, [15]).

We note that the pointwise rest frame may not be an inertial frame. Furthermore, the concepts of the gradient of a scalar field and the divergence or curl of a vector field become obscured in such a pointwise rest frame because to a pair of neighboring particles there correspond two different frames.

Fortunately, for slow speed approximations ($v \ll c$) as treated in this article, the Lorentzian invariance is not adhered to rigidly. Except for the Minkowski's formulation, we have no need to introduce the rest frame or pointwise rest frame in other formulations. For deformable bodies in motion, a rigorous relativistic theory of electrodynamics must be based on the *general principle of relativity* [29, 34].

4.1 The Minkowski Formulation (EBDH)

Based on the special principle of relativity, H. Minkowski proposed in 1908 a theory of electrodynamics for moving media [32]. His logic is very simple. Since the field equations, Eqs. (3.2), are known for a ponderable body at rest in the laboratory frame (the K-frame), the equations for the body moving with constant velocity $\mathbf{v}$ relative to the K-frame are also known. The latter have the same form as Eqs. (3.2) except that all field variables, coordinates, and time are in the rest frame (K'-frame). Symbolically, we attach a prime to all variables and coordinates in the equations to signify that they are in the K'-frame.

When these equations in the K'-frame are transformed back to those in the K-frame by the inverse Lorentz transformation, a system of equations for the moving body in the laboratory frame is obtained. This system of equations is identical to Eqs. (3.2) (see Section 34 of [16]).

We group these equations as Eqs. (I) where a roman numeral designates various formulation of the Maxwell equations for moving media.

$$\text{(I)}\quad \text{Eqs. (3.2A)–(3.2D).} \tag{4.1}$$

From the K'-frame to the K-frame, the field variables are transformed, in slow speed approximation, as

$$\begin{aligned} \mathbf{E}' &= \mathbf{E} + \mathbf{v} \times \mathbf{B}, & \mathbf{H}' &= \mathbf{H} - \mathbf{v} \times \mathbf{D}, \\ \mathbf{D}' &= \mathbf{D} + \mathbf{v} \times \mathbf{H}/c^2, & \mathbf{B}' &= \mathbf{B} - \mathbf{v} \times \mathbf{E}/c^2, \\ \mathbf{J}' &= \mathbf{J} - \sigma\mathbf{v}, & \sigma' &= \sigma - \mathbf{v} \cdot \mathbf{J}/c^2. \end{aligned} \tag{4.2}$$

Although the field equations are unchanged in form, except all primes are dropped, the constitutive equations are altered. Since Eqs. (3.3) are known constitutive equations for matter at rest, the constitutive equations for moving matter when expressed in the rest frame are

$$\mathbf{D}' = \varepsilon\mathbf{E}', \qquad \mathbf{B}' = \mu\mathbf{H}', \qquad \mathbf{J}' = \nu\mathbf{E}'. \tag{4.3A}$$

The same constitutive equations can be expressed in terms of the laboratory frame variables if we substitute Eqs. (4.2) into the above and then solve for **D** and **B**. In slow speed limit, these are

$$\begin{aligned} \mathbf{D} &= \varepsilon\mathbf{E} + (\varepsilon\mu - \varepsilon_0\mu_0)\mathbf{v} \times \mathbf{H}, \\ \mathbf{B} &= \mu\mathbf{H} + (\varepsilon_0\mu_0 - \varepsilon\mu)\mathbf{v} \times \mathbf{E}, \\ \mathbf{J} - \sigma\mathbf{v} &= \nu(\mathbf{E} + \mathbf{v} \times \mathbf{B}). \end{aligned} \tag{4.3B}$$

Equations (4.1) and (4.3B) complete the Minkowski formulation.

Minkowski's theory undoubtedly is the most widely known electrodynamic theory for moving media. Despite the original assumption that the body is rigid and moves with constant velocities, the theory has been applied without reservation in many cases to deformable bodies in motion (see a review by G. Paria [37]).

Corresponding to Eqs. (I), the forces on the moving body are calculated in Subsection 5.3 from the Minkowski energy–momentum tensor, and again in Subsection 6.2.

4.2 The Lorentz Formulation (EBPMv)

In 1892 H. A. Lorentz proposed the theory of electrons (see [30]). The theory ascribes all electromagnetic phenomena to the agency of moving electric charges. A material body is supposed to contain an immense number of extremely small charged particles called *electrons*.† The electrons interact with the fields which permeate the material medium and the fields satisfy the Maxwell equations *in vacuo*. Because the electrons are moving rapidly, the microscopic electric or magnetic fields are rapidly fluctuating in space and in time. However, the averaged value of these fields is a comparatively smooth function.

The averaging process is carried as follows: let ΔV be the infinitesimal region in the neighborhood of $\mathbf{x}$, and ΔT the short time interval around an instant t. The average of field quantity, $\mathbf{h}$, is defined as

$$\langle \mathbf{h}(\mathbf{x}, t) \rangle = \frac{1}{\Delta T \Delta V} \int_{\Delta T} d\tau \int_{\Delta V} \mathbf{h}(\mathbf{x} + \xi, t + \tau)\, d^3\xi. \tag{4.4}$$

Within ΔV, there are a large number of electrons. Thus the length scale for ΔV should be large when compared with the sizes of electrons and the distances they traverse, but still very small when compared with the dimension in phenomenological measurements. The averaged field, written as $\mathbf{H} \equiv \langle \mathbf{h} \rangle$, is the macroscopic field.

We do not repeat the derivation here as it is similar to that in the statistical formulation which is discussed in Section 7. A historical account of Lorentz's theory can be found in [19, Chapter XIII], and a modern treatment is given in [18, Chapter F III]. The complete set of the field equations of the Lorentz formulation are

† The "electron" in the Lorentz theory is different from the "electron" in the modern atomic theory. The latter is based on another theory of the electron by J. J. Thomson in 1898.

$$
\text{(II)} \begin{cases} \nabla \cdot \mathbf{B} = 0, & \text{(4.5A)} \\ \nabla \times \mathbf{E} + \partial \mathbf{B}/\partial t = 0, & \text{(4.5B)} \\ \varepsilon_0 \nabla \cdot \mathbf{E} = \sigma - \nabla \cdot \mathbf{P}, & \text{(4.5C)} \\ \mu_0^{-1} \nabla \times \mathbf{B} - \varepsilon_0\, \partial \mathbf{E}/\partial t = \mathbf{J} + \partial \mathbf{P}/\partial t + \nabla \times (\mathbf{P} \times \mathbf{v}) + \nabla \times \mathbf{M}. & \text{(4.5D)} \end{cases}
$$

The definitions for **P**, **M**, and other variables are discussed in Section 7 along with a comparison of (II) with the statistical formulation (III). Equations (4.5) differ from Eqs. (3.6) only in the term $\nabla \times (\mathbf{P} \times \mathbf{v})$ in the last equation. They will be identical if in Eq. (4.5D), the last two terms are replaced by $\nabla \times \mathbf{M}^S$ where

$$
\mathbf{M}^S = \mathbf{M} + \mathbf{P} \times \mathbf{v}. \tag{4.6}
$$

In the original memoir, Lorentz started with nonmagnetizable materials. Thus $\mathbf{M} = 0$, $\mathbf{B} = \mu_0 \mathbf{H}$ in Eq. (4.5) and the resulting equations are valid for electrically polarized material bodies moving at slow speed (see Chapter DIV, [1]). For magnetizable materials the validity of Eqs. (II) is questionable. In the statistical theory, it is shown (see Subsection 7.1) that Eq. (4.5D) is a special case of a more general result.

Forces and energy supply for the Lorentz formulation are discussed in Sections 7 and 8.

4.3 The Statistical Formulation (EBPM)†

With the advancing of the theory of statistical mechanics, many followers of Lorentz tried to modify the theory of electrons (see a review by P. Mazur [31]). A comprehensive treatment is contained in the recent monograph by de Groot and Suttorp [4]. In the new theory, the electrons are grouped into *stable groups* like atoms, ions, or molecules. The effect of electrons within each stable group is represented by microscopic electric and magnetic multipole moments (dipoles, quadrupoles,...). The statistical average of these multipole moments over a large number of stable groups are the polarization **P** and the magnetization **M**.

The nonrelativistic part of the theory will be discussed later in Subsection 7.1. It is sufficient to point out here that the final form of the electrodynamic equations for moving media is identical to Eqs. (3.6). They are designated as the Statistical Formulation (III),

† Many researchers contributed to the completion of this formulation, notably L. Rosenfeld, P. Mazur, S. R. deGroot, J. Vlieger, and L. F. Suttorp. See [4] for references.

(III) Eqs. (3.6A)–(3.6D). (4.7)

These equations were derived by S. R. de Groot and I. Vlieger [25] based on a covariant formulation of electrodynamics and are supposedly Lorentz invariant.

Forces and energy supply based on the statistical formulation are discussed in Section 7.

4.4 The Chu Formulation (EHPM)

In a recent text written for undergraduate students in electrical engineering, there contains another theory of electrodynamics for moving matter (Chapter 5, 9 [5]). The theory is credited to L. J. Chu (1960) and is extended in the monograph by Penfield and Haus [15].

The basic postulates in the Chu formulation are:

1. Material bodies, when moving or deforming, contribute to the electromagnetic fields by acting as sources (charges and currents) for these fields. The effect of these sources is equivalent to the effect they would have if located in the vacuum.
2. The sources are described by the distribution of free charges σ, free currents **J**, the electric polarization **P**, and the magnetization **M**, along with the corresponding surface sources.
3. The polarization and magnetization are modeled respectively by electric and magnetic charges, and dipoles. The postulates (1) and (2) are stated clearly in the text (p. 379). The postulate (3) is implied in the derivation that leads to the definition of **P** and **M**, and the final set of the field equations. We consider it a basic postulate for if the material sources were modeled by something other than the dipoles, the meaning of **P** and **M** as well as the resulting equations would be different.

We note that similar postulates are also made in the Lorentz formulation. The first postulate is essentially the foundation of the theory of electrons. The second is also common to both formulations. The only difference lies in the third postulate. Lorentz and his followers modeled the microscopic behavior of the material by the electrons and applied the statistical method to determine macroscopic behaviors. Chu directly modeled the macroscopic behavior by electric and magnetic charges and dipoles.

The two field variables in the vacuum are **E** and **H** in the Chu formulation, and the two source variables are the free charge σ and free current **J**. A polarizable material in the field introduces additional sources. The complete set of equations are

$$
\text{(IV)}\quad\begin{cases}
\nabla\cdot(\mu_0\mathbf{H}) = -\nabla\cdot(\mu_0\mathbf{M}), & (4.8\text{A})\\
\nabla\times\mathbf{E} + \mu_0\,\partial\mathbf{H}/\partial t = -\partial(\mu_0\mathbf{M})/\partial t - \nabla\times(\mu_0\mathbf{M}\times\mathbf{v}), & (4.8\text{B})\\
\varepsilon_0\nabla\cdot\mathbf{E} = \sigma - \nabla\cdot\mathbf{P}, & (4.8\text{C})\\
\nabla\times\mathbf{H} - \varepsilon_0\,\partial\mathbf{E}/\partial t = \mathbf{J} + \partial\mathbf{P}/\partial t + \nabla\times(\mathbf{P}\times\mathbf{v}). & (4.8\text{D})
\end{cases}
$$

In the above, polarization $\mathbf{P}$ and magnetization $\mathbf{M}$ are defined in terms of the electric and magnetic dipole moments which are modeled by two dipoles (see Eqs. (10.1) and (10.2)). On the right-hand sides of Eqs. (4.8), $-\nabla\cdot(\mu_0\mathbf{M})$ is the *magnetization charge*, $-\nabla\cdot\mathbf{P}$ the *polarization charge*, $\partial(\mu_0\mathbf{M})/\partial t + \nabla\times(\mu_0\mathbf{M}\times\mathbf{v})$ the *magnetization current*, and $\partial\mathbf{P}/\partial t + \nabla\times(\mathbf{P}\times\mathbf{v})$ the *polarization current*.

Note that if the relation (3.5) is adopted, Eqs. (IV) are the same as Eqs. (II) except the extra term, $-\nabla\times(\mu_0\mathbf{M}\times\mathbf{v})$, in Eq. (4.8B). For non-magnetizable materials ($\mathbf{M} = \mathbf{0}$) in slow motion, these two sets of equations, Eqs. (II) and (IV), are identical.

4.5 Global Laws for Electrodynamics

As shown in [35], the various formulations of the Maxwell equations for moving media can all be derived from the following global laws. They are postulated for electromagnetic fields in moving matter (Sects. 275, 276 [18])

Gauss–Faraday
$$\oint_S \mathbf{B}\cdot d\mathbf{S} = 0; \tag{4.9A}$$

Faraday
$$\oint_C \mathbf{E}_e\cdot d\mathbf{C} = -\frac{d}{dt}\int_S \mathbf{B}\cdot d\mathbf{S}; \tag{4.9B}$$

Gauss–Coulomb
$$\oint_S \mathbf{D}\cdot d\mathbf{S} = \int_V \sigma\, dV; \tag{4.9C}$$

Ampère–Maxwell
$$\oint_C \mathbf{H}_e\cdot d\mathbf{C} = \frac{d}{dt}\int_S \mathbf{D}\cdot d\mathbf{S} + \int_S \mathbf{J}_e\cdot d\mathbf{S}. \tag{4.9D}$$

In addition, there is the law of conservation of charges:

$$\oint_S \mathbf{J}_e\cdot d\mathbf{S} + \frac{d}{dt}\int_V \sigma\, dV = 0. \tag{4.9E}$$

All integrations are either taken over a material volume V which is enclosed by the surface S, or a material surface S bounded by the circuit C. The vector surface element $d\mathbf{S}$ is in the direction normal to the surface dS and the vector length element $d\mathbf{C}$ is along the tangent of the circuit C. Except for a subscript e, all field and source variables in Eq. (4.9) have the same meaning as those in Eq. (3.2). Note that Eqs. (4.9B), (4.9D), (4.9E) are in the form of the general balance equation (2.1).

The subscript e stands for "effective". Thus $\mathbf{E}_e$ is the *effective electric field intensity*, $\mathbf{H}_e$ the *effective magnetic field intensity*, and $\mathbf{J}_e$ the *effective current density*. For matter at rest, $\mathbf{E}_e = \mathbf{E}$, $\mathbf{H}_e = \mathbf{H}$ and $\mathbf{J}_e = \mathbf{J}$. When a material body is moving and deforming, $\mathbf{E}_e$ and $\mathbf{H}_e$ are the induced field intensities measured by an observer who moves with the material particles in S and C, the velocity being much slower than the speed of light.

Applying Eqs. (2.3) and (2.4) to Eq. (4.9), we obtain the local electrodynamic laws for moving media:

$$\nabla \cdot \mathbf{B} = 0, \tag{4.10A}$$
$$\nabla \times \mathbf{E}_e = -\overset{*}{\mathbf{B}} = -\partial \mathbf{B}/\partial t - \nabla \times (\mathbf{B} \times \mathbf{v}), \tag{4.10B}$$
$$\nabla \cdot \mathbf{D} = \sigma, \tag{4.10C}$$
$$\nabla \times \mathbf{H}_e = \overset{*}{\mathbf{D}} + \mathbf{J}_e = \partial \mathbf{D}/\partial t + \nabla \times (\mathbf{D} \times \mathbf{v}) + \sigma \mathbf{v} + \mathbf{J}_e, \tag{4.10D}$$
$$\nabla \cdot \mathbf{J}_e + \partial \sigma/\partial t + \nabla \cdot (\mathbf{v}\sigma) = 0. \tag{4.10E}$$

Note that Eqs. (4.9B) and (4.10B), which are the *Faraday law of induction for moving bodies*, were presented in Maxwell's *Treatise* (Section 598). This and other equations as contained in the original *Treatise* are discussed recently by C. T. Tai [41], using modern notations and interpretations. Examples illustrate the physical principle underlying this equation are given by D. Corson [23].

The entire set of equations (4.10) was proposed by H. Hertz (1890) as the electromagnetic field equations for moving bodies (see p. 328, [19]). The right-hand side of Eq. (4.10D) is what O. Heaviside (1892) called the *current of dielectric convection.*

For a moving body, a charge in the body which is stationary to an observer moving with the body appears as a convective current $\sigma\mathbf{v}$ to an observer in the laboratory frame. Thus in terms of laboratory frame variables,

$$\mathbf{J}_e \equiv \mathbf{J} - \sigma\mathbf{v}. \tag{4.11}$$

This relation reduces Eq. (4.10E) to the form of Eq. (3.2E), now valid also for moving bodies. When $\mathbf{v} = 0$, Eqs. (4.10) agree with Eqs. (3.2).

The effective fields $\mathbf{E}_e$ and $\mathbf{H}_e$ may be expressed in terms of the laboratory frame variables. However, the conversion is not unique because it is difficult to measure experimentally $\mathbf{E}$ or $\mathbf{H}$ field inside matter either in the rest frame or in the laboratory frame. Based on experimental facts for moving charges or current circuits in the vacuum, various interpretations may be assigned to the effective fields, resulting in a variety of formulations of the Maxwell equations for moving media.

A. The Minkowski Formulation (**EBDH**). If the effective fields are defined as

$$\mathbf{E}_e \equiv \mathbf{E} + \mathbf{v} \times \mathbf{B}, \qquad \mathbf{H}_e \equiv \mathbf{H} - \mathbf{v} \times \mathbf{D} \tag{4.12}$$

we obtain Eqs. (I). All variables are measured in the laboratory frame.

The first of Eq. (4.12) is based on the Lorentz force formula for a moving charge in vacuum. According to Eq. (3.30), a charge q which moves with velocity $\mathbf{v}$ experiences a net force

$$\mathbf{f} = q\mathbf{E} + q\mathbf{v} \times \mathbf{B} = q(\mathbf{E} + \mathbf{v} \times \mathbf{B}) = q\mathbf{E}_e.$$

Thus the effective electric field on the moving charge is $\mathbf{E}_e$. The second of Eq. (4.12) is based on a similar reasoning for a magnetic pole (see Chapters 1, 3, and 4, [3]). Comparing Eq. (4.12) with the first pair of Eq. (4.2), it is seen that the effective field is the same as the field in the rest frame for slow motions.

B. The Lorentz Formulation (**EPBMv**). Instead of $\mathbf{H}_e$ in Eq. (4.12), we may define

$$\mathbf{E}_e \equiv \mathbf{E} + \mathbf{v} \times \mathbf{B}, \qquad \mathbf{H}_e \equiv \mathbf{H} - \mathbf{v} \times \varepsilon_0\mathbf{E}. \tag{4.13}$$

Substituting the above into Eq. (4.10) and then converting **H** to **M** and **B**, and **D** to **E** and **P** according to Eq. (3.5), we obtain Eqs. (II).

C. The Statistical Formulation (**EBPM**). Recall that Eq. (3.6) can be directly deduced from Eq. (3.2) by a change of the field variables. Hence we can derive Eq. (III) from Eq. (4.10) by using the same definition as in Eq. (4.12) and then convert **H** and **D** to laboratory frame variables according to Eq. (3.5).

D. The Chu Formulation (**EHPMv**). Since in the free space $\mu_0\mathbf{H} = \mathbf{B}$ and $\varepsilon_0\mathbf{E} = \mathbf{D}$, we may change the definition of the effective fields in Eq. (4.12) to

$$\mathbf{E}_e \equiv \mathbf{E} + \mathbf{v} \times \mu_0\mathbf{H}, \qquad \mathbf{H}_e \equiv \mathbf{H} - \mathbf{v} \times \varepsilon_0\mathbf{E}. \tag{4.14}$$

Substituting the above to Eq. (4.10) and changing **B** and **D** to **E**, **H**, **P**, **M** according to Eq. (3.5), we obtain Eqs. (IV). The physical meaning for the effective fields in this case is discussed in [5] (p. 390).

It should become clear by comparing Eq. (4.14) with Eq. (4.12) that **E** and **H** in the Chu formulation and those in the Minkowski formulation are different. As first shown by Tai [40], they are related by

$$\mathbf{E}^{\mathrm{M}} = \mathbf{E}^{\mathrm{C}} + \mu_0 \mathbf{M}^{\mathrm{C}} \times \mathbf{v}, \qquad \mathbf{H}^{\mathrm{M}} = \mathbf{H}^{\mathrm{C}} - \mathbf{P}^{\mathrm{C}} \times \mathbf{v}, \tag{4.15}$$

where the superscripts M and C designate the Minkowski and the Chu formulation respectively.

Similarly, by comparing the Lorentz formulation (superscript L) and the statistical formulation (superscript S), we find

$$\mathbf{E}^{\mathrm{S}} = \mathbf{E}^{\mathrm{L}}, \qquad \mathbf{H}^{\mathrm{S}} = \mathbf{H}^{\mathrm{L}} + \mathbf{v} \times \mathbf{P}^{\mathrm{L}},$$

or equivalently,

$$\mathbf{P}^{\mathrm{S}} = \mathbf{P}^{\mathrm{L}}, \qquad \mathbf{M}^{\mathrm{S}} = \mathbf{M}^{\mathrm{L}} + \mathbf{P}^{\mathrm{L}} \times \mathbf{v}. \tag{4.16}$$

Except for the Lorentz formulation, detailed comparisons of various formulations of electrodynamics are made by Penfield and Haus (Chapter 7, [15]).

5 MAXWELL STRESS TENSOR AND MINKOWSKI ENERGY-MOMENTUM TENSOR

Based on the Maxwell–Minkowski equations, Eqs. (I), we begin our discussion on the stresses and forces in a material medium.

5.1 The Maxwell Stress Tensor

In Part IV, Chapter XI of the *Treatise* by Maxwell, the energy and stress in the electromagnetic field are discussed. The total energy density is shown to consist of two components, the electrostatic energy density $\frac{1}{2}\mathbf{E} \cdot \mathbf{D}$, and the magnetostatic energy density $\frac{1}{2}\mathbf{H} \cdot \mathbf{B}$. The magnetic force on a volume element is also made of two components: one, on the total current density in the element, $\mathbf{J}_t \times \mathbf{B}$, and the other on the magnetic moment density in the element, $(\nabla\mathbf{H}) \cdot \mu_0\mathbf{M}$. The former was based on the experiments conducted by Ampère and the latter was derived from the change of the magnetic energy due to the displacement of the element.

The total magnetic force per unit volume is

$$\mathbf{F} = (\nabla\mathbf{H}) \cdot \mu_0\mathbf{M} + \mathbf{J}_t \times \mathbf{B}. \tag{5.1}$$

In addition, there is the moment of force per unit volume

$$\mathbf{L} = \mathbf{B} \times \mathbf{H}. \tag{5.2}$$

Since $\mathbf{J}_t = \mathbf{J} + \partial\mathbf{D}/\partial t = \nabla \times \mathbf{H}$ and $\mu_0\mathbf{M} = \mathbf{B} - \mu_0\mathbf{H}$, Eq. (5.1) may be transformed to

$$\mathbf{F} = \nabla \cdot (\mathbf{B}\mathbf{H} - \tfrac{1}{2}\mu_0\mathbf{H}\cdot\mathbf{H}\underset{\sim}{\mathbf{I}}), \tag{5.3}$$

where $\underset{\sim}{\mathbf{I}}$ is the unit dyadic. The quantity in the parenthesis was identified by Maxwell as a stress, and it is now called the *Maxwell stress tensor*.

The earlier part of the *Treatise* also calculated forces and stresses on a system of electric charges in the vacuum by a different approach (see Section 9). The force, exerted by electric fields on polarized ponderable bodies, was calculated by W. H. Bragg in 1892 (p. 272, [19]).

5.2 Balance Laws of Electromagnetic Momentum and Energy

Local balance laws of electromagnetism in the form of Eq. (2.5) may be derived from the Maxwell equations. Consider first Eqs. (3.2) as applied to bodies at rest. By multiplying Eq. (3.2B) by $\mathbf{D}$ (vector product), and Eq. (3.2D) by $\mathbf{B}$, and then adding the two resulting equations, we obtain

$$-\nabla \cdot (\mathbf{D}\mathbf{E} + \mathbf{B}\mathbf{H}) + (\nabla\mathbf{E}) \cdot \mathbf{D} + (\nabla\mathbf{H}) \cdot \mathbf{B} + \partial(\mathbf{D} \times \mathbf{B})/\partial t = -\sigma\mathbf{E} - \mathbf{J} \times \mathbf{B}.$$

The above equation may be written compactly,

$$\begin{aligned}\nabla \cdot [(\mathbf{E}\cdot\mathbf{D} + \mathbf{H}\cdot\mathbf{B})\underset{\sim}{\mathbf{I}} - (\mathbf{D}\mathbf{E} + \mathbf{B}\mathbf{H})] + \partial(\mathbf{D} \times \mathbf{B})/\partial t \\ = -\sigma\mathbf{E} - \mathbf{J} \times \mathbf{B} + (\nabla\mathbf{D}) \cdot \mathbf{E} + (\nabla\mathbf{B}) \cdot \mathbf{H},\end{aligned} \tag{5.4}$$

or

$$\begin{aligned}\nabla \cdot [\tfrac{1}{2}(\mathbf{E}\cdot\mathbf{D} + \mathbf{H}\cdot\mathbf{B})\underset{\sim}{\mathbf{I}} - (\mathbf{D}\mathbf{E} + \mathbf{B}\mathbf{H})] + \partial(\mathbf{D} \times \mathbf{B})/\partial t \\ = -\sigma\mathbf{E} - \mathbf{J} \times \mathbf{B} + \tfrac{1}{2}[(\nabla\mathbf{D}) \cdot \mathbf{E} - (\nabla\mathbf{E}) \cdot \mathbf{D}] + \tfrac{1}{2}[(\nabla\mathbf{B}) \cdot \mathbf{H} - (\nabla\mathbf{H}) \cdot \mathbf{B}].\end{aligned} \tag{5.5}$$

Similarly, taking the dot product (scalar product) of Eq. (3.2B) with $\mathbf{H}$, and Eq. (3.2D) with $\mathbf{E}$, and subtracting one equation from the other, we obtain

$$\nabla \cdot (\mathbf{E} \times \mathbf{H}) + \left(\mathbf{H} \cdot \frac{\partial\mathbf{B}}{\partial t} + \mathbf{E} \cdot \frac{\partial\mathbf{D}}{\partial t}\right) = -\mathbf{J} \cdot \mathbf{E}. \tag{5.6}$$

Equation (5.6) is a balance law of electromagnetic energy and it was first derived by J. H. Poynting (1884) in global form. The term on the right of Eq. (5.6), $\mathbf{J} \cdot \mathbf{E}$, represents the supply of energy in the form of the Joule heat. From the second law of thermodynamics, it can be shown that this energy supply is an irreversible process. The term $\mathbf{H} \cdot \partial \mathbf{B}/\partial t + \mathbf{E} \cdot \partial \mathbf{D}/\partial t$ may be expressed as $\partial W/\partial t$ where W is the energy density stored in the electromagnetic field. In the first term, $\mathbf{E} \times \mathbf{H}$ is the Poynting vector of energy flow (Section 2.19, [17]).

Equation (5.5) may be expressed as

$$\nabla \cdot \underset{\sim}{\boldsymbol{\tau}}^{\mathrm{M}} + \frac{\partial}{\partial t}(\mathbf{D} \times \mathbf{B}) = -\mathbf{F}^{\mathrm{M}}, \tag{5.7}$$

where

$$\underset{\sim}{\boldsymbol{\tau}}^{\mathrm{M}} = \tfrac{1}{2}(\mathbf{D} \cdot \mathbf{E} + \mathbf{B} \cdot \mathbf{H})\underset{\sim}{\mathbf{I}} - (\mathbf{DE} + \mathbf{BH}), \tag{5.8}$$

$$\mathbf{F}^{\mathrm{M}} = \sigma \mathbf{E} + \mathbf{J} \times \mathbf{B} - \tfrac{1}{2}[(\nabla \mathbf{D}) \cdot \mathbf{E} - (\nabla \mathbf{E}) \cdot \mathbf{D}] - \tfrac{1}{2}[(\nabla \mathbf{B}) \cdot \mathbf{H} - (\nabla \mathbf{H}) \cdot \mathbf{B}]. \tag{5.9}$$

The magnetic part of $\underset{\sim}{\boldsymbol{\tau}}^{\mathrm{M}}$ resembles the terms inside the parentheses of Eq. (5.3) and we shall call $\underset{\sim}{\boldsymbol{\tau}}^{\mathrm{M}}$ the *Maxwell stress tensor*. The product $\mathbf{D} \times \mathbf{B}$ was identified as the *electromagnetic momentum* by J. J. Thomson (1893) and M. Abraham (1902) (p. 317 of [19]). The entire expression for $\mathbf{F}^{\mathrm{M}}$ represents the transfer of momentum to or from a unit volume of the field, and it is identified as a force.

Many texts (e.g. p. 159 of [17], p. 181 of [14]) present the balance law of momentum only for linear isotropic materials. In view of Eq. (3.3), Eq. (5.9) may be written

$$\mathbf{F}^{\mathrm{M}}(\text{linear}) = \sigma \mathbf{E} + \mathbf{J} \times \mathbf{B} - \tfrac{1}{2}E^2 \nabla \varepsilon - \tfrac{1}{2}H^2 \nabla \mu \tag{5.10}$$

and Eq. (5.6) may be written

$$\nabla \cdot (\mathbf{E} \times \mathbf{H}) + \partial W/\partial t = -\mathbf{J} \cdot \mathbf{E}, \tag{5.11}$$

where

$$W(\text{linear}) = \tfrac{1}{2}(\mathbf{E} \cdot \mathbf{D} + \mathbf{H} \cdot \mathbf{B}) - \tfrac{1}{2}(\varepsilon E^2 + \mu H^2). \tag{5.12}$$

The force expression in Eq. (5.10) was first derived by H. von Helmholz (1881), using a different approach, and it is known as the *Helmholtz force* (see Section 2.22 of [17]).

We have identified $\mathbf{F}^{\mathrm{M}}$ in Eqs. (5.9) or (5.10) as a force in electromagnetic fields. Whether we can now consider it as the body force in the mechanical momentum equation, Eq. (2.12), is far from certain.

First, we note that Eqs. (5.8) and (5.9) are not the unique expressions that can be inferred from the balance equations. Note that Eq. (5.4) can also be expressed in the form of Eq. (5.7) if we define a stress tensor and a force as follows:

$$\underset{\sim}{\tau}^{PH} = (\mathbf{D}\cdot\mathbf{E} + \mathbf{B}\cdot\mathbf{H})\underset{\sim}{\mathbf{I}} - (\mathbf{DE} + \mathbf{BH}), \tag{5.13}$$

$$\mathbf{F}^{PH} = \sigma\mathbf{E} + \mathbf{J}\times\mathbf{B} - (\nabla\mathbf{D})\cdot\mathbf{E} - (\nabla\mathbf{B})\cdot\mathbf{H}. \tag{5.14}$$

They are quite different from those in Eqs. (5.8), and (5.9).

Expressions like Eqs. (5.13) and (5.14) were proposed in [22] as the stress and force respectively in the field-material subsystem. Furthermore, the kinetic force acting on the material body as given in [15] (Eq. (7.42)) reduces to Eq. (5.14), but not to Eq. (5.9), when the body is at rest. However, the majority of the literature prefers $\underset{\sim}{\tau}^M$ and $\mathbf{F}^M$ as given by Eqs. (5.8) and (5.9) because they also occur naturally in the four-dimensional formulation of the Maxwell equations, which is discussed in the next subsection.

Secondly, the Helmholtz force Eq. (5.10) is restricted to linear isotropic materials (it can be generalized to include anisotropy). This is at variance with the modern approach in continuum physics that a balance law should be independent of the constitutive relations of materials. Furthermore, Eq. (5.10) reduces simply to $\mathbf{F}^M = \sigma\mathbf{E} + \mathbf{J}\times\mathbf{B}$ when the material is also homogeneous. Thus for an isotropic, homogeneous magnetic insulator ($\sigma = 0$, $\mathbf{J} = 0$), there would be no body force acting on the material according to Eq. (5.10). This seems to contradict experimental observations.

Finally, there remains the question on whether the momentum flow and force density in the electromagnetic field are equal and opposite to the momentum and force acting on the material body. We will reserve the discussion of this question for the next section.

5.3 The Minkowski Energy–momentum Tensor

In 1908 H. Minkowski formulated the laws of electromagnetism in a four-dimensional manifold of time-space [32]. He also showed that field equations for moving media can be derived from those for stationary matter by applying the principle of special relativity, which is discussed in Subsection 4.1. In a posthumous paper edited by M. Born [33], he extended the treatment to reconcile some differences between his theory and the Lorentz theory of electrons. The concept of an energy–momentum tensor in the four-dimensional manifold was initiated in these papers. Our discussion here follows A. Sommerfeld (Sections 31, 34, 35, [16]) and C. Møller (Chapters 5 and 6, [13]).

We shall use Latin indices to denote the components of a vector or tensor in the three-dimensional space (3-space) with Cartesian coordinates (x_1, x_2, x_3), and Greek indices to denote the components in the four-dimensional space–time manifold (4-space) with coordinates $(x_1, x_2, x_3, x_4 = ict)$. The metric of the 4-space is

$$ds^2 = dx_1^2 + dx_2^2 + dx_3^2 - d(ct)^2 = dx_\alpha dx_\alpha, \qquad \alpha = 1, 2, 3, 4. \tag{5.15}$$

Repeated indices imply the summing from 1 to 3 in Latin letters and 1 to 4 in Greek letters.

Define two antisymmetric tensors $F_{\mu\nu}$ and $G_{\mu\nu}$ as

$$[F_{\mu\nu}] = \begin{bmatrix} 0 & cB_3 & -cB_2 & -iE_1 \\ -cB_3 & 0 & cB_1 & -iE_2 \\ cB_2 & -cB_1 & 0 & -iE_3 \\ iE_1 & iE_2 & iE_3 & 0 \end{bmatrix},$$

$$[G_{\mu\nu}] = \begin{bmatrix} 0 & H_3 & -H_2 & -icD_1 \\ -H_3 & 0 & H_1 & -icD_2 \\ H_2 & -H_1 & 0 & -icD_3 \\ icD_1 & icD_2 & icD_3 & 0 \end{bmatrix} \tag{5.16}$$

and a vector in 4-space as

$$j_\mu = (\mathbf{J}, ic\sigma). \tag{5.17}$$

Since $F_{\mu\nu} = -F_{\nu\mu}$ and $G_{\mu\nu} = -G_{\nu\mu}$, out of sixteen components of each tensor only six are independent.

The Maxwell–Minkowski equations (3.2A)–(3.2D) can be written in terms of $F_{\mu\nu}$, $G_{\mu\nu}$, and $j_\mu (\lambda, \mu, \nu = 1, 2, 3, 4)$ as

$$\text{(I)} \quad \begin{cases} \partial_\lambda F_{\nu\mu} + \partial_\mu F_{\lambda\nu} + \partial_\nu F_{\mu\lambda} = 0, & (5.18) \\ \partial_\lambda G_{\mu\lambda} = j_\mu, & (5.19) \end{cases}$$

where $\partial_\lambda \equiv \partial/\partial x_\lambda$. Because of the skew-symmetry, Eq. (5.18) is nontrivial only when λ, ν, and μ are different from one another. We therefore obtain from it four scalar equations in accordance with the following trios of values: (2, 3, 4), (3, 4, 1), (4, 1, 2) and (1, 2, 3), the last giving rise to Eq. (3.2A) and the first three to Eq. (3.2B). The two Maxwell equations with sources are represented by Eq. (5.19); for $\mu = 1, 2, 3$ it yields Eq. (3.2D) and for $\mu = 4$ it yields Eq. (3.2C).

The formula for the force in Eq. (3.20) can be expressed by a vector f_ν in 4-space,

$$f_\nu = c^{-1}F_{\nu\mu}j_\mu. \tag{5.20}$$

The first three components are $\sigma\mathbf{E} + \mathbf{J} \times \mathbf{B}$ and the fourth component is $(i\mathbf{J} \cdot \mathbf{E}/c)$.

Taking the inner product of the tensor equation (5.19) with $F_{\nu\mu}$, we obtain (Section 7.7, [13])

$$\begin{aligned} F_{\nu\mu}j_\mu &= F_{\nu\mu}(\partial_\lambda G_{\mu\lambda}) \\ &= \partial_\lambda(F_{\nu\mu}G_{\mu\lambda}) - G_{\mu\lambda}(\partial_\lambda F_{\nu\mu}) \\ &= \partial_\lambda(F_{\nu\mu}G_{\mu\lambda}) - \tfrac{1}{2}(\partial_\lambda F_{\mu\nu} + \partial_\mu F_{\nu\lambda})G_{\lambda\mu} \\ &= \partial_\lambda(F_{\nu\mu}G_{\mu\lambda}) + \tfrac{1}{2}(\partial_\nu F_{\lambda\mu})G_{\lambda\mu}. \end{aligned} \tag{5.21}$$

In the last step use has been made of Eq. (5.18). Furthermore,

$$\tfrac{1}{2}G_{\lambda\mu}(\partial_\nu F_{\lambda\mu}) = \tfrac{1}{4}\partial_\nu(F_{\lambda\mu}G_{\lambda\mu}) + \tfrac{1}{4}[G_{\lambda\mu}(\partial_\nu F_{\lambda\mu}) - F_{\lambda\mu}(\partial_\nu G_{\lambda\mu})].$$

Thus

$$F_{\nu\mu}j_\mu = -\partial_\lambda(G_{\lambda\mu}F_{\nu\mu} - \tfrac{1}{4}\delta_{\lambda\nu}F_{\xi\mu}G_{\xi\mu}) + \tfrac{1}{4}[G_{\lambda\mu}(\partial_\nu F_{\lambda\mu}) - F_{\lambda\mu}(\partial_\nu G_{\lambda\mu})]. \tag{5.22}$$

The first term on the right of Eq. (5.22) is the divergence of a tensor in 4-space. Defining

$$S_{\lambda\nu}^{(e)} \equiv c^{-1}(G_{\lambda\mu}F_{\nu\mu} - \tfrac{1}{4}\delta_{\lambda\nu}F_{\xi\mu}G_{\xi\mu}), \tag{5.23}$$

we write Eq. (5.22) in the following form:

$$\frac{\partial S_{\lambda\nu}^{(e)}}{\partial x_\lambda} = -\frac{1}{c}F_{\nu\mu}j_\mu + \frac{1}{4c}\left(G_{\lambda\mu}\frac{\partial F_{\lambda\mu}}{\partial x_\nu} - F_{\lambda\mu}\frac{\partial G_{\lambda\mu}}{\partial x_\nu}\right). \tag{5.24}$$

The $S_{\lambda\nu}^{(e)}$ is called the *Minkowski energy–momentum tensor*.

In three dimensional space, the components of $S_{\lambda\nu}^{(e)}$ are

$$S_{\lambda\nu}^{(e)} = \left[\begin{array}{c:c} \tau_{jk}^{\mathrm{M}} & is_k/c \\ \hdashline icg_j & -W \end{array}\right], \tag{5.25}$$

where

$$\begin{aligned} \tau_{jk}^{\mathrm{M}} &= \tfrac{1}{2}(E_m D_m + H_m B_m)\delta_{jk} - (D_j E_k + B_j H_k) \\ g_j &= (\mathbf{D} \times \mathbf{B})_j \\ s_k &= (\mathbf{E} \times \mathbf{H})_k \\ W &= \tfrac{1}{2}(E_m D_m + H_m B_m), \qquad m, j, k = 1, 2, 3. \end{aligned} \tag{5.26}$$

Note that $\mathbf{s}$ is the Poynting vector and W the energy density in Eq. (5.11); $\underset{\sim}{\tau}^{\mathrm{M}}$ is the Maxwell stress tensor in Eq. (5.8) and $\mathbf{g}$ the electromagnetic momentum in Eq. (5.7).

Furthermore, the first term on the right of Eq. (5.24) is simply $-f_\nu$, and the second term is

$$\frac{1}{4c}\left(G_{\lambda\mu}\frac{\partial F_{\lambda\mu}}{\partial x_\nu} - F_{\lambda\mu}\frac{\partial G_{\lambda\mu}}{\partial x_\nu}\right)$$

$$= \frac{1}{2}\left(\frac{\partial \mathbf{B}}{\partial x_\nu}\cdot\mathbf{H} - \frac{\partial \mathbf{E}}{\partial x_\nu}\cdot\mathbf{D} - \mathbf{B}\cdot\frac{\partial \mathbf{H}}{\partial x_\nu} + \mathbf{E}\cdot\frac{\partial \mathbf{D}}{\partial x_\nu}\right). \tag{5.27}$$

Thus the spatial part ($\nu = 1, 2, 3$) of Eq. (5.24) is just the balance of momentum Eq. (5.7), and the temporal part ($\nu = 4$) of it is the balance of energy, Eq. (5.6).

In the aether, $\mathbf{D} = \varepsilon_0\mathbf{E}$ and $\mathbf{B} = \mu_0\mathbf{H}$, the two field tensors are related by

$$G_{\nu\mu} = (\varepsilon_0/\mu_0)^{1/2}F_{\nu\mu}. \tag{5.28}$$

Thus the second term on the right of Eq. (5.24) vanishes and

$$-\partial_\lambda S^{(e)}_{\lambda\nu}(\text{aether}) = c^{-1}F_{\nu\lambda}j_\lambda = f_\nu. \tag{5.29}$$

The spatial part on the right-hand side of this equation is the Lorentz force on a test body, Eq. (3.20).

Based on this fact and the notion that the transference of momentum to a volume element generates a body force on the element, Minkowski postulated that for a material body in uniform motion, there is a force, f^{M}_ν (in 4-space), acting on a volume element of the body; this force equals the negative of the divergence of the energy–momentum tensor,

$$f^{\mathrm{M}}_\nu \equiv -\partial_\lambda S^{(e)}_{\lambda\nu}. \tag{5.30}$$

From Eq. (5.24), we find

$$f^{\mathrm{M}}_\nu \equiv \frac{1}{c}F_{\nu\mu}j_\mu - \frac{1}{4c}\left(G_{\lambda\mu}\frac{\partial F_{\lambda\mu}}{\partial x_\nu} - F_{\lambda\mu}\frac{\partial G_{\lambda\mu}}{\partial x_\nu}\right). \tag{5.31}$$

The spatial part of f^{M}_ν is just the F^{M}_j in Eq. (5.9); the temporal part can be calculated from Eq. (5.27), with†

$$f^{\mathrm{M}}_j = \sigma E_j + (\mathbf{J}\times\mathbf{B})_j - \frac{1}{2}\left(\frac{\partial \mathbf{B}}{\partial x_j}\cdot\mathbf{H} - \frac{\partial \mathbf{H}}{\partial x_j}\cdot\mathbf{B}\right) - \frac{1}{2}\left(\frac{\partial \mathbf{D}}{\partial x_j}\cdot\mathbf{E} - \frac{\partial \mathbf{E}}{\partial x_j}\cdot\mathbf{D}\right), \tag{5.32}$$

† Linear isotropic materials were assumed in Minkowski's paper.

$$icf_4^{\mathrm{M}} = -\mathbf{J}\cdot\mathbf{E} - \frac{1}{2}\left(\frac{\partial\mathbf{B}}{\partial t}\cdot\mathbf{H} - \frac{\partial\mathbf{H}}{\partial t}\cdot\mathbf{B} + \frac{\partial\mathbf{D}}{\partial t}\cdot\mathbf{E} - \frac{\partial\mathbf{E}}{\partial t}\cdot\mathbf{D}\right). \tag{5.33}$$

So far as a stationary medium is concerned, nothing is gained by constructing the energy–momentum tensor in 4-space. For moving matter, Minkowski's analysis has a great advantage as the energy–momentum tensor so constructed is covariant with the electrodynamic equations (5.18) and (5.19). However, since these equations in a laboratory frame are the same as the Maxwell equations for stationary media, it is not surprising that Eqs. (5.24) are identical to the balance equations (5.6) and (5.7).

After constructing the energy–momentum tensor, $S_{\lambda\nu}^{(e)}$ for the electromagnetic field, Minkowski simply assumed that the force which is exerted by the field on the matter (rigid bodies) is $-\partial_\lambda S_\lambda^{(e)}$. Thus the question of how the matter interacts with the field is answered by an assumption. We shall return to this question in the next section.

Since the Minkowski tensor $S_{\lambda\nu}^{(e)}$ is not symmetric and its momentum component and energy component do not follow Planck's principle that **s** equals $c^2\mathbf{g}$, there were doubts about whether $-\partial_\lambda S_{\lambda\nu}^{(e)}$ indeed yields the "correct" expression for the force. Many other forms for the energy–momentum tensor have been proposed thereafter, notably the Abraham tensor and the Einstein–Laub tensor. These tensors were recently compared by I. Brevik [20]. By making additional assumptions such as linear constitutive laws and dipole models for polarizations, Brevik also derived the Minkowski's tensor from other first principles.

We have discussed only the energy–momentum tensor for the Minkowski formulation (I). The corresponding tensor in terms of **E**, **B**, **P**, **M** formulation III) was also presented by Minkowski [33] (see Chapter 18 of [14]), and that for the Chu formulation (IV) is given in the Appendix 1 of [5].

5.4 Interaction of Fields with Matter

At the time of Faraday and Maxwell, the model of aether was adopted to interpret the transmission of electromagnetic effects. The effect is exerted by one material object on another some distance away in space. The space was conceived as an unchangeable passive constituent of the universe. A hypothetical substance was assumed to permeate all space, including the volumes occupied by ordinary material substances, in which all dynamic actions took place. This hypothetical "immaterial medium" was called the *aether*. The dynamic action of electromagnetism in the aether was described by the Maxwell equations (3.2) with the "aetheral relation" Eq. (3.1).

The ordinary material substance (matter) was treated as if it were merely a modification of the aether, distinguished only by different constitutive laws. Thus if the material medium were at rest, there was no need to distinguish between the action of the electromagnetic field in the aether or that in the matter. On this basis, the Maxwell stress for stationary media as developed in the *Treatise* might be considered as the stresses acting on the matter.

However, some distinction between the aether and the matter must be made when the body is in motion. Maxwell presumed that material bodies when displaced carry the contained aether along with them. This supposition was not followed in later years because it is inconsistent with Fresnel's theory (1818) of the light propagation in a moving transparent body. Instead, electromagnetic stresses and forces correspond to those introduced by Maxwell are assigned to the aether, as distinct from material bodies. It is the aether which carries the dynamic actions as described by the Maxwell equations, including the stresses derived from them. The forces and stresses which act on the material bodies depend on how the matter interacts with the aether. This is the basis of the Lorentz theory of electrons (see Chapters 9, 10, 12, [19]) and it is widely accepted today.

Modern scientific literature tends to regard the aetherally pothesis unsound or at least unnecessary. Thus to specify a medium without material substances, the phrase *in the aether* is replaced by *in the vacuum* or *in vacuo*. The stresses generated by the electromagnetic fields *in the aether* are now simply the stresses *in the field*. With this understanding, we return to the question on how moving matter interacts with fields.

Neglecting the *deformation* which described the local change of the relative positions of material particles, the general motion of a *rigid body* is composed of a *translation* of an arbitrary particle in the body, known as the base point, and a *rotation* about this base point. Since the principle of special relativity applies to moving frames translating at constant velocities, the Minkowski theory as described in Subsection 5.3 pertains only to rigid bodies moving with constant velocities in electromagnetic fields, or to sources which generate electromagnetic fields moving with constant velocities relative to stationary matter. The interaction of the rigid body (matter) with the field is determined by constructing the energy-momentum tensor of the electromagnetic field, $S_{\mu\nu}^{(e)}$, and then postulating a force $f_\nu^{\mathrm{M}} = -\partial_\mu S_{\mu\nu}^{(e)}$ acting on the matter. Since the tensor $S_{\mu\nu}^{(e)}$ is covariant with the basic equations of electrodynamics, the f_ν^{M} so derived is invariant under the Lorentz transformation.

When a rigid body is in acceleration or in rotation, one could presumably follow the same procedure and logic to obtain a force (in 4-space) that acts on the body. However, the construction of $S_{\mu\nu}^{(e)}$ must be based on electrodynamic equations that follow the general principle of relativity.

For deformable media, the general motion is described by a transformation function $\hat{\mathbf{x}}(X, t)$ which maps a particle $\mathbf{X}$ in a reference configuration to the position $\mathbf{x}$ in the present configuration, as defined in Eq. (2.2). The velocity field $\mathbf{v} = \dot{\mathbf{x}} = \partial\hat{\mathbf{x}}(X, t)/\partial t$ satisfies the balance equations (2.11)–(2.14). In modern theories, the interaction of deformable matter with the electromagnetic field is prescribed by assuming particular forms for the force $\mathbf{F}$, couple $\mathbf{L}$, and energy source ϕ in these equations as functions of the electromagnetic variables. The interaction of a rigid body with the field is then a special case in which the function $\hat{\mathbf{x}}(X, t)$ is restricted to a rigid motion. The aforementioned theory of interaction was apparently originated by J. Lamor (1897) and G. H. Liven (1916) (see Sections 2.2102.23 of [17]). In Livens' theory (Sections 86 and 234 of [9]) one calculates the force exerted by the electromagnetic field on the polarized matter, and then these forces are introduced into the equation of elasticity to determine the deformation.

Livens' logic and procedure were criticized as follows. Since the force which causes the deformation depends on the polarization of the matter, and the deformation which depends on the magnitude of the force changes the polarization of the matter, two parts of the problem of interaction can not be handled separately (p. 146, [17]). Actually in this atricle, where both the balance equations of mechanics Eqs. (2.11)–(2.14) and the Maxwell equations are expressed in the present (deformed) configuration of the body, the two parts of the problem are treated simultaneously, not separately.

Prior to Lamor and Livens' work, D. J. Korteweg (1880) and H. von Helmholtz (1881) considered the effect of deformation by letting the dielectric constant ε in the linear constitutive law $\mathbf{D} = \varepsilon\mathbf{E}$ be a function of the strain. By calculating the change in free energy of the electrostatic field as a result of deformation, and equating this change of energy to the work done by electromagnetic forces, an expression for the body force in solid dielectrics is obtained (p. 145, [17])

$$\mathbf{F} = -\tfrac{1}{2}E^2\nabla\varepsilon + \tfrac{1}{2}\nabla\left(E^2\rho\,\frac{\partial\varepsilon}{\partial\rho}\right), \tag{5.34}$$

where ρ is the mass density. A similar expression can be obtained for magnetic forces. The additional term which depends on the mass density accounts for the *electrostrictive* effect or the *magnetostrictive effect.*

Helmholtz's treatment of the interaction had a dominant influence in physics literature, and his procedure became a standard one. In this procedure a free energy is first established for the electric or magnetic field in a body without deformation and another for the stress–strain field in the body neglecting electric or magnetic polarizations. The former gives rise to an expression similar to the first term in Eq. (5.34) and the latter yields the usual mechanical forces (not shown in Eq. (5.34)). The interaction is determined by prescribing an "interaction free energy" which would yield an expression similar to the second term in Eq. (5.34). The theory using this construction successfully accounts for many experimental results, but only when additional modifications are used (see a critical discussion by Brown, in Chapter 1 of [2]).

Both the Helmholtz and Liven treatments of interaction in deformed bodies are based on the linear theory of elasticity in which the distinction between the deformed configuration and undeformed configuration of a body is lost. Thus it is unclear as to whether the polarizations in the body are calculated on a volume element of the deformed configuration, or the undeformed configuration. For this reason the aforementioned criticism of Livens' treatment has some justification. The modern treatment using the finite deformation theory of continua was proposed by Toupin in 1956 [44]. His work has inspired many subsequent investigations.

One large question still remains. That is, how to specify the **F**, **L**, and ϕ in the balance equations in terms of the electromagnetic variables. Obviously additional postulates beyond those already contained in electromagnetic theories and finite deformation theories must be made. This accounts for the abundance of theories in literature, since different postulates, sometimes not even clearly stated, lead to different results. In this article, we try to sort out many schools of thought which can be grouped as follows:

1. Postulate an energy–momentum tensor for the combined system of the deformed media and the electromagnetic field. The interaction is determined from the conservation law of the total energy–momentum for the entire system. This is discussed in the next section.
2. Postulate that the interaction at the microscopic scale is characterized by the electrons (small charged particles) moving in the aether. The interaction on the macroscopic scale is then determined by applying a statistical averaging process. This follows the theory of electrons (Subsection 4.2). Details are discussed in Section 7.
3. Postulate that the force acting on the material by the field is given by a formula similar to that of Eq. (3.20). The charge distribution and current density to be substituted into the formula depend on the interpretation of various formulations of the Maxwell equations. This is discussed in Section 8.
4. Postulate the resultant force and moment on the entire material body

due to externally applied electromagnetic fields. The forces and moments on the interior elements of the body are then determined by evaluating the additional effect on the interior element due to its surrounding materials. This is discussed in Section 9.

5. Postulate models such as charges, poles, dipoles, and current circuits which characterize the macroscopic behavior of polarized materials. The interaction of the material with the electromagnetic field is then determined from these models. Various models are discussed in Section 10.

There are other approaches which are a variation or combination of the above approaches. We shall not delineate them here. Because of the difficulty of measuring the various electromagnetic field quantities inside the matter, the theories resulting from these postulates have not been confirmed experimentally. Early experimental works on electrodynamics of moving media are reviewed in Refs. [3] and [20].

6 TOTAL ENERGY–MOMENTUM TENSORS

In this section, we discuss the field–matter interaction based on the principle of conservation of total energy–momentum (in 4-space) in a closed system.

6.1 Closed Systems and Open Systems

A physical system is *closed* if the total momentum and energy within the system are conserved. Otherwise, it is an *open* system.

As an example, consider the motion of an elastic body under the influence of mechanical forces as one physical system. In the absence of the heat flux **Q** and the heat source ϕ_Q, the local forms of the balance equations (2.7) and (2.9) are

$$\frac{\partial}{\partial t}(\rho \mathbf{v}) + \nabla \cdot (\rho \mathbf{v}\mathbf{v} - \underset{\sim}{\tau}) = \mathbf{F}, \tag{6.1}$$

$$\frac{\partial \mathscr{E}}{\partial t} + \nabla \cdot (\mathbf{v}\mathscr{E} - \underset{\sim}{\tau} \cdot \mathbf{v}) = \mathbf{F} \cdot \mathbf{v} + \phi, \tag{6.2}$$

where

$$\mathscr{E} = \tfrac{1}{2}\rho \mathbf{v} \cdot \mathbf{v} + \rho U. \tag{6.3}$$

Note that Eq. (6.1) reduces to Eq. (2.12) because of Eq. (2.11),

$$\frac{\partial}{\partial t}(\rho \mathbf{v}) + \nabla \cdot (\rho \mathbf{v}\mathbf{v}) = \rho \frac{\partial \mathbf{v}}{\partial t} + \rho \mathbf{v} \cdot \nabla \mathbf{v} = \rho \frac{d\mathbf{v}}{dt}. \tag{6.4}$$

Equation (6.2) is equivalent to Eq. (2.14) because of Eq. (2.12) and

$$\frac{\partial}{\partial t}(\rho U) + \nabla \cdot (\mathbf{v}\rho U) = \rho \frac{dU}{dt}, \tag{6.5}$$

$$\frac{\partial}{\partial t}(\tfrac{1}{2}\rho \mathbf{v} \cdot \mathbf{v}) + \nabla \cdot (\tfrac{1}{2}\rho \mathbf{v}\mathbf{v} \cdot \mathbf{v}) = \rho \frac{d\mathbf{v}}{dt} \cdot \mathbf{v}. \tag{6.6}$$

In the above, $\mathscr{E}$ is the total energy (kinetic plus internal energy) of the system, $\rho\mathbf{v}$ the kinetic momentum, $(\rho\mathbf{v}\mathbf{v} - \underset{\sim}{\tau})$ the kinetic momentum flux, and $(\mathbf{v}\mathscr{E} - \underset{\sim}{\tau} \cdot \mathbf{v})$ the energy flux. The $\mathbf{F}$ in this example is a mechanical body force and ϕ a mechanical source of energy supply. Within this system, the total momentum and energy within the system are conserved if $\mathbf{F} = \mathbf{0}$ and $\phi = 0$. Thus the system is considered closed if both $\mathbf{F}$ and ϕ vanish, and it is open if either $\mathbf{F}$ or ϕ does not vanish.

For convenience Eqs. (6.1) and (6.2) can be combined into one tensorial equation in 4-space,

$$\partial_\mu S_{\mu\nu}^{(m)} = f_\nu^{(m)} \qquad \text{(open system)}, \tag{6.7}$$

where the nonrelativistic energy–momentum tensor, $S_{\mu\nu}^{(m)}$, and the force vector $f_\nu^{(m)}$, are defined as

$$S_{\mu\nu}^{(m)} = \left[\begin{array}{c:c} \rho v_j v_k - \tau_{jk} & \frac{i}{c}(\mathscr{E} v_k - \tau_{kn} v_n) \\ \hdashline ic\rho v_j & -\mathscr{E} \end{array}\right], \qquad f_\nu^{(m)} = \left[\begin{array}{c} F_j \\ \frac{i}{c}(\phi + F_n v_n) \end{array}\right] \tag{6.8}$$

The superscript (m) signifies that only the mechanical energy and momentum of the system are considered. If $f_\nu^{(m)} = 0$, the system is closed,

$$\partial_\mu S_{\mu\nu}^{(m)} = 0 \qquad \text{(closed system)}. \tag{6.9}$$

An energy–momentum tensor like $S_{\nu\mu}^{(m)}$ which is invariant under the Lorentz transformation could also be defined for the same system from the relativistic equations of mechanical momentum and energy (Chapter 6, [13]).

As another example, consider electromagnetic fields within a material medium as the physical system. The energy–momentum tensor for the entire system is given by $S_{\lambda\nu}^{(e)}$ in Eq. (5.23). From Eq. (5.24) it is seen that the system is, in general, open. However, for electromagnetic fields in the aether which is free from charges and currents, the f_ν in Eq. (5.29) vanishes, and the system is closed.

For any given physical system, we may postulate a *total energy–momentum tensor* $S_{\mu\nu}$. If the system is closed, $S_{\mu\nu}$ satisfies the two conservation equations (Chapters 6, 7, [13]),

$$\partial_\mu S_{\mu\nu} = 0, \tag{6.10}$$

$$S_{\mu\nu} = S_{\nu\mu}. \tag{6.11}$$

The first is the conservation of the energy–momentum within the system and the second the conservation of angular momentum. The angular momentum in 4-space is defined by a third-rank tensor

$$\Theta_{\lambda\mu\nu} = x_\mu S_{\lambda\nu} - x_\nu S_{\lambda\mu}. \tag{6.12}$$

The conservation of $\Theta_{\lambda\mu\nu}$ is expressed by

$$\partial_\lambda \Theta_{\lambda\mu\nu} = 0. \tag{6.13}$$

In view of Eq. (6.10) the above reduces to Eq. (6.11).

The entire physical system may be divided into two (or more) subsystems and the total tensor is then decomposed into two (or more) components,

$$S_{\mu\nu} = S_{\mu\nu}^{(1)} + S_{\mu\nu}^{(2)}. \tag{6.14}$$

Each tensor is related to one subsystem. The construction of the total energy–momentum tensor and the tensor for each subsystem is usually guided by physical principles and laws as illustrated by Eq. (6.8). Although the total energy momentum tensor is symmetric, the tensor for each subsystem may not be.

Even when the total system is closed, the subsystem may not be. If, for example, the second subsystem is open, we obtain from Eqs. (6.10) and (6.14)

$$\partial_\mu S_{\mu\nu}^{(2)} = -f_\nu^{(2)} = -\partial_\mu S_{\mu\nu}^{(1)}, \tag{6.15}$$

where $f_\nu^{(2)}$ is a force in 4-space acting on the second subsystem. Thus the negative of $f_\nu^{(2)}$ is acting on the first subsystem.

6.2 Total Energy–momentum Tensor

To investigate the interaction of electromagnetic fields with matter in motion, we will consider a physical system which is composed of the matter and the field. Excluding the thermal effects, the system may be decomposed into two subsystems: the mechanical and the electrical. Thermal or other effects can be included later by adding to the above a third subsystem.

Let the energy–momentum tensor of the entire system be $S_{\mu\nu}$, and those of the mechanical and electrical subsystems be $S_{\mu\nu}^{(m)}$ and $S_{\mu\nu}^{(e)}$, respectively,

$$S_{\mu\nu} = S_{\mu\nu}^{(m)} + S_{\mu\nu}^{(e)}, \qquad \mu, \nu = 1,2,3,4. \tag{6.16}$$

We now assume that $S_{\mu\nu}^{(m)}$ is given by Eq. (6.8) and $S_{\nu\mu}^{(e)}$ by Eq. (5.23), and the entire physical system is closed; that is,

$$\partial_\mu S_{\mu\nu} = \partial_\mu [S_{\mu\nu}^{(m)} + S_{\mu\nu}^{(e)}] = 0. \tag{6.17}$$

The spatial part, $\nu = 1,2,3$, and the temporal part, $\nu = 4$, of the above equations are,

$$\begin{aligned} &\partial_t(\rho v_j) + \partial_k(\rho v_j v_k) - \partial_k(\tau_{jk} - \tau_{jk}^{\mathrm{M}}) + \partial_t g_j = 0, \\ &\partial_t \mathscr{E} + \partial_k(\mathscr{E} v_k) - \partial_j(\tau_{jk} v_k) + \partial_k s_k + \partial_t W = 0, \end{aligned} \tag{6.18}$$

where

$$\partial_t \equiv \partial/\partial t \equiv ic\, \partial/\partial x_4, \qquad \partial_k \equiv \partial/\partial x_k, \qquad j,k = 1,2,3.$$

As mentioned in Section 2, the mechanical body force and mechanical source of energy supply, if any, are omitted from this investigation. Thus $f_\mu^{(m)}$ in Eq. (6.7) should be considered vanishing and the mechanical subsystem is closed with $\partial_\nu S_{\nu\mu}^{(m)} = 0$. The electrical subsystem is open with

$$\partial_\mu S_{\mu\nu}^{(e)} = f_\nu^{(e)}, \tag{6.19}$$

where $f_\mu^{(e)}$ represents the right-hand side of Eq. (5.24). The spatial and temporal parts of the above equation are

$$\begin{aligned} &\partial_k \tau_{kj}^{\mathrm{M}} + \partial_t g_j = f_j^{(e)}, \\ &\partial_k s_k + \partial_t W = -icf_4^{(e)}. \end{aligned} \tag{6.20}$$

Substituting Eq. (6.20) into Eq. (6.18) we obtain

$$\begin{aligned} &\partial_t(\rho v_j) + \partial_k(\rho v_k v_j) - \partial_k \tau_{kj} + f_j^{(e)} = 0, \\ &\partial_t \mathscr{E} + \partial_k(\mathscr{E} v_k) - \partial_k(\tau_{kj} v_j) - icf_4^{(e)} = 0. \end{aligned} \tag{6.21}$$

In view of Eqs. (6.4)–(6.6), the above can be further simplified

$$\rho d_t v_j = \partial_k \tau_{kj} - f_j^{(e)}, \tag{6.22}$$

$$\rho d_t U = \tau_{kj} \partial_k v_j + (icf_4^{(e)} + f_j^{(e)} v_j), \tag{6.23}$$

where $d_t \equiv \partial_t + v_k \partial_k$.

Equation (6.22) is in the same form of Eq. (2.12), and Eq. (6.23) is in the form of Eq. (2.14) when Q and ϕ_Q are neglected in the latter. Thus we may identify the body force and energy supply as

$$F_j = -f_j^{(e)}, \tag{6.24}$$

$$\phi = icf_4^{(e)} + f_j^{(e)} v_j. \tag{6.25}$$

Comparing Eq. (6.19) with Eq. (5.30), we find $f_j^{(e)} = -f_j^{\mathrm{M}}$ and $icf_4^{(e)} = -icf_4^{\mathrm{M}}$. Thus Eqs. (6.24) and (6.25) may be expressed as

$$(I)\left\{\begin{aligned} &\mathbf{F} = \sigma\mathbf{E} + \mathbf{J}\times\mathbf{B} - \tfrac{1}{2}[(\nabla\mathbf{B})\cdot\mathbf{H} - (\nabla\mathbf{H})\cdot\mathbf{B} - (\nabla\mathbf{E})\cdot\mathbf{D} + (\nabla\mathbf{D})\cdot\mathbf{E}], && (6.26)\\ &\phi = \mathbf{J}_e\cdot\mathbf{E}_e + \tfrac{1}{2}\left[\mathbf{H}\cdot\frac{d\mathbf{B}}{dt} - \mathbf{B}\cdot\frac{d\mathbf{H}}{dt} + \mathbf{D}\cdot\frac{d\mathbf{E}}{dt} - \mathbf{E}\cdot\frac{d\mathbf{D}}{dt}\right], && (6.27)\end{aligned}\right.$$

where $\mathbf{J}_e = \mathbf{J} - \sigma\mathbf{v}$ and $\mathbf{E}_e = \mathbf{E} + \mathbf{v}\times\mathbf{B}$ are the effective fields as defined in Eqs. (4.11) and (4.12) respectively, and $d/dt \equiv \partial/\partial t + \mathbf{v}\cdot\nabla$. The term $\mathbf{J}_e\cdot\mathbf{E}_e$ is the Joule heat supply due to currents in moving matter.

From the law of conservation of the total angular momentum, Eq. (6.11), we can determine the body couple η_{jk} in Eq. (2.13). The spatial part of $S_{\mu\nu}$ in Eq. (6.16) is $S_{jk} = \rho v_j v_k - \tau_{jk} + \tau_{jk}^{\mathrm{M}}$, $j, k = 1, 2, 3$. The condition of symmetry for S_{jk} gives rise to

$$\begin{aligned}\tau_{jk} - \tau_{kj} &= \tau_{jk}^{\mathrm{M}} - \tau_{kj}^{\mathrm{M}}\\ &= -(D_j E_k - D_k E_j) - (B_j H_k - B_k H_j),\end{aligned} \tag{6.28}$$

where the relation (5.8) has been substituted for τ_{jk}^{M}. Thus

$$\eta_{jk} = \tfrac{1}{2}(D_j E_k - D_k E_j) + \tfrac{1}{2}(B_j H_k - B_k H_j) = D_{[j}E_{k]} + B_{[j}H_{k]} \tag{6.29}$$

and the body couple vector $L_i = \varepsilon_{ijk}\eta_{jk}$ is

$$\text{(I)}: \quad \mathbf{L} = \mathbf{D}\times\mathbf{E} + \mathbf{B}\times\mathbf{H} = \mathbf{P}\times\mathbf{E} + \mu_0\mathbf{M}\times\mathbf{H}. \tag{6.30}$$

Because the mechanical energy–momentum tensor $S_{\mu\nu}^{(m)}$ in Eq. (6.16) is not relativistically invariant, the condition of symmetry for the temporal part of $S_{\mu\nu}$ is not satisfied (Section 289 [18]), that is

$$S_{j4} \neq S_{4j}, \qquad j = 1, 2, 3. \tag{6.31}$$

Equations (6.26), (6.27), and (6.30) complete the nonrelativistic specification of the matter–field interaction in moving matter based on the Maxwell–Minkowski equation (I).

6.3 The Principle of Virtual Power

The success of the application of the conservation equation for the total energy–momentum tensor depends upon the knowledge of the energy–momentum tensor for each subsystem. The latter is often constructed on the basis of the balance laws of physics formulated in three-dimension space. Therefore, we proceed directly with the balance equations in 3-space.

Consider a total system which is divided into N subsystems which may be open. The conservation equation for an energy–momentum tensor of the ith subsystem is

$$\partial_\mu S_{\mu\nu}^{(i)} = f_\nu^{(i)}, \qquad i = 1, 2, \ldots, N. \tag{6.32}$$

Analogous to Eq. (6.19), the spatial and temporal parts of the above are

$$\partial_k \tau_{kj}^{(i)} + \partial_t g_j^{(i)} = f_j^{(i)}, \tag{6.33}$$

$$\partial_k S_k^{(i)} + \partial_t W^{(i)} = \psi^{(i)}, \qquad j, k = 1, 2, 3, \tag{6.34}$$

where $\psi^{(i)} = icf_4^{(i)}$. Here $\tau_{kj}^{(i)}$ is a momentum flux, $g_j^{(i)}$ the momentum density, $S_k^{(i)}$ an energy flux, and W an energy density. The two source terms $f_j^{(i)}$ and $\psi^{(i)}$ vanish if the ith subsystem is closed.

Instead of invoking the total energy–momentum tensor, we can base the entire analysis on the laws of balance of momentum, Eq. (6.33), and balance of energy, Eq. (6.34), for subsystems. Let the first subsystem be a kinetic system representing the motion of a body, and the remainder be the electromagnetic subsystem, dissipative subsystem (temperature–heat flow), etc. The $f_j^{(1)}$ and $\psi^{(1)}$ of the first system are then the body force and energy source which we want to determine in the field–matter interaction. Since the total system is closed, we have

$$\sum_{i=1}^{N} f_j^{(i)} = 0, \qquad \sum_{i=1}^{N} \psi^{(i)} = 0. \tag{6.35}$$

From these two equations, $f_j^{(1)}$ and $\psi^{(1)}$ can be calculated if the rest of the source terms are known.

In the remaining subsystems $i = 2, 3, \ldots, N$, $f_j^{(i)}$ and $\psi^{(i)}$ can be calculated from Eqs. (6.33) and (6.34) respectively when the members on the left-hand sides of these equations are known. For a given physical system, the quantities $S_k^{(i)}$, $W^{(i)}$, and $\psi^{(i)}$ which are associated with the energy equation are generally easier to determine than those associated with the momentum equation. If indeed they are found for the subsystem, the momentum flux $\tau_{jk}^{(i)}$ and density $g_j^{(i)}$ may be determined by applying the *principle of virtual power*. The essence of the principle which was proposed first by Penfield and Haus in 1961 (see [15]) is as follows.

In nonrelativistic theory, the two balance equations are invariant under a Glilean transformation (Sections 4.1, 4.2, [15]). To the body in motion and deformation, we attach a *pointwise rest frame*, one to each particle, which moves with velocity v_j. The quantities τ_{jk}, g_j, s_j, W (superscripts (i) are hereafter dropped) refer to a laboratory frame K are then related to those in the pointwise rest frame K' by

$$\begin{aligned} s_j &= s'_j + \tau'_{jk}v_k + v_jW', \\ W &= W' + g'_k v_k, \qquad \text{etc.}, \end{aligned} \tag{6.36}$$

where a prime denotes a quantity measured in the frame K'. Substituting Eq. (6.36) into Eq. (6.34) and letting the K'-frame approach the K-frame so that the terms containing v_j (but not the derivatives of v_j) vanish, one obtains

$$\partial_k s'_k + \partial_t W' + W'\partial_k v_k - \psi' = -\tau'_{jk}\partial_j v_k - g'_k\partial_t v_k. \tag{6.37}$$

Note that the momentum flux tensor τ'_{ik} and momentum density g'_k which we set out to determine appear on the right-hand side of Eq. (6.37) and the three quantities s'_k, W', and ψ' appear on the left. It is understood that these three energy-related quantities in the rest frame are known or can be ascertained for the subsystem.

Since the K'-frame can be brought infinitesimally close to the K-frame, $\partial_i v_k$ is regarded as the virtual rate of change of deformation and $\partial_t v_k$ the virtual velocity (hence the name of virtual power). When all terms in Eq. (6.37) are properly assembled those without the factor $\partial_t v_k$ or $\partial_i v_k$ should jointly vanish, for otherwise energy would not be conserved for stationary, rigid materials. Since $\partial_t v_k$ and $\partial_j v_k$ are arbitrary and independent, the remaining terms which are grouped with one of them as the common factor should also jointly vanish. Hence a knowledge of ψ', W', and s'_k enables the determination of τ'_{jk} and g'_k in Eq. (6.37).

Based on the principle of virtual power and the Maxwell–Minkowski equations (I), Chu, Penfield, and Haus calculated the body force and energy supply to a moving body. Details are contained in Refs. [22] (Section VII) and [15] (Section 7.3). Their end results (nonrelativistic) differ from those given in the previous subsection. These results have been criticized by Brevik (p. 60, Part 2 of [20]) who also raised serious question about the principle of virtual power.

Nevertheless the principle which is extensively applied in Ref. [15] has been a useful tool in obtaining results for all other formulations of electrodynamics. The results for the Boffi and Chu formulations given in the monograph (Sections 7.2 and 7.5 of [15]) are identical to those obtained using different principles and methods. However, none of the relativistic results summarized in Chapter 7 of Ref. [15] have been verified by other principles and methods of derivation.

6.4 Discussion

The concept of a closed system and the principle of conservation of the total energy–momentum in 4-space, or the conservations of total momentum, and energy in 3-space, provide a rational basis to determine the field–matter interaction. However, the end results depend upon the construction of the energy–momentum tensor for each subsystem. It is no coincidence that the final result presented in Subsection 6.2 agrees with that in Subsection 5.3; Eq. (6.26) and Eq. (5.32) are identical. The agreement is due to, first, the fact that only two subsystems are assumed. Secondly, the components of the energy–momentum tensor in the electrical subsystem are chosen from those in the momentum and energy balance equations of the previous section. In fact $S^{(e)}_{\mu\nu}$ in Eq. (6.17) is taken from Eq. (5.25).

Had a different energy–momentum tensor been chosen for each subsystem, the resulting expressions for **F**, **L**, and ϕ would have been different. This is possible because we have shown, in Subsection 5.2, that two sets of balance equations (Eqs. (5.8) and (5.9) versus Eqs. (5.13) and (5.14)) may be derived from the same Maxwell–Minkowski equations. Thus a new energy–momentum tensor may be constructed for the electrical subsystem. This tensor may not be covariant with the electrodynamic equations, but this is not critical in the slow speed approximation for deformable bodies.

Furthermore, the stress tensor τ_{jk} assumed in the mechanical subsystem may be dependent on the electromagnetic fields through constitutive relations. It is then possible to assume a mechanical tensor, $\overline{S}^{(m)}_{\mu\nu}$, which is completely independent of electromagnetic fields. The difference $S^{(m)}_{\mu\nu} - \overline{S}^{(m)}_{\mu\nu}$ is absorbed into the electrical tensor $S^{(e)}_{\mu\nu}$ to form a new tensor $\overline{S}^{(e)}_{\mu\nu}$. The total energy–momentum tensor is unchanged,

$$S_{\mu\nu} = S^{(m)}_{\mu\nu} + S^{(e)}_{\mu\nu} = \overline{S}^{(m)}_{\mu\nu} + \overline{S}^{(e)}_{\mu\nu},$$

but the end results for the mechanical subsystem would be altered.

This nonuniqueness of decomposition of the total energy–momentum tensor into many tensors of the subsystems is the source of controversies in the literature. As previously noted, various tensors for the electromagnetic subsystem have been proposed in the literature and they are compared by Brevik [20]. A brief discussion is also given in Møller's monograph (Section 7.7, [13]). Both authors favor the using of the Minkowski tensor.

7 THE THEORY OF ELECTRONS AND STATISTICAL MECHANICS

As discussed in Subsection 4.2, H. A. Lorentz proposed in 1892 the theory of electrons. Although Faraday and Maxwell advanced the supposition that a moving electrified body is equivalent to an electric current, it was not experimentally confirmed until 1876 by H. A. Rowland. If a small body with a charge q moves with velocity $\mathbf{u}$, the *convective current* is $q\mathbf{u}$. When moving in a magnetic induction $\mathbf{B}$, a charged body should experience a force, $q\mathbf{u} \times \mathbf{B}$, exactly the same as the conduction current in a circuit (Eq. (3.12)). This was first proposed by O. Heaviside in 1889.

In the original memoir, Lorentz assumed a Lagrangian function for a charge q moving with velocity $\mathbf{u}$ in an electric field $\mathbf{e}$, and magnetic field $\mathbf{h}$. These two microscopic fields satisfy the Maxwell equations *in vacuo*. From the Lagrangian equation, he obtained a force acting on the charged particle which was called *electron*,†

$$\mathbf{f} = q\mathbf{e} + q\mathbf{u} \times (\mu_0 \mathbf{h}). \tag{7.1}$$

This is now known as the *Lorentz force* (p. 395, [19]).

For electrified particles moving in a vacuum, Eq. (7.1) has been confirmed by experiments. However, in the electron theory of matter, the microscopic particles are assumed to move inside the matter, in which the field and the force acting on them cannot be measured with certainty. The application of the formula (7.1) to moving matter is thus another basic postulate in the theory of electrons. This was stated clearly by Lorentz (Sections 7, 8, [10]).

Another basic postulate in the electron theory is that the moving electrons do not interact directly with each other, but with the aetheral medium in which they are embedded. Furthermore, the aether is assumed to remain at rest all the time, and a moving body which carries electrons can not communicate its motion to the aether. The dynamic action in the aether is described completely by the Maxwell equations *in vacuo*.

Thus the Lorentz theory of electrons establishes, through postulates and assumptions, how matter interact with the aether and the field. The logic

† The terms *electron* and *atom* used here are not the same as those in the modern theory of atoms.

is as follows: the matter is composed of electrons. The electrons move in the matter as if they are isolated charges moving in the aether, in which the electromagnetic fields are governed by the Maxwell equations *in vacuo*. The aether exerts a force **f** as given by Eq. (7.1) on the electrons. The total field inside the matter is the field in the aether plus the microscopic fields generated by the electrons as sources. On the macroscopic scale, the field is obtained from the microscopic fields by an averaging process.

We mentioned in Subsection 4.2 that the macroscopic field equations (II) were derived by Lorentz with the application of the averaging formula (4.4). The next step would seem to be the application of the same formula to the microscopic forces **f** to derive the macroscopic force on a material body. However, the derivation is not as straightforward because the expression for **f** in Eq. (7.1) involves the product of two microscopic quantities, and the average of the product of two quantities does not always equal the product of the averages of each quantity.

It should be noted that the atomic theory of matter and statistical mechanics were not fully developed in Lorentz's time for correct determination of these forces. The necessary advances have been accomplished in recent years, and the results are comprehensively presented in the monograph by deGroot and Suttorp [4]. In this section, we will summarize the nonrelativistic results that are contained in the first two chapters of the monograph.

7.1 Microscopic and Macroscopic Field Equations

One improvement in the modern theory is the precise specification of the distribution of the charged particles (electrons) by using the delta function. As mentioned in Subsection 4.2 the charged particles in the material medium are gathered into *stable groups* which can simply be termed *atoms*. The location of the ith electron in the kth atom is denoted by (Fig. 4)

$$\mathbf{x}_{ki} = \mathbf{x}_k + \xi_{ki}, \tag{7.2}$$

where $\mathbf{x}_k$ is the position vector of the center of mass of the stable group and ξ_{ki} the internal coordinate within the atom. Thus at an observation point $\mathbf{x}$, the microscopic fields $\mathbf{e}(\mathbf{x})$ and $\mathbf{b}(\mathbf{x})$ are generated by a large number of atoms, each containing many electrons. The fields are governed by the Maxwell equations in aether,

$$\nabla \cdot \mathbf{b} = 0, \tag{7.3A}$$

$$\nabla \times \mathbf{e} + \partial \mathbf{b}/\partial t = 0, \tag{7.3B}$$

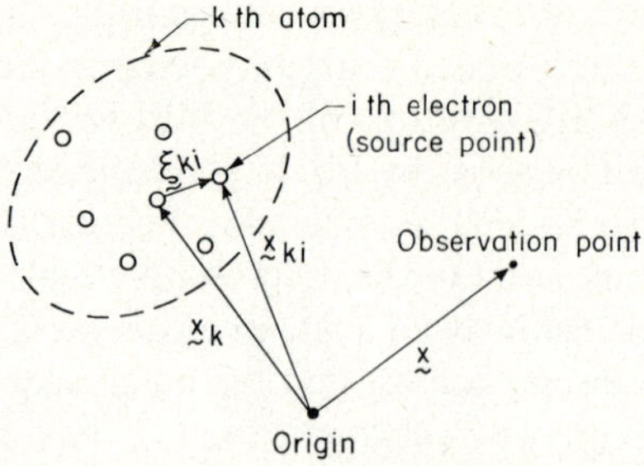

Fig. 4 Electrons (charged particles) and atoms (stable groups).

$$\varepsilon_0 \nabla \cdot \mathbf{e} = \sum_k \sum_i q_{ki} \delta(\mathbf{x}_{ki} - \mathbf{x}), \tag{7.3C}$$

$$\mu_0^{-1} \nabla \times \mathbf{b} - \varepsilon_0 \, \partial \mathbf{e}/\partial t = \sum_k \sum_i q_{ki} \dot{\mathbf{x}}_{ki} \delta(\mathbf{x}_{ki} - \mathbf{x}). \tag{7.3D}$$

Here q_{ki} is the charge of the ith electron in the kth atom, $q_{ki}\dot{\mathbf{x}}_{ki}$ is the convective current where $\dot{\mathbf{x}}_{ki} = d\mathbf{x}_{ki}/dt = \partial \mathbf{x}_{ki}/\partial t$. The delta function $\delta(\mathbf{x}_{ki} - \mathbf{x})$ specifies the location of the concentrated charge.

Since, physically, we are interested in the fields outside the stable groups, the delta functions may be developed into power series of the internal coordinates ξ_{ki} according to

$$\delta(\mathbf{x}_{ki} - \mathbf{x}) = \sum_{n=0}^{\infty} \frac{(-1)^n}{n!} (\xi_{ki} \cdot \nabla_k)^n \delta(\mathbf{x}_k - \mathbf{x}), \tag{7.4}$$

where $\nabla_k \equiv \partial/\partial x_k$. The expansion in Eq. (7.4) when multiplied by charges q_{ki} or current $q_{ki}\dot{x}_{ki}$ is known as the multipole expansion. We shall in this article retain the terms up to ξ_{ki} and neglect terms with higher order. Thus the series in Eq. (7.3C) is reduced to

$$\sum_k \sum_i q_{ki}[\delta(\mathbf{x}_k - \mathbf{x}) + \xi_{ki} \cdot \nabla_k \delta(\mathbf{x}_k - \mathbf{x}) + \tfrac{1}{2}\xi_{ki}\xi_{ki} : \nabla_k \nabla_k \delta(\mathbf{x}_k - \mathbf{x}) + \cdots]$$

$$\cong \sum_k [q_k \delta(\mathbf{x}_k - \mathbf{x}) + \boldsymbol{\mu}_k \cdot \nabla_k \delta(\mathbf{x}_k - \mathbf{x})] = \sigma_q - \nabla_x \cdot \mathbf{p},$$

where the following definitions for the charge q_k, electric dipole moment $\boldsymbol{\mu}_k$, and magnetic moment $\boldsymbol{\nu}_k$ have been applied:

$$q_k = \sum_i q_{ki}, \qquad \boldsymbol{\mu}_k = \sum_i q_{ki}\xi_{ki}, \qquad \boldsymbol{\nu}_k = \tfrac{1}{2} \sum_i q_{ki} \xi_{ki} \times \dot{\xi}_{ki}. \tag{7.5A}$$

The last step of the above derivation has made use of the relation

$$\boldsymbol{\mu}_k \cdot \nabla_k \delta(\mathbf{x}_k - \mathbf{x}) = -\boldsymbol{\mu}_k \cdot \nabla_x \delta(\mathbf{x}_k - \mathbf{x}) = -\nabla_x \cdot [\boldsymbol{\mu}_k \delta(\mathbf{x}_k - \mathbf{x})],$$

where $\nabla_x \equiv \partial/\partial x$, and

$$\sigma_q = \sum_k q_k \delta(\mathbf{x}_k - \mathbf{x}), \qquad \mathbf{p} = \sum_k \boldsymbol{\mu}_k \delta(\mathbf{x}_k - \mathbf{x}), \tag{7.5B}$$

The double-summed series in Eq. (7.3D) can be similarly reduced to $\mathbf{j} + \partial \mathbf{p}/\partial t + \nabla_x \times \mathbf{m}$ with the following definitions:

$$\mathbf{j} = \sum_k q_k \dot{\mathbf{x}}_k \delta(\mathbf{x}_k - \mathbf{x}), \tag{7.6A}$$

$$\mathbf{m} = \sum_k (\boldsymbol{\mu}_k \times \dot{\mathbf{x}}_k + \boldsymbol{\nu}_k)\delta(\mathbf{x}_k - \mathbf{x}). \tag{7.6B}$$

In this derivation we first note $\dot{\mu}_k = \sum q_{ki}\dot{\xi}_{ki}$. Additional use is then made of the relation

$$\frac{\partial \mathbf{p}}{\partial t} = \frac{\partial}{\partial t} \sum_i q_{ki}[\xi_{ki}\delta(\mathbf{x}_k - \mathbf{x}) - \tfrac{1}{2}\xi_{ki}(\xi_{ki} \cdot \nabla_k)\delta(\mathbf{x}_k - \mathbf{x})],$$

where the higher order term in ξ_{ki} is kept in the definition for $\mathbf{p}$ so that after taking the time differentiation, a term of the first order in ξ_{ki} (in the form of $\xi_{ki}\dot{\xi}_{ki}$) will emerge. The vector identity $\nabla \times (\mathbf{w} \times \mathbf{v}) = \nabla \cdot (\mathbf{vw} - \mathbf{wv})$ is also used. Equations (7.3) now assume the following form:

$$\begin{aligned} &\nabla \cdot \mathbf{b} = 0, \\ &\nabla \times \mathbf{e} + \partial \mathbf{b}/\partial t = 0, \\ &\varepsilon_0 \nabla \cdot \mathbf{e} = \sigma_q - \nabla \cdot \mathbf{p}, \\ &\mu_0^{-1} \nabla \times \mathbf{b} - \varepsilon_0\, \partial \mathbf{e}/\partial t = \mathbf{j} + \partial \mathbf{p}/\partial t + \nabla \times \mathbf{m}. \end{aligned} \tag{7.7}$$

The σ_q and $\mathbf{j}$ are the distribution functions for microscopic charges and currents respectively at the point $\mathbf{x}$; $\mathbf{p}$ and $\mathbf{m}$ are the microscopic polarization and magnetization respectively.

As evidenced by the presence of the delta functions, the microscopic fields $\mathbf{e}$ and $\mathbf{b}$ fluctuate rapidly in space. However, the physical dimension of the phenomenological laws are much larger than the size of each atom. This macroscopic field quantities may be defined in terms of the statistical average of the microscopic fields over regions that contain a large number of atoms.

In statistical mechanics the averaging formula (4.4) is replaced by the following statistical formula:

$$\mathbf{H}(\mathbf{x}, t) = \langle \mathbf{h} \rangle = \int \mathbf{h}(\mathbf{x}; \mathbf{r}) f(t; \mathbf{r})\, d\mathbf{r}, \tag{7.8}$$

where $\mathbf{r}$ represents the ensemble $(\mathbf{x}_k, \dot{\mathbf{x}}_k, \xi_{ki}, \dot{\xi}_{ki})$, and $d\mathbf{r} = d\mathbf{x}_k\, d\dot{\mathbf{x}}_k \Pi(d\xi_{ki}\, d\dot{\xi}_{ki})$ is an element in the *fluxion space*. The product $f\, d\mathbf{r}$ represents the probability to find $\mathbf{h}$ in the fluxion space element $d\mathbf{r}$. From Eq. (7.8) it can be shown that $\langle \nabla \mathbf{h} \rangle = \nabla \langle \mathbf{h} \rangle$ and $\langle d\mathbf{h}/dt \rangle = \partial \langle \mathbf{h} \rangle / \partial t$.

We now define the macroscopic field variables,

$$\begin{aligned} &\langle \mathbf{e} \rangle = \mathbf{E}, \qquad \langle \mathbf{b} \rangle = \mathbf{B}, \qquad \langle \mathbf{p} \rangle = \mathbf{P}. \\ &\langle \mathbf{m} \rangle = \mathbf{M}, \qquad \langle \sigma_q \rangle = \sigma, \qquad \langle \mathbf{j} \rangle = \mathbf{J}. \end{aligned} \tag{7.9}$$

We express Eqs. (7.7) in terms of these variables by taking the statistically averaged value, as follows:

$$\text{(III)} \quad \begin{cases} \nabla \cdot \mathbf{B} = 0, & \text{(7.10A)} \\ \nabla \times \mathbf{E} + \partial \mathbf{B}/\partial t = 0, & \text{(7.10B)} \\ \varepsilon_0 \nabla \cdot \mathbf{E} = \sigma - \nabla \cdot \mathbf{P}, & \text{(7.10C)} \\ \mu_0^{-1} \nabla \times \mathbf{B} - \varepsilon_0\, \partial \mathbf{E}/\partial t = \mathbf{J} + \partial \mathbf{P}/\partial t + \nabla \times \mathbf{M}. & \text{(7.10D)} \end{cases}$$

Note that Eqs. (7.10) are not the same as the original version derived by Lorentz (Eqs. (4.5)). The difference can be traced back to the definition for $\mathbf{p}$ in Eq. (7.5B) and $\mathbf{m}$ in Eq. (7.6B). If all atoms near $\mathbf{x}$ have approximately the same velocity $\mathbf{v} = d\mathbf{x}/dt$, the $\dot{\mathbf{x}}_k$ in Eq. (7.6B) can be replaced by $\mathbf{v}$. Hence

$$\mathbf{m} \cong \sum_k (\boldsymbol{\mu}_k \times \mathbf{v} + \boldsymbol{\nu}_k)\delta(\mathbf{x}_k - \mathbf{x}) = \mathbf{p} \times \mathbf{v} + \bar{\mathbf{m}}, \tag{7.11}$$

where

$$\bar{\mathbf{m}} = \sum_k \boldsymbol{\nu}_k \delta(\mathbf{x}_k - \mathbf{x}). \tag{7.6C}$$

Equation (7.3D) now reduces to

$$\mu_0^{-1} \nabla \times \mathbf{b} - \varepsilon_0\, \partial \mathbf{e}/\partial t = \mathbf{j} + \partial \mathbf{p}/\partial t + \nabla \times (\mathbf{p} \times \mathbf{v} + \bar{\mathbf{m}}). \tag{7.12}$$

Defining $\langle \bar{\mathbf{m}} \rangle = \mathbf{M}^{\mathrm{L}}$ and leaving all other definitions unchanged, we obtain, in place of Eq. (7.10D),

$$\mu_0^{-1} \nabla \times \mathbf{B} - \varepsilon_0\, \partial \mathbf{E}/\partial t = \mathbf{J} + \partial \mathbf{P}/\partial t + \nabla \times (\mathbf{P} \times \mathbf{v}) + \nabla \times \mathbf{M}^{\mathrm{L}}$$

which is the last equation of the Lorentz formulation (II). Note that the average of Eq. (7.11) is just Eq. (4.16).

7.2 Momentum Equation for Composite of Particles

Let a composite of electrons move in an externally applied electric field $\mathbf{E}_0$, and magnetic induction $\mathbf{B}_0$. The external fields must satisfy the Maxwell equations in free space,

$$\begin{aligned} \nabla \cdot \mathbf{B}_0 &= 0, & \nabla \times \mathbf{E}_0 + \partial \mathbf{B}_0/\partial t &= \mathbf{0}, \\ \varepsilon_0 \nabla \cdot \mathbf{E}_0 &= 0, & \mu_0 \nabla \times \mathbf{B}_0 - \varepsilon_0\, \partial \mathbf{E}_0/\partial t &= \mathbf{0}. \end{aligned} \tag{7.13}$$

Both $\mathbf{B}_0$ and $\mathbf{E}_0$ are functions of $\mathbf{x}$ and time t.

The fields at $\mathbf{x}$ generated by a single electron q_{lj} at $\mathbf{x}_{lj}$ may be calculated by solving Eqs. (7.3C) and (7.3D) with

$$\begin{aligned} \mathbf{e}_{lj}(\mathbf{x}) &= -\nabla_x[q_{lj}/(4\pi\varepsilon_0 |\mathbf{x} - \mathbf{x}_{lj}|)], \\ \mathbf{b}_{lj}(\mathbf{x}) &= \nabla_x \times [q_{lj}\dot{x}_{lj}/(4\pi\varepsilon_0 c^2 |\mathbf{x} - \mathbf{x}_{lj}|)]. \end{aligned} \tag{7.14}$$

The total electric field and magnetic induction at the location $\mathbf{x}_{ki}$ occupied by the ith electron in the kth atom are (p. 11, [4])

$$\mathbf{e}_t(\mathbf{x}_{ki}) = \sum_{j(\neq i)} \mathbf{e}_{kj}(\mathbf{x}_{ki}) + \sum_{l(\neq k)} \sum_{j} \mathbf{e}_{lj}(\mathbf{x}_{ki}) + \mathbf{E}_0(\mathbf{x}_{ki}), \tag{7.15A}$$

$$\mathbf{b}_t(\mathbf{x}_{ki}) = \sum_{j(\neq i)} \mathbf{b}_{kj}(\mathbf{x}_{ki}) + \sum_{l(\neq k)} \sum_{j} \mathbf{b}_{lj}(\mathbf{x}_{ki}) + \mathbf{B}_0(\mathbf{x}_{ki}). \tag{7.15B}$$

The first term on the right is the *intra-atomic field* (field by electrons within the kth atom) and the second the *inter-atomic field*.

The force action on the ith electron is given by the Lorentz formula (7.1) and the kinetic equation of motion for the electron with mass ρ_{ki} is

$$\rho_{ki}\ddot{\mathbf{x}}_{ki} = q_{ki}[\mathbf{e}_t(\mathbf{x}_{ki}) + \dot{\mathbf{x}}_{ki} \times \mathbf{b}_t(\mathbf{x}_{ki})]. \tag{7.16}$$

For the composite of electrons within the kth atom, we sum equations such as (7.16) over the index i, and introduce the center of mass $\mathbf{x}_k$ with

$$\sum_i \rho_{ki}\mathbf{x}_{ki} = \rho_k \mathbf{x}_k, \qquad \rho_k = \sum_i \rho_{ki}. \tag{7.17}$$

When Eqs. (7.15) are substituted into Eq. (7.16), the sum of all intra-atomic forces vanishes; $\mathbf{b}_t$ in Eq. (7.15B) may be approximated by $\mathbf{B}_0$ because the product $\dot{\mathbf{x}}_{ki} \times \mathbf{b}_{lj}(\mathbf{x}_{ki})$ is of the order $(\dot{x}_{ki}^2/c^2)$, which is neglected in the non-relativistic approximation. Thus, the motion of the center of mass of the kth atom is described by

$$\rho_k \ddot{\mathbf{x}}_k = \sum_i q_{ki} \left[\sum_{l(\neq k)} \sum_j \mathbf{e}_{lj}(\mathbf{x}_{ki}) + \mathbf{E}_0(\mathbf{x}_{ki}) + \dot{\mathbf{x}}_{ki} \times \mathbf{B}_0(\mathbf{x}_{ki}) \right]. \tag{7.18}$$

Since $\mathbf{e}_{lj}$, $\mathbf{E}_0$ and $\mathbf{B}_0$ all are functions of the position vector $\mathbf{x}_{ki} = \mathbf{x}_k + \xi_{ki}$, they can be expanded into a power series of the internal coordinates ξ_{ki}. Retaining only the effect of dipole moments, the result is (see Eq. (50), p. 13 of [4])

$$\rho_k \ddot{\mathbf{x}}_k = \mathbf{f}_k^{(\mathrm{L})} + \mathbf{f}_k^{(\mathrm{S})}, \tag{7.19}$$

where

$$\mathbf{f}_k^{(\mathrm{L})} = (q_k + \boldsymbol{\mu}_k \cdot \nabla_k)[\mathbf{E}_0(\mathbf{x}_k) + \dot{\mathbf{x}}_k \times \mathbf{B}_0(\mathbf{x}_k)] + (\dot{\boldsymbol{\mu}}_k + \mathbf{v}_k \times \nabla_k) \times \mathbf{B}_0(\mathbf{x}_k) + \mathbf{f}'_k, \tag{7.20}$$

$$\mathbf{f}_k^{(\mathrm{S})} = \sum_i \sum_{l(\neq k)} \sum_j q_{ki} \mathbf{e}_{lj}(\mathbf{x}_{ki}) - \mathbf{f}'_k, \tag{7.21}$$

and where

$$\mathbf{f}'_k = \sum_{l(\neq k)} (1 + q_l \mathbf{v}_k \cdot \nabla_k + q_k \mathbf{v}_l \cdot \nabla_l) \nabla_k \frac{1}{4\pi\varepsilon_0 |\mathbf{x}_k - \mathbf{x}_l|}.$$

Here $\mathbf{f}'_k$ is the power series expansion of the inter-atomic force about $\mathbf{x}_k$; and $\mathbf{f}_k^{(\mathrm{L})}$ is the *long-range* microscopic force mainly due to the action of the externally applied field on the kth atom, which also contains f'_k which is the long-range inter-atomic force. The term $\mathbf{f}_k^{(\mathrm{S})}$ is the *short-range* force.

Because $\nabla \cdot \mathbf{B}_0 = 0$, $\nabla \times \mathbf{E}_0 = -\partial \mathbf{B}_0/\partial t$, Eq. (7.20) can be expressed as (see Eq. (54), p. 13 of [4]),

$$\mathbf{f}_k^{(\mathrm{L})} = q_k(\mathbf{E}_0 + \mathbf{v}_k \times \mathbf{B}_0) + (\nabla_k \mathbf{E}_0) \cdot \boldsymbol{\mu}_k + (\nabla_k \mathbf{B}_0) \cdot (\mathbf{v}_k + \boldsymbol{\mu}_k \times \mathbf{v}_k) + (d/dt)(\boldsymbol{\mu}_k \times \mathbf{B}_0) + \mathbf{f}'_k, \tag{7.22}$$

where $\mathbf{v}_k = \dot{\mathbf{x}}_k$, and $\mathbf{E}_0$ and $\mathbf{B}_0$ are functions of $\mathbf{x}_k$ and t.

The energy equation and angular momentum equation for the kth atom can be established analogously. We omit their derivations here (see Section I.5, [4]).

7.3 Equations of Statistical Mechanics

We have already seen how the macroscopic field equations (7.10) are derived from the atomic field equations by applying the statistical formula (7.8). The same procedure can be applied to derive the balance equations. In case that two atoms at different locations are involved, the probability function f in Eq. (7.8) should be replaced by a two-point distribution function.

The macroscopic *mass density* $\rho(\mathbf{x}, t)$ is defined as

$$\rho(\mathbf{x}, t) = \left\langle \sum_k \rho_k \delta(\mathbf{x}_k - \mathbf{x}) \right\rangle. \tag{7.23}$$

Introducing the *local barycentric velocity* $\mathbf{v}(\mathbf{x}, t)$, we define the macroscopic momentum $\rho\mathbf{v}$ as

$$\rho(\mathbf{x}, t)\mathbf{v}(\mathbf{x}, t) = \left\langle \sum_k \rho_k \mathbf{v}_k \delta(\mathbf{x}_k - \mathbf{x}) \right\rangle. \tag{7.24}$$

The difference between $\mathbf{v}$ and $\mathbf{v}_k = \dot{\mathbf{x}}_k$ is called the *local velocity fluctuation* $\hat{\mathbf{v}}_k$,

$$\hat{\mathbf{v}}_k(\mathbf{x}, t) = \mathbf{v}_k - \mathbf{v}(\mathbf{x}, t). \tag{7.25}$$

Differentiating Eq. (7.23) with respect to time, we find

$$\partial\rho(x, t)/\partial t = -\nabla \cdot \left\langle \sum_k \rho_k \mathbf{v}_k \delta(\mathbf{x}_k - \mathbf{x}) \right\rangle, \tag{7.26}$$

where use has been made of the following result:

$$\frac{d}{dt}\,\delta(\mathbf{x}_k - \mathbf{x}) = \dot{\mathbf{x}}_k \cdot \nabla_k \delta(\mathbf{x}_k - \mathbf{x}) = -\nabla \cdot [\mathbf{v}_k \delta(\mathbf{x}_k - \mathbf{x})].$$

Here ∇_k operates on $\mathbf{x}_k$ and ∇ operates on $\mathbf{x}$. Comparing Eq. (7.24) to Eq. (7.26) we obtain the balance equation of mass Eq. (2.11)

$$\partial\rho/\partial t + \nabla \cdot (\rho\mathbf{v}) = 0. \tag{7.27}$$

The momentum equation is obtained if we take the time derivative of Eq. (7.24),

$$\partial(\rho\mathbf{v})/\partial t = -\nabla \cdot \left\langle \sum_k \rho_k \mathbf{v}_k \mathbf{v}_k \delta(\mathbf{x}_k - \mathbf{x}) \right\rangle + \left\langle \sum_k \rho_k \dot{\mathbf{v}}_k \delta(\mathbf{x}_k - \mathbf{x}) \right\rangle. \tag{7.28}$$

Replacing $\mathbf{v}_k$ in the first term of Eq. (7.28) by $\hat{\mathbf{v}}_k + \mathbf{v}$ as in Eq. (7.25), and $\rho_k \dot{\mathbf{v}}_k$ in the second by $\mathbf{f}_k^{(L)} + \mathbf{f}_k^{(S)}$ in Eq. (7.19), one obtains the balance equation of momentum

$$\partial(\rho\mathbf{v})/\partial t = -\nabla \cdot (\rho\mathbf{v}\mathbf{v} + \underset{\sim}{\tau}^{(K)}) + \mathbf{F}^{(L)} + \mathbf{F}^{(S)}, \tag{7.29}$$

where the statistical average of $\rho_k \hat{\mathbf{v}}_k$ vanishes and

$$\mathbf{F}^{(\mathrm{L})} = \left\langle \sum_k \mathbf{f}_k^{(\mathrm{L})} \delta(\mathbf{x}_k - \mathbf{x}) \right\rangle,$$

$$\mathbf{F}^{(\mathrm{S})} = \left\langle \sum_k \mathbf{f}_k^{(\mathrm{S})} \delta(\mathbf{x}_k - \mathbf{x}) \right\rangle, \tag{7.30}$$

$$\underset{\sim}{\tau}^{(\mathrm{K})} = \left\langle \sum_k \rho_k \hat{\mathbf{v}}_k \hat{\mathbf{v}}_k \delta(\mathbf{x}_k - \mathbf{x}) \right\rangle.$$

$\underset{\sim}{\tau}^{(\mathrm{K})}$ is the *kinetic pressure tensor* and $\underset{\sim}{\mathbf{F}}^{(\mathrm{L})}$ and $\underset{\sim}{\mathbf{F}}^{(\mathrm{S})}$ are the *long-range force* and *short-range force* respectively.

Substituting Eq. (7.22) into Eq. (7.30A) and making use of the definition for σ, $\mathbf{J}$, $\mathbf{P}$, and $\mathbf{M}$ in Eq. (7.9), we find

$$\begin{aligned} \mathbf{F}^{(\mathrm{L})} &= \sigma \mathbf{E}_0 + \mathbf{J} \times \mathbf{B}_0 + (\nabla \mathbf{E}_0) \cdot \mathbf{P} + (\nabla \mathbf{B}_0) \cdot \mathbf{M} + \partial(\mathbf{P} \times \mathbf{B}_0)/\partial t \\ &\quad + \nabla \cdot (\mathbf{v}\mathbf{P} \times \mathbf{B}) + \nabla \cdot \left\langle \sum_k \hat{\mathbf{v}}_k \boldsymbol{\mu}_k \times \mathbf{B}_0 \delta(\mathbf{x}_k - \mathbf{x}) \right\rangle + \left\langle \sum_k \mathbf{f}_k' \delta(\mathbf{x}_k - \mathbf{x}) \right\rangle. \end{aligned} \tag{7.31}$$

The external fields $\mathbf{E}_0$ and $\mathbf{B}_0$ can be expressed in terms of the statistically averaged fields $\mathbf{E}$, $\mathbf{B}$, etc., as follows. Note that $\mathbf{E}_0$ and $\mathbf{B}_0$ satisfy Eq. (7.13) which is the homogeneous part of Eq. (7.10). Regard $\sigma - \nabla \cdot \mathbf{P}$ as a source function for the inhomogeneous equation Eq. (7.10C), and obtain the complete solution for $\mathbf{E}$ as (p. 117, [4])

$$\mathbf{E}(\mathbf{x}) = \mathbf{E}_0(\mathbf{x}) - \nabla \int [\sigma(\mathbf{x}') - \nabla' \cdot \mathbf{P}(\mathbf{x}')] \frac{1}{4\pi\varepsilon_0 |\mathbf{x} - \mathbf{x}'|} \, d\mathbf{x}'. \tag{7.32}$$

Similarly, the complete solution for $\mathbf{B}$ can be obtained from Eq. (7.10D),

$$\mathbf{B}(\mathbf{x}) = \mathbf{B}_0(\mathbf{x}) + \nabla \times \int \left[\mathbf{J}(\mathbf{x}') + \frac{\partial \mathbf{P}(\mathbf{x}')}{\partial t} + \nabla' \times \mathbf{M}(\mathbf{x}') \right] \frac{\mu_0}{4\pi |\mathbf{x} - \mathbf{x}'|} \, d\mathbf{x}'. \tag{7.33}$$

In the above formulas, all field variables may depend explicitly or implicitly upon the time variable t. In the same order of slow speed approximation as in Eq. (7.18), we take

$$\mathbf{B}(\mathbf{x}) \cong \mathbf{B}_0(\mathbf{x}). \tag{7.34}$$

Substituting Eqs. (7.32) and (7.34) into Eq. (7.31), we obtain by a long and complicated reduction (p. 40, [4])

$$\mathbf{F}^{(L)} = \sigma\mathbf{E} + \mathbf{J} \times \mathbf{B} + (\nabla\mathbf{E}) \cdot \mathbf{P} + (\nabla\mathbf{B}) \cdot \mathbf{M} + \partial(\mathbf{P} \times \mathbf{B})/\partial t + \nabla \cdot (\mathbf{v}\mathbf{P} \times \mathbf{B}) - \nabla \cdot \boldsymbol{\tau}^{(F)} + \mathbf{F}^{(C)}. \tag{7.35}$$

The term $\mathbf{F}^{(C)}$ is the correlation contribution from the inter-atomic forces and $\boldsymbol{\tau}^{(F)}$ is a contribution to the pressure tensor due to the action of the **B** field on the electric dipoles.

For fluids and solids, $\mathbf{F}^{(C)}$ in Eq. (7.35) and the short-range force $\mathbf{F}^{(S)}$ in Eq. (7.29) can be expressed as the divergence of the pressure tensors $\boldsymbol{\tau}^{(C)}$ and $\boldsymbol{\tau}^{(S)}$ respectively. Substituting Eq. (7.35) into (7.29) and combining all pressure tensors, we finally obtain (p. 56, [4])

$$\partial(\rho\mathbf{v})/\partial t + \nabla \cdot (\rho\mathbf{v}\mathbf{v}) = \nabla \cdot \underset{\sim}{\boldsymbol{\tau}} + \mathbf{F}, \tag{7.36}$$

where

$$\underset{\sim}{\boldsymbol{\tau}} = -(\underset{\sim}{\boldsymbol{\tau}}^{(K)} + \underset{\sim}{\boldsymbol{\tau}}^{(F)} + \underset{\sim}{\boldsymbol{\tau}}^{(C)} + \underset{\sim}{\boldsymbol{\tau}}^{(S)}), \tag{7.37}$$

$$\text{(III)}: \quad \mathbf{F} = \sigma\mathbf{E} + \mathbf{J} \times \mathbf{B} + (\nabla\mathbf{E}) \cdot \mathbf{P} + (\nabla\mathbf{B}) \cdot \mathbf{M} + \partial(\mathbf{P} \times \mathbf{B})/\partial t + \nabla \cdot (\mathbf{v}\mathbf{P} \times \mathbf{B}). \tag{7.38}$$

Equation (7.36) is the balance equation of momentum where the body force **F** has been expressed in terms of the macroscopic fields which satisfy Eq. (7.10). The stress tensor $\underset{\sim}{\boldsymbol{\tau}}$ in Eq. (7.36) is composed of the kinetic pressure tensor $\underset{\sim}{\boldsymbol{\tau}}^{(K)}$ and other components which are the result of electric and magnetic fields acting on the dipole moments.

In view of Eq. (7.10), the first four terms in **F** can be expressed as (p. 49, [4])

$$\sigma\mathbf{E} + \mathbf{J} \times \mathbf{B} + (\nabla\mathbf{E}) \cdot \mathbf{P} + (\nabla\mathbf{B}) \cdot \mathbf{M} \equiv -\nabla \cdot \underset{\sim}{\boldsymbol{\tau}}^{S} - \partial(\mathbf{D} \times \mathbf{B})/\partial t, \tag{7.39}$$

where $\mathbf{D} = \varepsilon_0\mathbf{E} + \mathbf{P}$, $\mathbf{B} = \mu_0(\mathbf{H} + \mathbf{M})$ and

$$\underset{\sim}{\boldsymbol{\tau}}^{S} = (\tfrac{1}{2}\varepsilon_0\mathbf{E} \cdot \mathbf{E} + \tfrac{1}{2}\mu_0^{-1}\mathbf{B} \cdot \mathbf{B} - \mathbf{M} \cdot \mathbf{B})\underset{\sim}{\mathbf{I}} - (\mathbf{D}\mathbf{E} + \mathbf{B}\mathbf{H}). \tag{7.40}$$

Hence the formula (7.38) can be written as

$$\text{(III)}: \quad \mathbf{F} = -\nabla \cdot \underset{\sim}{\boldsymbol{\tau}}^{S} - \partial(\varepsilon_0\mathbf{E} \times \mathbf{B})/\partial t + \nabla \cdot (\mathbf{v}\mathbf{P} \times \mathbf{B}). \tag{7.41}$$

Here $\underset{\sim}{\boldsymbol{\tau}}^{S}$ is another form of stress tensor in the material medium.

The macroscopic balance equation of energy can also be derived from the energy equation for each electron

$$\rho_{ki}(\dot{\mathbf{x}}_{ki} \cdot \ddot{\mathbf{x}}_{ki}) = q_{ki}(\dot{\mathbf{x}}_{ki} \cdot \mathbf{e}_t). \tag{7.42}$$

The above equation is obtained by taking the scalar product of Eq. (7.16) with $\dot{\mathbf{x}}_{ki}$. Without giving further details, we list here the final form of the equation of energy for fluids or solids (p. 57, [4])

$$\partial(\tfrac{1}{2}\rho v^2 + \rho U)/\partial t + \nabla\cdot[\mathbf{v}(\tfrac{1}{2}\rho v^2 + \rho U)]$$
$$= \nabla\cdot(\underset{\sim}{\tau}\cdot\mathbf{v} - Q) + \mathbf{J}\cdot\mathbf{E} + \frac{\partial\mathbf{P}}{\partial t}\cdot\mathbf{E} - \mathbf{M}\cdot\frac{\partial\mathbf{B}}{\partial t} + \nabla\cdot(\mathbf{vP}\cdot\mathbf{E}). \quad (7.43)$$

U is the total internal energy which is the sum of three parts $U^{(K)}$, $U^{(C)}$ and $U^{(S)}$, the first being the internal energy of a pure mechanical system. **Q** represents the heat flow which also consists of three parts, the kinetic, correlation, and short-range parts. If Eq. (7.43) is converted to the global form Eq. (2.9), we can identify the last four terms as the energy supply ($\phi_Q = 0$ in Eq. 2.9),

$$\mathbf{F}\cdot\mathbf{v} + \phi = \mathbf{J}\cdot\mathbf{E} + \frac{\partial\mathbf{P}}{\partial t}\cdot\mathbf{E} - \mathbf{M}\cdot\frac{\partial\mathbf{B}}{\partial t} + \nabla\cdot(\mathbf{vP}\cdot\mathbf{E}). \quad (7.44)$$

Substituting Eq. (7.38) into the above, we find after a long manipulation of vector identities,

$$\text{(III)}:\quad \phi = \left[\mathbf{J} - \sigma\mathbf{v} + \frac{\partial\mathbf{P}}{\partial t} + \nabla\cdot(\mathbf{vP})\right]\cdot(\mathbf{E} + \mathbf{v}\times\mathbf{B})$$
$$- (\mathbf{M} + \mathbf{v}\times\mathbf{P})\left(\frac{\partial\mathbf{B}}{\partial t} + \mathbf{v}\cdot\nabla\mathbf{B}\right). \quad (7.45)$$

Introducing the effective current $\mathbf{J}_e$ in Eq. (4.11) and the effective electric field $\mathbf{E}_e$ in Eq. (4.12), plus the effective magnetization $\mathbf{M}_e$,

$$\mathbf{M}_e = \mathbf{M} + \mathbf{v}\times\mathbf{P}, \qquad \mathbf{E}_e = \mathbf{E} + \mathbf{v}\times\mathbf{B}, \quad (7.46)$$

we obtain from Eq. (7.45),

$$\text{(III)}:\quad \phi = \mathbf{J}_e\cdot\mathbf{E}_e + \rho\frac{d}{dt}\left(\frac{\mathbf{P}}{\rho}\right)\cdot\mathbf{E}_e - \mathbf{M}_e\cdot\frac{d\mathbf{B}}{dt}. \quad (7.47)$$

In view of Eq. (2.11), we have used the identity $\rho(d/dt)(\mathbf{P}/\rho) = \partial\mathbf{P}/\partial t + \nabla\cdot(\mathbf{vP})$.

From the angular momentum equation, a body couple is derived (p. 63, [4])

$$\text{(III)}:\quad \mathbf{L} = \mathbf{P}\times\mathbf{E} + \mathbf{M}\times\mathbf{B} + \mathbf{v}\times(\mathbf{P}\times\mathbf{B}). \quad (7.48)$$

Equations (7.38), (7.47), and (7.48) complete the specification of the interactions of moving matter with fields based on the Lorentz theory of electrons and the statistical theory.

DeGroot and Suttorp's monograph [4] also presents thermodynamic equations and a complete relativistic statistical theory. The results for **F**, ϕ, and **L** are given on p. 278 and p. 286 of Ref. [4].

In Subsection 10.3 we shall show that these final results agree with those based on the macroscopic dipole-current circuit model.

7.4 Discussion

The statistical theory as summarized in this section is the most comprehensive treatment on the matter–field interactions. The basic postulates are in the theory of electrons, and the remainder is derived from the principle of statistical mechanics. The theory clearly separates the forces that are generated by the external electromagnetic fields (field–matter interaction) from the interatomic forces (matter–matter interaction within a body). It identifies the nature of the long-range force and that of the short-range force. It shows that the stress tensor in the final form of the balance equation of linear momentum, Eq. (7.36), is composed of the usual kinetic stress tensor and other stress tensors due to electromagnetic forces.

A slight uncertainty in the derivation is the grouping of the total electromagnetic force into a body force **F** and a surface force which are represented by $\underset{\sim}{\tau}^{(F)} + \underset{\sim}{\tau}^{(C)} + \underset{\sim}{\tau}^{(S)}$ in Eq. (7.36). The part $\underset{\sim}{\tau}^{(S)}$ is due to the short-range force, but the other two parts are not. If the term $\mathbf{F}^{(C)}$ in Eq. (7.35) were not converted to $\nabla \cdot \underset{\sim}{\tau}^{(C)}$ the final form for the body force **F** in Eq. (7.36) would be different. This nonuniqueness in the separation of the total force into a body force and a surface force is inherent in the theory of electromagnetic forces in continua. The statistical theory does not completely resolve this uncertainty.

Since the statistical theory is built upon the physics of microscopic particles, it does not establish jump conditions or boundary conditions for the macroscopic variables. This is rather critical here for deciding which component of the total force could be expressed as a divergence of a pressure tensor. Changing $\mathbf{F}^{(C)} + \mathbf{F}^{(S)}$ to $\nabla \cdot (\underset{\sim}{\tau}^{(C)} + \underset{\sim}{\tau}^{(S)})$ implies that within a continuum some of the field quantities in $\mathbf{F}^{(C)} + \mathbf{F}^{(S)}$ are singular at a surface of discontinuity. This is not evidenced by the statistical theory.

Note that Eqs. (7.10) can be brought into agreement with Eqs. (3.2) by the substitution of $\mathbf{D} = \varepsilon_0\mathbf{E} + \mathbf{P}$ and $\mathbf{B} = \mu_0(\mathbf{H} + \mathbf{M})$ (Eq. (3.5)). However, this agreement is superficial. The Minkowski formulation is valid for rigid bodies moving with constant velocities in electromagnetic fields whereas the statistical formulation is supposed valid for deformable bodies. In the former the constitutive equations for **D**, **H**, and **J** are rigidly specified by Eq. (4.3) but in the latter the constitutive equations for **P**, **M** and **J** remain unspecified.

8 MACROSCOPIC MAXWELL–LORENTZ FORCES

Instead of working with the microscopic fields and accepting the assumption of Eq. (7.1) for the force on electrons, many followers of Maxwell and Lorentz simply extended Eq. (3.20) and postulated the macroscopic body force (per unit volume) as

$$\mathbf{F} = \sigma_t \mathbf{E} + \mathbf{J}_t \times \mathbf{B}, \tag{8.1}$$

where σ_t is the *total charge* and J_t is the *total current.* Note that a similar formula was postulated by Maxwell in deriving the stress tensors (Subsection 5.1); hence **F** in Eq. (8.1) will be termed *the Maxwell–Lorentz force.*

Analogously, the energy supply is assumed as

$$\phi = \mathbf{J}_t \cdot \mathbf{E} - \mathbf{F} \cdot \mathbf{v}. \tag{8.2}$$

The first term is an extension of the Joule heating due to the free current, $\mathbf{J} \cdot \mathbf{E}$; the second term is the rate of work done by the body force.

The total charge and total current are usually identified from the source terms in the electromagnetic field equations. The results for various formulations are given in the following two subsections.

8.1 The EBPMv and EBPM Formulations

The field equations of the Lorentz formulation (II) are given in Section 4. The right-hand side of Eq. (4.5C) is the sum of the free charge and polarization charge, and that of Eq. (4.5D) is the sum of free, polarization, and magnetization currents. Assuming in Eq. (8.1)

$$\text{(II)}: \quad \begin{cases} \sigma_t = \sigma - \nabla \cdot \mathbf{P}, & (8.3) \\ \mathbf{J}_t = \mathbf{J} + \partial \mathbf{P}/\partial t + \nabla \times (\mathbf{P} \times \mathbf{v}) + \nabla \times \mathbf{M}, & (8.4) \end{cases}$$

we find

$$\begin{aligned} \text{(II)}: \quad \mathbf{F} &= \sigma \mathbf{E} + \mathbf{J} \times \mathbf{B} - (\nabla \cdot \mathbf{P})\mathbf{E} + [\partial \mathbf{P}/\partial t + \nabla \times (\mathbf{P} \times \mathbf{v}) + \nabla \times \mathbf{M}] \times \mathbf{B} \\ &= \varepsilon_0 (\nabla \cdot \mathbf{E})\mathbf{E} - \mathbf{B} \times (\nabla \times \mathbf{B}/\mu_0 - \varepsilon_0\, \partial \mathbf{E}/\partial t). \end{aligned} \tag{8.5}$$

When the field equations (4.5) are applied, we obtain

$$\mathbf{F} = \varepsilon_0 (\nabla \cdot \mathbf{E})\mathbf{E} - \mathbf{B} \times [\nabla \times (\mathbf{B}/\mu_0) - \varepsilon_0\, \partial \mathbf{E}/\partial t].$$

Furthermore, we may insert $(\nabla \cdot \mathbf{B})\mathbf{B} (= 0)$ into the above expression. Noting that

$$\varepsilon_0 \mathbf{B} \times \frac{\partial \mathbf{E}}{\partial t} = -\varepsilon_0 \frac{\partial(\mathbf{E} \times \mathbf{B})}{\partial t} - \varepsilon_0 \mathbf{E} \times (\nabla \times \mathbf{E}),$$

$$\mathbf{E} \times (\nabla \times \mathbf{E}) = (\nabla \mathbf{E}) \cdot \mathbf{E} - \mathbf{E} \cdot \nabla \mathbf{E}, \quad \text{etc.,}$$

we finally obtain another expression for the body force,

$$\text{(II)}: \quad \mathbf{F} = -\nabla \cdot \underset{\sim}{\tau}^{\mathrm{L}} - \varepsilon_0\, \partial(\mathbf{E} \times \mathbf{B})/\partial t, \tag{8.6}$$

where

$$\underset{\sim}{\tau}^{\mathrm{L}} = \tfrac{1}{2}(\varepsilon_0 \mathbf{E} \cdot \mathbf{E} + \mathbf{B} \cdot \mathbf{B}/\mu_0)\underset{\sim}{\mathbf{I}} - (\varepsilon_0 \mathbf{E}\mathbf{E} + \mathbf{B}\mathbf{B}/\mu_0). \tag{8.7}$$

$\underset{\sim}{\tau}^{\mathrm{L}}$ is the electromagnetic stress tensor for moving matter based on the Lorentz formulation. It is identical in form to the Maxwell stress tensor *in vacuo* (Eq. (5.8)).

From Eqs. (8.2), (8.4), and (8.5) we find

$$\begin{aligned}\text{(II)}: \quad \phi &= (\mathbf{J} - \sigma \mathbf{v}) \cdot \mathbf{E} + \mathbf{J} \cdot (\mathbf{v} \times \mathbf{B}) + (\overset{*}{\mathbf{P}} + \nabla \times \mathbf{M}) \cdot \mathbf{E} \\ &\quad + \left[\frac{\partial \mathbf{P}}{\partial t} + \nabla \times (\mathbf{P} \times \mathbf{v}) + \nabla \times \mathbf{M}\right] \cdot (\mathbf{v} \times \mathbf{B}) \\ &= \mathbf{J}_e \cdot \mathbf{E}_e + (\overset{*}{\mathbf{P}} + \nabla \times \mathbf{M}) \cdot \mathbf{E}_e,\end{aligned} \tag{8.8}$$

where according to Eq. (2.4)

$$\overset{*}{\mathbf{P}} = \frac{\partial \mathbf{P}}{\partial t} + \mathbf{v}(\nabla \cdot \mathbf{P}) + \nabla \times (\mathbf{P} \times \mathbf{v}),$$

and $\mathbf{J}_e = \mathbf{J} - \sigma \mathbf{v}$ and $\mathbf{E}_e = \mathbf{E} + \mathbf{v} \times \mathbf{B}$ are the effective current and field respectively, as defined in Eqs. (4.11) and (4.13).

For nonmagnetizable material in slow motion, we set $\mathbf{M} = \mathbf{0}$ in the above formulas. The end results completely agree with those assumed by R. Toupin. He called $\mathbf{E}_e$ the *electromotive intensity* and neglected free charges and currents in the final results (Eqs. $(5.1)_3$ and $(5.1)_9$ of [45]).

In the **EBPM** formulation (III), the total charge and current can be identified from Eqs. (3.6) as

$$\text{(III)}: \begin{cases} \sigma_t = \sigma - \nabla \cdot \mathbf{P}, & (8.9) \\ \mathbf{J}_t = \mathbf{J} + \partial \mathbf{P}/\partial t + \nabla \times \mathbf{M}. & (8.10) \end{cases}$$

These forms are identical to those in Eqs. (8.3) and (8.4) except that in the latter, there is an extra term $\nabla \times (\mathbf{P} \times \mathbf{v})$. However, the physical meanings for $\mathbf{M}$ in these two formulations are different. Substituting the above expressions into Eqs. (8.1) and (8.2), we find

$$\mathbf{F} = -\nabla \cdot \underset{\sim}{\tau}^{\mathrm{SL}} + \varepsilon_0\, \partial(\mathbf{E} \times \mathbf{B})/\partial t, \tag{8.11}$$

$$\phi = \mathbf{J}_e \cdot \mathbf{E}_e + (\partial \mathbf{P}/\partial t + \nabla \times \mathbf{M}) \cdot \mathbf{E}_e, \tag{8.12}$$

where $\underset{\sim}{\tau}^{SL} \equiv \underset{\sim}{\tau}^{L}$ which is given in Eq. (8.7). The effective fields, $\mathbf{E}_e$, in these two formulations are the same.

In the **EBDH** formulation (I), it is difficult to identify the total charge and total current from the field equations (4.1). However, when they are transformed to the **EBPM** representation by Eqs. (3.5), Eqs. (I) are then identical to Eqs. (III). Thus the total charge and total current are given by Eqs. (8.9) and (8.10) respectively. This has been proposed by P. Poincelot [38] as an alternative basis on which to construct an energy–momentum tensor [39].

It is interesting to note that the same expression for $\underset{\sim}{\tau}^{L}$ (or $\underset{\sim}{\tau}^{SL}$) also emerges from the balance equations of electromagnetic momentum in the Lorentz (or statistical) formulation. Repeating the procedure that leads to Eq. (5.7) we obtain from Eqs. (II) (or Eqs. III) the following balance equation:

$$\text{(II), (III)}: \quad \nabla \cdot \underset{\sim}{\tau}^{L} + \varepsilon_0\, \partial(\mathbf{E} \times \mathbf{B})/\partial t = -\,(\sigma_t \mathbf{E} + \mathbf{J}_t \times \mathbf{B}), \tag{8.13}$$

where $\underset{\sim}{\tau}^{L}$ (or $\underset{\sim}{\tau}^{SL}$) is given by Eq. (8.7), and σ_t and J_t are respectively the total charge and current appropriate for each formulation. The left-hand side of Eq. (8.13) represents the net change of momentum density within a unit volume. The right-hand side is the supply of electromagnetic momentum to and from the unit volume. This transfer of momentum which is equivalent to the action of a force, occurs in the electromagnetic field. Hence the right-hand side of Eq. (8.13) is interpreted as a force action *on the field*; see discussions in Subsection 5.2.

On the other hand, Eq. (8.1) is a formula to calculate the force acting *on the material body*. By the assumption of Eqs. (8.3) and (8.4) (or Eqs. (8.9) and (8.10)) we have made the force in the kinetic subsystem (the material body) equal to the negative of the force in the electromagnetic subsystem (the field).

8.2 The Chu Formulation

In the Chu formulation (IV), the field variables in the free space are chosen to be **E** and **H** and the sources in the material medium can be identified from the right-hand sides of Eqs. (4.8),

$$\text{(IV)}: \quad \begin{aligned} \sigma_t &= \sigma - \nabla \cdot \mathbf{P}, & \mathbf{J}_t &= \mathbf{J} + \partial \mathbf{P}/\partial t + \nabla \times (\mathbf{P} \times \mathbf{v}), \\ \sigma_t^* &= -\nabla \cdot (\mu_0 \mathbf{M}), & \mathbf{J}_t^* &= \partial(\mu_0 \mathbf{M})/\partial t + \nabla \times (\mu_0 \mathbf{M} \times \mathbf{v}). \end{aligned} \tag{8.14}$$

σ_t^* and $\mathbf{J}_t^*$ are the total magnetic charge and magnetic currents respectively.

Instead of Eq. (8.1), the formula for the body force is postulated as (Eq. (10.7) of [5])

$$\mathbf{F} = \sigma_t\mathbf{E} + \mathbf{J}_t \times (\mu_0\mathbf{H}) + \sigma_t^*\mathbf{H} - \mathbf{J}_t^* \times (\varepsilon_0\mathbf{E}). \tag{8.15}$$

The end result is of the form (Eq. (10.10) of [5])

$$\mathbf{F} = -\nabla \cdot \underset{\sim}{\tau}^{\text{CL}} - \varepsilon_0\mu_0\, \partial(\mathbf{E} \times \mathbf{H})/\partial t, \tag{8.16}$$

where

$$\underset{\sim}{\tau}^{\text{CL}} = \tfrac{1}{2}(\varepsilon_0\mathbf{E} \cdot \mathbf{E} + \mu_0\mathbf{H} \cdot \mathbf{H})\underset{\sim}{\mathbf{I}} - (\varepsilon_0\mathbf{E}\mathbf{E} + \mu_0\mathbf{H}\mathbf{H}). \tag{8.17}$$

This stress tensor is in the same form as the Maxwell stress tensor in the vacuum. In addition, the energy supply is given by (Eq. (10.30) of [5])

$$\begin{aligned} \phi &= \mathbf{J}_t \cdot \mathbf{E} + \mathbf{J}_t^* \cdot \mathbf{H} - \mathbf{F} \cdot \mathbf{v} \\ &= (\mathbf{E} + \mathbf{v} \times \mu_0\mathbf{H}) \cdot (\mathbf{J}_t - \sigma_t\mathbf{v}) + (\mathbf{H} - \mathbf{v} \times \varepsilon_0\mathbf{E}) \cdot (\mathbf{J}_t^* - \sigma_t^*\mathbf{v}). \end{aligned} \tag{8.18}$$

The above results were later considered incomplete [15]. We shall discuss this formulation again in Section 10.

8.3 Discussion

Without question, the postulation of a macroscopic Lorentz force in the form of (8.1) or (8.15) is the simplest approach to calculate the body force. The end results are compatible with those emerging from the balance equation of electromagnetic momentum. Both expressions for the stress tensors, $\underset{\sim}{\tau}^{\text{L}}$ (or $\underset{\sim}{\tau}^{\text{SL}}$) and $\underset{\sim}{\tau}^{\text{C}}$, are the same as the Maxwell stress tensor in the aetheral medium (τ^{M} in Eq. (5.8) with $\mathbf{D} = \varepsilon_0\mathbf{E}$ and $\mathbf{B} = \mu_0\mathbf{H}$).

The simplicity of this approach is also the root of its weakness. Conceptually the postulation of (8.1) together with the assumptions (8.3) and (8.4), or their equivalent, means that the electromagnetic momentum transferred in the field is completely converted to a kinetic momentum which affects the motion of the body. Thus the force acting on the field is equal and opposite to that acting on the matter. We have already mentioned in Subsection 5.4 that this concept is untenable for moving media.

Once the Eq. (8.1) is postulated, the identification of the σ_t and $\mathbf{J}_t$ from the field equations actually requires additional information or assumptions. For instance, given the field equations (I), one may want to consider the right-hand side of Eqs. (3.2) as the "sources" and to assume $\sigma_t = \sigma$ and $\mathbf{J}_t = \mathbf{J} + \partial\mathbf{D}/dt$. But this is known to be incomplete (see Eqs. (5.1) to (5.3)). Instead, Eqs. (I) are first transformed to Eqs. (III) and the assumptions (8.9) and (8.10) are adopted. This manipulation is guided by facts not implied in the original assumptions.

The fact that $\underset{\sim}{\tau}^{SL}$ in Eq. (8.11) differs from $\underset{\sim}{\tau}^{S}$ in Eq. (7.40), both being based on the same field equations (III), casts further doubts about this approach. The same can be said for $\underset{\sim}{\tau}^{CL}$ in Eq. (8.17) as it differs from $\underset{\sim}{\tau}^{C}$ in Section 10, both being derived from the Chu formulation. Nevertheless, criticism that is based solely on the comparisons of end results is not totally justified here because the nature of the electromagnetic surface forces, in this case, is not specified. Equation (8.1) does not assume any form for the surface forces. If some of them can be converted to an expression such as a volume force, the final expressions for the resultant body force are quite different. We shall illustrate this point by reporting the following results given by Livens (Section 235 of [9]).

According to Eqs. (III) and Eqs. (8.1), (8.9), and (8.10), the body force per unit volume is

$$\mathbf{F} = (\sigma - \nabla \cdot \mathbf{P})\mathbf{E} + (\mathbf{J} + \partial \mathbf{P}/\partial t + \nabla \times \mathbf{M}) \times \mathbf{B}. \tag{8.19}$$

For an arbitrary volume V bounded by the surface S, the volume force is

$$\mathbf{F}_r^{(v)} = \int_V (\sigma - \nabla \cdot \mathbf{P})\mathbf{E}\, dV + \int_V \left(\mathbf{J} + \frac{\partial \mathbf{P}}{\partial t} + \nabla \times \mathbf{M}\right) \times \mathbf{B}\, dV, \tag{8.20}$$

where we have separated the electric part from the magnetic part. On the surface S there are additional tractions due to surface polarization $(\mathbf{n} \cdot \mathbf{P})$ and surface magnetization current $(-\mathbf{n} \times \mathbf{M})$,

$$\mathbf{F}_r^{(s)} = \int_S (\mathbf{n} \cdot \mathbf{P})\mathbf{E}\, dS + \int_S (-\mathbf{n} \times \mathbf{M}) \times \mathbf{B}\, dS. \tag{8.21}$$

Adding Eqs. (8.20) and (8.21) and converting the surface integral by the divergence theorem for dyadics (see derivations leading to Eqs. (9.16B) and (9.16C)), we obtain

$$\mathbf{F}_r = \mathbf{F}_r^{(v)} + F_r^{(s)} = \int_V (\sigma + \mathbf{P} \cdot \nabla)\mathbf{E}\, dV + \int_V \left[\left(\mathbf{J} + \frac{\partial \mathbf{P}}{\partial t}\right) \times \mathbf{B} + (\nabla \mathbf{B}) \cdot \mathbf{M}\right] dV. \tag{8.22}$$

The body force per unit volume is then

$$\begin{aligned}\mathbf{F} &= \sigma\mathbf{E} + \mathbf{J} \times \mathbf{B} + \mathbf{P} \cdot \nabla\mathbf{E} + (\nabla\mathbf{B}) \cdot \mathbf{M} + \frac{\partial \mathbf{P}}{\partial t} \times \mathbf{B} \\ &= \sigma\mathbf{E} + \mathbf{J} \times \mathbf{B} + (\nabla\mathbf{E}) \cdot \mathbf{P} + (\nabla\mathbf{B}) \cdot \mathbf{M} + \frac{\partial}{\partial t}(\mathbf{P} \times \mathbf{B}).\end{aligned} \tag{8.23}$$

Note that Eq. (8.23) differs from Eq. (8.11) but agrees with Eq. (7.38) when $\mathbf{v} = \mathbf{0}$.

9 MAGNETOSTATIC FORCES ON A WHOLE BODY

In Maxwell's *Treatise*, the mechanical action between two systems of electrical charges (p. 153, vol. 1, [12]) is discussed in terms of the resultant force on each system. For a system of charges in a vacuum, the resultant force acting on the entire system by an externally applied field $\mathbf{E}_0$ is (Eq. (3.9))

$$\mathbf{F}_r = \int_V \sigma \mathbf{E}_0(\mathbf{x})\, dV.$$

The basis for this formula, as discussed in Subsection 3.2, is Coulomb's law, and the transmission of such a force is an *action at a distance*. Extension of this type of force to dielectric medium was mentioned (p. 166, vol. 1, [12]) but not completed in the *Treatise*. However, when Maxwell discussed the forces in magnetized material medium, an entirely different approach was adopted. Instead of evaluating the resultant force, a formula similar to Eq. (5.1) was used for calculating the force density, and the *action* was assumed to transmit *through contact* of aetheral medium (p. 276, vol. 2, [12]).

The same approach of evaluating the resultant force was adopted by W. F. Brown recently to determine the forces in magnetized materials. In his monograph [2], the magnetoelastic interactions are discussed in terms of the basic principles of magnetostatics and modern theory of continuum mechanics. We shall summarize his treatment on magnetic forces and stresses in a deformable body in this section (Chapter 2 and 5 of [2]).

9.1 Pole, Dipole, and Current–circuit Models of Magnetizations

In Subsection 3.2 we discussed that on the basis of the Mitchell–Coulomb law, the magnetic field intensity at the position $\mathbf{x}$ due to a magnetic pole of strength p' at the position $\mathbf{x}'$ is (Eq. (3.11))

$$\mathbf{H}(\mathbf{x}) = \frac{p'}{4\pi\mu_0} \frac{\mathbf{x} - \mathbf{x}'}{|\mathbf{x} - \mathbf{x}'|^3} = -\nabla\left[\frac{p'}{4\pi\mu_0 |\mathbf{x} - \mathbf{x}'|}\right]. \tag{9.1}$$

If a second pole of strength $-p'$ is placed at $\mathbf{x}' + \xi$, the total field at $\mathbf{x}$ is the

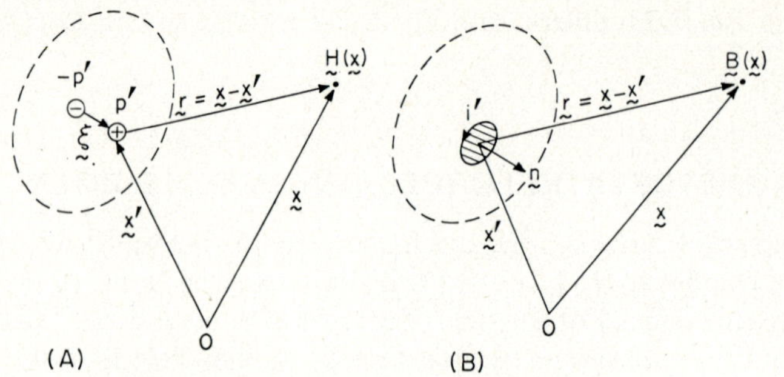

Fig. 5 (A) Magnetic dipoles; (B) current circuits.

sum of the contributions from both poles. In the limit $\xi \to 0$ and $|p'\xi|$ approaches a finite limit, the field due to the dipole is (Fig. 5)†

$$\mathbf{H}_2(\mathbf{x}) = -\nabla\phi_2, \qquad \phi_2 = \frac{\mathbf{m}' \cdot (\mathbf{x} - \mathbf{x}')}{4\pi|\mathbf{x} - \mathbf{x}'|^3}, \tag{9.2}$$

where the magnetic dipole moment **m** (prime omitted) is defined (Fig. 5A)

$$\mathbf{m} = \lim_{\xi \to 0} (p\xi/\mu_0). \tag{9.3}$$

From Eq. (3.12), the magnetic induction **B**(**x**) generated by a closed circuit with current i' is (Fig. 3)

$$\mathbf{B}(\mathbf{x}) = \frac{\mu_0}{4\pi}\oint_{C'} \frac{i'\, d\mathbf{C}' \times (\mathbf{x} - \mathbf{x}')}{|\mathbf{x} - \mathbf{x}'|^3} = \nabla \times \oint_{C'} \frac{\mu_0 i'\, d\mathbf{C}'}{4\pi|\mathbf{x} - \mathbf{x}'|}. \tag{9.4}$$

When the distance $|\mathbf{x} - \mathbf{x}'|$ is large in comparison with circuit dimensions, the line integral may be evaluated by expanding $|\mathbf{x} - \mathbf{x}'|^{-1}$ about the center of the circuit C'. In the limit of zero dimension for the circuit, the magnetic induction is (Fig. 5B)

$$\mathbf{B}_2(\mathbf{x}) = \nabla \times \mathbf{A}_2, \qquad \mathbf{A}_2 = \frac{\mu_0 \mathbf{m}' \times (\mathbf{x} - \mathbf{x}')}{4\pi|\mathbf{x} - \mathbf{x}'|^3}, \tag{9.5}$$

where the magnetic dipole moment is defined (primes omitted)

$$\mathbf{m} = \lim_{S_0 \to 0} (iS_0\mathbf{n}). \tag{9.6}$$

S_0 is the area of the circuit, and **n** a unit vector normal to the plane of the circuit (Fig. 6B).

† The subscript 2 denotes a field associated with the dipole.

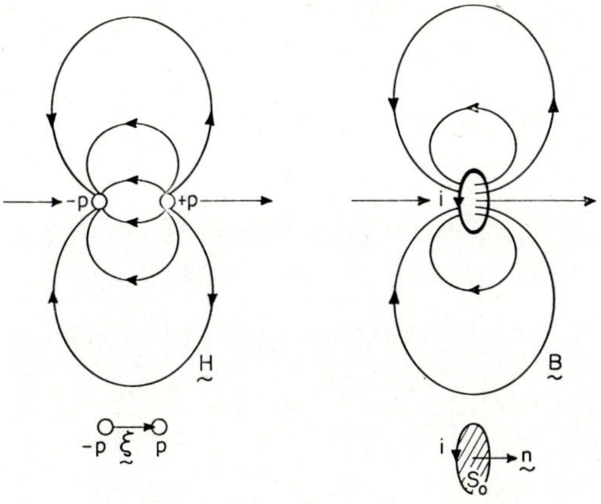

(A) (B)

Fig. 6 (A) Magnetic dipole and magnetic field strength **H**; (B) amperian current circuit and magnetic induction **B**.

With either Eqs. (9.3) or (9.6) as the definition for magnetic dipole moments, we find

$$\begin{aligned}\mathbf{H}_2 &= -\nabla\phi_2 = \frac{1}{4\pi}\left[-\frac{\mathbf{m}'}{r^3}+\frac{3(\mathbf{m}'\cdot\mathbf{r})\mathbf{r}}{r^5}\right],\\ \mathbf{B}_2 &= \nabla\times\mathbf{A}_2 = \frac{\mu_0}{4\pi}\left[-\frac{\mathbf{m}'}{r^3}+\frac{3(\mathbf{m}'\cdot\mathbf{r})\mathbf{r}}{r^5}\right],\end{aligned} \tag{9.7}$$

where $\mathbf{r} = \mathbf{x} - \mathbf{x}'$ and $r = |\mathbf{x} - \mathbf{x}'|$. Thus at the observing point $\mathbf{x}$, $\mathbf{B}(\mathbf{x}) = \mu_0\mathbf{H}(\mathbf{x})$ as it should. Shown in Fig. 6 are the magnetic field lines generated by a dipole source and a current circuit source. The lines of field are the same only when $\mathbf{x}$ is at a point far away from the sources, or when $\xi \to 0$ and $S_0 \to 0$.

Equations (9.2) and (9.5) can be generalized to a distribution of magnetic dipole moments within a volume V. Defining the magnetization $\mathbf{M}$ as the magnetic moment per unit volume, we have

$$\phi_2(\mathbf{x}) = \int \frac{\mathbf{M}'\cdot(\mathbf{x}-\mathbf{x}')}{4\pi|\mathbf{x}-\mathbf{x}'|^3}\,dV', \tag{9.8}$$

$$\mathbf{A}_2(\mathbf{x}) = \mu_0 \int \frac{\mathbf{M}'\times(\mathbf{x}-\mathbf{x}')}{4\pi|\mathbf{x}-\mathbf{x}'|^3}\,dV'. \tag{9.9}$$

Since

$$\nabla \frac{1}{|\mathbf{x}-\mathbf{x}'|} = -\nabla' \frac{1}{|\mathbf{x}-\mathbf{x}'|} = -\frac{\mathbf{x}-\mathbf{x}'}{|\mathbf{x}-\mathbf{x}'|^3},$$

and

$$\nabla \times (\mathbf{M}/r) = (\nabla \times \mathbf{M})/r - \mathbf{M} \times \nabla(1/r),$$
$$\nabla \cdot (\mathbf{M}/r) = (\nabla \cdot \mathbf{M})/r + \mathbf{M} \cdot \nabla(1/r),$$

where $r = |\mathbf{x} - \mathbf{x}'|$, the above expressions may be written as

$$\phi_2 = \frac{1}{4\pi}\left[\int \frac{-\nabla' \cdot \mathbf{M}'}{|\mathbf{x}-\mathbf{x}'|}\,dV' + \int \frac{\mathbf{n} \cdot \mathbf{M}'}{|\mathbf{x}-\mathbf{x}'|}\,dS'\right], \tag{9.10}$$

$$\mathbf{A}_2 = \frac{\mu_0}{4\pi}\left[\int \frac{\nabla' \times \mathbf{M}'}{|\mathbf{x}-\mathbf{x}'|}\,dV' + \int \frac{-\mathbf{n} \times \mathbf{M}'}{|\mathbf{x}-\mathbf{x}'|}\,dS'\right]. \tag{9.11}$$

In Eq. (9.8) or (9.10), ϕ_2 is the scalar potential for the magnetic field strength $\mathbf{H}_2(= -\nabla\phi_2)$ which is generated by a continuous distribution of magnetic dipole moments inside a body of volume V and bounding surface S. Similarly $\mathbf{A}_2$ is the vector potential for the magnetic induction $\mathbf{B}_2(= \nabla \times \mathbf{A}_2)$ which is generated by the same distribution of dipole moments. In the scalar potential ϕ_2, the magnetization may be interpreted by the model of magnetic dipoles (Eq. (9.3), whereas in $\mathbf{A}_2$, it is interpreted by the model of current circuits (Eq. (9.6)). Comparing Eq. (9.10) with the magnetic pole formula (9.1), we find that $-\nabla \cdot (\mu_0\mathbf{M})$ corresponds to a magnetic pole, and $\mathbf{n} \cdot (\mu_0\mathbf{M})$ to a surface magnetic pole. Similarly, a comparison of Eq. (9.11) with Eq. (9.4), or its three-dimensional version Eq. (3.14), shows that $\nabla \times \mathbf{M}$ corresponds to the current density $\mathbf{J}$, and $-\mathbf{n} \times \mathbf{M}$ is equivalent to a surface current density $\mathbf{K}$.

When the observation point $\mathbf{x}$ is outside the volume V, $\mathbf{B}_2(\mathbf{x}) = \nabla \times \mathbf{A}_2 = -\mu_0\nabla\phi_2 = \mu_0\mathbf{H}_2(\mathbf{x})$. If on the other hand the point $\mathbf{x}$ is inside the material volume V, the integrand becomes singular and $\nabla \times \mathbf{A}_2$ (Eq. (9.11)) does not equal $-\mu_0\nabla\phi_2$ (Eq. (9.10)). The difference is (p. 17, [2])

$$\mathbf{B}_2 - \mu_0\mathbf{H}_2 = \mu_0\mathbf{M}. \tag{9.12}$$

Thus the current loop and dipole model representations of magnetized bodies lead to the same magnetic field ($\mathbf{B}$ or $\mu_0\mathbf{H}$) at external points, but not at internal points. This result is compatible with the definition in Eq. (3.5).

From Eqs. (9.4) and (9.5), it can be concluded that $\mathbf{B}$ field is *solenoidal*; from (9.1) and (9.2), $\mathbf{H}$ is *irrotational* in the absence of conduction current in the body. If there are conduction currents and, possibly, magnetization,

$$\nabla \cdot \mathbf{B} = 0, \qquad \nabla \times \mathbf{H} = \mathbf{J}; \qquad \mathbf{B} = \mu_0(\mathbf{H} + \mathbf{M}). \tag{9.13}$$

These are the time-independent Maxwell equations (3.2). At the surface of discontinuity, the normal components of **B**, and the tangential components of **H** are continuous. If (+) and (−) denote the positive and negative sides respectively of the surface of discontinuity, we write

$$\mathbf{n} \cdot (\mathbf{B}^+ - \mathbf{B}^-) = 0, \qquad \mathbf{n} \times (\mathbf{H}^+ - \mathbf{H}^-) = \mathbf{K}, \tag{9.14}$$

where **n** is the unit normal to the surface, pointing toward the positive side.

9.2 Body Forces and Surface Forces

Consider a magnetizable body which is placed in an externally applied induction $\mathbf{B}_0(= \mu_0\mathbf{H}_0)$. Within the region V occupied by the body,

$$\nabla \cdot \mathbf{B}_0 = 0, \qquad \nabla \times \mathbf{B}_0 = \mathbf{0}. \tag{9.15}$$

The first equation is always true for any magnetic induction; the second is true because $\mathbf{B}_0$ is assumed stationary (magnetostatic case) and the current circuit sources are outside V.

To calculate the resultant force acting on the entire body, the magnetizations are modeled by current circuits (Amperian currents). As discussed in the previous subsection, $\nabla \times \mathbf{M}$ is equivalent to **J**, and $-\mathbf{n} \times \mathbf{M}$ is equivalent to a surface current. Applying Eqs. (3.13) or (3.15), we obtain the formula for the resultant force on the whole body of volume V and bounding surface S (Section 5.3 of [2]),

$$\mathbf{F}_r = \int_S (-\mathbf{n} \times \mathbf{M}) \times \mathbf{B}_0 \, dS + \int_V (\nabla \times \mathbf{M}) \times \mathbf{B}_0 \, dV. \tag{9.16A}$$

In addition, there is the resultant torque

$$\mathbf{L}_r = \int_S \mathbf{x} \times [(-\mathbf{n} \times \mathbf{M}) \times \mathbf{B}_0] \, dS + \int_V \mathbf{x} \times [(\nabla \times \mathbf{M}) \times \mathbf{B}_0] \, dV. \tag{9.17A}$$

By applying vector indentities, Eqs. (9.16A) and (9.17A) may be transformed into various forms. First, the surface integral may be transformed into a volume integral.

$$\begin{aligned}\int (-\mathbf{n} \times \mathbf{M}) \times \mathbf{B}_0 \, dS &= \int [\mathbf{n}(\mathbf{B}_0 \cdot \mathbf{M}) - \mathbf{n} \cdot \mathbf{B}_0\mathbf{M}] \, dS \\ &= \int [\nabla(\mathbf{B}_0 \cdot \mathbf{M}) - \nabla \cdot (\mathbf{B}_0\mathbf{M})] \, dV.\end{aligned}$$

Next, we note

$$(\nabla \times \mathbf{M}) \times \mathbf{B}_0 = (\mathbf{M} \cdot \nabla)\mathbf{B}_0 + (\mathbf{B}_0 \cdot \nabla)\mathbf{M} + \mathbf{M} \times (\nabla \times \mathbf{B}_0) - \nabla(\mathbf{M} \cdot \mathbf{B}_0).$$

Substituting the above into Eq. (9.16A), we find

$$\begin{aligned} \mathbf{F}_r &= \int_V [(\mathbf{M} \cdot \nabla)\mathbf{B}_0 + \mathbf{M} \times (\nabla \times \mathbf{B}_0) - \mathbf{M}\nabla \cdot \mathbf{B}_0]\, dV \\ &= \int_V (\mathbf{M} \cdot \nabla)\mathbf{B}_0\, dV. \end{aligned} \tag{9.16B}$$

In the last step, use has been made of the Eqs. (9.15).

Furthermore, since

$$\nabla \cdot (\mathbf{M}\mathbf{B}_0) = (\nabla \cdot \mathbf{M})\mathbf{B}_0 + (\mathbf{M} \cdot \nabla)\mathbf{B}_0,$$

we have from Eq. (9.15B)

$$\mathbf{F}_r = \int_S (\mathbf{n} \cdot \mathbf{M})\mathbf{B}_0\, dS + \int_V (-\nabla \cdot \mathbf{M})\mathbf{B}_0\, dV. \tag{9.16C}$$

Equations (9.16 A, B, C) are equally valid expressions for the resultant force on the entire body.

In an analogous manner, we find from Eq. (9.17A),

$$\mathbf{L}_r = \int_V \mathbf{x} \times (\mathbf{M} \cdot \nabla\mathbf{B}_0)\, dV + \int_V \mathbf{M} \times \mathbf{B}_0\, dV \tag{9.17B}$$

$$= \int_S \mathbf{x} \times (\mathbf{n} \cdot \mathbf{M}\mathbf{B}_0)\, dS + \int_V \mathbf{x} \times (-\nabla \cdot \mathbf{M})\mathbf{B}_0\, dV. \tag{9.17C}$$

Since the external magnetic field can also be expressed in terms of $\mathbf{H}_0$ with $\mathbf{B}_0 = \mu_0\mathbf{H}_0$ in MKSA units, all $\mathbf{B}_0$'s in Eqs. (9.15) and (9.16) may be replaced by $\mu_0\mathbf{H}_0$. As discussed in previous subsection, $\mathbf{n} \cdot \mathbf{M}$ and $-\nabla \cdot \mathbf{M}$ in Eq. (9.16C) are associated with the pole model for magnetications. Traditionally, the magnetic fields in connection with poles and dipoles are denoted by $\mathbf{H}$. Henceforth, we shall write Eqs. (9.16B) and (9.17B) as

$$\mathbf{F}_r = \mu_0 \int_V (\mathbf{M} \cdot \nabla)\mathbf{H}_0\, dV, \tag{9.18}$$

$$\mathbf{L}_r = \mu_0 \int_V \mathbf{x} \times (\mathbf{M} \cdot \nabla\mathbf{H}_0)\, dV + \mu_0 \int_V \mathbf{M} \times \mathbf{H}_0\, dV. \tag{9.19}$$

These two formulas are regarded as fundamental in Brown's monograph

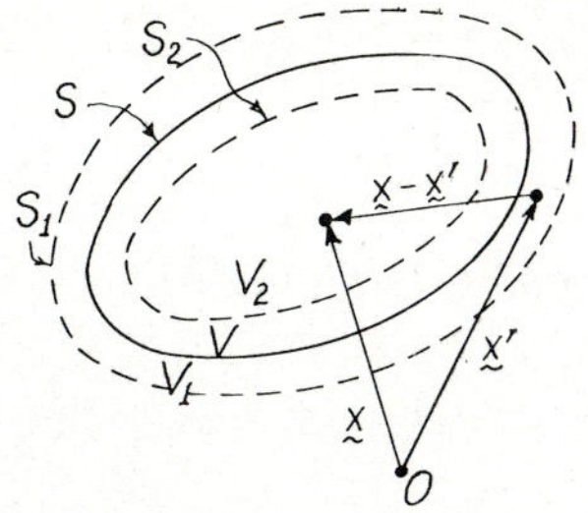

Fig. 7 An arbitrary volume V (inside the solid line) within a magnetized body.

(p. 55, [2]). They are also cited as the forces and torque on a body in a magnetic field by Landau and Lifshitz (Section 34, [8]).

On the basis of Eq. (9.18), we can not conclude that the magnetic force on a volume element dV is $\mathbf{M} \cdot \nabla \mathbf{H}_0 \, dV$, because the rest of the magnetized body also generates a magnetic field at dV. The forces due to the internal fields have been cancelled during the process of summing all forces acting on the entire body. To calculate the force on an arbitrary volume V bounded by a surface S inside the magnetized body, consider a closed surface S_2 inside S, and a surface S_1 outside S (Fig. 7). The volumes bounded by S_2 and S_1 are V_2 and V_1 respectively.

Let $\mathbf{B}_0 = \mu_0 \mathbf{H}$ be the field intensity of all sources outside S_1. Inside S_2, the total field intensity is

$$\mathbf{H} = \mathbf{H}_0 + \mathbf{H}_{12}, \tag{9.20}$$

where $\mathbf{H}_{12}$ is the part of the field intensity contributed by the matter in V_1, calculated at points $\mathbf{x}$ inside V_2. Applying Eq. (9.18) to the volume V_2, we obtain

$$\mathbf{F}_r = \mu_0 \int_{V_2} (\mathbf{M} \cdot \nabla)\mathbf{H} \, dV + \mathbf{F}_{12}, \tag{9.21}$$

where

$$\begin{aligned} \mathbf{F}_{12} &= -\mu_0 \int_{V_2} (\mathbf{M} \cdot \nabla)\mathbf{H}_{12} \, dV \\ &= -\mu_0 \int_{S_2} (\mathbf{n} \cdot \mathbf{M})\mathbf{H}_{12} \, dS - \mu_0 \int_{V_2} (-\nabla \cdot \mathbf{M})\mathbf{H}_{12} \, dV. \end{aligned} \tag{9.22}$$

The $\mathbf{H}_{12}$ can be evaluated by applying Eqs. (9.2) and (9.10) to the volume V_1

$$\mathbf{H}_{12} = -\frac{1}{4\pi}\left[\int_{S_1} (\mathbf{n}' \cdot \mathbf{M}') \frac{\mathbf{x} - \mathbf{x}'}{|\mathbf{x} - \mathbf{x}'|^3} dS' + \int_{V_1} (-\nabla' \cdot \mathbf{M}') \frac{\mathbf{x} - \mathbf{x}'}{|\mathbf{x} - \mathbf{x}'|^3} dV'\right]. \tag{9.23}$$

Insertion of Eq. (9.23) into Eq. (9.22) leads to

$$\mathbf{F}_{12} = \mu_0 \int_{S_1} (\mathbf{n}' \cdot \mathbf{M}')\mathbf{H}_{21}\, dS' + \mu_0 \int_{V_1} (-\nabla' \cdot \mathbf{M}')\mathbf{H}_{21}\, dV', \tag{9.24}$$

where $\mathbf{H}_{21}$ is the field intensity contributed by the matter in V_2, calculated at points $\mathbf{x}'$ in V_1.

We first add the two expressions for $\mathbf{F}_{12}$ and divide the sum by 2, and then take the limit of $S_1, S_2 \to S$. In the limit, the volume integrals in Eqs. (9.22) and (9.24) cancel each other but the surface integrals do not, because $\mathbf{H}_{12}$ is always evaluated inside S_2, whereas $\mathbf{H}_{21}$ is always evaluated outside it. In the limit, the contribution from the two surface integrals is

$$\mathbf{F}_{12} = \frac{\mu_0}{2} \int_S (\mathbf{n} \cdot \mathbf{M})(\mathbf{H}^+ - \mathbf{H}^-)\, dS, \tag{9.25}$$

where $\mathbf{H}$ is magnetic field intensity of the poles associated with the magnetized matter in V, and $(+)$ and $(-)$ denote the values just outside and just inside S.

From Eq. (9.14) we know that across a surface of discontinuity, the normal components of $\mathbf{H}$ are discontinuous. Since $\mathbf{n} \cdot (\mathbf{B}^+ - \mathbf{B}^-) = 0$ and $\mathbf{B} = \mu_0(\mathbf{H} + \mathbf{M})$, we find

$$\mathbf{n} \cdot (\mathbf{H}^+ - \mathbf{H}^-) = \mathbf{n} \cdot \mathbf{M} = M_n.$$

Thus

$$\mathbf{H}^+ - \mathbf{H}^- = \mathbf{n} M_n, \tag{9.26}$$

where $M_n = \mathbf{n} \cdot \mathbf{M}$ is the normal component of $\mathbf{M}$. The formula (9.25) can now be written

$$\mathbf{F}_{12} = \int_S \tfrac{1}{2}\mathbf{n}(\mu_0 M_n^2)\, dS. \tag{9.27}$$

As $V_2 \to V$, Eq. (9.21), when combined with the above, reads

$$\mathbf{F}_r = \int_V (\mu_0 \mathbf{M} \cdot \nabla)\mathbf{H}\, dV + \int_S \tfrac{1}{2}\mathbf{n}(\mu_0 M_n^2)\, dS. \tag{9.28}$$

The volume V is an arbitrary volume bounded by the surface S inside the body, and $\mathbf{H}$ and $\mathbf{M}$ satisfy Eq. (9.13).

The above derivation indicates that the long-range force exerted by one component of the magnetized body on the other can be expressed in terms of a body force (the volume integral in Eq. (9.28) and a surface force in the surface integral). Furthermore, this resolution is not unique. If we started with Eqs. (9.16 A, C), instead of Eq. (9.16B), different expressions for the resultant force would emerge. These expressions are given in Ref. [2, p. 57].

A similar calculation gives rise to the resultant torque on an arbitrary volume V inside a body,

$$\mathbf{L}_r = \int_V \mathbf{x} \times [(\mu_0 \mathbf{M} \cdot \nabla)\mathbf{H}]\, dV + \int_S \mathbf{x} \times \mathbf{n}(\tfrac{1}{2}\mu_0 M_n^2)\, dS + \int_V (\mu_0 \mathbf{M} \times \mathbf{H})\, dV. \tag{9.29}$$

The last term is a body couple which corresponds to the $\int \mathbf{L}\, dV$ term in Eq. (2.9).

9.3 Various Stress Tensors and Momentum Equations

Since V in Eq. (9.28) is an arbitrary volume inside a body and $\mathbf{F}_r$ is the resultant force on the material inside the surface S, the balance equation of momentum Eq. (2.7) now reads

$$\frac{d}{dt}\int_V \rho \mathbf{v}\, dV = \int_S (\mathbf{t}^{(n)} + \tfrac{1}{2}\mu_0 M_n^2 \mathbf{n})\, dS + \int_V (\mu_0 \mathbf{M} \cdot \nabla)\mathbf{H}\, dV. \tag{9.30}$$

Note that in addition to the usual surface traction $\mathbf{t}^{(n)}$, there is the force $\frac{1}{2}\mu_0 M_n^2 \mathbf{n}$ at the same surface.

Defining a new surface traction vector

$$\mathbf{t}_e^{(n)} = \mathbf{t}^{(n)} + \tfrac{1}{2}\mu_0 (\mathbf{M} \cdot \mathbf{n})^2 \mathbf{n}, \tag{9.31}$$

and a stress tensor $\underset{\sim}{\tau}$ such that

$$\mathbf{t}_e^{(n)} = \mathbf{n} \cdot \underset{\sim}{\tau}, \tag{9.32}$$

we obtain from Eq. (9.30) the local balance law for linear momentum,

$$\rho \dot{\mathbf{v}} = \nabla \cdot \underset{\sim}{\tau} + \mu_0 (\mathbf{M} \cdot \nabla)\mathbf{H}. \tag{9.33}$$

Similarly, for the balance of angular momentum, we have

$$\varepsilon_{ijk} \tau_{jk} = \mu_0 (\mathbf{M} \times \mathbf{H})_i. \tag{9.34}$$

Since $\mathbf{M} \times \mathbf{H}$ does not vanish in general, the stress tensor so introduced is asymmetric. Equations (9.33) and (9.34), which correspond to Eqs. (2.12) and (2.13A) respectively, are the balance equations of momenta for stationary, magnetized media. $\mathbf{M}$ and $\mathbf{H}$ satisfy the magnetostatic equations (9.13).

The last volume integral in Eq. (9.30) can be converted into the sum of a volume integral and a surface integral by noting the vector identity

$$(\mu_0 \mathbf{M} \cdot \nabla)\mathbf{H} = \nabla \cdot (\mu_0 \mathbf{M}\mathbf{H}) - (\mu_0 \nabla \cdot \mathbf{M})\mathbf{H}. \tag{9.35}$$

A new global equation for the balance of linear momentum is thus obtained. Introducing a new stress tensor $\bar{\tau}_{ij}$ such that

$$\mathbf{n} \cdot \underset{\sim}{\bar{\tau}} = \mathbf{t}^{(n)} + \tfrac{1}{2}\mu_0(\mathbf{M} \cdot \mathbf{n})^2\mathbf{n} + \mu_0(\mathbf{n} \cdot \mathbf{M})\mathbf{H}, \tag{9.36}$$

we obtain another set of momentum equations

$$\rho\dot{\mathbf{v}} = \nabla \cdot \underset{\sim}{\bar{\tau}} + \mu_0(-\nabla \cdot \mathbf{M})\mathbf{H}, \tag{9.37}$$

$$\bar{\tau}_{ij} = \bar{\tau}_{ji}. \tag{9.38}$$

The new stress tensor $\bar{\tau}_{ij}$ is symmetric.

A third form of balance equation is established by Brown (p. 60, [2]) if the current circuit model is adopted (Eq. (9.16A)). The local equations are

$$\rho\dot{\mathbf{v}} = \nabla \cdot \underset{\sim}{\bar{\tau}}' + (\nabla \times \mathbf{M}) \times \mathbf{B}, \tag{9.39}$$

$$\bar{\tau}'_{ij} = \bar{\tau}'_{ji}. \tag{9.40}$$

The stress tensor $\bar{\tau}'_{ij}$ is also symmetric, and $\mathbf{M}$ and $\mathbf{B}$ satisfy Eq. (9.13). It should now become clear that by applying other vector identities to transform the volume and surface integrals in global balance laws, other forms of local equations may be derived.

As noted by Brown, Eq. (9.39) could be directly established from the Amperian current model for the magnetization. Since $\nabla \times \mathbf{M}$ corresponds to a current $\mathbf{J}$, there is a body force per unit volume $\mathbf{F} = \mathbf{J} \times \mathbf{B} = (\nabla \times \mathbf{M}) \times \mathbf{B}$ acting on each volume element. The effect of the surface magnetization current which appears as a surface integral in Eq. (9.16A) is combined with the usual mechanical surface traction. The combined effect is represented by the new stress tensor $\bar{\tau}'_{ij}$.

Similarly, if the magnetizations are represented by the pole model, Eq. (9.37) could be established directly from that $-\nabla \cdot \mathbf{M}$ represents a magnetic pole and the force on a magnetic pole, according to Eq. (3.11), is $(-\nabla \cdot \mathbf{M})\mathbf{H}$. Again, the effect of surface magnetization which is represented by the first integral in Eq. (9.16C) is combined into the stress tensor $\bar{\tau}_{ij}$.

Finally, we note that the body force term $(\mu_0 \mathbf{M} \cdot \nabla)\mathbf{H}$ in Eq. (9.33) can be directly obtained when the magnetizations are modeled by dipoles. Further discussion of forces on dipoles is given in the next section.

9.4 Discussion

Brown's monograph also contains discussions on the boundary conditions, energy equations, and constitutive equations for paramagnetic and ferromagnetic materials. It is a very comprehensive, modern treatise on the subject of magnetoelastic interactions.

His approach to the determination of magnetic forces and torque as summarized in this section is free from what the author called "superfluous postulations." The magnetostatic equations in Subsection 9.1 are based on the Michell–Coulomb law and Ampère's law, which in turn were founded on experiments. The concept of magnetic moments and magnetizations are well defined and generally accepted. The pole, dipole, and current circuit models are useful in interpreting the results at various stages but they are not essential in developing the basic equations and formulas. The formula (Eq. 9.16) for the resultant forces on the whole body are established without additional postulates beyond those already contained in the theory of magnetostatics.

Although the discussions in the monograph are confined to magnetic materials, they can be applied equally well to dielectrics. If **P** is substituted for $\mu_0\mathbf{M}$, and **E** for **H**, Eqs. (9.33) and (9.34) or Eqs. (9.37) and (9.38) are the dynamic equations for dielectrics.

However, Brown's theory can not be easily generalized to moving bodies and nonstatic fields. First the magnetostatic equations (9.13) must be replaced by the complete set of Maxwell equations, and it is not clear which of the several formulations mentioned in Section 4 should be adopted. Next the formulas (9.16) and (9.17) for the resultant force and torque must be modified. At this point great difficulty would be encountered if the polarization charge, polarization current, magnetization charge, and magnetization current were all present in moving matter. Thirdly the deduction of the energy balance equation which is rather complicated already in the statical case (Chapter 7, [2]) may become even more intricate. Finally, the transformation from the resultant force on the entire body to the force on an arbitrary volume (from Eq. (9.18) to Eq. (9.28)) cannot be easily evaluated for a body in motion.

This last step (from Eq. (9.18) to Eq. (9.28)) could have been omitted if one had modeled the magnetizations by poles, dipoles, or American current circuits. As mentioned before, once these models are accepted, one can establish the momentum equation (9.33), (9.37), or (9.39) immediately. Recalling that the magnetic field equations (9.13) are based on either the model of magnetic poles, or current circuits, we see that the adoption of one of these two models for magnetization may not be totally superfluous. This, however, has been carefully avoided in Brown's treatment.

In the next and in the final sections of this article, we discuss the magnetoelastic interactions for a body in motion based on the pole and dipole models. We shall show that Eq. (9.33) is a special case of the more general equations.

10 MODELS FOR FIELD–MATTER INTERACTIONS

In a recent paper [35], Y. H. Pao and K. Hutter proposed determination of the matter–field interaction on the basis of the Chu formulation of electrodynamics. The basic postulates in the Chu formulation are listed in Subsection 4.4, of which the first two were stated in the original text and the third one is our assertion. We believe in that the third postulate—the polarization and magnetization are modeled respectively by electric and magnetic charges, and dipoles—must be adjoined in order to arrive at the final form of the field equations (4.8). The precise meanings of these dipoles are discussed in the following subsection. For convenience, we shall refer to this assumption as the *two-dipole model.*

Following these postulates and the concept of the Lorentz force on electric charges together with the corresponding forces on magnetic poles, Pao and Hutter calculated the body force, body couple, and energy supply to the moving media. Their results agree with those obtained by Chu, Penfield and Haus (slow speed approximation), which are deduced from the principle of virtual power [22].

A commonly favored model for the magnetic interaction is the Amperian current circuit. In this model, the previous role of the dipole is played by a small circuit with circulating current, known as the *Amperian current circuit*, or simply *current circuit*. We already discussed how the magnetic moment is defined in terms of the limiting value of a current circuit (Eq. (9.6)). A combination of electric dipoles for electric polarization and Amperian current circuits for magnetization will be referred to as the *dipole–current circuit model.*

Based on the dipole–current circuit model, many researchers have investigated the matter–field interactions. In 1957 L. V. Boffi investigated the electrodynamics of moving media. His results are contained in an unpublished doctoral thesis [21]. A summary of the results is given by Penfield and Haus under the heading "the Boffi formulation" (Section 7.5 of [15]). Recently, H. F. Tiersten and C. F. Tsai investigated the case of deformable media [43]. In both cases, the field equations and the expressions for the body force agree with those in the statistical formulation as presented in the Section 7.

10.1 Electric and Magnetic Dipoles and Current-circuits

The role of free charges in dielectrics and free currents in conductor as sources for electromagnetic field is well understood. Under the influence of applied electric and magnetic fields, additional sources are induced in material media. The effects of these induced sources may be represented by two dipoles, one electric and one magnetic, or a combination of an electric dipole and Amperian current circuit.

Each dipole is composed of a pair of negative and positive electric charges, $\pm q$, or magnetic poles, $\pm p$. If two electric charges $\pm q$ are separated at a distance ξ, the electric dipole moment is defined as $q\xi$ where the vector is traditionally pointing from the negative charge to the positive. The polarization density $\mathbf{P}$ which is the electric dipole moment per unit volume is then defined as

$$\mathbf{P} = \lim_{\xi \to 0} Nq\xi, \tag{10.1}$$

where N is the number of dipoles per unit volume and $q\xi$ is assumed to remain finite in the limit.

Similarly, the magnetic dipole moment is defined as $p\xi$ where ξ separates the pole p from $-p$. The magnetization which is the magnetic dipole moment per unit volume is (Fig. 6A)

$$\mu_0\mathbf{M} = \lim_{\xi \to 0} N^*p\xi, \tag{10.2}$$

where N^* is the number of magnetic dipoles per unit volume.

With these definitions and the basic laws of electromagnetism as applied to the charges or poles in motion, Chu was able to establish the field equations for moving media, Eqs. (4.8).

Since the electric and magnetic dipoles are assumed to be contained in the material particles, the number of dipoles $\mathbf{N}(\mathbf{x}, t)$ of either type is conserved. When the particle moves with velocity $\mathbf{v}$, we have

$$\partial N/\partial t + \nabla \cdot (\mathbf{v}N) = 0. \tag{10.3}$$

This is analogous to the equation of conservation of mass Eq. (2.11). For any scalar or vector function $\Phi(\mathbf{x}, t)$ Eq. (10.3) leads to the following identity:

$$N\frac{d}{dt}\left(\frac{\Phi}{N}\right) = \frac{d\Phi}{dt} - \frac{\Phi}{N}\left(\frac{\partial N}{\partial t} + \mathbf{v}\cdot\nabla N\right) = \frac{\partial \Phi}{\partial t} + \nabla \cdot (\mathbf{v}\Phi). \tag{10.4}$$

The magnetic dipole moment may also be defined in terms of a small circuit C with circulating current i. If N^* is the density of circuits per unit volume, the magnetization is

$$\mathbf{M} = \lim_{S_0 \to 0} (N^* i S_0 \mathbf{n}), \tag{10.5}$$

where S_0 is the area enclosed by the plane circuit C and $\mathbf{n}$ is a unit vector normal to the plane (Fig. 6B).

The magnetic induction field generated by a circuit is *solenoidal* and is traditionally denoted by $\mathbf{B}$ with $\nabla \cdot \mathbf{B} = 0$. The field generated by a dipole is traditionally denoted by $\mathbf{H}$ which is *irrotational.* The magnitudes of the fields due to each source are evaluated in Subsection 9.1; Eqs. (9.2), (9.5), and (9.7).

10.2 Force, Couple, and Energy Supply of the Two-dipole Model

To determine the interactions of matter and electromagnetic fields, a fourth postulate is added to the Chu formulation [35]:

4. The forces acting on the electric charges and magnetic poles are, respectively,

$$\mathbf{f}_q = q\mathbf{E} + q\mathbf{v} \times (\mu_0 \mathbf{H}) = q\mathbf{E}_e, \tag{10.6}$$

$$\mathbf{f}_p = p\mathbf{H} - p\mathbf{v} \times (\varepsilon_0 \mathbf{E}) = p\mathbf{H}_e, \tag{10.7}$$

where $\mathbf{v}$ is the velocity of the material particle, and $\mathbf{E}$ and $\mathbf{H}$ satisfy Eqs. (10.3).

The two effective fields in the above equations, $\mathbf{E}_e = \mathbf{E} + \mathbf{v} \times \mu_0 \mathbf{H}$ and $\mathbf{H}_e = \mathbf{H} - \mathbf{v} \times \varepsilon_0 \mathbf{E}$ are defined in Eq. (4.14). The force $\mathbf{f}_q$ is similar to the $\mathbf{f}$ in Eq. (7.1) which is known as the Lorentz force. The Lorentz force is assumed for a charge moving in the vacuum, and it can be measured directly in a laboratory. A generalization of it to a charge moving in a material medium as in Eq. (10.6) is not subject to direct experimental confirmation. Thus it is considered as an assumption.

The first part of $\mathbf{f}_p$ follows the Michell–Coulomb law of magnetostatics

(Eq. (3.11)). The assumption of $\mathbf{f}_p = -\varepsilon_0 p\mathbf{v} \times \mathbf{E}$ has a weaker foundation. One verification is given in Section 7.1 of Ref. [5], based on the calculation of torque on a polarized body. Another verification is based on the theory of relative motion, where the force on a magnetic pole moving with constant velocity $\mathbf{v}$ is $p(\mathbf{H} - \mathbf{v} \times \mathbf{D})$ (Chapters 1, 4 of [3]), and $\mathbf{D}$ equals $\varepsilon_0\mathbf{E}$ in the vacuum. Since Eq. (10.7) will always be applied to a pair of positive and negative poles, the assumption of $\mathbf{f}_p$ should be viewed as the force acting on magnetic dipoles.

Once the forces on electric charges and magnetic poles are assumed, the body force and couple on an arbitrary volume element can be calculated by applying a standard averaging and limiting process which is discussed in great detail in the text by Mason and Weaver [11]. The energy supply per unit volume must be evaluated in conjunction with the balance law of energy (Eq. (2.9)).

(A) *Body Forces.* Consider an assemblage of material particles with continuous distribution of mass density ρ. To each assemblage are attached two types of dipoles: the electric and magnetic. In the present configuration, the assemblage occupies the region dV surrounding the point $\mathbf{x}$.

Consider first the electric dipoles. Without the loss of generality, we place the negative charge, $-q$, of a dipole at $\mathbf{x}$, and the positive charge, $+q$, at $\mathbf{x} + \xi$. The respective velocities of the charges are

$$\mathbf{v}^- = \mathbf{v} = \dot{\mathbf{x}}, \qquad \mathbf{v}^+ = \mathbf{v} + \dot{\xi}. \tag{10.8}$$

Due to the fields $\mathbf{E}$ and $\mathbf{H}$, the force on each charge is, according to Eq. (10.6),

$$\begin{aligned} \mathbf{f}_q^- &= -q\mathbf{E}(\mathbf{x}) - q\mathbf{v} \times \mu_0\mathbf{H}(\mathbf{x}) \\ \mathbf{f}_q^+ &= q\mathbf{E}(x + \xi) + q[(\mathbf{v} + \dot{\xi}) \times \mu_0\mathbf{H}(\mathbf{x} + \xi)]. \end{aligned} \tag{10.9}$$

Expanding the field $\mathbf{E}(\mathbf{x} + \xi)$ and $\mathbf{H}(\mathbf{x} + \xi)$ into Taylor series,

$$\mathbf{E}(\mathbf{x} + \xi) = \mathbf{E}(\mathbf{x}) + \xi \cdot \nabla\mathbf{E}(\mathbf{x}) + \cdots, \tag{10.10}$$

we find the total force on the dipole,

$$\mathbf{f}_q = \mathbf{f}_q^- + \mathbf{f}_q^+ = q\xi \cdot \nabla\mathbf{E}(\mathbf{x}) + q\mathbf{v} \times [\mu_0\xi \cdot \nabla\mathbf{H}(\mathbf{x})] + \mu_0 q\dot{\xi} \times \mathbf{H}(\mathbf{x}) + \cdots. \tag{10.11}$$

We now consider one unit of the volume element which contains $N(\mathbf{x}, t)$ dipoles. If a dipole is separated by the surface of the volume, it is excluded from the calculation because the force on one charge of the broken dipole by the companion charge is a type of surface force. All forces acting on the surface of the volume element are included in the stress vector $\mathbf{t}^{(n)}$ in Eq. (2.7). The total number of charges of opposite sign can be assumed equal because forces on any isolated positive or negative charge will be considered separately, see Eq. (10.17).

Multiplying Eq. (10.11) by N and taking the limit $\xi \to 0$, we find first

$$Nq\dot{\xi} = N\frac{d}{dt}\left(\frac{Nq\xi}{N}\right) = \frac{\partial}{\partial t}(Nq\xi) + \nabla\cdot(\mathbf{v}Nq\xi), \tag{10.12}$$

$$\lim_{\xi\to 0}(Nq\dot{\xi}) = \frac{\partial \mathbf{P}}{\partial t} + \nabla\cdot(\mathbf{vP}) = \rho\frac{d}{dt}\left(\frac{\mathbf{P}}{\rho}\right). \tag{10.13}$$

In the above equations, use has been made of Eq. (10.4) and the definition for $\mathbf{P}$ as given by Eq. (10.1). The last step of Eq. (10.13) is obtained by applying the conservation equation of mass with

$$\frac{\partial \mathbf{P}}{\partial t} + \nabla\cdot(\mathbf{vP}) = \frac{\partial \mathbf{P}}{\partial t} + \mathbf{v}\cdot\nabla\mathbf{P} + (\nabla\cdot\mathbf{v})\mathbf{P} = \frac{d\mathbf{P}}{dt} - \frac{1}{\rho}\frac{d\rho}{dt}\mathbf{P} = \rho\frac{d}{dt}\left(\frac{\mathbf{P}}{\rho}\right).$$

Next we define the body force per unit volume due to electric polarization as

$$\mathbf{F}_q = \lim_{\xi\to 0}(N\mathbf{f}_q). \tag{10.14}$$

From Eqs. (10.11) and (10.3), we obtain

$$\mathbf{F}_q = \mathbf{P}\cdot\nabla\mathbf{E} + \mathbf{v}\times(\mathbf{P}\cdot\nabla)(\mu_0\mathbf{H}) + [\partial\mathbf{P}/\partial t + \nabla\cdot(\mathbf{vP})]\times(\mu_0\mathbf{H}) \tag{10.15A}$$

$$= \mathbf{P}\cdot\nabla\mathbf{E} + \mathbf{v}\times(\mathbf{P}\cdot\nabla)(\mu_0\mathbf{H}) + \rho\frac{d}{dt}\left(\frac{\mathbf{P}}{\rho}\right)\times(\mu_0\mathbf{H}). \tag{10.15B}$$

The body force per unit volume on magnetic dipoles can be calculated in an analogous manner. With $\mathbf{F}_p = \lim N^*(\mathbf{f}_p^+ + \mathbf{f}_p^-)$ where $\mathbf{f}_p$ is given by Eq. (10.7), we find

$$\mathbf{F}_p = \mu_0\mathbf{M}\cdot\nabla\mathbf{H} - \mathbf{v}\times(\mu_0\mathbf{M}\cdot\nabla)(\varepsilon_0\mathbf{E}) - [\partial(\mu_0\mathbf{M})/\partial t + \nabla\cdot(\mu_0\mathbf{vM})]\times\varepsilon_0\mathbf{E} \tag{10.16A}$$

$$= \mu_0\mathbf{M}\cdot\nabla\mathbf{H} - \mathbf{v}\times(\mu_0\mathbf{M}\cdot\nabla)(\varepsilon_0\mathbf{E}) - \rho\frac{d}{dt}\left(\frac{\mu_0\mathbf{M}}{\rho}\right)\times(\varepsilon_0\mathbf{E}). \tag{10.16B}$$

The total body force per unit volume is the sum of $\mathbf{F}_q$ and $\mathbf{F}_p$, plus the force on the free charge, $\sigma\mathbf{E}$, and on the free current, $\mathbf{J}\times(\mu_0\mathbf{H})$. The final result is

$$\begin{aligned}(\text{IV}): \quad \mathbf{F} &= \sigma\mathbf{E} + \mathbf{J} \times (\mu_0\mathbf{H}) \\ &+ \mathbf{P}\cdot\nabla\mathbf{E} + \mathbf{v} \times (\mathbf{P}\cdot\nabla)(\mu_0\mathbf{H}) + [\partial\mathbf{P}/\partial t + \nabla\cdot(\mathbf{vP})] \times (\mu_0\mathbf{H}) \\ &+ \mu_0\mathbf{M}\cdot\nabla\mathbf{H} - \mathbf{v} \times (\mu_0\mathbf{M}\cdot\nabla)(\varepsilon_0\mathbf{E}) \\ &- [\partial(\mu_0\mathbf{M})/\partial t + \nabla\cdot(\mu_0\mathbf{vM})] \times (\varepsilon_0\mathbf{E}).\end{aligned} \tag{10.17}$$

The above result agrees with that given by Penfield and Haus (Eq. (7.25) of [15]) for the nonrelativistic case. For the quasi-static case, $\mathbf{v} = \mathbf{0}$ but $\mathbf{E}$, $\mathbf{H}$, $\mathbf{P}$, $\mathbf{M}$ may still be time dependent, Eq. (10.17) reduces to

$$\mathbf{F} = \sigma\mathbf{E} + \mathbf{J} \times (\mu_0\mathbf{H}) + \mathbf{P}\cdot\nabla\mathbf{E} + \mu_0\mathbf{M}\cdot\nabla\mathbf{H}. \tag{10.18}$$

The term $\mu_0\mathbf{M}\cdot\nabla\mathbf{H}$ is identical to the body force in Eqs. (9.28) and (9.33). The last two terms in the above are generally known as the *Kelvin force*.

A stress tensor may be derived from $\mathbf{F}$. Note that in Eq. (10.17),

$$\begin{aligned}\mathbf{P}\cdot\nabla\mathbf{E} + \mathbf{v} \times (\mathbf{P}\cdot\nabla)(\mu_0\mathbf{H}) &= \mathbf{P}\cdot\nabla\mathbf{E}_e + \mu_0\mathbf{H} \times (\mathbf{P}\cdot\nabla\mathbf{v}) \\ \partial\mathbf{P}/\partial t + \nabla\cdot(\mathbf{vP}) + \mathbf{J} &= \partial\mathbf{P}/\partial t + \nabla \times (\mathbf{P} \times \mathbf{v}) + \mathbf{J} + \nabla\cdot(\mathbf{Pv}) \\ &= \nabla \times \mathbf{H} - \varepsilon_0\, \partial\mathbf{E}/\partial t + \nabla\cdot(\mathbf{Pv}),\end{aligned}$$

where use has been made of Eqs. (10.3). Two similar expressions can be obtained for the magnetization force. Hence

$$\begin{aligned}\mathbf{F} &= \nabla\cdot(\mathbf{PE}_e) + (\nabla \times \mathbf{H} - \varepsilon_0\, \partial\mathbf{E}/\partial t) \times (\mu_0\mathbf{H}) - (\nabla\cdot\mathbf{P})\mathbf{E} + \sigma\mathbf{E} \\ &+ \nabla\cdot(\mu_0\mathbf{MH}_e) + (\nabla \times \mathbf{E} + \mu_0\, \partial\mathbf{H}/\partial t) \times (\varepsilon_0\mathbf{E}) - (\nabla\cdot\mu_0\mathbf{M})\mathbf{H},\end{aligned}$$

or

$$(\text{IV}): \quad \mathbf{F} = -\nabla\cdot\underset{\sim}{\tau}^{C} - c^{-2}\, \partial(\mathbf{E} \times \mathbf{H})/\partial t \tag{10.19}$$

where

$$\underset{\sim}{\tau}^{C} = \tfrac{1}{2}(\varepsilon_0 E^2 + \mu_0 H^2)\underset{\sim}{\mathbf{I}} - (\varepsilon_0\mathbf{EE} + \mu_0\mathbf{HH} + \mathbf{PE}_e + \mu_0\mathbf{MH}_e) \tag{10.20}$$

is the stress tensor in the Chu formulation.

(B) *Body Couples*. The forces acting on the charges of a dipole produce a moment about the point $\mathbf{x}$. Since we have assumed, without the loss of generality, the negative charge is at $\mathbf{x}$ and the positive at $\mathbf{x} + \xi$, the moment on the electric dipole is

$$\xi \times \mathbf{f}_q^{+} = \xi \times [q\mathbf{E}(\mathbf{x}) + q\mathbf{v} \times \mu_0\mathbf{H}(\mathbf{x}) + \cdots].$$

Multiplying it by N and taking the limit $\xi \to 0$, we obtain the body couple per unit volume,

$$\mathbf{L}_q = \mathbf{P} \times (\mathbf{E} + \mathbf{v} \times \mu_0 \mathbf{H}) = \mathbf{P} \times \mathbf{E}_e. \tag{10.21}$$

Similarly, the body couple on magnetic dipoles is

$$\mathbf{L}_p = \mu_0 \mathbf{M} \times (\mathbf{H} - \mathbf{v} \times \varepsilon_0 \mathbf{E}) = \mu_0 \mathbf{M} \times \mathbf{H}_e. \tag{10.22}$$

The total body couple per unit volume is

$$\text{(IV)}: \quad \mathbf{L} = \mathbf{L}_q + \mathbf{L}_p = \mathbf{P} \times \mathbf{E}_e + \mu_0 \mathbf{M} \times \mathbf{H}_e. \tag{10.23}$$

This body couple is intrinsic in a polarizable and magnetizable medium. It is independent of the material properties and constitutive equations.

For quasi-static case, the magnetic part of Eq. (10.23) agrees with the body couple in Eq. (9.34).

(C) *Supplies of Energy*. To determine the energy source term ϕ in Eqs. (2.9) and (2.14) we first calculate the rate of work done on the electric dipole during motion. For a pair of charges at $\mathbf{x}$ and $\mathbf{x} + \xi$, the rate of work done is

$$\begin{aligned} \mathbf{f}_q^- \cdot \mathbf{v}^- + \mathbf{f}_q^+ \cdot \mathbf{v}^+ &= -q[\mathbf{E}(\mathbf{x}) + \mathbf{v} \times \mu_0 \mathbf{H}(\mathbf{x})] \cdot \mathbf{v} \\ &\quad + q[\mathbf{E}(\mathbf{x} + \xi) + (\mathbf{x} + \dot{\xi}) \times \mu_0 \mathbf{H}(\mathbf{x} + \xi)] \cdot (\mathbf{v} + \dot{\xi}) \\ &= q\mathbf{E}(\mathbf{x}) \cdot \dot{\xi} + q(\xi \cdot \nabla)\mathbf{E}(\mathbf{x}) \cdot \mathbf{v} + q\xi \cdot \nabla \mathbf{E}(\mathbf{x}) \cdot \dot{\xi} + \cdots. \end{aligned} \tag{10.24}$$

Multiplying the above by the dipole density and taking the limit $\xi \to 0$, we find from the first two terms the rate of work done per unit volume

$$W_q = \rho \mathbf{E} \cdot \frac{d}{dt}\left(\frac{\mathbf{P}}{\rho}\right) + (\mathbf{P} \cdot \nabla \mathbf{E}) \cdot \mathbf{v},$$

where use has been made of Eq. (10.13). A similar expression can be derived for the rate of work done on the magnetic dipoles. The total rate of work done which includes the part due to the free current, $\mathbf{J} \cdot \mathbf{E}$, is

$$W = \mathbf{J} \cdot \mathbf{E} + \rho \mathbf{E} \cdot \frac{d}{dt}\left(\frac{\mathbf{P}}{\rho}\right) + (\mathbf{P} \cdot \nabla \mathbf{E}) \cdot \mathbf{v} + \rho \mathbf{H} \cdot \frac{d}{dt}\left(\frac{\mu_0 \mathbf{M}}{\rho}\right) + (\mu_0 \mathbf{M} \cdot \nabla \mathbf{H}) \cdot \mathbf{v}. \tag{10.25}$$

In Eq. (2.9), ϕ_Q represents the energy supply from heat sources. The remaining terms of the last integrand, $\mathbf{F} \cdot \mathbf{v} + \phi$, are the power supplied from the electromagnetic field. When the polarized material is modeled by the two dipoles, this power supply equals the rate of work done on the electric and magnetic dipoles as given by Eq. (10.25),

$$\mathrm{W} = \mathbf{F} \cdot \mathbf{v} + \phi.. \tag{10.26}$$

Substituting W (Eq. (10.25)) and $\mathbf{F}$ (Eq. (10.17)) into the above, we obtain

$$\text{(IV)}: \quad \phi = \mathbf{J}_e \cdot \mathbf{E}_e + \rho \mathbf{E}_e \cdot \frac{d}{dt}\left(\frac{\mathbf{P}}{\rho}\right) + \rho \mathbf{H}_e \cdot \frac{d}{dt}\left(\frac{\mu_0 \mathbf{M}}{\rho}\right), \tag{10.27}$$

where $\mathbf{J}_e = \mathbf{J} - \sigma \mathbf{v}$. Equation (10.27) agrees with the energy-supply expression assumed in Ref. [15] (Eq. (7.35)).

The expressions Eqs. (10.17), (10.23), and (10.27) specify the body force $\mathbf{F}$, body couple $\mathbf{L}$ and the energy supply ϕ in Eqs. (2.12), (2.13A), and (2.14) respectively. From the expression for $\mathbf{F}$, a stress tensor $\boldsymbol{\tau}^C$ is derived in Eq. (10.20).

10.3 The Dipole–current Circuit Model

When the dipole–current circuit model is postulated for the interaction, the field equations (3.2) or (3.6) are usually assumed. With the relations $\mathbf{B} = \mu_0(\mathbf{M} + \mathbf{H})$ and $\mathbf{D} = \varepsilon_0 \mathbf{E} + P$, these two sets of field equations which are designated as (I) and (III) respectively are exactly equivalent. They are valid equations of electrodynamics for moving matter if either Minkowski's logic, or the Lorentz theory of electrons and the statistical theory are followed. However, many authors never stated the reasons for adopting these equations as the governing field equations. Hence, it is difficult to examine whether the subsequent assumptions and derivations in these papers are consistent with either Minkowski's or Lorentz's postulates.

In addition to the postulation of the dipole–circuit model, the forces acting on these models must also be postulated. Generally, it is assumed that

$$\begin{aligned} \mathbf{f}_q &= q\mathbf{E} + q\mathbf{v} \times \mathbf{B}, \\ d\mathbf{f}_i &= i\, d\mathbf{C} \times \mathbf{B}. \end{aligned} \tag{10.28}$$

The first formula is similar to Lorentz's original assumption except that the $\mathbf{E}$ and $\mathbf{B}$ are the fields inside the material body, not those in the vacuum generated by applied sources. It is analogous to Eq. (10.6). The second formula is based on Ampère's law, Eq. (3.12), which was originally established for isolated current circuits in the vacuum.

In Boffi's investigation [21] the field equations in the form of (III) are established from the elementary principles and the assumed material sources of electric and magnetic fields. The body force, as given in [15] (Eq. (7.112), nonrelativistic case), is

$$\mathbf{F} = \sigma\mathbf{E} + \mathbf{J} \times \mathbf{B} + \mathbf{P} \cdot \nabla\mathbf{E} + \mathbf{P} \times (\nabla \times \mathbf{E}) + \mathbf{M} \cdot \nabla\mathbf{B} + \mathbf{M} \times (\nabla \times \mathbf{B}) + \partial(\mathbf{P} \times \mathbf{B})/\partial t + \nabla \cdot (\mathbf{v}\mathbf{P} \times \mathbf{B}) \qquad (10.29)$$

Since

$$(\nabla\mathbf{B}) \cdot \mathbf{M} = \mathbf{M} \cdot \nabla\mathbf{B} + \mathbf{M} \times (\nabla \times \mathbf{B}),$$
$$(\nabla\mathbf{E}) \cdot \mathbf{P} = \mathbf{P} \cdot \nabla\mathbf{E} + \mathbf{P} \times (\nabla \times \mathbf{E}),$$

the above result is identical to that in Eq. (7.38).

In the paper by Tiersten and Tsai [43], a more sophisticated model is proposed to account for the effects of electronic polarization resonance and magnetic spin resonance in a deformable insulator. The part of the model that represents the effects of electric polarization and magnetizations which are considered in this article, is essentially a dipole–current circuit. The body force, couple and energy supply are found as (Eqs. (4.12), (3.48), (5.7), and Section 8 of [43])†

$$\mathbf{F} = \sigma\mathbf{E} + \mathbf{J} \times \mathbf{B} + \mathbf{P} \cdot \nabla\mathbf{E} + (\nabla\mathbf{B}) \cdot \mathbf{M}_e + \mathbf{v} \times (\mathbf{P} \cdot \nabla\mathbf{B}) + \left[\frac{\partial\mathbf{P}}{\partial t} + \nabla \cdot (\mathbf{v}\mathbf{P})\right] \times \mathbf{B}, \qquad (10.30)$$

$$\mathbf{L} = \mathbf{B} \times \mathbf{M}_e + \mathbf{P} \times \mathbf{E}_e, \qquad (10.31)$$

$$\mathbf{F} \cdot \mathbf{v} + \phi = \mathbf{E} \cdot \rho \frac{d}{dt}\left(\frac{\mathbf{P}}{\rho}\right) + \mathbf{P} \cdot \nabla\mathbf{E} \cdot \mathbf{v} - \mathbf{M}_e \cdot \frac{\partial\mathbf{B}}{\partial t} + \mathbf{J} \cdot \mathbf{E}. \qquad (10.32)$$

In the above equations, $\mathbf{M}_e$ and $\mathbf{E}_e$ are the magnetization and electric field in the rest frame which according to Minkowski's theory are related to the laboratory frame variables,

$$\mathbf{M}_e = \mathbf{M} + \mathbf{v} \times \mathbf{P} \qquad \mathbf{E}_e = \mathbf{E} + \mathbf{v} \times \mathbf{B}. \qquad (10.33)$$

These relations are the low velocity limits of the relativistic transformations; see Eq. (4.2).

In Eq. (10.30), noting that

$$\mathbf{P} \cdot \nabla\mathbf{E} + (\partial\mathbf{P}/\partial t) \times \mathbf{B} = (\nabla\mathbf{E}) \cdot \mathbf{P} + \partial(\mathbf{P} \times \mathbf{B})/\partial t,$$
$$\nabla\mathbf{B} \cdot \mathbf{v} \times \mathbf{P} + \mathbf{v} \times (\mathbf{P} \cdot \nabla\mathbf{B}) = \mathbf{P} \times (\mathbf{v} \cdot \nabla\mathbf{B}),$$
$$\nabla \cdot (\mathbf{v}\mathbf{P} \times \mathbf{B}) = (\nabla \cdot \mathbf{v})(\mathbf{P} \times \mathbf{B}) + (\mathbf{v} \cdot \nabla\mathbf{P}) \times \mathbf{B} + \mathbf{P} \times (\mathbf{v} \cdot \nabla\mathbf{B}),$$

† The original results are for insulators. We have inserted $\sigma\mathbf{E} + \mathbf{J} \times \mathbf{B}$ in (10.30) and $\mathbf{J} \cdot \mathbf{E}$ in (10.32) for conductors.

we can thus deduce another expression for the body force,

$$\text{(III)}: \quad \mathbf{F} = \sigma\mathbf{E} + \mathbf{J} \times \mathbf{B} + (\nabla\mathbf{E}) \cdot \mathbf{P} + (\nabla\mathbf{B}) \cdot \mathbf{M} + \frac{\partial}{\partial t}(\mathbf{P} \times \mathbf{B}) + \nabla \cdot (\mathbf{v}\mathbf{P} \times \mathbf{B}). \tag{10.34}$$

This is the same as Eq. (7.38). Similarly, by using the vector identities

$$(\mathbf{v} \times \mathbf{P}) \times \mathbf{B} + \mathbf{P} \times (\mathbf{v} \times \mathbf{B}) = \mathbf{v} \times (\mathbf{P} \times \mathbf{B}),$$
$$\mathbf{v} \times \mathbf{P} \cdot \nabla \times \mathbf{E} = \mathbf{v} \cdot (\nabla\mathbf{E}) \cdot \mathbf{P} - (\mathbf{P} \cdot \nabla\mathbf{E}) \cdot \mathbf{P},$$

we deduce from Eqs. (10.31) and (10.32) the body couple and energy supply,

$$\text{(III)}: \begin{cases} \mathbf{L} = \mathbf{M} \times \mathbf{B} + \mathbf{P} \times \mathbf{E} + \mathbf{v} \times (\mathbf{P} \times \mathbf{B}), & (10.35) \\ \mathbf{F} \cdot \mathbf{v} + \phi = \mathbf{E} \cdot \dfrac{\partial \mathbf{P}}{\partial t} - \mathbf{M} \cdot \dfrac{\partial \mathbf{B}}{\partial t} + \nabla \cdot (\mathbf{v}\mathbf{P} \cdot \mathbf{E}) + \mathbf{J} \cdot \mathbf{E}. & (10.36) \end{cases}$$

They agree with Eqs. (7.45) and (7.44) respectively.

We shall not repeat the derivations that lead to Eqs. (10.30)–(10.32) except to point out that the calculation of magnetic forces on a moving current circuit is not an easy task. In [43], this difficulty is circumvented by attaching the current circuit to a rest frame (thus the result $\mathbf{M}_e$ in Eq. (10.30)). For deformable media, this *rest frame* must be interpreted as a *pointwise rest frame*, a concept previously discussed in Section 4.

10.4 Discussion

In this section, we have discussed the determination of matter–field interactions by specifying models which represent the effects of electric and magnetic moments. Except for the modeling of magnetization, the postulates and basic principles of the two approaches, one based on the two-dipoles model, and the other on the dipole–current circuit model, are the same. However, the end results, Eqs. (10.17), (10.32), and (10.27) of formulation IV versus Eqs. (10.34), (10.35), and (10.36) of formulation III (see also results Eqs. (7.38), (7.48), and (7.47)) are quite different.

Before making a detailed comparison, it should be noted that although the same symbols are used in these two formulations, they may have different meanings for moving media [40]. For purpose of comparison, both formulations should first be converted to **EBDH** according to $\mathbf{B} = \mu_0(\mathbf{H} + \mathbf{M})$ and $\mathbf{D} = \varepsilon_0\mathbf{E} + \mathbf{P}$. The **B** and **D** in both formulations are the same, but the **E** and **H** are not. They are related by

$$\mathbf{E}^{\mathrm{S}} = \mathbf{E}^{\mathrm{C}} - \mu_0 \mathbf{v} \times \mathbf{M}^{\mathrm{C}}, \tag{10.37}$$

$$\mathbf{H}^{\mathrm{S}} = \mathbf{H}^{\mathrm{C}} + \mathbf{v} \times \mathbf{P}^{\mathrm{C}}, \tag{10.38}$$

where the superscripts S and C designate the statistical formulation (III) and the Chu formulation (IV), respectively.

These two sets of variables and field equations are identical when $\mathbf{v} = \mathbf{0}$. However, even for the case of $\mathbf{v} = \mathbf{0}$, the expressions for $\mathbf{F}$ are different. With $\mathbf{v} = \mathbf{0}$ and $\mathbf{B} = \mu_0(\mathbf{H} + \mathbf{M})$, Eq. (10.34) reduces to

$$\begin{aligned}(\text{III—Stationary}):\quad \mathbf{F} = \sigma\mathbf{E} + \mathbf{J} \times \mu_0\mathbf{H} + \mathbf{P}\cdot\nabla\mathbf{E} + \frac{\partial \mathbf{P}}{\partial t} \times (\mu_0\mathbf{H}) \\ + \mu_0\mathbf{M}\cdot\nabla\mathbf{H} + \mu_0(\nabla\mathbf{M})\cdot\mathbf{M} - \varepsilon_0 \frac{\partial \mathbf{E}}{\partial t} \times (\mu_0\mathbf{M}),\end{aligned} \tag{10.39}$$

whereas Eq. (10.17) reduces to

$$\begin{aligned}(\text{IV—Stationary}):\quad \mathbf{F} = \sigma\mathbf{E} + \mathbf{J} \times \mu_0\mathbf{H} + \mathbf{P}\cdot\nabla\mathbf{E} + \frac{\partial \mathbf{P}}{\partial t} \times (\mu_0\mathbf{H}) \\ + \mu_0\mathbf{M}\cdot\nabla\mathbf{H} + \mu_0 \frac{\partial \mathbf{M}}{\partial t} \times (\varepsilon_0\mathbf{E}).\end{aligned} \tag{10.40}$$

The difference between the last terms of these two equations can not be reconciled without introducing the effect of surface forces.

The fact that the dipole–current circuit model leads to a result (Eq. (10.34)) which is in total agreement with that of the statistical theory (Eq. (7.38)), may create an impression that the dipole-circuit is a better representation than the two-dipole model. However, the magnetic force in Eq. (10.39), which is a static case of Eq. (10.34), does not reduce to any one of the three expressions in Brown's theory (the body force in Eq. (9.33), (9.37), or (9.38)). But the corresponding force in the two-dipole model (Eq. (10.40)) does. Comparison of these two models is discussed in [2] and [15].

Both models (two-dipole, dipole–current circuit) discussed in this section, and the model of electrons in Section 7 are postulated to represent and to interpret macroscopic phenomena. They are never meant to describe the microscopic nature of atoms in matter. The charged particles assumed in the theory of electrons very closely resemble the electrons and protons of the atomic theory of matter. However, they are not true representations of atomic behavior. In the theory of electrons, the concepts and results of electric dipole moment and polarization agree very closely with the results in the atomic theory of dielectrics, but the corresponding magnetic concepts do not agree at all with the quantum theory of magnetism. The concept of electrons in orbital motion, which generates a magnetic dipole moment (Eqs. (7.5), (7.6)), is only useful to interpret the diamagnetism. Paramagnetism and ferromagnetism are primarily due to electron spins of quantum origin. An introductory exposition of atomic theories of dielectrics and magnetism can be found in the text by Feymann (Chapters 11, 34, 35, 36 of [6]).

Since neither the two-dipole model nor the dipole-circuit model truly represents the atomic behavior of magnetism, they cannot be contrasted or evaluated on an atomic basis. The conceptual and observational bases for the postulation of either model are equally sound, and there are long historical lineages of both models. It is difficult at this stage to make a choice of these two models objectively.

11 SUMMARY OF ELECTROMAGNETIC FORCES AND ENERGY

In the previous sections, we have reviewed four different formulations of the Maxwell equations for moving media: (I) the Minkowski, (II) the Lorentz, (III) the Statistical, and (IV) the Chu formulation. Each is based on a different principle and postulation. For each formulation of field equations, there is a corresponding set of expressions for the body force, body couple, and energy supply, which when combined with mechanical balance equations specifies the interaction of the fields with matter in motion. Only non-relativistic theories are considered here.

Aside from σ, $\mathbf{J}$, and $\mathbf{v}$, only four of the six field variables $\mathbf{E}$, $\mathbf{B}$, $\mathbf{D}$, $\mathbf{H}$, $\mathbf{P}$, $\mathbf{M}$ are used in each formulation. The remaining two are defined in terms of the other four by

$$\mathbf{D} = \varepsilon_0 \mathbf{E} + \mathbf{P}, \qquad \mathbf{B} = \mu_0(\mathbf{H} + \mathbf{M}). \tag{3.5}$$

With these relations, all four formulations reduce to an identical form when $\mathbf{v} = \mathbf{0}$, the Maxwell equations for stationary media (Eqs. (3.2) or (3.6)).

For moving media, the Maxwell equations in the Minkowski formulation (Eqs. (I), or Eqs. (4.1)) and the statistical formulation (Eqs. (III), or Eqs. (4.7)) should be viewed as exactly equivalent. The equations in the Chu formulation (Eqs. (IV), or Eqs. (4.8)) which have been proposed in recent times are quite different. The equations in the Lorentz formulation (Eqs. (II) or Eqs. (4.5)) which were developed originally only for electrically polarizable materials are, in our view, superseded by Eqs. (III). Thus only two sets of the Maxwell equations, Eqs. (I) or (III), and Eqs. (IV), are to be considered.

Based on these forms of the Maxwell equations, various theories of field–

matter interactions have been proposed. Each theory determines the interactions by specifying the body force **F**, couple **L** and energy supply ϕ (Eqs. (2.12)–(2.14)) in terms of the electromagnetic variables. A total of five theories of which two were originated from Maxwell, one was proposed by Minkowski, and two were based upon the Lorentz theory of electrons, have been reviewed in Sections 5, 6, 7, 8, 9, and 10.

Of the two theories of interaction originated from Maxwell's *Treatise*, one is based on the concept of total charge and total current in a volume element of the matter. This is discussed in Subsection 5.1, its extension in Section 8. More will be said later about this theory. The other is based on the concept of the resultant force on the entire body. This is discussed in Subsection 3.2 for electric charges in the vacuum, and extended in Section 9 for static interactions of deformable bodies with magnetic fields. Generalization of the concept of the resultant force to moving bodies is not easy. Thus only one of the two theories pertains to the dynamical interaction of non-stationary fields with moving and deformable matter (Sections 5 and 8).

The theory proposed by Minkowski is based on the principle of relativity. It is discussed in Subsections 5.3 and 6.2. Minkowski postulated an energy-momentum tensor in four-dimensional space such that the divergence of this tensor is the force (in 4-space) acting on the matter (Subsection 5.3). This postulation may be replaced by applying the principle of conservation of the total energy-momentum in 4-space for a closed physical system, which is another postulation (Eq. (6.10)). However, the results as obtained in Subsection 6.2 indicate that the force acting in the electromagnetic field (in the aether) is equal and opposite to that acting on the matter. This conclusion was conceptionally unacceptable for moving matter (Subsection 5.4). Furthermore, the adherence to the special theory of relativity fixes the constitutive laws for the electromagnetic fields (Eqs. (4.3)), which may not be valid for deformable bodies. Hence the Minkowski theory should be viewed as valid for *rigid bodies moving at constant velocities.* Whether the expressions for **F**, **L**, and ϕ in Eqs. (6.26), (6.27), and (6.30) respectively are applicable to dynamical interactions of fields with deformable matter remains questionable.

The remaining two dynamic theories which are originated from the Lorentz theory of electrons postulate models to represent the source of charges, currents, electric polarizations, and magnetizations. These sources are embedded in, and move with the matter. The interaction of the matter with the field is determined by the forces that act on these sources by the field.

The first of the two theories is presented in Section 7. A *microscopic* model, the electron, is postulated and the microscopic force, the Lorentz force, is assumed in terms of the *microscopic* fields, $\mathbf{b}$ and $\mathbf{e}$ (Eq. (7.1)). The macroscopic fields and forces are then calculated by applying the theory of statistical mechanics. This theory is a departure from Maxwell's original approach (Section 8) which postulates two *macroscopic* models, the total charge σ_t and total current $\mathbf{J}_t$, and the Maxwell–Lorentz forces in terms of *macroscopic* field (Eq. (8.1)). Two versions of the second theory are presented in Section 10, in which *microscopic* models of dipoles and current-circuits are postulated and forces acting on these sources are assumed in terms of the *macroscopic* fields. The two-dipole model (Subsection 10.2) gives rise to a set of expressions for electromagnetic forces different from those of all other theories. The electromagnetic forces based on the dipole–current circuit model (Subsection 10.3) are the same as those derived from statistical theory in Section 7.

Comparison of the assumptions made in the theory of the Maxwell–Lorentz force with those of the Lorentz theory of electrons indicates that the former are more restrictive. First, the σ_t and $\mathbf{J}_t$ in Eq. (8.1) are macroscopic sources whereas the q in Eqs. (7.1) or (10.6) is a microscopic source to which the Lorentz force formula is known to be applicable. Next, the σ_t and $\mathbf{J}_t$ must be identified from the field equations. This is usually based on other theories or additional assumptions. Furthermore, the stress tensors $\underset{\sim}{\tau}^{\mathrm{L}}$ or $\underset{\sim}{\tau}^{\mathrm{LS}}$ (Eq. (8.7)) and $\underset{\sim}{\tau}^{\mathrm{CL}}$ (Eq. (8.17)) thus derived are identical to the Maxwell stress tensor in vacuous media. Even the revised result by Livens, Eq. (8.23), agrees with Eq. (7.38) only when $\mathbf{v} = \mathbf{0}$ (stationary media). Hence we have serious doubts on whether the theory in Section 8 can be applied to moving media.

By elimination, we are now left with two theories, and two sets of equations of electrodynamics for moving and deformable media. One set is based on the Lorentz theory of electrons and statistical mechanics (Section 7), this set being the same as that based on the dipole–current circuit model (Subsection 10.3). The other set is based upon the formulation of the two-dipole model (Subsection 10.2).

For the convenience of further discussions, we summarize these two sets in the following, the key equations being identified by (III) or (IV) for the set III or set IV respectively.

The *balance equations of mechanics*, Eqs. (2.11)–(2.14), apply to both sets:

$$
\text{(III), (IV)} \quad
\begin{cases}
\partial\rho/\partial t + \nabla\cdot(\mathbf{v}\rho) = 0, & (11.1)\\
\rho\, d\mathbf{v}/dt = \nabla\cdot\underset{\sim}{\tau} + \mathbf{F}, & (11.2)\\
\varepsilon_{ijk}\tau_{jk} = -L_i, & (11.3)\\
\rho\, dU/dt = \underset{\sim}{\tau} : \nabla\mathbf{v} - \nabla\cdot\mathbf{Q} + \phi_Q + \phi, & (11.4)
\end{cases}
$$

where $d/dt = \partial/\partial t + \mathbf{v}\cdot\nabla$. In addition, there is the classical *second law of thermodynamics*,

$$\rho\, dS/dt + \nabla\cdot(\mathbf{Q}/T) - \phi_Q/T \geq 0, \tag{11.5}$$

where S is the entropy density and T the absolute temperature.

Although the same symbols are used, the definitions for $\mathbf{E}_e$, $\mathbf{M}_e$, and $\mathbf{H}_e$ in the following two formulations, and the physical meanings for $\mathbf{E}$, $\mathbf{P}$, and $\mathbf{M}$ in two sets of equations are different.

1. *The Two-dipole Model Formulation.*

Electromagnetic field equations (Eqs. (4.8), (4.10E)):

$$\text{(IV)}\quad \begin{cases} \nabla\cdot(\mu_0\mathbf{H}) = -\nabla\cdot(\mu_0\mathbf{M}), \\ \nabla\times\mathbf{E} + \mu_0\,\partial\mathbf{H}/\partial t = -\partial(\mu_0\mathbf{M})/\partial t - \nabla\times(\mu_0\mathbf{M}\times\mathbf{v}), \\ \varepsilon_0\nabla\cdot\mathbf{E} = \sigma - \nabla\cdot\mathbf{P}, \\ \nabla\times\mathbf{H} - \varepsilon_0\,\partial\mathbf{E}/\partial t = \mathbf{J} + \partial\mathbf{P}/\partial t + \nabla\times(\mathbf{P}\times\mathbf{v}), \\ \nabla\cdot\mathbf{J} + \partial\sigma/\partial t = 0. \end{cases} \tag{11.6}$$

Body force (Eqs. (10.17), (10.19)):

$$\begin{aligned} \text{(IV)}\quad \mathbf{F} &= \sigma\mathbf{E} + \mathbf{J}\times(\mu_0\mathbf{H}) + \mathbf{P}\cdot\nabla\mathbf{E} + \mathbf{v}\times(\mathbf{P}\cdot\nabla)(\mu_0\mathbf{H}) + \frac{d}{dt}\left(\frac{\mathbf{P}}{\rho}\right)\times(\mu_0\mathbf{H}) \\ &\quad + \mu_0\mathbf{M}\cdot\nabla\mathbf{H} - \mathbf{v}\times(\mu_0\mathbf{M}\cdot\nabla)(\varepsilon_0\mathbf{E}) - \rho\frac{d}{dt}\left(\frac{\mathbf{M}}{\rho}\right)\times(\varepsilon_0\mathbf{E}), \qquad (11.7) \\ &= -\nabla\cdot\underset{\sim}{\tau}^{C} - c^{-2}\,\partial(\mathbf{E}\times\mathbf{H})/\partial t, \end{aligned}$$

$$\text{(IV)}\quad \underset{\sim}{\tau}^{C} = \tfrac{1}{2}(\varepsilon_0\mathbf{E}\cdot\mathbf{E} + \mu_0\mathbf{H}\cdot\mathbf{H})\underset{\sim}{\mathbf{I}} - (\varepsilon_0\mathbf{E}\mathbf{E} + \mu_0\mathbf{H}\mathbf{H} + \mathbf{P}\mathbf{E}_e + \mu_0\mathbf{M}\mathbf{H}_e). \tag{11.8}$$

Body couple (Eq. (10.23)):

$$\text{(IV)}\quad \mathbf{L} = \mathbf{P}\times\mathbf{E}_e + \mu_0\mathbf{M}\times\mathbf{H}_e. \tag{11.9}$$

Energy supply (Eq. (10.27)):

$$\text{(IV)}\quad \phi = \mathbf{J}_e\cdot\mathbf{E}_e + \rho\mathbf{E}_e\cdot\frac{d}{dt}\left(\frac{\mathbf{P}}{\rho}\right) + \rho\mathbf{H}_e\cdot\frac{d}{dt}\left(\frac{\mu_0\mathbf{M}}{\rho}\right), \tag{11.10}$$

where

$$\mathbf{J}_e = \mathbf{J} - \sigma\mathbf{v}, \qquad \mathbf{E}_e = \mathbf{E} + \mathbf{v}\times\mu_0\mathbf{H}, \qquad \mathbf{H}_e = \mathbf{H} - \mathbf{v}\times\varepsilon_0\mathbf{E}, \tag{4.14}$$

$$\rho \frac{d}{dt}\left(\frac{\mathbf{M}}{\rho}\right) = \frac{\partial \mathbf{M}}{\partial t} + \nabla \cdot (\mathbf{vM}), \qquad \rho \frac{d}{dt}\left(\frac{\mathbf{P}}{\rho}\right) = \frac{\partial \mathbf{P}}{\partial t} + \nabla \cdot (\mathbf{vP}). \tag{10.13}$$

2. *The Electron Model and Dipole–current Circuit Model Formulation.*

Electromagnet field equations (Eqs. (4.7), (4.10E)):

$$\text{(III)} \begin{cases} \nabla \cdot \mathbf{B} = 0, \\ \nabla \times \mathbf{E} + \partial \mathbf{B}/\partial t = 0, \\ \varepsilon_0 \nabla \cdot \mathbf{E} = \sigma - \nabla \cdot \mathbf{P}, \\ \mu_0^{-1} \nabla \times \mathbf{B} - \varepsilon_0\, \partial \mathbf{E}/\partial t = \partial \mathbf{P}/\partial t + \nabla \times \mathbf{M} + \mathbf{J}, \\ \nabla \cdot \mathbf{J} + \partial \sigma/\partial t = 0. \end{cases} \tag{11.11}$$

Body force (Eqs. (7.38), (7.40), (10.34)):

$$\begin{aligned} \text{(III)} \quad \mathbf{F} &= \sigma \mathbf{E} + \mathbf{J} \times \mathbf{B} + (\nabla \mathbf{E}) \cdot \mathbf{P} + (\nabla \mathbf{B}) \cdot \mathbf{M} + \partial(\mathbf{P} \times \mathbf{B})/\partial t + \nabla \cdot (\mathbf{vP} \times \mathbf{B}) \\ &= -\nabla \cdot \underset{\sim}{\tau}^S - \partial(\varepsilon_0 \mathbf{E} \times \mathbf{B})/\partial t + \nabla \cdot (\mathbf{vP} \times \mathbf{B}), \end{aligned} \tag{11.12}$$

$$\text{(III)} \quad \underset{\sim}{\tau}^S = \tfrac{1}{2}(\varepsilon_0 \mathbf{E} \cdot \mathbf{E} + \mathbf{B} \cdot \mathbf{B}/\mu_0 - 2\mathbf{M} \cdot \mathbf{B})\underset{\sim}{\mathbf{I}} - (\varepsilon_0 \mathbf{EE} + \mathbf{PE} + \mathbf{BB}/\mu_0 - \mathbf{BM}). \tag{11.13}$$

Body couple (Eqs. (7.48), (10.35)):

$$\text{(III)} \quad \mathbf{L} = \mathbf{P} \times \mathbf{E} + \mathbf{M}_e \times \mathbf{B}. \tag{11.14}$$

Energy supply (Eqs. (7.47), (10.36)):

$$\text{(IV)} \quad \phi = \mathbf{J}_e \cdot \mathbf{E}_e + \rho \frac{d}{dt}\left(\frac{\mathbf{P}}{\rho}\right) \cdot \mathbf{E}_e - \mathbf{M}_e \cdot \frac{d\mathbf{B}}{dt}, \tag{11.15}$$

where

$$\mathbf{J}_e = \mathbf{J} - \sigma \mathbf{v}, \qquad \mathbf{E}_e = \mathbf{E} + \mathbf{v} \times \mathbf{B}, \qquad \mathbf{M}_e = \mathbf{M} + \mathbf{v} \times \mathbf{P}. \tag{4.11, 12}$$

Equations (11.1)–(11.4) plus Eqs. (11.6)–(11.10) or Eqs. (11.11)–(11.15) complete the field equations (balance equations) for the interaction of electromagnetic fields with moving and deformable bodies. The equations are valid relativistically up to the order of v^2/c^2. All field variables are referred to the present configuration of the deformed body.

The above set of equations are indeterminate in the sense that the number of dependent variables exceeds the number of equations. Additional equations are supplied from the constitutive laws of the material media, which together with the boundary conditions are discussed in the next section.

12 CONSTITUTIVE EQUATIONS AND BOUNDARY CONDITIONS†

The constitutive equations can be established consistent with the first and second law of thermodynamics. We mentioned in Section 7 that the balance equation for energy Eq. (7.43), which is a form of the first law of thermodynamics, can be derived from a statistical average of microscopic equations. However, the entropy inequality, Eq. (11.5), which is a form of the second law of thermodynamics, cannot be so derived. Thus the entire question of constitutive equations must be approached from a continuum theory which postulates Eqs. (11.4) and (11.5) as the first and second fundamental laws of thermodynamics [46, Sections 14 and 20].

The derivation of boundary conditions for a continuum follows an entirely different logic. The starting point for the derivation is the global forms of the balance laws of mechanics, Eqs. (2.6)–(2.9), and electrodynamics, Eqs. (4.9A)–(4.9E). Originally, these global laws are assumed for field quantities which are continuous within a given volume or an area, resulting in the corresponding local balance equations, Eqs. (2.11)–(2.14), and various formulations of the Maxwell equations. We now assume that these global laws are valid even when some field quantities are discontinuous within the given volume or area. The transport equations like Eq. (2.4) should then be modified to account for the discontinuity of the field quantity across a moving singular surface or a line. They are then applied to the global laws for a pillbox enclosing the singular surface or line. In the limit of vanishing volume or area for the pillbox, the jump conditions are derived. By taking the singular surface to be the material surface, we obtain the boundary conditions (p. 525 of Ref. [18]).

The constitutive laws and boundary conditions corresponding to the first formulation of the preceding section are given in Ref. [35], those to the second formulation are discussed in Refs. [43] and [47]. Only elastic solids are considered here. Constitutive equations for viscous fluids are treated in Ref. [35].

† At the suggestion of the editor, this section was added after the completion of the previous sections which are devoted primarily to the theory of electromagnetic forces and energy. A few references were also added.

12.1 Constitutive Equations for the Two-dipole Formulation

The complete set of balance equations for the two-dipole formulation plus the second law of thermodynamics are summarized in the previous section as Eqs. (11.1)–(11.10). Eliminating ϕ_Q, the heat source function, form Eqs. (11.4) and (11.5), we obtain an energy–entropy inequality:

$$\rho\dot{U} - \underset{\sim}{\tau}:\nabla\mathbf{v} + \nabla\cdot\mathbf{Q} - \phi - \rho T\dot{S} - T\nabla\cdot(\mathbf{Q}/T) \leq 0, \tag{12.1}$$

where $\dot{U} = dU/dt$, etc., and U is a hitherto unspecified integral energy function.

For investigating the deformation of a solid, it is convenient to introduce the deformation gradient $\underset{\sim}{\xi}$, and its inverse $\underset{\sim}{\xi}^{-1}$,

$$\underset{\sim}{\xi} = \partial\mathbf{x}/\partial\mathbf{X} \qquad \text{and} \qquad \underset{\sim}{\xi}^{-1} = \partial\mathbf{X}/\partial\mathbf{x}, \tag{12.2}$$

or

$$\xi_{iJ} = \partial x_i/\partial X_J = x_{i,J} \qquad \text{and} \qquad \xi_{Ji}^{-1} = \partial X_J/\partial x_i = X_{J,i}.$$

The present position $\mathbf{x}$ of a particle is related to its reference coordinates $\mathbf{X}$ by Eq. (12.2).

The deformation tensor $\underset{\sim}{C}$ is defined by

$$\underset{\sim}{C} = \underset{\sim}{\xi}^T\cdot\underset{\sim}{\xi} \qquad \text{or} \qquad C_{JK} = \xi_{iJ}\xi_{iK} = x_{i,J}x_{i,K}, \tag{12.3}$$

where the superscript T denotes the transpose of a tensor. If a displacement vector $\mathbf{u}$ is introduced as $\mathbf{u} = \mathbf{x} - \mathbf{X}$, then (Fig. 1)

$$C_{JK} = \frac{\partial u_J}{\partial X_K} + \frac{\partial u_K}{\partial X_J} + \frac{\partial u_L}{\partial X_J}\frac{\partial u_L}{\partial X_K} + \delta_{JK}. \tag{12.4}$$

The Lagrangian strain tensor is

$$\underset{\sim}{\Sigma} = \tfrac{1}{2}(\underset{\sim}{C} - \underset{\sim}{I}) \qquad \text{or} \qquad \Sigma_{JK} = \tfrac{1}{2}(C_{JK} - \delta_{JK}). \tag{12.5}$$

For fluid medium, we use the symmetric and antisymmetric parts of $\nabla\mathbf{v}$ to describe its rate of deformation. With $\underset{\sim}{\xi}$, we can express the second term in Eq. (12.1) as

$$\begin{aligned}\underset{\sim}{\tau}:\nabla\mathbf{v} &= \tau_{ij}v_{j,i} = \tau_{ij}v_{j,K}X_{K,i} = \tau_{ij}\dot{x}_{j,K}X_{K,i} \\ &= \xi_{Ki}^{-1}\tau_{ij}\dot{\xi}_{jK} = (\underset{\sim}{\xi}^{-1}\cdot\underset{\sim}{\tau})^T:\dot{\underset{\sim}{\xi}}.\end{aligned} \tag{12.6}$$

With Eq. (12.6) and the ϕ in Eq. (11.10), it is natural to regard U in Eq. (12.1) as a function of the independent variables $\underset{\sim}{\xi}$, $\mathbf{M}/\rho$, $\mathbf{P}/\rho$, and the entropy S. However, the variables $\mathbf{H}_e$, $\mathbf{E}_e$, and T are preferred. We thus introduce the free energy F as

$$F = U - TS - \mathbf{E}_e \cdot (\mathbf{P}/\rho) - \mathbf{H}_e \cdot (\mu_0 \cdot \mathbf{M}/\rho), \tag{12.7}$$

and assume F to be a function of ξ, $\mathbf{E}_e$, $\mathbf{H}_e$, T, and ∇T. Substituting Eqs. (12.7), (12.6), and (11.10) into Eq. (12.1), we find

$$\rho(\dot{F} + S\dot{T}) - (\underset{\sim}{\xi}^{-1} \cdot \underset{\sim}{\tau})^T : \dot{\underset{\sim}{\xi}} + \mathbf{P} \cdot \dot{\mathbf{E}}_e + \mu_0 \mathbf{M} \cdot \dot{\mathbf{H}}_e + \nabla T \cdot (\mathbf{Q}/T) - \mathbf{J}_e \cdot \mathbf{E}_e \leq 0. \tag{12.8}$$

This reduced free energy–entropy inequality forms the basis for deducing the constitutive equations.

For polarizable and magnetizable thermoelastic solids, we assume the existence of a free energy function $\mathscr{F}$ such that

$$F = \mathscr{F}(\xi, \mathbf{E}_e, \mathbf{H}_e, T, \mathbf{g}) \tag{12.9}$$

and

$$\dot{F} = \frac{\partial \mathscr{F}}{\partial \underset{\sim}{\xi}} : \dot{\underset{\sim}{\xi}} + \frac{\partial \mathscr{F}}{\partial \mathbf{E}_e} \cdot \dot{\mathbf{E}}_e + \frac{\partial \mathscr{F}}{\partial \mathbf{H}_e} \cdot \dot{\mathbf{H}}_e + \frac{\partial \mathscr{F}}{\partial T} \dot{T} + \frac{\partial \mathscr{F}}{\partial \mathbf{g}} \cdot \dot{\mathbf{g}}, \tag{12.10}$$

where $\mathbf{g} = \nabla T$. Substitution of (12.10) into (12.8) yields

$$\left[\rho \frac{\partial \mathscr{F}}{\partial \xi} - (\underset{\sim}{\xi}^{-1} \cdot \underset{\sim}{\tau})^T\right] : \dot{\underset{\sim}{\xi}} + \left(\rho \frac{\partial \mathscr{F}}{\partial \mathbf{E}_e} + \mathbf{P}\right) \cdot \dot{\mathbf{E}}_e + \left(\rho \frac{\partial \mathscr{F}}{\partial \mathbf{H}_e} + \mu_0 \mathbf{M}\right) \cdot \dot{\mathbf{H}}_e$$
$$+ \rho \left(\frac{\partial \mathscr{F}}{\partial T} + S\right) \dot{T} + \rho \frac{\partial \mathscr{F}}{\partial \mathbf{g}} \cdot \dot{\mathbf{g}} + \frac{1}{T} \mathbf{Q} \cdot \mathbf{g} - \mathbf{J}_e \cdot \mathbf{E}_e \leq 0. \tag{12.11}$$

For this inequality to be valid at all times and for all admissible thermodynamic processes, the following conditions are sufficient:

$$\mathbf{Q} \cdot (\nabla T)/T - \mathbf{J}_e \cdot \mathbf{E}_e \leq 0, \tag{12.12}$$

$$\partial \mathscr{F}/\partial \mathbf{g} = \mathbf{0}, \tag{12.13}$$

$$\begin{cases} S = -\,\partial \mathscr{F}/\partial T, \\ \mu_0 \mathbf{M}/\rho = -\,\partial \mathscr{F}/\partial \mathbf{H}_e, \\ \mathbf{P}/\rho = -\,\partial \mathscr{F}/\partial \mathbf{E}_e, \\ \underset{\sim}{\tau} = \rho \underset{\sim}{\xi} \cdot (\partial \mathscr{F}/\partial \underset{\sim}{\xi})^T. \end{cases} \tag{12.14}$$

Equation (12.13) states that $\mathscr{F}$ is independent of $\mathbf{g} = \nabla T$. Thus, in lieu of Eq. (12.9), we assume

$$F = \mathscr{F}(\xi, \mathbf{E}_e, \mathbf{H}_e, \mathbf{T}). \tag{12.15}$$

Once the free energy function is known, S, $\mu_0\mathbf{M}/\rho$, $\mathbf{P}/\rho$ and $\underset{\sim}{\tau}$ can all be calculated from $\mathscr{F}$ according to Eq. (12.14). The heat flux $\mathbf{Q}$ and the effective conduction current $\mathbf{J}_e$ remain general functions of the ensemble (ξ, $\mathbf{E}_e$, $\mathbf{H}_e$, T, $\mathbf{g}$), so long as they satisfy the inequality (12.12).

Substituting Eq. (12.14), (12.10), and (12.7) into Eq. (11.4) one obtains†

$$\text{(IV)} \quad \rho T\dot{S} = \mathbf{J}_e \cdot \mathbf{E}_e - \nabla \cdot \mathbf{Q} + \phi_Q. \tag{12.16}$$

This equation replaces the balance of energy, Eq. (11.4), in future applications.

The constitutive equations (12.14) and (12.15) as they stand do not satisfy the principle of material objectivity. Through a consideration of invariance under an orthogonal transformation which represents a rigid rotation of the spatial frame, one can show that Eq. (12.15) is invariant if (p. 1019 of [35]),

$$\text{(IV)} \quad F = \mathscr{F}(\underset{\sim}{\boldsymbol{\Sigma}}, \hat{\mathbf{E}}, \hat{\mathbf{H}}, T), \tag{12.17}$$

where $\underset{\sim}{\boldsymbol{\Sigma}}$ is the Lagrangian strain tensor in Eq. (12.5), and

$$\begin{aligned} \hat{\mathbf{E}} &= \xi^T \cdot \mathbf{E}_e \quad \text{or} \quad \hat{E}_K = x_{j,K} E_j^e, \\ \hat{\mathbf{H}} &= \xi^T \cdot \mathbf{E}_e \quad \text{or} \quad \hat{H}_K = x_{j,K} H_j^e. \end{aligned} \tag{12.18}$$

Note that the components of the new variables, $\boldsymbol{\Sigma}$, $\hat{\mathbf{H}}$, $\hat{\mathbf{E}}$, are referred to the reference (undeformed) configuration, which are more appropriate for defining constitutive laws for solids, whereas all physical quantities in Eqs. (11.1)–(11.15) pertain to the body at the present (deformed) configuration. Note also that we have written E_j^e for $(\mathbf{E}_e)_j$ and H_j^e for $(\mathbf{H}_e)_j$.

In terms of the new free energy function, the constitutive equations are

$$\text{(IV)} \quad \begin{cases} \tau_{ij} = \rho \dfrac{\partial x_i}{\partial X_J} \dfrac{\partial \mathscr{F}}{\partial \Sigma_{JK}} \dfrac{\partial x_j}{\partial X_K} - P_i E_j^e - \mu_0 M_i H_j^e, \\ P_i = -\rho \dfrac{\partial x_i}{\partial X_J} \dfrac{\partial \mathscr{F}}{\partial \hat{E}_J}, \\ \mu_0 M_i = -\rho \dfrac{\partial x_i}{\partial X_J} \dfrac{\partial \mathscr{F}}{\partial \hat{H}_J}, \\ S = -\partial \mathscr{F}/\partial T. \end{cases} \tag{12.19}$$

The objective constitutive equations for $\mathbf{Q}$ and $\mathbf{J}_e$ are

$$Q_i = (\partial x_i/\partial X_J)\mathscr{Q}_J(\Sigma_{KL}, \hat{E}_K, \hat{H}_K, T, T,_K), \tag{12.20}$$

$$J_i^e = (\partial x_i/\partial X_J)\mathscr{J}_J(\Sigma_{KL}, \hat{E}_K, \hat{H}_K, T, \hat{T},_K), \tag{12.21}$$

where $J_i^e \equiv (\mathbf{J}_e)_i$ and

† Like in the previous section, the key equations are identified by (IV) for set IV, and (III) for set III.

$$\hat{T},_K = x_{j,K} T,_j. \tag{12.22}$$

$\mathcal{Q}_J$ and $\mathcal{J}_J$ are two general vector functions governed by the inequality (12.12).

Since Σ_{JK} is a symmetric tensor, the antisymmetric part of τ_{ij} in Eq. (2.19) is

$$\text{(IV)} \quad \tau_{[ij]} = -P_{[i}E^e_{j]} - \mu_0 M_{[i}H^e_{j]} \tag{12.23}$$

This is identical to Eq. (11.3) with the substituting of Eqs. (11.9) for **L**. Hence the balance equation for the angular momentum is redundant and will not be needed when an objective constitutive equation (12.19) is adopted. The stress tensor in this formulation is, in general, nonsymmetric.

Because the functions $\mathcal{F}$, $\mathcal{Q}_J$, $\mathcal{J}_J$ are given in a general form, the constitutive equations (12.19)–(12.21) are rather complicated. For many real materials, they can be expressed as polynomials of the constitutive variables [43, 42, 36], or other elementary functions of them. The following polynomials may be assumed for a large class of materials:

$$\mathcal{F} = \mathcal{F}^\Sigma + \mathcal{F}^E + \mathcal{F}^H + \mathcal{F}^Q, \tag{12.24}$$

and

$$\begin{aligned}
\mathcal{F}^\Sigma &= (2\rho_0)^{-1} C_{IJKL}\Sigma_{IJ}\Sigma_{KL},\\
\mathcal{F}^E &= (\rho_0/2)\chi_{JK}\hat{E}_J\hat{E}_K + \rho_0\beta^E_{IJKL}\Sigma_{IJ}\hat{E}_K\hat{E}_L + \theta_{IJK}\Sigma_{IJ}\hat{E}_K + \lambda_J\hat{E}_J T,\\
\mathcal{F}^H &= (\rho_0/2)\mu_0\alpha_{JK}\hat{H}_J\hat{H}_K + (\rho_0/4)\mu_0\alpha_{IJKL}\hat{H}_I\hat{H}_J\hat{H}_K\hat{H}_L\\
&\quad + \rho_0\mu_0\beta^H_{IJKL}\Sigma_{IJ}\hat{H}_K\hat{H}_L,\\
\mathcal{F}^Q &= \tfrac{1}{2}CT^2 + m_{JK}\Sigma_{JK}T.
\end{aligned} \tag{12.25}$$

$\mathcal{F}^\Sigma$ is the free energy function due to elastic strain, ρ_0 being the mass density at the reference configuration and C_{IJKL} being the elastic constants. $\mathcal{F}^E$ is due primarily to electric polarizations, χ_{IJ}, β^E_{IJKL}, θ_{IJK}, and λ_J being the reciprocal susceptibility, electrostrictive, piezoelectric, and pyroelectric constants respectively. $\mathcal{F}^H$ is due primarily to magnetizations, α_{JK}, α_{IJKL}, and β^H_{IJKL} being the anisotropy, fourth-order anisotropy, and magnetostrictive constants respectively. The inverse of the anisotropy constant is related to the magnetic susceptibility. All preceding material constants can also be a function of the temperature T. $\mathcal{F}^Q$ is due primarily to thermal effect, C and m_{JK} being the heat capacity and thermoelastic constants respectively.

For materials with a center of symmetry, all odd-order tensor material constants, e.g. λ_J, θ_{IJK} in Eq. (12.25), must vanish. The constitutive equations (12.20) and (12.21) become particularly simple for such materials [28],

$$\mathscr{Q}_J = \gamma^E_{JK}\hat{E}_K - \kappa_{JK}\hat{T},_K, \tag{12.26}$$

$$\mathscr{J}_J = \nu_{JK}\hat{E}_K + \gamma^Q_{JK}\hat{T},_K. \tag{12.27}$$

κ_{JK} and ν_{JK} are thermal and electrical conductivity respectively; γ^E_{JK} and γ^Q_{JK} are two thermoelectric constants. Their values must satisfy the inequality (12.12). All material constants in the preceding equations must eventually be determined by experiments.

The material constants delineated in Eqs. (12.25)–(12.27) represent various mechanical, thermal, and electromagnetic properties of materials. For any given material under investigation, not all properties are of equal significance. If we retain only the predominant part, the constitutive equations as well as the electromagnetic forces and energies will be greatly simplified. We mention here a few commonly encountered materials:

(A) *Non-ferrous metals.* For non-ferrous metals in an electromagnetic field, the effects of electric polarization and magnetization can be neglected, but the thermal and electric conductions are significant. We thus set, in Eq. (12.25), $\mathscr{F}^E = 0$ and $\mathscr{F}^H = 0$, which amounts to omitting all terms containing **P** and **M** in the field equations. This version of magneto-thermo-elasticity is the basis for the theory in Ref. [37].

(B) *Dielectrics.* Since the effects of magnetization and electrical conduction are comparatively insignificant for dielectrics we set $\mathscr{F}^H = 0$, which amounts to the omission of **M** in all equations. For the functions $\mathscr{Q}_J$ and $\mathscr{J}_J$, we set γ^E_{JK}, γ^Q_{JK}, and ν_{JK} zero. κ_{JK} is retained along with the thermal free energy $\mathscr{F}^Q$ to account for the thermoelastic effect [47]. Soft ferroelectric materials may be treated as a dielectric. However, for certain hard ferroelectric materials, or when dealing with the surface phenomenon of dielectrics, the free energy function may be also dependent on the polarization gradient $\nabla\mathbf{P}$, or equivalently $\nabla\mathbf{E}_e$. Furthermore, another balance equation of intra-molecular force must be added to the field equations. This aspect of the theory of elastic dielectrics, which is discussed in Refs. [44], and [48], is not included in this article.

(C) *Ferrous metals.* For paramagnetic and soft ferromagnetic materials, we omit the effect of electric polarization by setting $\mathscr{F}^E = 0$ (dropping all **P**'s) [36]. The effect of conduction current which is a major source of heating and magnetic damping is usually retained. Soft ferrites may be analyzed as soft ferromagnetic materials except that they are poor conductors for heat and electric currents, $\nu_{JK} = \nu^{Q}_{JK} = \gamma^{E}_{JK} = 0$. For hard ferromagnetic materials, the polynomial representation of the free energy function $\mathscr{F}^H$ is not appropriate. In fact, the existence of $\mathscr{F}^H$ is in question when the hysteresis loss is pronounced. Continuum theory for polycrystalline hard-ferromagnetic materials has not yet been fully developed.

(D) *Ferrimagnetic insulator.* A certain hard ferrimagnetic crystalline material like yttrium iron garnet (YIG) has very low hysteresis loss and mechanical damping. It can maintain the propagation of elastic waves of ultrahigh frequencies ($\sim 10^9$ hertz), which interacts with spin waves under magnetic fields. As an important material for making microwave devices, it has stimulated in recent years the research on magnetoelastic interactions [2, 42, 49]. Spin waves is of quantum origin and can be described in terms of the continuum theory by adding the magnetization gradient ($\nabla \mathbf{M}$ or equivalently $\nabla \mathbf{H}_e$) as a constitutive variable in the free-energy function. However, the addition of this variable must be accompanied by the addition of a balance equation for magnetization, which is not considered in this article.

There are many other materials with special electronic properties like semiconductors and superconductors, which are not considered here.

12.2 Constitutive Equations for the Dipole–current Circuit Formulation

Starting from Eq. (12.1) while ϕ is given in Eq. (11.15), we may establish the constitutive equations for the dipole–current circuit formulation in analogous to those of the two-dipole formulation. Instead of the Legendre transformation, Eq. (12.7), we assume a free energy F,

$$F = U - TS - \mathbf{E}_e \cdot (\mathbf{P}/\rho). \tag{12.28}$$

This accomplishes the purpose of changing the independent variables from S and $(\mathbf{P}/\rho)$ to T and $\mathbf{E}_e$. The reduced free energy–entropy inequality becomes

$$\rho(\dot{F} + S\dot{T}) - (\underset{\sim}{\xi}^{-1} \cdot \underset{\sim}{\tau})^T : \dot{\underset{\sim}{\xi}} + \mathbf{P} \cdot \dot{\mathbf{E}}_e + \mathbf{M}_e \cdot \dot{\mathbf{B}} + \nabla T \cdot (\mathbf{Q}/T) - \mathbf{J}_e \cdot \mathbf{E}_e \leq 0. \tag{12.29}$$

In analogous to Eq. (12.17), we define a free energy function $\mathscr{F}$ with (Eq. (8.15) of Ref. [43]),

$$\text{(III)} \quad F = \mathscr{F}(\underset{\sim}{\Sigma}, \hat{\mathbf{E}}, \hat{\mathbf{B}}, T), \tag{12.30}$$

where

$$\hat{E}_K = x_{j,K} E^e_j, \qquad \hat{B}_K = x_{j,K} B^e_j. \tag{12.31}$$

The inequality (12.29) is satisfied by the following objective constitutive equations:

$$\text{(III)} \begin{cases} \tau_{ij} = \rho \dfrac{\partial x_i}{\partial X_J} \left(\dfrac{\partial \mathscr{F}}{\partial \Sigma_{JK}} \right) \dfrac{\partial x_j}{\partial X_K} - P_i E^e_j - M^e_i B_j, \\ P_i = - \rho \dfrac{\partial x_i}{\partial X_J} \left(\dfrac{\partial \mathscr{F}}{\partial \hat{E}_J} \right), \\ M^e_i = - \rho \dfrac{\partial x_i}{\partial X_J} \left(\dfrac{\partial \mathscr{F}}{\partial \hat{B}_J} \right), \\ S = - \partial \mathscr{F} / \partial T. \end{cases} \tag{12.32}$$

The objective constitutive equations for Q_i and J^e_i are analogous to Eqs. (12.20) and (12.21),

$$\text{(III)} \begin{cases} Q_i = (\partial x_i/\partial X_J) \mathscr{Q}_J(\Sigma_{KL}, \hat{E}_K, \hat{B}_K, T, \hat{T}_{,K}), \\ J^e_i = (\partial x_i/\partial X_J) \mathscr{J}_J(\Sigma_{KL}, \hat{E}_K, \hat{B}_K, \hat{T}, \hat{T}_{,K}), \end{cases} \tag{12.33}$$

where

$$\hat{T}_{,K} = x_{j,K} T_{,j}.$$

Polynomial representations for the functions $\mathscr{F}$, $\mathscr{Q}_J$, $\mathscr{J}_J$ may be assumed as in Eqs. (12.24)–(12.27). However, the $\hat{\mathbf{H}}$ in $\mathscr{F}^H$ should be replaced by $\hat{\mathbf{B}}$.

Like Eq. (12.16), the reduced energy equation for this formulation is

$$\text{(III)} \quad \rho T \dot{S} = \mathbf{J}_e \cdot \mathbf{E}_e - \nabla \cdot \mathbf{Q} + \phi_Q. \tag{12.34}$$

Note again, the antisymmetric part of the stress tensor in Eq. (12.32) is

$$\text{(III)} \quad t_{[ij]} = - P_{[i} E^e_{j]} - M^e_{[i} B_{j]}. \tag{12.35}$$

The above result is compatible with the balance law for angular momentum and the body couple $\mathbf{L}$ in Eq. (11.14).

12.3 Boundary Conditions

The starting point for deriving the boundary conditions is to postulate that the balance laws in global form for both mechanical and electromagnetic fields are valid even when the field quantities are discontinuous within a material volume or area. The derivation of the mechanical boundary conditions is well known [18], and that of the electromagnetic boundary conditions for a moving boundary is given by Costen and Adamson [50] for the Minkowski formulation. Derivations for other formulations are essentially the same (see Section 13 of Ref. [46]). We omit all details and list the final form of the boundary conditions.

In the following, **n** denotes a unit vector normal to the material surface of a body in the present configuration. The surface which moves with the material particle velocity **v** divides the material volume into a plus (+) side and a minus (−) side while **n** pointing toward (+) side. The symbol $[\![\psi]\!] = \psi^+ - \psi^-$ denotes the jump of the tensor quantity ψ, from the positive (+) side to the negative (−) side of the surface. We also use the symbol σ' and $\mathbf{J}'$ to denote the surface electric charge and surface electric current respectively on the material surface.

(A) *The Two-dipole Model Formulation.* From Eqs. (2.7) and (2.9), we obtain [35],

$$
\text{(IV)} \quad \begin{cases} \mathbf{n}\cdot[\![\underset{\sim}{\tau} + \underset{\sim}{\tau}^C]\!] = \mathbf{0}, & (12.36) \\ \mathbf{n}\cdot[\![(\underset{\sim}{\tau} + \mathbf{PE} + \mu_0\mathbf{MH})\cdot\mathbf{v} + \tfrac{1}{2}(\varepsilon_0 E^2 + \mu_0 H^2)\mathbf{v} - (\mathbf{Q} + \mathbf{E}\times\mathbf{H})]\!] = 0. & (12.37) \end{cases}
$$

The first is the boundary condition for the stress tensor $\underset{\sim}{\tau}$ while $\underset{\sim}{\tau}^C$ is given in Eq. (11.18). The second is for the heat flux vector **Q**. No useful boundary conditions result from the global laws for mass and angular momentum.

From Eqs. (4.9A)–(4.9E) and Eq. (4.14) we find

$$
\text{(IV)} \quad \begin{cases} \mu_0\mathbf{n}\cdot[\![\mathbf{M} + \mathbf{H}]\!] = 0, \\ \mathbf{n}\times[\![\mathbf{E} + \mu_0\mathbf{v}\times\mathbf{H}]\!] = \mathbf{0}, \\ \mathbf{n}\cdot[\![\varepsilon_0\mathbf{P} + \mathbf{E}]\!] = \sigma', \\ \mathbf{n}\times[\![\mathbf{H} - \varepsilon_0\mathbf{v}\times\mathbf{E}]\!] = \mathbf{J}' - \sigma'\mathbf{v}, \end{cases} \tag{12.38}
$$

and

$$
\text{(IV)} \quad [\![\mathbf{J}]\!]\cdot\mathbf{n} - v_n[\![\sigma]\!] + \mathbf{n}\cdot\nabla\times(\mathbf{n}\times\mathbf{J}') + v_n\nabla\cdot(\mathbf{n}\sigma') = -\frac{\partial\sigma'}{\partial t}, \tag{12.39}
$$

where $v_n = \mathbf{n}\cdot\mathbf{v}$ is the normal component of the surface velocity **v**. The four conditions in Eq. (12.38) reduce to the familiar conditions for fixed boundaries ($\mathbf{v} = \mathbf{0}$) if one recalls that $\mu_0(\mathbf{M} + \mathbf{H}) = \mathbf{B}$ and $\varepsilon_0\mathbf{P} + \mathbf{E} = \mathbf{D}$.

Note that although the equation of continuity for the electric current, either in global form of Eq. (4.9E) or in local form of Eq. (4.10E), may be derived from the other four Maxwell equations, the jump conditions (12.39) is independent of the four corresponding jump conditions in Eq. (12.38).

(B) *The Dipole–Current Circuit Model Formulation.* In analogous to the boundary conditions of part (A), we have for stresses and heat flux [43, 47]:

$$\text{(III)} \begin{cases} \mathbf{n} \cdot [\![\underset{\sim}{\tau} + \underset{\sim}{\tau}^{S}]\!] = \mathbf{0}, & (12.40) \\ \mathbf{n} \cdot [\![(\underset{\sim}{\tau} + \mathbf{PE} + \mathbf{MB}) \cdot \mathbf{v} + \tfrac{1}{2}(\varepsilon_0 E^2 + B^2/\mu_0)\mathbf{v} - (\mathbf{Q} + \varepsilon_0 \mathbf{E} \times \mathbf{B})]\!] = 0, & (12.41) \end{cases}$$

where $\underset{\sim}{\tau}^{S}$ is given in Eq. (11.13).

Similarly, corresponding to Eqs. (11.11), we have [43, 50],

$$\text{(III)} \begin{cases} \mathbf{n} \cdot [\![\mathbf{B}]\!] = 0, \\ \mathbf{n} \times [\![\mathbf{E} + \mathbf{v} \times \mathbf{B}]\!] = \mathbf{n} \times [\![\mathbf{E}]\!] - v_n [\![\mathbf{B}]\!] = \mathbf{0}, \\ \mathbf{n} \cdot [\![\varepsilon_0 \mathbf{P} + \mathbf{E}]\!] = \sigma', \\ \mathbf{n} \times [\![\mathbf{B}/\mu_0 - \mathbf{M}]\!] + v_n [\![\varepsilon_0 \mathbf{P} + \mathbf{E}]\!] = \mathbf{J}', \end{cases} \tag{12.42}$$

and

$$\text{(III)} \quad [\![\mathbf{J}]\!] \cdot \mathbf{n} - v_n [\![\sigma]\!] + \mathbf{n} \cdot \nabla \times (\mathbf{n} \times \mathbf{J}') + v \nabla \cdot (\mathbf{n}\sigma') = -\frac{\partial \sigma'}{\partial t}. \tag{12.43}$$

The fourth of Eq. (12.42) is derived from $\mathbf{n} \cdot [\![\mathbf{H} - \mathbf{v} \times \mathbf{D}]\!] = \mathbf{J}' - \sigma' \mathbf{v}$; and Eq. (12.43) is identical to Eq. (12.39).

12.4 Summary and Conclusion

With the supplement of constitutive equations, we have two determinate sets of equations, one for each theory. The balance equation of mass (11.1) contains four unknowns, ρ and $\mathbf{v}(=\dot{\mathbf{x}})$. The three balance equations of linear momentum (11.2) contains ρ, $\mathbf{v}$, and nine unknown stress components $\underset{\sim}{\tau}$, the body force in them has been expressed in terms of the electromagnetic variables in (11.7) or (11.12). Since the balance equations of angular momentum (11.3) are satisfied identically when objective constitutive equations are adopted, they are redundant and will not be needed. The balance equation of energy (11.4) is superseded by the energy equation (12.16) or (12.34), which contains ρ and additional unknowns $S, T, \mathbf{J}_e, \mathbf{E}_e$, and $\mathbf{Q}$. In these equations, the heat supply ϕ_Q is a prescribed quantity and should not be treated as an unknown.

In either set of the Maxwell equations, Eqs. (11.6) or (11.11), only seven of the ten equations are independent. For dynamical problems, it is convenient to take the second, fourth, and fifth equation as the independent ones.

In Set IV, Eqs. (11.6) contains the electromagnetic field variables **E**, **P**, **H**, **M**, **J**, σ and the velocity **v**. Since $\underset{\sim}{\tau}$, **P**, **M**, and S are related in Eq. (12.19) to the free energy function $\mathscr{F}$ which is assumed known for a given material, and $\mathbf{J}_e$ and **Q** can be expressed in terms of other variables by Eqs. (12.20) and (12.21), they are eliminated as unknowns by the constitutive equations. Thus only ρ, **v**(or **x**), T, **E**, **H** and σ remain as the unknown variables. They are governed by the complete set of equations: Eqs. (11.1), (11.2), (11.6) and (12.16) of Set IV, a total of twelve independent equations for twelve unknowns. The boundary conditions for this set are given in Eqs. (12.36)–(12.39).

In Set III, the constitutive equations (12.32) and (12.33) have related $\underset{\sim}{\tau}$, **P**, **M**, S, **Q** and $\mathbf{J}_e$ to other dependent variables, and we are left with ρ, **v**(or **x**), T, **E**, **B**, and σ as the twelve unknown variables. They are governed by Eqs. (11.1), (11.2), (11.11), and (12.34). The boundary conditions are specified by Eqs. (12.40)–(12.43).

The complete set of equations of either theory, even with the simplified constitutive equations as assumed in Eqs. (12.24)–(12.27), are highly non-linear and complex. It is unlikely that an exact, nontrivial solution to these equations will ever be found. In order to derive approximate solutions with a predetermined accuracy, these equations are simplified by perturbation or other procedures. A procedure which allows the linearization of these equations systematically is presented in [35], [36], [51].

After this long and ponderous account of the history, logic, and theory of electrodynamics of deformable media, one question still remains: Of these two sets of theories and equations, which one is "correct" or "more accurate"? This cannot be answered until more experiments are conducted to test the results which are derived from these equations. Although the electromagnetic fields **E**, **B**, **D**, **H**, **P**, and **M** inside moving bodies cannot be measured directly, precise measurements of the velocity field **v**, or its gradient $\nabla\mathbf{v}$, are possible. Since the major difference in these two sets lies in the representation of the induced magnetization, experiments of the propagation of elastic (acoustic) waves in magnetized materials may provide a test of these theories.

Acknowledgment. This research was supported in part by the National Science Foundation through a grant to the Material Science Center of Cornell University.

13 REFERENCES

Books and Monographs

1. Becker, Richard, *Electromagnetic Fields and Interactions, vol. 1 Electromagnetic Theory and Relativity* (edited by F. Sauter and translated from German by A. W. Knudson), Blackie, London (1964).
2. Brown, William F., Jr., *Magnetoelastic Interactions*, Springer-Verlag, New York (1966).
3. Cullwick, E. G., *Electromagnetism and Relativity*, Longmans, Green & Co., London (1957).
4. DeGroot, S. R. and Suttorp, L. G., *Foundations of Electrodynamics*, North-Holland Publishing Co., Amsterdam (1972).
5. Fano, R. M., Chu, L. J. and Adler, R. B., *Electromagnetic Fields, Energy, and Forces*, John Wiley & Sons, Inc., New York (1960). (Reprinted by the M.I.T. Press.)
6. Feynman, R. P., Leighton, R. B. and Sands, M., *The Feynman Lectures on Lectures on Physics*, Vol. 2, Addison-Wesley, Reading, Mass. (1964).
7. Jackson, John D., *Classical Electrodynamics*, John Wiley & Sons, Inc., New York (1962).
8. Landau, L. D. and Lifshitz, E. M., *Electrodynamics of Continuous Media*, Addison-Wesley, Mass. (1960).
9. Livens, George H., *The Theory of Electricity*, 2nd edition, Cambridge University Press, London (1926).
10. Lorentz, Hendrik A., *The Theory of Electrons and Its Applications to the Phenomena of Light and Radiant Heat*, 2nd edition, Verlag von B. G. Teubner, Leipzig (1916).
11. Mason, M. and Weaver, W., *The Electromagnetic Field*, Dover Publications, New York (reprint of 1929 edition).
12. Maxwell, James Clerk, *A Treatise on Electricity and Magnetism*, Vols. 1 and 2, Dover Publications, Inc., New York (1954) (reprint of the 3rd edition, 1891).
13. Møller, C., *The Theory of Relativity*, 2nd edition, Oxford University Press, London (1972).
14. Panofsky, W. K. H. and Phillips, M., *Classical Electricity and Magnetism*, 2nd edition, Addison-Wesley Publishing Co., Reading, Mass. (1962).
15. Penfield Jr., Paul and Haus, Hermann A., *Electrodynamics of Moving Media*, M.I.T. Press, Cambridge, Mass. (1967).
16. Sommerfeld, Arnold, *Electrodynamics* (Lectures on Theoretical Physics, Vol. III), Academic Press, New York (1952).
17. Stratton, Julius A., *Electromagnetic Theory*, McGraw-Hill Book Co., New York (1941).
18. Truesdell, C. and Toupin, R. A., *The Classical Field Theories*, in *Handbuch der Physik*, Vol. III/1, S. Flügge (ed.), Springer-Verlag, Berlin (1960).
19. Whittaker, Edmund, *A History of the Theories of Aether and Electricity*, Vol. 1, *The Classical Theories*, Vol. 2, *The Modern Theories*, Harper & Brothers, New York. (Reprint of 1951 and 1953 edition.)

Research and Review Papers

20. Brevik, I., "Electromagnetic Energy-momentum Tensor within Material Media, Part I, Minkowski's Tensor, Part II. Discussion of Various Tensor Form," *Royal Danish Academy of Sciences and Letters* (*Det Kongelige Danske Videnskabernes Selskab, Matema tisk-fysiske Meddelelser*) **37**, Nos. 11 and 13 (1970).
21. Boffi, L. V., "Electrodynamics of Moving Media," Sc.D. Thesis, Department of Electrical Engineering, M.I.T., Cambridge, Mass. (Aug. 1957).

22. Chu, L. J., Haus, H. A. and Penfield, P., Jr., "The Force Density in Polarizable and Magnetizable Fluids," *Proc. IEEE* **54** (1966) 920.
23. Corson, Dale R., "Electromagnetic Induction in Moving Systems," *Amer. J. Phys.* **24** (1956) 126.
24. DeGroot, S. R. and Suttorp, L. G., "The Relativistic Energy-momentum Tensor in Polarized Media, Parts 1–7," *Physica* **37** (1967) 284, 297; **39** (1968) 28, 41, 61, 77, 84.
25. DeGroot, S. R. and Vlieger, J., "Deviation of Maxwell's Equations—the statistical theory of the macroscopic equations," *Physica* **31** (1965) 254.
26. Einstein, A., "Zur Elektrodynamik bewegter Körper," *Annalen der Physik* **17** (1905) 891. (English translation by W. Perrett and G. B. Jeffrey, in the *Principle of Relativity*, 1923, Dover Publications, New York, N.Y.).
27. Grot, Richard A. and Eringen, A. Cemal, "Relativistic Continuum Mechanics, Part 1. Mechanics and Thermodynamics, Part 2. Electromagnetic Interactions with Matter," *Int. J. Engng Sci.* **4** (1966) 611, 639.
28. Hutter, Kolumban and Pao, Yih-Hsing, "A Dynamic Theory for Magnetizable Elastic Solids with Thermal and Electrical Conduction," *J. Elasticity* **4** (1974) 89.
29. Lianis, G., "Formulation and Application of Relativistic Constitutive Equations of Deformable Electromagnetic Materials," *Il Nuovo Cimento* **16B** (1973) 1.
30. Lorentz, Hendrik A., "Weiterbildung der Maxwellschen Theorie, Elektronentheorie," *Enzyklopädie der mathematischen Wissenschaften*, Bd. V-2, Art. 13, 14 (1902).
31. Mazur, P., "On Statistical Mechanics and Electromagnetic Properties of Matter," *Adv. Chem. Phys.* **1** (1958) 309.
32. Minkowski, Hermann, "Die Grundgleichungen für die elektromagnetischen Vorgänge in bewegten Körpern," *Mathemat. Annal.* **68** (1910) 472.
33. Minkowski, Hermann and Born, Max, "Ein Ableitung der Grundgleichungen für die elektromagnetischen Vorgänge in bewegten Körper vom Standpunkte der Electronentheorie," *Mathemat. Annal.* **68** (1910) 526.
34. Mo, Tse Chin, "Theory of Electrodynamics in Media in Noninertial Frames and Applications," *J. Math. Phys.* **11** (1970) 2589.
35. Pao, Y. H. and Hutter, K., "Electrodynamics for Moving Elastic Solids and Viscous Fluids," *Proc. IEEE*, **63** (1975) 1011.
36. Pao, Yih-Hsing and Yeh, Chau-Shioung, "A Linear Theory for Soft Ferromagnetic Elastic Solids," *Int. J. Engng Sci.* **11** (1973) 415.
37. Paria, Gundahar, "Magneto-elasticity and Magneto-thermo-elasticity," *Adv. Appl. Mech.* **10** (1967) 73 (Academic Press, N.Y., edited by G. Kuerti).
38. Poincelot, Paul, "Sur l'expression de la densité d'énergie électromagnétique," *Comptes Rendus,* l'Académie des Sciences, Paris, **264B** (1967) 1064.
39. Poincelot, Paul, "Sur le tenseur électrodynamique," *Comptes Rendus*, l'Académie des Sciences, **264B** (1967) 1179, 1225–1560.
40. Tai, C. T., "Present Views on Electrodynamics of Moving Media," *Radio Sci.* **2** (1967) 245.
41. Tai, Chen-To, "On the Presentation of Maxwell's Theory," *Proc. IEEE* **60** (1972) 936.
42. Tiersten, H. F., "Coupled Magnetomechanical Equations for Magnetically Saturated Insulators," *J. Math. Phys.* **5** (1964) 1298.
43. Tiersten, H. F. and Tsai, C. F., "On the Interaction of the Electromagnetic Field with Heat Conducting Deformable Insulators," *J. Math. Phys.* **13** (1972) 361.
44. Toupin, R. A., "The Elastic Dielectrics," *J. Rational Mech. Anal.* **5** (1956) 849.
45. Toupin, R. A., "A Dynamical Theory of Elastic Dielectrics," *Int. J. Engng Sci.* **1** (1963) 101.
46. Jaunzemis, W., *Continuum Mechanics*, MacMillan, New York (1967).
47. Tiersten, H. F., "On the Nonlinear Equations of Thermo-electroelasticity," *Int. J. Engng Sci.* **9** (1971) 587.
48. Mindlin, R. D., *Polarization Gradient in Elastic Dielectrics*, Course and Lectures at Udine, Springer-Verlag, New York (1970).
49. LeCraw, R. C. and Comstock, R. L., "Magnetoelastic Interactions in Ferromagnetic

Insulators," in *Physical Acoustics* **3B** (1965); also Strauss, W., "Magnetoelastic Properties of Yttrium-Iron Garnet," in *Physical Acoustics* **4B**, Academic Press, New York (1968).

50. Costen, R. C. and Adamson, D., "Three Dimensional Derivation of the Electrodynamic Jump Conditions and Moment-energy Laws at a Moving Boundary," *Proc. IEEE* **53** (1965) 1181.
51. Van de Ven, A. A. F., *Interaction of Electromagnetic and Elastic Fields in Solids*, Dr. of Science Thesis, Technichische Hogeschool te Eindhoven, the Netherlands (1975).

V

Problems in Magneto-solid Mechanics

Francis C. Moon†

Cornell University, Ithaca, New York

1 INTRODUCTION

Conventional devices in which large magnetic fields and electric currents are present include switches, motors, and electromagnets. Also structures are sometimes called upon to withstand the forces of transient current pulses of over 10^4 amperes such as occur in lightning strikes or magnetomotive tools. With recent advances in superconductivity, new devices will be required to withstand the forces due to high currents such as superconducting motors, transmission lines, inductance storage devices, rings for magnetically levitated trains, and superconducting toroidal magnets for fusion reactors.

The purpose of this paper is to present to mechanicians some interesting magnetoelastic problems with potential application. We explore methods to calculate the effects of magnetic fields and currents on the stresses, stability, and dynamic behavior of structural members. Experiments are also described with which the author has observed nontrivial effects of magnetic fields and electric currents on the mechanical behavior of beams, plates, rings, and shells.

To obtain a physical "feeling" for what is a "high" magnetic field or current one is reminded that if B is the magnetic field in webers/meters^2, and μ_0 ($4\pi 10^{-7}$ in MKS units) the magnetic permeability of vacuum, then the quantity

$$\left[\frac{B^2}{2\mu_0}\right] = \frac{\text{newtons}}{\text{meter}^2}$$

† Formerly with Princeton University, Princeton, New Jersey.

has the units of stress. At a magnetic field of 1 wb/m^2, about the limit for permanent magnets, $B^2/2\mu_0 = 40\ N/cm^2$. Also if I represents the electric current in amperes, then the quantity

$$[\mu_0 I^2] = \text{newtons}$$

has the units of force. Only for currents above 10^3 amps does this quantity become greater than one newton. These quantities will appear in the calculations involving the deformation of a body in a magnetic field or carrying current. In general, for engineering sized structures, the stresses or deformations will only be large for magnetic fields of the order of 1 wb/m^2 or currents greater than 1 kiloamp.

Magneto-solid mechanics has its roots in the nineteenth century, especially in the study of magneto-striction. The effect of mechanical strains on the magnetic properties of iron and steel was also studied in the last century by Wiedemann, Matteucci, and Wertheim. These and other problems concerning elastic solids and magnetic fields have been surveyed by Todhunter and Pearson [1] in their history of elasticity.

By the late nineteenth century the concepts of magnetic forces and their relation to magnetic energy through a variational principal were known and may be found in Maxwell's famous treatise on electromagnetics [2]. These ideas led to the now familiar expression for magnetic body forces in terms of the divergence of a magnetic stress tensor. Modern treatments of this concept may be found in Jackson [3] and Stratton [4]. This concept of magnetic forces, Maxwell attributes to Faraday, who imagined tension along the magnetic field lines and pressure at right angles to them. When the body containing the magnetized material or current is cut, the mechanical stresses must equilibrate the above-mentioned electromagnetic tensions and pressures.

Thus the basic tenets of magnetoelasticity would appear to have been established by the late nineteenth century, were it not for the fact that for magnetizable or ferromagnetic materials, the representation of the total forces by the magnetic stress tensor was not unique. Modern authors have shown that to uniquely determine the internal forces in a magnetized body the constitutive relations between strain, stress, and magnetic fields must be rigorously formulated using an energy approach (see, for example, Brown [5], Penfield and Haus [6], and Pao [7]).

In the last decade intense interest in nonlinear continuum mechanics and the axiomatic method led researchers to a reexamination of the interaction of electromagnetic fields with solids, as illustrated by Toupin [8] and Eringen [9] in their theory of elastic dielectrics. Many of these works were motivated primarily by an interest in the effects of electric and magnetic fields on the constitutive or stress–strain relations and did not arise out of a specific engineering application. The reader is referred to the article by Pao (Chapter IV, this volume), for a review of these theoretical questions.

This article will focus instead on the applications of magneto-solid mechanics. In spite of the interest in magnetomechanics by early mechanicians, study of the forces and stresses in electromagnetic devices was mainly calculated by electrical engineers or applied physicists. Forces in rotating electrical machinery, for example, may be found in a text by Hague [10]. And forces between conductors or electric coils carrying currents are found in a number of papers by Dwight [11]. A similar example is a paper by Higgins [12] calculating the forces between channel beams when the conductors carry short-circuit high currents. Such problems are usually important in electrical power distribution systems where lightning can cause transient currents with high magnetic forces.

Another problem involving magnetic forces of importance to electrical engineering is the electrical contact. This problem involves elastic and plastic deformation, thermoelastic stresses due to heating, melting, and magnetic forces, when electrical current passes between two bodies with limited surface contact. A survey of some of the early research in this area may be found in the book by Holm [13].

In many of these problems knowledge of the magnetic field determines the body forces and hence the deformation and stresses in the solid. Thus the magnetic field problem can be decoupled from the stress analysis. In the last decade, however, many problems were investigated involving mutual coupling of field and deformation. Knopoff [14], for example, studied the effect of the earth's magnetic field on seismic waves. Another was the work of Tiersten and others [15, 16] on waves in saturated magnetoelastic materials which paralleled the development of ultrasonic devices using magnetoelastic waves.

Similar papers appeared on the study of elastic and plastic wave motion in the presence of a large static magnetic field many of these originating in the Polish journals by Kaliski and coworkers [17]. These problems usually involved the calculation of perturbations in the elastic motion due to the magnetic field and many were not directed toward a specific technical or engineering problem. A review of many of these papers has been given by Paria [18].

The reader will find in the present review many references to devices and applications. This is because analysis of magneto-solid problems have been almost exclusively carried out with a particular device in mind. The author has sought to study applications where the magnetic forces can produce significant deformation, in contrast to those problems where small signals for electronic communication signals are important. This survey is aimed at a class of magnetoelastic problems whose origins derive from devices such as: high field magnets, pulsed magnetic coils for magnetic forming, superconducting coils for fusion reactors, and magnetic levitation and propulsion magnets for high-speed trains and short-circuit problems in transmission lines. Common to most of these problems are high magnetic fields, forces, or currents.

An important class of problems not covered in this article is the interaction of laser beams or electron beams with solids. While engineering devices to drill, weld and machine solids with lasers are well developed, little study of the electromagneto-mechanics has been published beyond a few thermoelastic analyses.

2 METHODS

There are two basic methods with which one can calculate the effects of magnetic fields in solids. One is the magnetic force-field method. In this method the magnetic field is determined from the current distribution in the solid and magnetic body forces are used in the classical equations of mechanics. In the second method the magnetic energy of the currents is incorporated in a Lagrangian. This allows one to construct approximate solutions from a variational method. Both these methods are illustrated below.

In the examples below only methods for nonferromagnetic conductors are treated. This is because a unique description of magnetic stresses in magnetizable materials is not available. While self-consistent theories of magnetoelasticity of magnetizable materials can be formulated, such as Brown [5], Tiersten [15], and Penfield and Haus [6], these theories contain constitutive relations between magnetic field and strain that have yet to be determined for specific structural materials. Such magnetostrictive data that does exist has not been correlated with the specific continuum theories mentioned above.

2.1 Magnetic Forces—Field Method

When magnetization is not present in a conducting solid, the body force on an element with charge density q and current density $\mathbf{J}$ is given by [4]

$$\mathbf{f} = q\mathbf{E} + \mathbf{J} \times \mathbf{B} \tag{2.1}$$

where **E** is the electric field and **B** the magnetic field.

In many problems the magnetic field can be assumed to be quasi-static and satisfies the equation†

$$\nabla \times \mathbf{B} = \mu_0 \mathbf{J}, \tag{2.2}$$

$$\nabla \cdot \mathbf{B} = \mathbf{0}. \tag{2.3}$$

The current **J** depends on the velocity of the solid **v**, and the electric and magnetic fields, i.e.

$$\mathbf{J}' = \sigma \mathbf{E}'. \tag{2.4}$$

The primes indicate values measured relative to the material. When the material moves this relation becomes

$$\mathbf{J} = \sigma(\mathbf{E} + \mathbf{v} \times \mathbf{B}) + q\mathbf{v}.$$

In good conductors the last term can often be dropped. Also where the material is electrically anisotropic the electrical conductivity σ becomes a second-order tensor.

In superconducting materials the relation (2.4) is replaced by the equation [20]

$$\mathbf{E} = \mu_0 \lambda^2 \dot{\mathbf{J}}, \tag{2.4a}$$

where λ is the penetration depth of a static magnetic field into the conductor.

The electric field **E** can either be supplied by an external source, or induced in the conductor by a time varying magnetic field, as in magnetic forming, or magnetic levitation of vehicles. The electric and magnetic fields are then coupled by Faraday's law

$$\nabla \times \mathbf{E} + \frac{\partial \mathbf{B}}{\partial t} = \mathbf{0}. \tag{2.5}$$

In the force method the body force is incorporated in the elasticity equations (assuming $q = 0$),

$$\nabla \cdot \mathbf{t} + \mathbf{J} \times \mathbf{B} = \rho \ddot{\mathbf{u}}, \tag{2.6}$$

where **t** is the stress tensor. When the conductor is in the form of a plate the magnetic forces are integrated through the thickness

$$D\nabla_1^4 w + \rho h \frac{\partial^2 w}{\partial t^2} = F + \mathbf{n} \cdot \nabla \times \mathbf{C}, \tag{2.7}$$

† Meter-kilogram-second-Coulomb units are used in this paper, $\mu_0 = 4\pi\ 10^{-7}$.

where

$$\mathbf{F} = \int \mathbf{n} \cdot (\mathbf{J} \times \mathbf{B})\, dz,$$

$$\mathbf{C} = \int z\mathbf{n} \times (\mathbf{J} \times \mathbf{B})\, dz,$$

where w is the transverse plate displacement; $\mathbf{n}$ is the unit vector normal to the plate surface; $D = Eh^3/12(1 - \nu^2)$; E is Young's modulus; ρ is the plate density; and $\nabla_1^2 \equiv (\partial^2/\partial x^2) + (\partial^2/\partial y^2)$.

The above equation neglects the inplane or membrane stresses. The average membrane stresses are determined by the equation

$$\nabla \cdot \mathbf{t} + \int (\mathbf{J} \times \mathbf{B}) \cdot [\mathbf{I} - \mathbf{nn}]\, dz = \rho h \frac{\partial^2 \mathbf{U}}{\partial t^2}. \tag{2.8}$$

Here $\mathbf{I}$ is the identity dyadic and $\mathbf{U}$ the average inplane displacements.

This formulation neglects forces due to electric charges. For a conductor electric charges will move to the surfaces in a very short time. Thus if charges are present they will reside on the surface and provide a force $Q\mathbf{E}$, where Q is the surface charge density.

Also if the membrane stresses $\mathbf{t}$ are large they will introduce moments in Eq. (2.7).

A. *Steady Stresses in Thin Current-carrying Conductors.* To calculate stresses in a conductor one must add body forces to the equations of elastic equilibrium. The body forces for non-ferromagnetic materials are given by $\mathbf{J} \times \mathbf{B}$ where $\mathbf{J}$ is the current density vector and $\mathbf{B}$ the magnetic field. While calculation of magnetic fields of currents and solution of the equations of elasticity are usually well-understood mathematical problems, exact stress distributions due to currents in conductors for problems of interest to engineering designers are few. Further, while the formal mathematical problems are understood, actual determination of fields and stresses for particular geometries is a nontrivial problem. This results from the fact that simplification of the elasticity problem may complicate the field problem as is illustrated in the example chosen.

Aside from the stresses, magnetoelastic problems involving the stability of the elastic conductor under the electromagnetic body forces will also interest the structural mechanician. Buckling of conducting strips under high currents has been observed by the author in the laboratory and is also

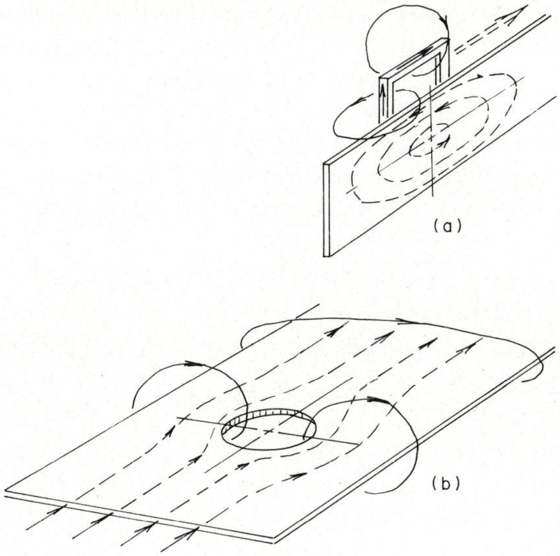

Fig. 1 (a) Induced eddy current patterns in a strip due to a moving magnet. (b) Magnetic "stress concentration" due to a hole in a conducting strip.

observed in lightning-strike damage. The buckling current is dependent on the induced stresses in the conductor.

One of the simplest classes of problems in elasticity is the case of plane stress in thin plate-like structures (Fig. 1). However, for thin plates, calculation of the magnetic field near the edges becomes a problem. In the first section we introduce a stream function for the current and derive an integral expression over the plate to calculate the self-magnetic field. Finally we use the calculated fields to determine the stress distribution.

B. *Statement of the Problem.* We consider here cases for which the conductor is a thin flat plate, where the current distribution across the plate thickness can be assumed to be uniform. Further we assume that the currents are either steady or quasi-steady as in the case of oscillating currents. If the currents are oscillatory, we assume that the skin depth (see, for example, [4]), which is inversely proportional to the square root of the frequency, is large compared with the plate thickness Δ. We are primarily concerned here with the so-called self-field and self-stresses of the current on the conductor. The net force and moments on the plate due to these self-fields are zero. Under these assumptions the equilibrating stress system is planar, i.e. no bending occurs in the plate. It might be noted here, that if a stability analysis of the plate were made, the effect of small out-of-plane deformations on the field would be made and bending stresses would become important. The present problem is then the precursor to the stability problem or can be referred to as the initial stress problem without bending deformations.

Since the current across the thickness is uniform we define

$$\mathbf{I} = \Delta\mathbf{J} = \sigma\Delta\mathbf{E}, \tag{2.9}$$

where $\mathbf{E}$ is the electric field.

For steady-state or low-frequency currents we have a continuity condition

$$\nabla \cdot \mathbf{I} = \nabla \cdot \mathbf{J} = 0 \tag{2.10}$$

which suggests the use of a "stream potential" which is found in hydrodynamics. A similar approach has been used by Silvester and Popovic to calculate induced eddy currents in conducting sheets [21]. One sets

$$\mathbf{I} = \nabla \times (\psi\mathbf{n}) = -\mathbf{n} \times \nabla\psi,$$

where $\mathbf{n}$ is normal to the plate. The concomitant magnetic field to the current distribution is

$$\mathbf{B}^1(\mathbf{r}) = \frac{\mu_0}{4\pi\Delta}\int_{\text{plate}} \frac{\mathbf{I}(\mathbf{r}) \times (\mathbf{r} - \mathbf{r}')}{|\mathbf{r} - \mathbf{r}'|^3}\, dv'. \tag{2.11}$$

In general, where $\mathbf{I}$ is induced by time-varying fields $\mathbf{B}^0$, outside the plate, the stream function is determined by Faraday's law [4],

$$\nabla^2\psi = \sigma\Delta\frac{\partial}{\partial t}(B_z^0 + B_z^1), \tag{2.12}$$

where x, y are in the plane, and z has origin at plate is midsurface. When steady currents exist $\nabla^2\psi = 0$ and the solutions are analogous to those in hydrodynamics. Lines of constant ψ will give the direction of current flow while lines of constant $|\nabla\psi|$ will give constant $\mathbf{J} \cdot \mathbf{J}$ lines of energy deposition.

The inplane stresses satisfy the equations

$$\frac{\partial t_{xx}}{\partial x} + \frac{\partial t_{xy}}{\partial y} + J_y(B_z^0 + B_z^1) = \rho\frac{\partial^2 u_x}{\partial t^2},$$

$$\frac{\partial t_{xy}}{\partial x} + \frac{\partial t_{yy}}{\partial y} - J_x(B_z^0 + B_z^1) = \rho\frac{\partial^2 u_y}{\partial t^2}.$$

Note that the planar component of the body force $\mathbf{I} \times \mathbf{B}$ is $-B_z \nabla\psi$, so that $\nabla\psi$ gives the direction of the electromagnetic force in the plate.

The determination of the stresses has three steps: (i) find ψ for given $\mathbf{B}^0$ and boundary conditions; (ii) calculate B_z^1 from ψ; and (iii) integrate the equations of motion.

C. *Self-field of Currents in a Thin Sheet.* The current stream function ψ is assumed to be defined throughout three-dimensional space, and constant across the sheet thickness. Let $\mathbf{R} = \mathbf{r} - \mathbf{r}'$ in Eq. (2.11). Then the Biot–Savart law reads (Proof by W. Markiewicz, Intermagnetics, Guilderland, N.Y.)

$$B_z^1 = \frac{\mu_0}{4\pi\Delta} \int \left[I_x \frac{\partial}{\partial y'}\left(\frac{1}{|\mathbf{R}|}\right) - I_y \frac{\partial}{\partial x'}\left(\frac{1}{|\mathbf{R}|}\right)\right] dv'.$$

The integrand can be written in the form

$$\nabla_1\psi \cdot \nabla_1 \frac{1}{|\mathbf{R}|} - \frac{\partial\psi}{\partial z'}\left(\frac{\partial}{\partial z'}\left(\frac{1}{|\mathbf{R}|}\right)\right),$$

where ∇_1 is the gradient and operates on the $\mathbf{r}'$ variables. If the integral is carried out within the boundaries of the plate then $\partial\psi/\partial z' = 0$ and the second term of the preceding expression is zero. Using an identity to replace the first term, one obtains the expression for B_z^1 as

$$B_z^1 = \frac{\mu_0}{4\pi\Delta} \int \left\{ \nabla_1 \cdot \left[\psi \nabla_1 \frac{1}{|\mathbf{R}|}\right] - \psi \nabla_1^2 \frac{1}{|\mathbf{R}|}\right\} dv',$$

and hence for values of B_z^1 in the plate,

$$B_z^1 = \frac{\mu_0 \psi}{\Delta} + \frac{\mu_0}{4\pi\Delta} \int \nabla_1 \cdot \left[\psi \nabla_1 \frac{1}{|\mathbf{R}|}\right] dv'.$$

The next step is to use the divergence theorem on the second integral where we set $\psi = 0$ on the narrow edges of the plate;

$$B_z^1 = \frac{\mu_0\psi}{\Delta} + \frac{\mu_0}{4\pi\Delta} \int_{\text{top}} \psi \frac{\partial}{\partial z} \frac{1}{|\mathbf{R}|} da' - \frac{\mu_0}{4\pi\Delta} \int_{\text{bottom}} \psi \frac{\partial}{\partial z'} \frac{1}{|\mathbf{R}|} da'.$$

Since we are interested in stresses in the midplane of the plate we set $z = 0$ in the above expression to obtain

$$B_z^1(x, y, 0) = \frac{\mu_0 \psi}{\Delta} - \frac{\mu_0}{4\pi} \int \frac{\psi(x', y')\, da'}{[(x - x')^2 + (y - y')^2 + \Delta^2/4]^{3/2}}. \tag{2.13}$$

Since ψ is only determined to an additive constant it may appear that the above expression for B_z^1 is not completely determined by ψ. However, one can show that when $\psi = C$, a constant everywhere, or $I = 0$, then

$$B_z^1 = \frac{\mu_0 C}{\Delta}\left[1 - \frac{\Delta}{4\pi}\int_0^\infty \int_0^{2\pi} \frac{2\pi r\, dr\, d\theta}{[r^2 + \Delta^2/4]^{3/2}}\right] = 0.$$

Also when I is independent of the variable x, then the self field becomes

$$B_z^1(0, y, 0) = \frac{\mu_0 \psi}{\Delta} - \frac{\mu_0}{2\pi}\int \frac{\psi(y')\, dy'}{(y' - y)^2 + \Delta^2/4}. \tag{2.14}$$

Example: *Uniform Current in a Strip*

Consider a thin plate of width $2a$ in the y-direction and uniform current flowing in the x-direction. The stream function and normal magnetic field then are

$$\psi = I_0 y/2a,$$

$$B_z^1 = \frac{\mu_0 I_0}{4a}\left[\bar{y} - \frac{1}{\pi}\int_{-2a/\Delta}^{2a/\Delta} \frac{\eta\, d\eta}{(\eta - \bar{y})^2 + 1}\right] = \frac{\mu_0 I_0}{4a} F(\bar{y}), \tag{2.15}$$

where $\bar{y} = 2y/\Delta$, and $F(\bar{y})$ is given by

$$F(\bar{y}) = \left[\bar{y} - \frac{1}{2\pi}\ln\frac{(A - \bar{y})^2 + 1}{(A + \bar{y})^2 + 1} - \frac{\bar{y}}{\pi}\tan^{-1}(A - \bar{y}) - \tan^{-1}(A + \bar{y})\right]. \tag{2.16}$$

For current distributions which are independent of position along the strip, the internal stress t_{yy} may be directly integrated assuming zero stress on the edge, i.e.

$$t_{yy} = \int_a^y \frac{\partial \psi}{\partial \eta} B_z^1(\eta)\, d\eta.$$

Integration of the electromagnetic body force was carried out for the case of constant current distribution across a strip and the resulting stresses are shown in Fig. 2.

For current distributions which vary along the strip, direct integration of the stresses is not possible due to the existence of shear stresses.

From Fig. 2 it is clear that uniform current in a thin plate produces compressive stresses. Compressive stresses in thin plates can sometimes induce *buckling*. Thus one would expect that for large enough currents the plate will collapse under inplane compressive stresses in analogy to Euler

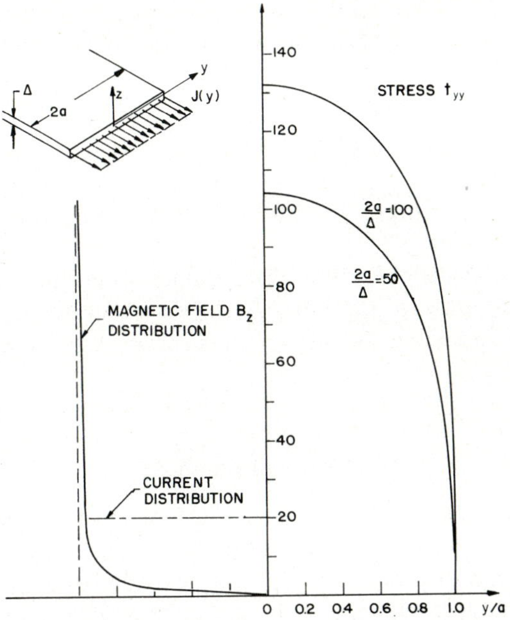

Fig. 2 Distribution of magnetic field and stress in a thin conducting strip.

Fig. 3 Photograph of the buckling of an aluminum strip under several kiloamps of electric current.

buckling of a column. Providing that temperature effects due to Joule heating are not significant, the critical current for buckling should be given by a relation of the form

$$\mu_0 I^2 = kY\Delta^2\left(\frac{\Delta}{a}\right)^2,$$

where Δ/a is the thickness to width ratio of the conductor, and k is a non-dimensional constant.

Experimental evidence for this effect is often found in structures subject to high currents due to lightning. A simple illustration of the magnetic collapse of a conducting foil under transient currents of up to 4000 A is shown in Fig. 3 [22].

2.2 Magnetic Forces—Energy Method

Consider a structure with a strain energy function V and let the structure have N degrees of freedom each of which has a generalized coordinate U_k. These coordinates may represent, for example, the bending amplitude of a beam or the change in radius of a circular magnet. Whatever the interpretation the strain energy is assumed to be a known function of the U_k.

Now imagine that electric generators pump M independent currents I_k through the structure and that the voltage at each generator is e_k. Thus the total work done in a small time dt is

$$I_k e_k \, dt,$$

where repeated indices are summed.

From Maxwell's equations the rate of change of flux ϕ_k through the kth circuit is given by

$$\frac{\partial \phi_k}{\partial t} = e_k,$$

where the circuit is assumed to be nonresistive.

The stored magnetic energy in the circuit is assumed to be a known function of the fluxes and displacements, i.e. $W = W(\phi_n, U_n)$.

Then the variational principal states that the virtual changes in elastic and magnetic energies is equal to the virtual work, i.e.

$$I_k \, d\phi_k = \frac{\partial W}{\partial \phi_k} d\phi_k + \frac{\partial W}{\partial U_k} dU_k + \frac{\partial V}{\partial U_k} dU_k. \tag{2.17}$$

Since the dU_k and $d\phi_k$ are independent we have

$$I_k = \frac{\partial W}{\partial \phi_k}, \qquad \frac{\partial W(\phi_i, U_i)}{\partial U_k} + \frac{\partial V}{\partial U_k} = 0. \tag{2.18}$$

Often, however, the magnetic energy is known as a function of the currents. We can make a Legendre transformation by defining a coenergy

$$W^*(I_k) = I_k \phi_k - W(\phi_i).$$

Thus if I_i, ϕ_i are considered independent variables,

$$dW = I_k\, d\phi_k + \phi_k\, dI_k - dW^*.$$

Substituting into (2.18) the principal of virtual work becomes

$$0 = \phi_k\, dI_k - dW^* + dV$$

which leads to the equations

$$\phi_k = \frac{\partial W^*}{\partial I_k},$$

$$\frac{\partial W^*(I_i, U_i)}{\partial U_k} = \frac{\partial V}{\partial U_k}. \tag{2.19}$$

Thus if $Q_k^e = -(\partial V/\partial U_k)$ is called the generalized elastic force then the equilibrium between elastic and magnetic forces is

$$\frac{\partial W}{\partial U_k} + Q_k^e = 0,$$

or the generalized magnetic force is given by

$$Q_k^m = \left.\frac{\partial W^*}{\partial U_k}\right|_{I \text{ constant}} = -\left.\frac{\partial W}{\partial \phi_k}\right|_{\phi \text{ constant}} \tag{2.20}$$

For a single linear magnetic current $W = W^* = LI^2/2$ and $\phi = LI$ where L is the inductance and is assumed to be dependent on the deformation U_k. Then the magnetic force is

$$Q_k^m = \frac{I^2}{2}\frac{\partial L}{\partial U_k} = -\frac{\phi^2}{2}\frac{\partial 1/L}{\partial U_k}. \tag{2.21}$$

A. *Energy Method–Dynamic Effects.* To model the mechanics of elastic structures with electric currents, we incorporate the magnetic energy of the currents, W, into Hamilton's principle (see, for example, Crandall *et al.* [23]) where

$$W = \frac{\mu_0}{4\pi} \int \frac{\mathbf{J} \cdot \mathbf{J}'}{|\mathbf{r} - \mathbf{r}'|} \, dr \, dr'. \tag{2.22}$$

If the currents are confined to a filamentary path through which I amperes flow, then

$$W = \tfrac{1}{2} L I^2,$$

where L is called the self-inductance of the device. For a device with N current paths and N currents I_k,

$$W = \tfrac{1}{2} L_{ij} I_i I_j. \tag{2.23}$$

Suppose then we can represent the displacements of a structure by either nodal or modal amplitudes $A_n(t)$. For example, in a plate we could write the transverse displacements w as

$$w(x_1, x_2) = A_n(t) f_n(x_1, x_2). \tag{2.24}$$

Using generalized masses and stiffnesses for these modes, the kinetic and potential energies, T and V, can be written in terms of the modal amplitudes $A_n(t)$. Likewise the magnetic energy can also be written in terms of the variables A_n. Thus

$$T = \tfrac{1}{2} m_n A_n^2(t), \qquad V = \tfrac{1}{2} k_n A_n^2(t), \qquad W = W(I_k, A_n).$$

If we are satisfied with a finite number of modes as an approximation, a Lagrangian can be constructed as

$$L = T - V + W. \tag{2.25}$$

The independent variables are A_n and q_k, where

$$I_k = \frac{dq_k}{dt}.$$

The usual variational method leads to the set of equations

$$\ddot{A}_n(t) + \Omega_n^2 A_n - \frac{1}{m_n} \frac{\partial W}{\partial A_n} = \frac{F_n}{m_n}, \qquad \frac{d}{dt} \frac{\partial W}{\partial I_k} + R_k I_k = E_k, \tag{2.26}$$

where F_n are generalized external forces on the structure and Ω_n are the natural frequencies. Also R_k are circuit resistances and E_k are circuit voltages which maintain the currents.

For superconducting devices operation in the persistent mode $E_k = 0$ and $R_k = 0$. Thus the magnetic flux is conserved, i.e.

$$\phi_k = \frac{\partial W}{\partial I_k} = \text{constant}.$$

For linear circuits the magnetic flux is given by

$$\phi_k = L_{kj} I_j.$$

For one circuit $\phi_0 = LI$,

$$\ddot{A}_n(t) + \Omega_n^2 A_n - \frac{1}{m_n}\frac{\phi_0^2}{L^2}\frac{\partial L}{\partial A_n} = \frac{F_n}{m_n} \quad \text{(no sum on } n\text{)}. \tag{2.27}$$

Consider the case of one modal amplitude, A, where the inductance can be written as a Taylor series in A,

$$L = L_0 + L'A + \frac{L''}{2}A^2 + \cdots \tag{2.28}$$

If A defines deviations from an equilibrium point such that

$$F = -\frac{\phi_0^2}{L^2}L',$$

then the vibrations of the system are determined from

$$\ddot{A}(t) + \left(\Omega^2 - \frac{\phi_0^2}{mL^2}[L'' - 2(L')^2]\right)A = 0. \tag{2.29}$$

for the constant flux case; or

$$\ddot{A}(t) + \left(\Omega^2 - \frac{I^2 L''}{m}\right)A = 0 \tag{2.30}$$

for the constant current case.

Another example is the rigid-body vibration of a moving superconducting ring levitated at a height h above a conducting track. When the velocity is high enough, the magnetic field is screened out by the conductor and the force on the coil may be viewed as due to an image coil beneath the surface (see Fig. 4); see, for example, Reitz [24] for a discussion of forces on magnetic levitation magnets. The mutual inductance for a circular coil of radius R is approximately ($h \ll R$)

$$L = \mu_0 R\left(\ln\frac{4R}{h} - 2\right)$$

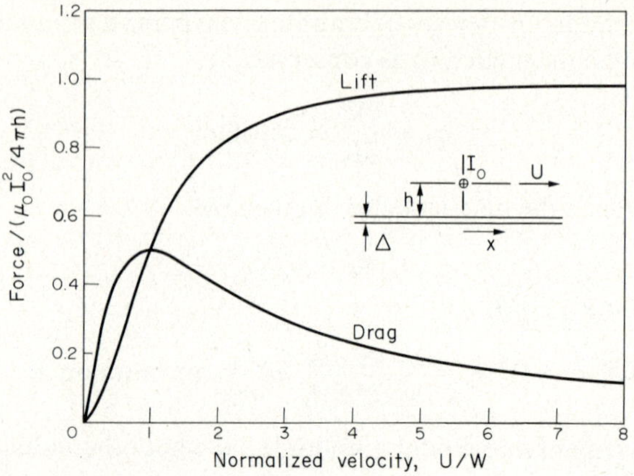

Fig. 4 Forces due to a superconducting coil moving over a conducting sheet track.

so that the magnetic energy of the coil and conducting sheet guideway is

$$W = -\tfrac{1}{2}LI^2. \tag{2.31}$$

If we perturb the height by an amount A, i.e. $h = h_0 + A$, the inductance can be expanded in a Taylor series in A,

$$L = L_0 - \mu_0 R\left(\frac{A}{h_0}\right) + \frac{\mu_0}{2} R\left(\frac{A}{h_0}\right)^2 + \cdots \tag{2.32}$$

Substituting (2.31, 32) into (2.26), where the mechanical stiffness is non-existent, i.e. $\Omega^2 = 0$, we obtain the equation of motion as

$$\ddot{A} + \frac{\mu_0 I^2}{m} \frac{R}{h_0^2} A = 0.$$

Since the current must be adjusted so that the repulsive magnetic force equilibrates the gravitational force $m\,g$, it is easy to show that the *frequency is independent of the mass*, i.e. for $A = A_0\,e^{i\omega t}$, $\omega^2 = g/h_0$, so that the vehicle vibrates like a pendulum of length h_0.

3 STABILITY OF FERROELASTIC STRUCTURES IN MAGNETIC FIELDS

3.1 Magnetoelastic Buckling of Beam-plates

From a research point of view the most interesting problems in magneto-solid mechanics are those in which the magnetic field and strain or deformation are coupled in some nontrivial way. Two examples of coupling are mutual effects of strain and magnetic field on the constitutive relations and coupling through boundary conditions. Both types of problems have been investigated, particularly those for which the problem can be linearized about an initial strain or magnetic field state $\mathbf{B}_0$. Values of the initial magnetic field $\mathbf{B}$ for which there exist multiple solutions are known as critical values and the solid or structure is said to be unstable or can buckle in analogy to Euler column buckling.

Formulation of the stability problem of a saturated ferromagnetic body in an initial uniform magnetic field has been given by Brown [5]. Initially the magnetization vector $\mathbf{M}$, which has a constant magnitude, lies along the direction of the applied field $\mathbf{B}_0$. The field is lowered until the magnetization, $\mathbf{M}$, switches to a direction other than parallel to $\mathbf{B}_0$. This value of the field is called the *nucleation field* [5]. Solutions to this problem have been found only for the rigid solid case by Brown [5], Frei *et al.* [25], and Aharoni [26] for ellipsoids of revolution. They have found three modes for the direction cosines of $\mathbf{M}$ which they call curling, coherent rotation, and "buckling". In these papers, however, deformation has been assumed to be zero.

Physical buckling of magnetized ferromagnetic whiskers in forms of elastica has been observed by DeBlois [27], as shown in Fig. 5. In his experiment the initial magnetization lies along the rod so there exist magnetic poles of strength ρ_m on each end. The whisker is then acted on by forces $\rho_m \mathbf{B}_0$ at the ends, if the pole model for magnetic forces is used. When

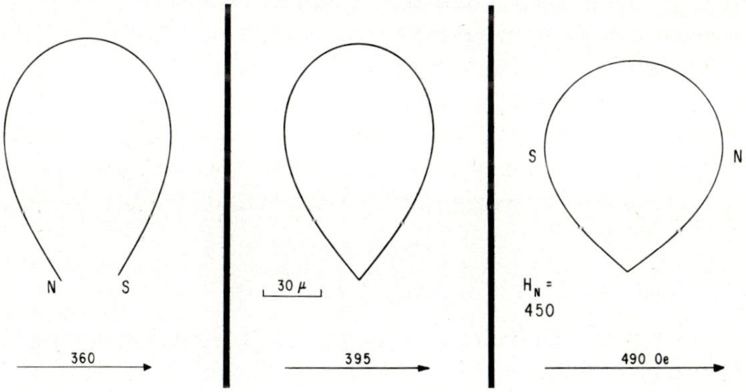

Fig. 5 Photograph of the buckling of a ferromagnetic whisker in a longitudinal magnetic field (personal communication from R. W. DeBlois).

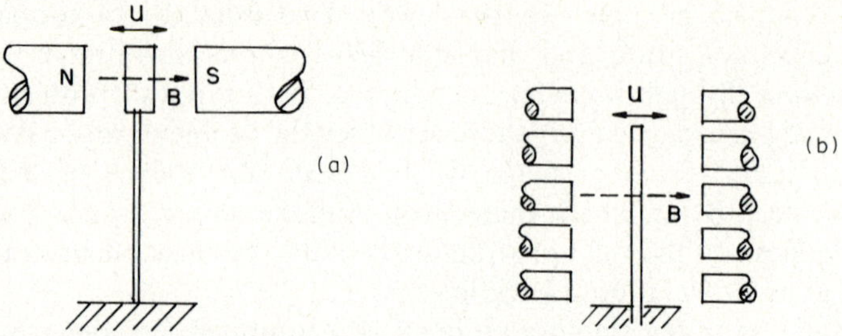

Fig. 6 (a) Vibration of a ferromagnetic mass between the poles of an electromagnet. (b) Vibration of ferromagnetic beam between a series of magnetic poles.

these forces are large enough the whisker buckles into the classic shape of the elastica.

On a macroscopic scale one of the early papers on buckling of a solid in a magnetic field was by Mozniker [28] in 1959. He placed a ferromagnetic solid on the end of a cantilevered beam, between two electro-magnets (Fig. 6a). The equation of motion for the first mode was given as

$$\frac{d^2u}{dt^2} + k\frac{du}{dt} + \omega_0^2 u - KB_0^2 \frac{Gu}{(G^2 - u^2)^2} = 0, \tag{3.1}$$

where G is the gap between the magnet and beam and u is the transverse displacement. The linearized solution (neglecting damping, $k = 0$) predicts a natural frequency which decreases with current, i.e.

$$\omega^2 = \omega_0^2(1 - KB_0^2/G^3\omega_0^2). \tag{3.2}$$

Mozniker measured such a decrease in natural frequency. At the value of B_0 for which $\omega \to 0$ the beam will move to a new equilibrium position. This is referred to as *magnetoelastic buckling.* This effect was not studied in any detail by Mozniker, however.

Panovko and Gubanova [29] extended this problem to a beam situated midway between a set of magnet poles (Fig. 6b). This led them to postulate a negative foundation type magnetic force distribution in the beam equation

$$EJ\frac{\partial^4 u}{\partial x^4} = \frac{KB_0^2}{G_0^3}u, \tag{3.3}$$

where G_0 is the gap between the poles and B_0 is the magnetic field. For a plate-like beam of thickness $2h$, and length L, this theory leads to a critical magnetic field relation of the form,

$$B_c \sim (h/L)^2.$$

However, the physics of this problem is buried in the phenomenological constant K relating the force to the deflection. Also no experiments were reported in [29].

In 1967 the author presented a dissertation [30] which reexamined this problem. This involved the theoretical and experimental study of a ferromagnetic beam-plate in a uniform magnetic field. These results were presented in two papers by Moon and Pao [31, 32]. Their treatment differs from Panovko and Gubanova in the assumption of the nature of the magnetic forces. Rather than a magnetic force proportional to the displacement, they recognized that in a uniform magnetic field there can be no net force on a magnetized body. However, the local magnetization in the plate $\mathbf{M}$ can interact with the applied magnetic field $\mathbf{B}_0$ to produce a couple $\mathbf{M} \times \mathbf{B}_0$. This leads to a plate equation for sinusoidal deformations, of wavelength $\lambda = 2\pi/k$, given below

$$D \frac{d^3u}{dx^3} + b \frac{du}{dx} = 0, \tag{3.4}$$

where

$$b = \frac{2\chi^2 B_0^2}{\mu_0 \mu_r} \frac{\sinh kh}{k\Delta},$$

and

$$\Delta = \mu_r \sinh kh + \cosh kh, \qquad \mu_r = \chi + 1.$$

Since $\chi = \mu_r - 1$, then in the limit where $\chi \sim \mu_r$, and $\mu_r h/L \gg 1$ the critical field for a periodically pinned beam-plate is found to be [31]

$$\frac{B_0^2}{\mu_0 E} \simeq \frac{\pi^3}{24(1-\nu^2)} \left(\frac{2h}{L}\right)^3. \tag{3.5}$$

In this theory the perturbed magnetic field is calculated for a deformed plate where the rotated plate normal $\mathbf{n}^*$ is related to the original normal $\mathbf{n}$ by the rotation on the surface,

$$\mathbf{n}^* = \mathbf{n} + \boldsymbol{\Omega} \times \mathbf{n},$$
$$\boldsymbol{\Omega} = \nabla_1 \times u(x, y)\mathbf{n},$$

and ∇_1 is the gradient operator in the plane. The continuity boundary conditions on the magnetic field are written as

$$[\mathbf{H}] \times \mathbf{n}^* = 0,$$
$$[\mathbf{B}] \cdot \mathbf{n}^* = 0,$$

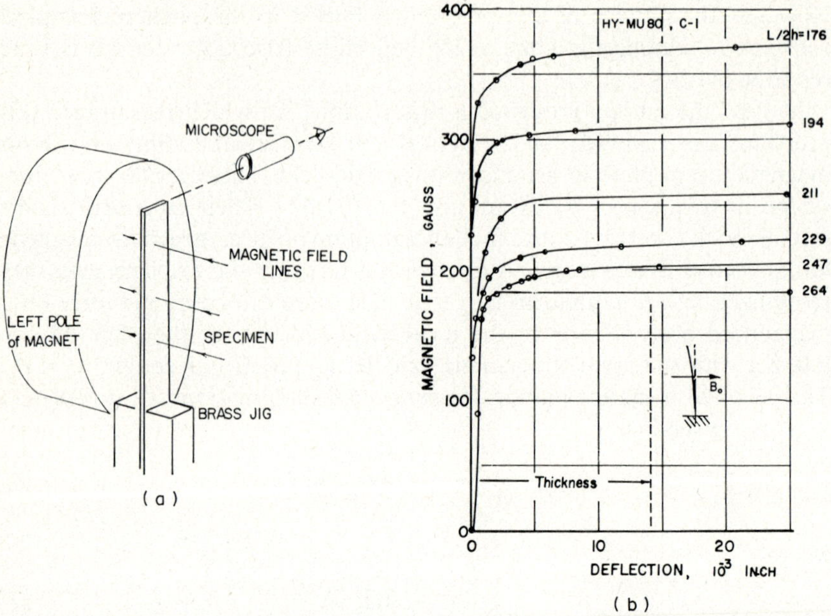

Fig. 7 (a) Diagram of ferromagnetic plate in a transverse magnetic field. (b) Experimental buckling curves for a ferroelastic plate in a transverse magnetic field (from F. C. Moon and Y.-H. Pao, "Magnetoelastic Buckling of a Thin Plate," *J. Appl. Mech.* **35**, No. 1 (1969) 1–9).

where [ ] indicates a jump in the function across the plate surface. The total magnetic field is the sum of the initial and perturbed fields,

$$\mathbf{B} = \mathbf{B}_0 + \mathbf{b}, \qquad \mathbf{H} = \mathbf{H}_0 + \mathbf{h}.$$

The theory has been extended to two-dimensional plate problems by this writer [33]. The equation to be satisfied is given by

$$D\nabla_1^4 u + 2d\rho \frac{\partial^2 u}{\partial t^2} = q + \mathbf{n} \cdot \nabla_1 \times \mathbf{C}. \tag{3.6}$$

In a uniform external magnetic field the magnetic body force q is zero, while the couple per unit area $\mathbf{C}$ is given by

$$\mathbf{C} = \chi \int_{-d}^{d} \mathbf{h} \times \mathbf{B}_0 \, dz$$

when the initial field is normal to the plate.

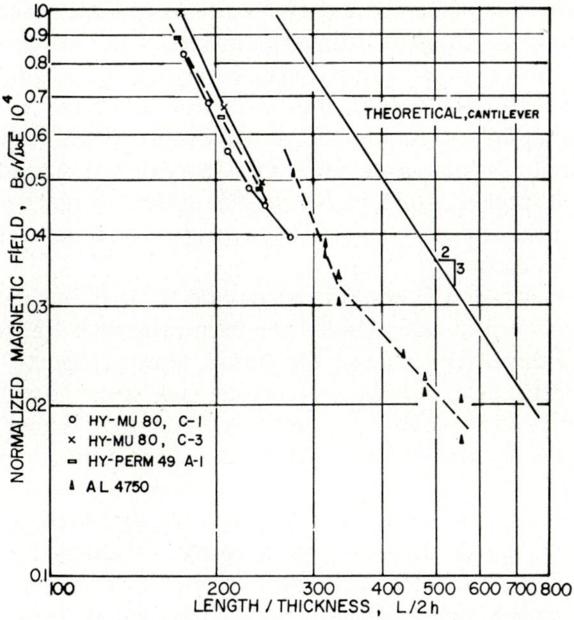

Critical magnetic field for buckling of beam-plates

Fig. 8 Critical magnetic field for buckling of beam-plates versus length/thickness ratio (from F. C. Moon and Y.-H. Pao, "Magnetoelastic Buckling of a Thin Plate," *J. Appl. Mech.* **35**, No. 1 (1969) 1–9).

Experimental verification of this phenomena is shown in Figs. 7 and 8. As can be seen, the dependence of the critical field on the 3/2 power of the thickness ratio seems to confirm the assumption of a magnetic couple across the plate in comparison to the Panovko and Gubanova theory which leads to a second power law. As is evident, however, the absolute magnitude of the critical field is lower by a factor of about 1.8 than the theoretical. This was surprising since B_0 is independent of the permeability for large μ_r. A search for this factor has led to a number of additional papers on the subject which are reviewed below.

3.2 Comparison of Buckling Theory and Experiment

One question the reader might ask concerning this problem is whether it is of any general importance to mechanics or whether it is simply an ad hoc problem with difficult boundary conditions for which exact solutions cannot be found to compare with experiments. The answer is partially affirmative to both options; namely, magnetoelastic buckling might offer a way to test macroscopic theories of interaction between the magnetic field and deformation. This is because it leads to a linearized eigenvalue problem with no dynamics to complicate either the theory or experiment. However, experiments on specimens of finite size present difficult boundary conditions for the theoretician and can complicate the comparison of different formulations of magnetoelasticity.

For the past decade a number of axiomatic theories of the interaction of electromagnetic fields and solids have been proposed. Kaliski [34] and Dunkin and Eringen [35] studied the elastic, linear, magnetic plate in a uniform magnetic field but did not predict any instability phenomena. These theories included conduction currents, which were ignored by Moon and Pao, but the deletion of eddy currents should not affect static stability phenomena.

Tiersten [15] also constructed a self-consistent magnetoelastic theory for saturated ferromagnetic materials which correctly predicted some highly applicable spin wave–acoustic wave interactions. However, no macroscopic stability phenomena were predicted or studied using this theory.

One of the basic differences in these theories has been the assumption of the form of the magnetic force distribution in the body. Brown [5] has shown that the total force on a magnetizable body can be expressed in terms of different body force distributions. These are reviewed here briefly. If the total field is written as the sum of the field created by the external sources $\mathbf{B}_0$ and the magnetized body alone $\mathbf{B}_1$, the force in a magnetized body is given by (see Chapter IV by Pao in this volume):

Pole model;

$$\mathbf{F} = \int (-\nabla \cdot \mathbf{M})\mathbf{B}_0 \, dv + \int_A (\mathbf{n} \cdot \mathbf{M})\mathbf{B}_0 \, da, \tag{3.7}$$

Ampere current model;

$$\mathbf{F} = \int (\nabla \times \mathbf{M}) \times \mathbf{B}_0 \, dv + \int (\mathbf{M} \times \mathbf{n}) \times \mathbf{B}_0 \, da, \tag{3.8}$$

Dipole model;

$$\mathbf{F} = \int (\mathbf{M} \cdot \nabla)\mathbf{B}_0 \, dv, \tag{3.9}$$

with body couple

$$\mathbf{C} = \int \mathbf{M} \times \mathbf{B}_0 \, dv. \tag{3.10}$$

These volume integrals can be written in terms of the total field $\mathbf{B} = \mathbf{B}_0 + \mathbf{B}_1$ and for linear magnetic materials (i.e. $\mathbf{M} = \chi\mathbf{H}$) can be transformed into integrals of the form

$$\mathbf{F} = \int \mathbf{f} \, dv + \int [\mathbf{T}] \cdot \mathbf{n} \, da,$$

where we have:

Pole model;

$$\begin{aligned} \mathbf{f} &= \mu_0(\nabla \cdot \mathbf{n})\mathbf{H}, \\ \mathbf{T} &= \frac{1}{\mu_0}(\mathbf{HH} - \tfrac{1}{2}\mathbf{IH} \cdot \mathbf{H}), \end{aligned} \tag{3.11}$$

Ampere current model;

$$\begin{aligned} \mathbf{f} &= (\nabla \times \mathbf{M}) \times \mathbf{B}, \\ \mathbf{T} &= \frac{1}{\mu_0}(\mathbf{BB} - \tfrac{1}{2}\mathbf{IB} \cdot \mathbf{B}), \end{aligned} \tag{3.12}$$

Maxwell model;

$$\begin{aligned} \mathbf{f} &= \frac{1}{2}\left(\mu_r \mathbf{f}_{\text{pole}} + \frac{1}{\mu_r}\mathbf{f}_{\text{amp}}\right), \\ \mathbf{T} &= \mathbf{HB} - \tfrac{1}{2}\mathbf{I}(\mathbf{B} \cdot \mathbf{H}). \end{aligned} \tag{3.13}$$

For a linear magnetic material $\nabla \cdot \mathbf{B} = 0$ implies $\mathbf{f}_{\text{pole}} = \nabla \cdot \mathbf{H} = 0$. Thus the force distribution in the pole model exists where discontinuities in magnetic fields occur, as on the surface. Hence free mechanical traction boundary condition may still have nonzero "magnetic stress" components on the surface.

Similar remarks hold for the Ampere model, since for no conduction currents, $\mathbf{J} = 0$, and $\nabla \times \mathbf{H} = 0$ implies $\nabla \times \mathbf{M} = 0$ for linear magnetic materials. The conclusion that follows these observations is that different magnetic force models lead to different "magnetic stress" boundary conditions. Resolution of this dilemma has been recognized by many researchers as requiring different constitutive relations between stress, strain, and magnetization for each model (e.g. [5], [6]). Determination of these relations must be based on linearization of the nonlinear equations derived from continuum theory. In this way a self-consistent model can be found. Magnetic stress boundary conditions in one model will appear as constitutive terms in the stress–strain–magnetization relation.

Thus use of a particular "magnetic stress" boundary condition and unmodified or conventional stress–strain relations, in predicting magnetoelastic buckling can only be considered a guess. In the Moon–Pao theory the Amperian or dipole models were used with the usual Hooke's law and were chosen because they led to instability. However, in the Dunkin–Eringen analysis, a Maxwell model or pole model was effectively used with Hooke's law and could not lead to instability of the plate in a magnetic field as formulated unless the problem was linearized about an initial magnetic, in-plane compressional stress.

In an attempt to resolve the Moon–Pao experiments with theory, Pao and Yeh constructed a self-consistent linear ferroelastic continuum theory and reexamined the stability problem [36]. If $\bar{M}$, $\bar{H}$ represent initial fields perpendicular to the plate surface, and **m**, **h** represent the perturbed fields, the equilibrium equations and boundary conditions used by Pao and Yeh become

$$\begin{aligned} \nabla \cdot \mathbf{t} + \mu_0(\bar{\mathbf{M}} \cdot \nabla \bar{\mathbf{H}} + \bar{\mathbf{M}} \cdot \nabla \mathbf{h} + \mathbf{m} \cdot \nabla \bar{\mathbf{H}}) &= 0, \\ t_{ij} = \bar{t}_{ij} + C_{ijkl} u_{k,l} + 2\beta_{ijkl} \bar{M}_l m_k &+ \mu_0(\bar{H}_j m_i + \bar{H}_i m_j), \\ \bar{t}_{ij} = \mu_0 \chi^{-1} \bar{M}_i \bar{M}_j &+ \tfrac{1}{2}\mu_0 \beta_{ijkl} \bar{M}_k \bar{M}_l, \\ \mathbf{n} \cdot [\mathbf{t}] = \tfrac{1}{2}\mu_0 \mathbf{n}[\bar{\mathbf{M}}^2] &+ \mu_0 \mathbf{n}[\bar{\mathbf{M}} \cdot \mathbf{m}]. \end{aligned} \tag{3.14}$$

Here the β_{ijkl} act as magnetostrictive constants. Two sets of solutions for symmetric and antisymmetric displacements about the midplane of a plate in a normal magnetic field were found.

The lowest buckling field for long wavelength, sinusoidal bending (symmetric displacement) was found to be identical to that found by Moon in the limit as $\mu_r \to \infty$, $\mu_r k\Delta \gg 1$.

In contrast to the above equations, Moon [30] used the conventional elasticity theory with the Ampere stress model to obtain the set of equations

$$\nabla \cdot \mathbf{t} = 0, \qquad t_{ij} = C_{ijkl} U_{k,l}, \qquad [\mathbf{t}] \cdot \mathbf{n} = (\mathbf{M} \times \mathbf{n}^*) \times \mathbf{B}_0, \tag{3.15}$$

where $\mathbf{n}^*$ is the rotated plate normal. This formulation again led to the Moon–Pao buckling field in the high μ_r limit.

Yeh [37] also conducted buckling experiments and again obtained experimental critical fields below the theoretical values by 17–40%. However, he found that as the plate width to thickness ratio became larger, for a fixed

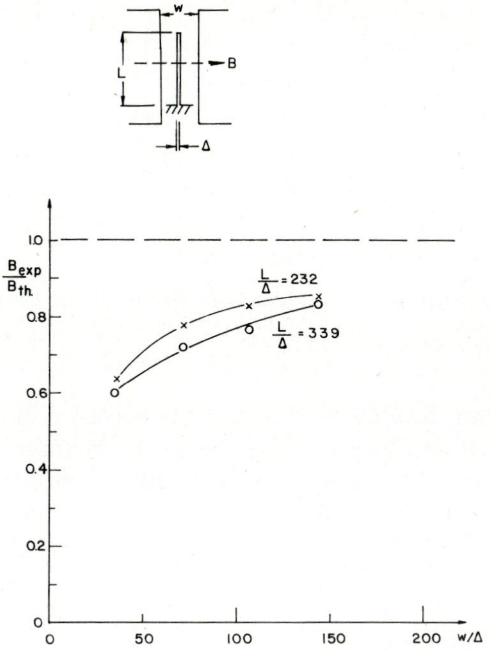

Fig. 9 Critical magnetic field for buckling of beam-plates versus width/thickness ratio (based on data from Yeh [37]).

material, and $L/2h$ ratio, the measured critical field approached the theoretical value; Fig. 9.

Thus the hope that a more logical development of the continuum theory would resolve the experiment and the theory was dimmed. It is paradoxical that in the limit where magnetic effects become nontrivial vis-à-vis the mechanics, i.e. $\mu_r \to \infty$, the difference between formulations seems to be washed out. Of course, Pao and Yeh did not use the magnetostrictive constants β_{ijkl}. However, in the experiments of Moon [30] and Yeh [37] different transformer-type plate materials seem to yield the same B_c vs. $2h/L$ curves for the same width to thickness ratio.

Another effort to resolve the discrepancies was the work of Wallerstein, Peach, and coworkers [38–40]. They adopted the basic Moon–Pao model but tried to find a more accurate magnetic field distribution by including the width effect. Based on numerical calculations Wallerstein and Peach [38] concluded that the actual field at the surface of the undeformed plate in a normal magnetic field was 86% higher than the applied field $\mathbf{B}_0$, where the

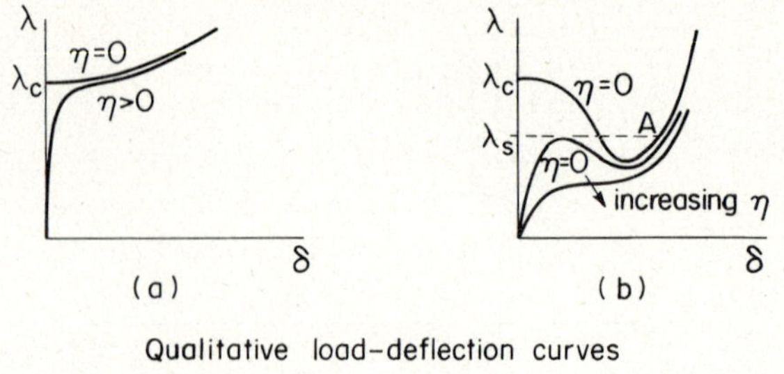

Fig. 10 (a) Imperfection insensitive buckling curves. (b) Imperfection sensitive buckling curves (from C. H. Popelar and C. O. Bast, "An Experimental Study of the Postbuckling Behavior of a Beam," *Exp. Mech.* **12**, No. 12 (1972)).

infinite plate theory admits no change in the normal field due to the plate. Dalrymple, Peach, and Viegelahn [39] have also determined experimentally that the buckling field increases as the width to thickness of the plates increased. They reason that for a finite plate, the flux near the edge is concentrated, thus raising the average flux across the plate above the external field which is assumed in the Moon–Pao model.

A third attempt to explain the difference between magnetoelastic buckling theory and experiments was the investigation of Popelar [41] into the sensitivity of the critical magnetic field to misalignment and initial curvature in the beam-plate. A post-buckling theory was proposed and experimental evidence presented by Popelar and Bast [42] to support the thesis that the instability was imperfection sensitive.

Illustration of imperfection sensitive buckling is shown in Fig. 10 where the loading factor, $\lambda = B_0^2$, and the ordinate is the tip beam deflection. Misalignment of the field or initial curvature is represented by η. Popelar's theory suggests that for η small the instability is a snap phenomenon and the measured critical field can be significantly lower than the critical value for $\eta = 0$. This seems to be borne out in experiments [42] in which they systematically varied the misalignment of the field (Fig. 11). This theory, which is based on nonlinear bending of the beam, predicts that the critical field is more sensitive to misalignment of the field with the plate normal than initial curvature of the beam.

3.3 Magnetoelastic Stability of Circular Rods

(a) *Theory.* It should be noted that the destabilizing effect of the magnetic fields occurs at values of the field common in many magnetic environments.

PHYSICAL PROPERTIES

Material	AL 4750
Relative Permeability at 200 gauss	15,000
Initial Permeability	3,500
Maximum Permeability	40,000 to 130,000
Induction at Maximum Permeability (gauss)	5,000
Saturation Induction (gauss)	15,500
Modulus of Elasticity (psi)	22 x 10^6

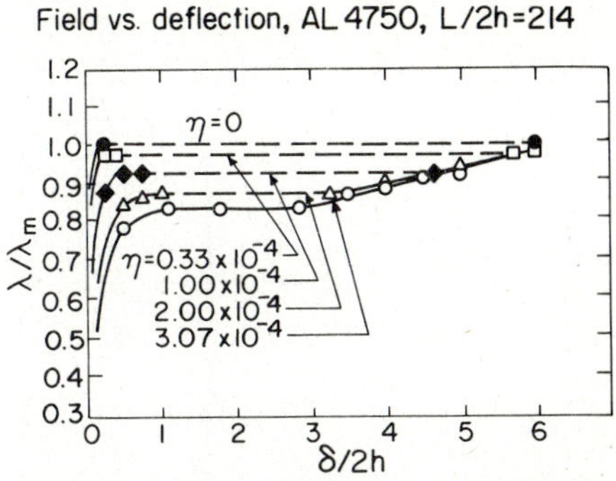

Fig. 11 Magnetic buckling field versus misalignment for a beam-plate in a transverse magnetic field (from C. H. Popelar and C. O. Bast, "An Experimental Study of the Postbuckling Behavior of a Beam," *Exp. Mech.* **12**, No. 12 (1972)).

Buckling of beams with $L/2h$ ratios of 200–300 occurred at around 300 gauss whereas fields in superconducting magnet environments for MHD or fusion reactors reach 5 Wb/m^2 (50,000 gauss). Also while the magnetic field may be below critical, it still may affect the natural frequencies of the structure.

With a view to application of these ideas to more complex structural elements, this writer has examined other shapes namely rods, plates, and shells in magnetic fields.

The problem of a ferroelastic rod in a transverse magnetic field, which will be discussed next, was first treated by this writer for the analog of a dielectric fiber in an electric field [43]. The analysis below follows Ref. [43] very closely. But whereas the buckling analysis of a dielectric fiber leads to a very high electric field—too high for easy laboratory tests, extension to a ferromagnetic rod leads to easily attainable magnetic fields—similar to the beam-plate experiments. Experimental evidence has been obtained for the instability of the rod of circular cross-section in a magnetic field.

While the problem appears similar to that of beam-plate analysis, there are two important differences. First the magnetic boundary conditions on the edges, which were a problem in the beam-plate analysis, are satisfied explicitly on the circular boundary of the rod. Second, the rod has two directions of bending: one along the field and the other transverse to the field. With zero magnetic field these two bending modes are degenerate, i.e. the frequencies are the same. The experiments show however that in the presence of a field, transverse to the rod, the modes are split with the *mode along the field decreasing in frequency* and the *mode normal to the field increasing slightly in frequency*.

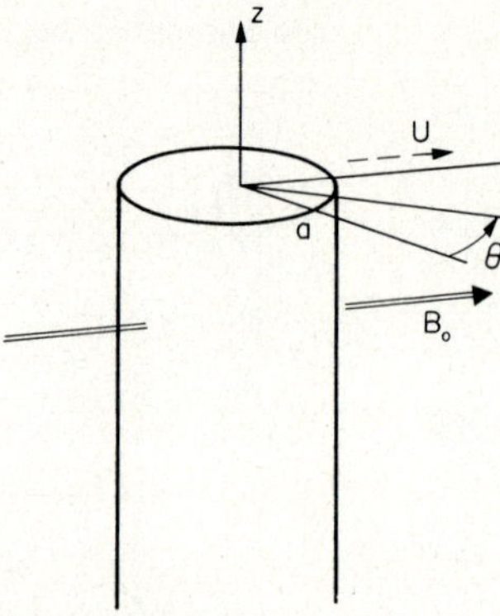

Fig. 12 Diagram of a circular ferroelastic rod in a transverse magnetic field.

Consider a circular elastic cylinder of radius a in a transverse magnetic field $\mathbf{B}_0$ (Fig. 12). If the cylinder is made of linear magnetic material, the magnetic field due to the presence of the cylinder in the otherwise uniform field is given by

$$B_r^0 = B_0 \sin\theta[1 + \beta a^2/r^2],$$
$$B_\theta^0 = B_0 \cos\theta[1 - \beta a^2/r^2], \qquad r \geq a,$$
$$\mathbf{B}^0 = \frac{2\mu_r}{(\mu_r + 1)} B_0 \mathbf{e}_y, \qquad r < a,$$

where

$$\beta = (\mu_r - 1)/(\mu_r + 1).$$

Now suppose the rod is subject to a sinusoidal displacement of its axis in the direction of the applied field, i.e. the center displacement is given by

$$\mathbf{u} = u\mathbf{e}_y = u_0\, e^{ikz}\,(\cos\theta\mathbf{e}_\theta + \sin\theta\mathbf{e}_r).$$

This displacement will cause the rod surface to rotate thereby inducing a magnetization vector $\mathbf{M}$ with a component along the axis. Using the dipole model for magnetic forces, we observe that the magnetic body force is zero while a distributed body couple will act on the rod; this body couple is given by

$$\mathbf{C} = C\mathbf{e}_x = \mu_0 \int_0^a \int_0^{2\pi} \mathbf{M} \times \mathbf{H}_0 r\, dr\, d\theta, \qquad (3.16)$$

where $\mathbf{H}_0 = B_0\mathbf{e}_y/\mu_0$, and $\mathbf{M} = \chi\mathbf{H} = \chi(\mathbf{H}^0 + \mathbf{H}_1)$. Again neglecting magnetostrictive effects, and self-forces between one part of the bar and another, we use the classical bending theory of rods with a body couple

$$YJ\frac{d^4u}{dz^4} - \frac{dC}{dz} = 0, \tag{3.17}$$

where Y is the Young's modulus and J the moment of inertia.

Since this is a magnetostatic problem with zero current, one finds the perturbed magnetic field $\mathbf{H}_1$ using potential theory and standard perturbation techniques, i.e.

$$\mathbf{H}_1 = \nabla\phi, \qquad \nabla^2\phi = \mathbf{0},$$
$$\phi = \Phi(r, \theta)\, e^{ikz},$$

so that

$$\nabla^2\Phi - k^2\Phi = 0. \tag{3.18}$$

We obtain the boundary or jump conditions on Φ, applied at the deformed surface (indicated by [ ]*), by expanding the fields in a Taylor series. Thus $[\mathbf{B}\cdot\mathbf{n}]^* = 0$ becomes

$$[\mathbf{B}_1\cdot\mathbf{n} + (\mathbf{u}\cdot\nabla)\mathbf{B}^0\cdot\mathbf{n}] = 0.$$

Also $[\mathbf{H}\times\mathbf{n}]^* = \mathbf{0}$ leads to $[\phi]^* = 0$ which becomes

$$[\phi] + [\mathbf{u}\cdot\mathbf{H}^0] = 0,$$

where the brackets [ ] (without the superposed star) indicate jumps at the undeformed surface.

These conditions lead to

$$\left[\mu_r\frac{\partial\Phi}{\partial r}\right] + \frac{2\beta U_0 B_0}{a}\cos 2\theta = 0,$$
$$[\Phi] + \tfrac{1}{2}U_0\beta H_0(1 - \cos 2\theta) = 0. \tag{3.19}$$

Solutions to the Helmholtz equation (3.18) may be written in terms of modified Bessel functions, and take the form

$r < a$

$$\Phi = A_1 I_0(kr) + A_2 I_2(kr)\cos 2\theta,$$

$r > a$

$$\Phi = B_1 K_0(kr) + B_2 K_2(kr)\cos 2\theta.$$

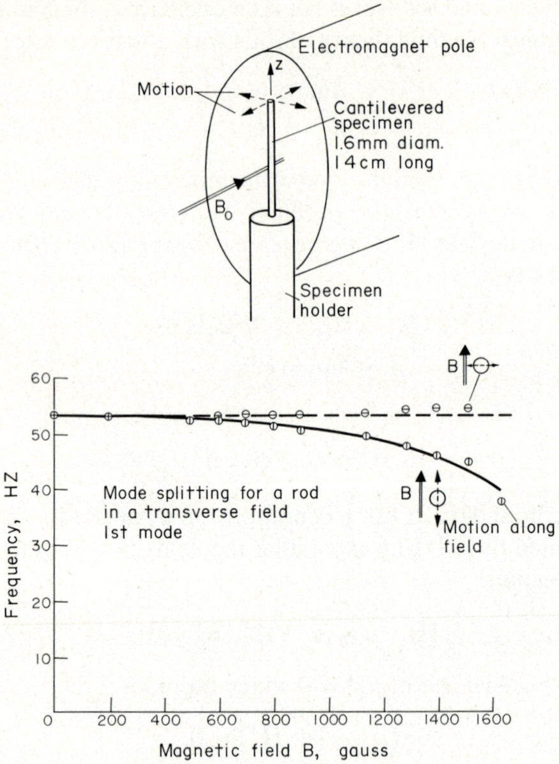

Fig. 13 Natural frequency of a ferroelastic rod of circular cross-section in a transverse magnetic field (from F. C. Moon, "Mechanics of Structures with High Electric Currents," Princeton University Report AMS 1111 (July 1973)).

The four constants are determined from the boundary conditions (3.19), but the couple **C** only depends on A_1, i.e.

$$C = -2\pi a \chi B_0 \, e^{ikz} \, A_1 i I_1(ka).$$

where

$$A_1 = \frac{U_0 H_0 \beta K_0'(ka)}{(I_0 K_0' - \mu_r K_0 I_0')}.$$

The Bessel functions are evaluated at $r = a$ and primes indicate differentiation.

For long wavelength disturbance, $ka \ll 1$, the critical buckling field is given by

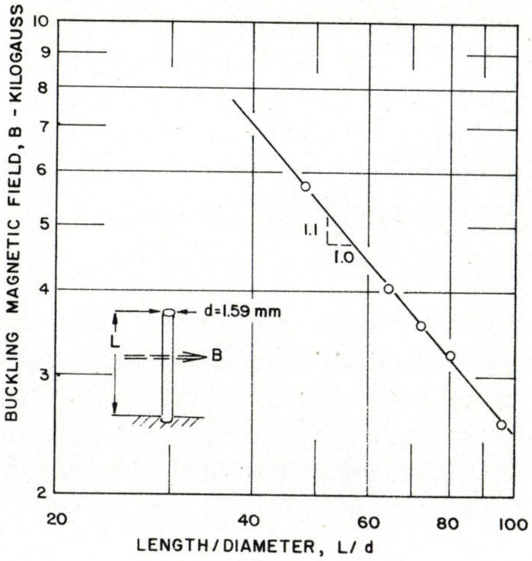

Fig. 14 Magnetic buckling field versus length/diameter ratio for a cold rolled steel rod in a transverse magnetic field.

$$\left(\frac{B_0^2}{2\mu_0}\right) = \frac{(\mu_r + 1)}{\chi^2}\frac{r}{2}(ka)^2[1 + \mu_r(ka)^2\,|\ln ka|]. \tag{3.20}$$

For vibrations in the direction of the magnetic field, the analysis, neglecting induced currents, leads to a decrease in natural frequency with magnetic field for a periodically pinned rod, i.e. $\omega^2 = \omega_0^2(1 - B^2/B_c^2)$.

(b) *Experiments: Circular Rod in a Magnetic Field.* A few experiments have been performed on a rod in a magnetic field. Using a 17.8-cm diameter magnet with a 2.5-cm gap, a 1.59-mm diameter cold rolled-steel rod was tested using various lengths. Vibration tests were performed and show the predicted decrease in frequency with field (Fig. 13) for vibrations in the direction of the field. However, for vibrations normal to the initial field lines slight increases in frequency were found. The slopes of the ω^2 vs. B^2 curves give the critical magnetic field values, which are shown in Fig. 14 for a few length to diameter ratios. The slope of this curve on log paper indicates a dependence on (L/a) to a power greater than one. This is reasonable since the log term in (3.20) is of order unity for these experiments and $\mu_r(ka)^2 \leq 1$.

A hollow tube of similar dimensions was also tested and showed the same behavior as the solid rod.

It seems reasonable to expect that more complicated structural shaped columns such as I-beams, U, or angle cross-sections will exhibit buckling in a transverse magnetic field environment. From an engineering design point of view then, the effective bending stiffness of ferromagnetic beams in transverse magnetic fields might be lowered by the fields and should be checked when applications call for such configurations.

(c) *Post-buckling of a Circular Rod—Energy Method.* An interesting limiting case of magnetoelastic buckling is the high length to diameter limit where the magnetization becomes proportional to the rotation or slope of the rod axes. If Θ is the angle the tangent to the rod axis makes with the undeformed rod axis, the magnetization in the rod has the following form for bending in the field direction (magnetic field initially normal to the rod axis):

$$M_1 \sim \cos \Theta,$$
$$M_{11} \sim \sin \Theta,$$

where M_1 is the component normal to the rod axis. The magnetic energy of the rod (neglecting self energies) is

$$W = -\frac{1}{2} \int \mathbf{M} \cdot \mathbf{B}_0 \, dv.$$

Then the equation of equilibrium of the deformed rod can be obtained from the variational condition $\delta U = 0$, where the functional U is given by

$$U = \int_0^L \left[\frac{YJ}{2} \left(\frac{d\Theta}{dS} \right)^2 - \frac{\pi a^2}{2} \mathbf{M} \cdot \mathbf{B}_0 \right] dS; \tag{3.21}$$

in which S measures length along the rod. The resulting linearized equation is identical to (3.17) and the nonlinear equation has the form

$$\frac{d^2\Theta}{dS^2} + \alpha^2 \sin 2\Theta = 0, \tag{3.22}$$

where

$$\alpha^2 = \frac{B_0^2}{2\mu_0} \frac{\pi a^2}{YJ} \mu_r.$$

This equation is similar to the equation of the elastica; see, for example, Love [44].

3.4 Plate Vibrations in a Magnetic Field

A natural extension of the one-dimensional bending of a beam in a magnetic field is to consider the effects on a two-dimensional plate. It is often said there is nothing much new under the sun and this is true also of plate vibrations in magnetic fields. The problem was studied first by Matteucci [45] in 1845, who was a friend of Faraday. He excited vibrations in square plates of glass, brass, and iron when they were placed in a magnetic field. He reported, however, no effect on the vibration or Chladni patterns.

It was shown by Moon and Pao [32] that the vibration and stability problem were closely connected, in that the frequency decreased as the magnetic field increased:

$$\omega^2 = \omega_0^2(1 - B^2/B_c^2). \tag{3.23}$$

By measuring the slope of this relation, one could experimentally determine B_c without buckling the plate. This in fact was done in [32] and also by Yeh [37] who found the dynamic B_c to be about 2–10% above the value found by the static buckling tests.

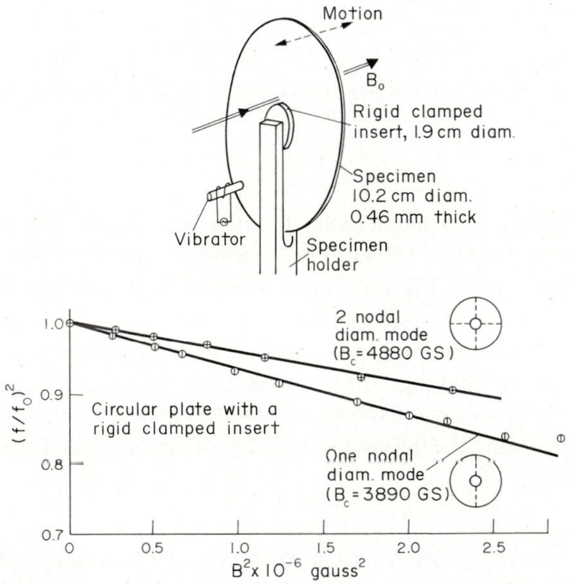

Fig. 15 Natural frequency of circular cold rolled steel plates in a transverse magnetic field (from F. C. Moon, "Mechanics of Structures with High Electric Currents," Princeton University Report 1111 (July 1973)).

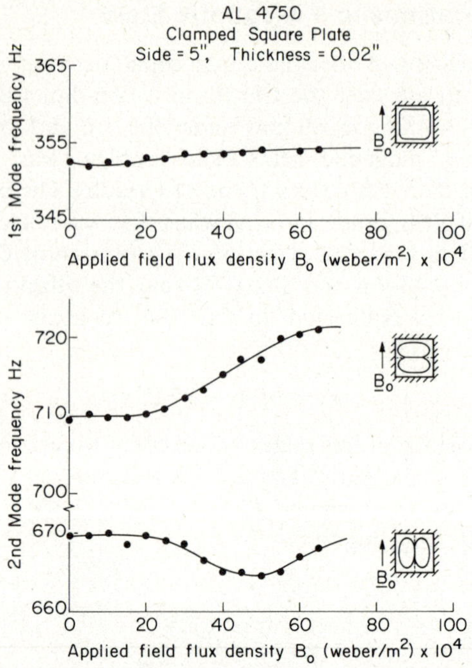

Fig. 16 Natural frequencies for various modes of a square steel plate in an inplane magnetic field (from S. Srinivasan, "Vibration and Stability of Beams and Plates in the Presence of Electromagnetic Fields," Ph.D. Thesis, Ohio State University (1970)).

A two-dimensional magnetoelastic buckling theory is more difficult because the sinusoidal deformation assumption is not useful, as for the case of circular plates. This writer, however, extended the beam-plate theory to two-dimensional plates in oblique magnetic fields using the Fourier transform. It was found that the magnetic field has to be close to the plate normal to achieve buckling and a decrease in natural frequency with field. For fields at an angle with the plate normal, greater than a characteristic value for each mode, the frequency increases with magnetic field for a simply supported plate [33].

(a) *Two-dimensional Plates Experiments.* As often happens in mechanics, boundary conditions which simplify the theory are not easy to obtain experimentally and vice versa. This writer tested annular steel plates, clamped at the center and free at the outer edges (Fig. 15). This is the analog to the cantilevered beam, and allowed the deflection of the plate to be monitored. Thus the two-dimensional theory could not be tested quantitatively. The test results do confirm the fact that the circular plate can buckle in a transverse magnetic field. This was evidenced by the decrease in frequency of the mode with one modal diameter, which appears to have the lowest critical field; see Fig. 15.

These results suggest an application to sound-producing devices using plates as pistons. The mechanical impedance of the plate can be changed by varying an applied static magnetic field in the vicinity of the plate.

Thus for more complex ferromagnetic structures an applied DC magnetic field may change the effective stiffness. An application of this idea has been applied to an ultrasonic generator by Birr and Koryu Ishii [46]. A transformer core is excited by a coil and emits ultrasound in the 20-kHz range. The frequency of this ultrasound was found to change by as much as 100% by application of a small static magnetic field of the order of 200 gauss.

Srinivasan [47] in a dissertation studied the vibrations of a ferroelastic beam in a *magnetic field parallel to the plane of the plate.* He found that the magnetic field *increased the natural frequencies* in contrast to the normal field case; Fig. 16.

4 MECHANICS OF ELASTIC CONDUCTORS

4.1 Introduction

As described in the introduction the flow of electric current can create large forces and stresses in the solids that contain them. To appreciate the order of magnitude of these forces one need only examine the quantity $\mu_0 I^2$, which has dimensions of force, in the following table:

I-Amperes	$F = \mu_0 I^2$ Newtons	(lbf)	Device
1	$1.25\ 10^{-6}$	$(0.28\ 10^{-6})$	Low-power electronics
10^2	$1.25\ 10^{-2}$	$(0.28\ 10^{-2})$	Power motors
10^3	1.25	(0.28)	Electric welders
10^4	125	(28)	High-power switches Lightning
10^5	12,500	(2800)	Magnets for levitating trains
10^6	$12.5\ 10^5$		Magnets for fusion rectors

In many problems in which stresses due to currents must be calculated, the determination of the magnetic fields and currents is independent of the deformation and motion of the solid. In such problems the current–field constitutive relation (Ohms law) is independent of stress and the stress–strain relations are assumed to be decoupled from the magnetic field or current. These include nonferromagnetic materials where the motion of the solid is small. When the motion of the solid is appreciable then coupling can occur in Ohm's law through the relation

$$\mathbf{J} = \sigma(\mathbf{E} + \mathbf{v} \times \mathbf{B}),$$

where $\mathbf{v}$ is the local velocity of the solid.

To solve such uncoupled problems one calculates the applied body forces $\mathbf{J} \times \mathbf{B}$ and then determines the stresses using conventional stress analysis. In normal conductors the temperature distribution in the structure must also be calculated due to the Joule heating which is proportional to $\mathbf{J} \cdot \mathbf{J}/\sigma$. In some problems thermoelastic stresses due to $\mathbf{J} \cdot \mathbf{J}$ may in fact be more important than those due to the direct magnetic forces.

While such decoupled problems seem straightforward from a conceptual point of view, exact solutions of stress and current fields for particular problems are few. A simple example is a circular wire with a uniform current distribution across the radius. The magnetic field is circumferential and linear across the conductor producing a radial body force and a stress state [48] similar to that in a rotating disc, i.e.

$$\begin{aligned} \sigma_r &= -\frac{1}{16\pi^2}\frac{\mu_0 I_0^2}{R^2}\left(1 - \frac{r^2}{R^2}\right)\frac{(3-2\nu)}{(1-\nu)}, \\ \sigma_\theta &= -\frac{1}{16\pi^2}\frac{\mu_0 I_0^2}{R^2}\left(1 - \frac{1+2\nu}{(3-2\nu)}\right)\frac{(3-2\nu)}{(1-\nu)} \end{aligned} \tag{4.1}$$

where I_0 is the total current, ν is Poissons ratio, R is the radius of the wire and the stress along the wire is assumed to be zero. For a superconducting wire of 1 mm diameter carrying 10^3 A current, the stress at the center is $\sigma_r = \sigma_\theta \simeq 10\ \text{N/cm}^2$. Thus while an exact solution can be found for this problem, in practice the stresses produced are not very high.

In addition to the magnetic body forces, the current will produce heat and hence thermoelastic stresses. Yuan [48] has treated these effects for a circular rod.

4.2 Continuum Models

If ferroelastic materials can be thought of as the interaction between an elastic continuum and a spin or magnetization continuum, then problems of elastic conductors can be seen as the interaction between an elastic continuum and an electron gas continuum. While this view of current carrying solids may be necessary to understand certain high-frequency phenomenon such as laser–solid interactions at a surface, the relaxation time for relative displacements of metal actions and their electron gas is so short that forces acting on the free electrons can for most problems be considered to act directly on the elastic or solid continuum.

A number of continuum or phenomenological theories of the two continuum models have been presented in the literature. In the field of solid-state theory interactions of the current and deformation are treated from the viewpoint of phonons and a quantized electron gas; see, for example, [49]. A treatment closer to classical continuum theories is that illustrated in Steele and Vural [50].

This writer has presented a nonlinear theory of an elastic continuum and an electron gas [51, 52]. This model uses a single temperature mixture theory of Green and Naghdi [53]. Demiray [54] in a similar treatment uses a multi-temperature mixture theory of Eringen. Eringen also has applied his micromorphic or polar continuum models to superconducting solids [55].

In an attempt to model the superconducting–normal state transition, Tiersten [56] has presented a two-fluid continuum theory consisting of normal conducting gas and a superconducting electron gas.

Pipkin and Rivlin [57] have used a single continua model to formulate the constitutive relations for electrical conductivity in deformed solids. In the linearized form the generalized Ohm's law for isotropic material is

$$\mathbf{J} = \left[(S_1 + S_2 Tr\varepsilon + S_3 \mathbf{E} \cdot \varepsilon \cdot \mathbf{E})\mathbf{I} + S_4 \varepsilon\right] \cdot \mathbf{E},$$

where ε is the usual strain tensor and $\mathbf{I}$ the identity tensor. In two other papers [58, 59] they also investigate steady current flow in rigid rods for materials which obey a more general Ohm's law of the form $\mathbf{J} = \mathbf{f}(\mathbf{E}, \mathbf{B})$.

4.3 Stresses in High Current Magnets and Coils

One of the major applications of magneto-solid mechanics is in the design of high current–high magnetic field devices. Such devices, originally designed for experimental research in physics, now find potential application in MHD machines, inductive storage devices, and magnets for fusion reactors.

The basic problem is shown in Fig. 17. Currents are made to flow in cylindrical paths creating axial and radial components of the magnetic field. The current-carrying elements thus have radial and axial body forces which

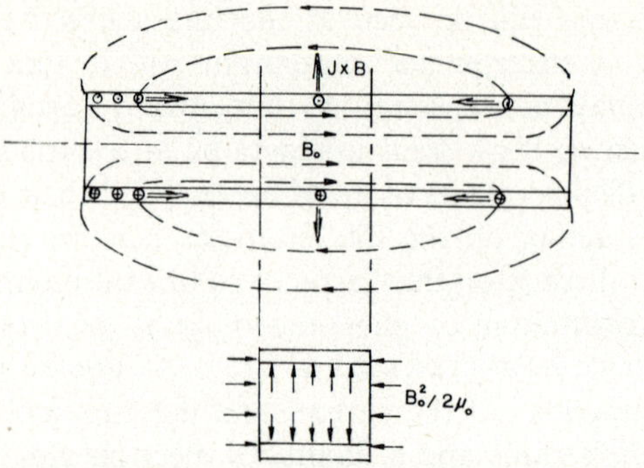

Fig. 17 Magnetic forces and stresses in a long thin solenoidal coil.

must be balanced by internal stresses in the conducting elements themselves or by external reinforcing structures.

In the limit of a long thin-walled solenoid (Fig. 17) the magnetic forces can be approximated in the middle section by an internal pressure creating tensile hoop stresses and axial compressive forces proportional to the square of the mean magnetic field (see, for example, [61]),

$$t_{\theta\theta} = \frac{B_0^2}{2\mu_0}\frac{R}{\Delta},$$
$$t_{zz} = -\tfrac{1}{2}t_{\theta\theta}, \tag{4.2}$$

where R/Δ is the radius to thickness ratio of the shell.

In the limit of a single turn of wire the average circumferential stresses are given by [62] (Fig. 18),

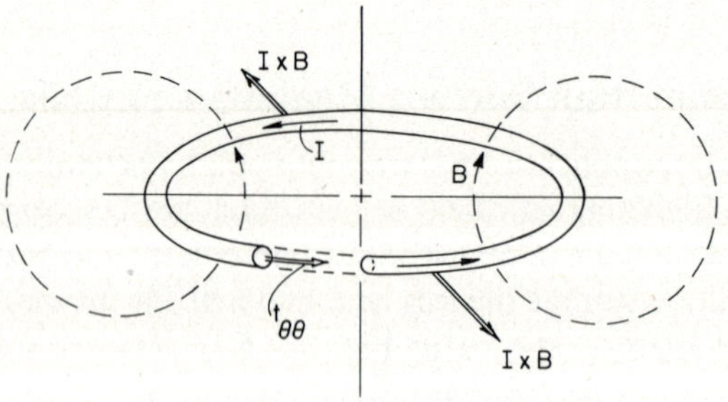

Fig. 18 Magnetic forces and stresses in a circular ring magnet.

$$t_{\theta\theta} = \frac{\mu_0 J_0^2 a^2}{4}\left[\ln\left(\frac{8R}{a}\right) - \frac{3}{4}\right], \tag{4.3}$$

where J_0 is the current density in the coil, R is the major radius and "a" the radius of the circular cross-section. One can observe that if J_0 is fixed, then the circumferential stress increases with cross-sectional area of the coil. Thus for a superconducting coil with $J_0 = 3 \times 10^4$ amps/cm^2, $R = 1$ meter, $a = 2$ cm, the stress is about 6 10^3 N/cm^2 (8700 psi).

It is the intermediate cases that present the most difficulty. An early analysis was given by Cockroff [63] in 1928. He used thick cylinder elasticity theory in which the body forces were calculated using the inductance—or energy method, i.e.,

$$\begin{aligned} \frac{\partial}{\partial r}(rt_{rr}) - t_{\theta\theta} = f_r = \frac{J^2}{2\pi D}\frac{\partial M}{\partial r}, \\ \frac{\partial t_{zz}}{\partial z} = f_z = \frac{J^2}{2\pi Dr}\frac{\partial M}{\partial z}, \end{aligned} \tag{4.4}$$

where J is the current density, D is the number of turns per cross-section, and M is the mutual induction between the whole coil and a single turn. While the method is clever, the calculation of M is tedious.

High field solenoid magnets of the steady and pulsed field type have been analyzed using different models, some of which are discussed by Brechna [60] and Montgomery [61] in monographs on magnet design. Middleton and Trowbridge [64] take a particular magnet and use three different models for analysis. One model uses a homogeneous isotropic cylinder analysis (Fig. 19) found in elementary elasticity texts, the second model places spacers between each turn and uses a series of concentric shells. The third incorporates the helical nature of the windings of the magnets, accounting for winding friction.

Kilb and Westendorp [65] tried to use a composite type analysis to account for the reinforcing steel bands between the copper conducting strips. They use an effective modulus calculated from a mixture rule

$$E_m = f_c E_c + f_s E_s,$$

where f_c, f_s are the volume fractions of copper and steel respectively. However, their model uses E_m in an isotropic elasticity analysis. While E_m might be valid for an effective circumferential modulus, the radial modulus should be different, i.e.

$$\frac{1}{E_m'} = \frac{f_c}{E_c} + \frac{f_s}{E_s},$$

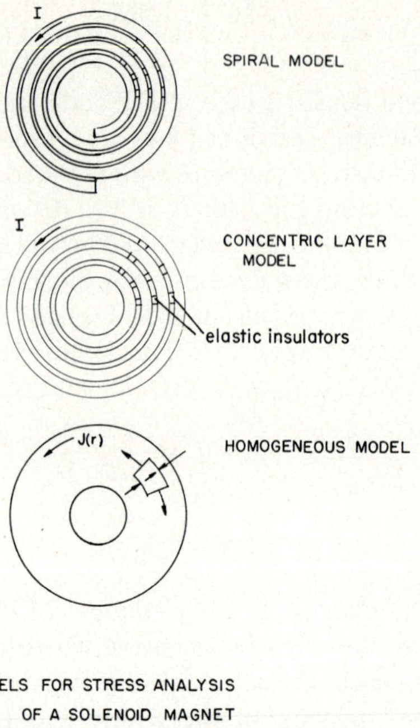

Fig. 19 Models for stress analysis of a solenoid magnet.

and an anisotropic elastic model should be used. In another paper Westendorp considers the effect of initial winding stresses [66].

4.4 Virial Theorem and Force-free Magnets

In an attempt to produce high magnetic energy systems with low mechanical stresses the concept of the force-free magnet, or force-reduced magnet, has been investigated. Since the body forces are $\mathbf{J} \times \mathbf{B}$ a force-free magnet is one in which the current is wound such that

$$\mathbf{J} = \alpha \mathbf{B} \tag{4.5}$$

so that

$$\nabla \times \mathbf{B} = \mu_0 \alpha \mathbf{B}. \tag{4.6}$$

Such a concept has been discussed by Furth *et al.* [68] and Wakefield [69]. For a solenoid the forces can be reduced by winding the currents in a helical path on the cylinder. Wells and Mills [70] experimented with force-reduced magnets in the shape of toroids.

While one can reduce forces in one part of a magnetic system by choosing windings which satisfy (4.5, 6), one cannot eliminate all the forces in an isolated magnetic system. This follows from a theorem similar to the Virial theorem in classical mechanics; see, for example, Goldstein [71]. One considers the equilibrium of a solid under magnetic body forces and examines the integral

$$\int (\nabla \cdot \mathbf{t} + \mathbf{J} \times \mathbf{B}) \cdot \mathbf{r}\, dv = 0. \tag{4.7}$$

The body force is replaced by $\nabla \cdot \mathbf{T}^m$ where

$$\mu_0 T_{ij}^m = B_i B_j - \tfrac{1}{2} B^2 \delta_{ij}. \tag{4.8}$$

Application of the divergence theorem and the fact that for an isolated magnetic system

$$\lim_{\mathbf{r} \to \infty} \int \mathbf{r} \cdot \mathbf{T}^m \cdot \mathbf{n}\, da \to 0$$

one can show that

$$W_m \equiv \int_{\text{space}} \frac{B^2}{2\mu_0}\, dv = \int \mathrm{Tr}(\mathbf{t})\, dv, \tag{4.9}$$

where $\mathrm{Tr}(\mathbf{t})$ is the trace of the stress tensor $\mathbf{t}$.

This idea was discussed by Parker [72] and Levy [73]. In the latter the theorem is used to choose a lower bound on the structural mass required to sustain stresses in a superconducting magnet with energy W_m in connection with possible space applications.

The theorem being quite general states that while some parts of the system might be designed to be stress free, other parts of the structure must have the product of structural volume and trace of the stress tensor proportional to the stored magnetic energy. Thus, as has been suggested by Wakefield [69], the toroidal magnets for a fusion reactor might be force-free or force reduced with a large ring outside of the reactor area taking up the stress (Fig. 20).

4.5 Superconducting Magnets

With the recent development of high field high current density superconductors, high ampere-turn magnetic devices have been built for levitation

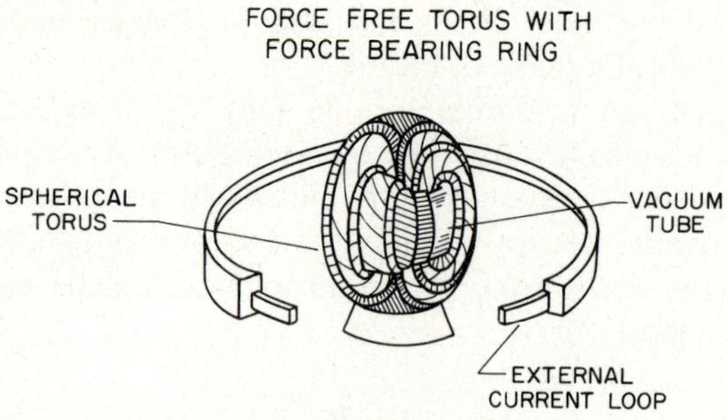

Fig. 20 Force free torus with a force bearing ring (from K. E. Wakefield, *Design of Force-free Toroidal Magnets*, Princeton University, N.J. (Mar. 1964)).

of vehicles, bubble chambers, power transmission lines, and rotating machinery [74]. However, even larger magnets are planned for proposed fusion reactors with a dozen or more 6-meter diameter magnets, each carrying up to 7×10^6 ampere-turns of currents. In such large devices structural over-design must be at a minimum. Using a system analysis on the material constraints on the design of fusion reactors of the toroidal type, DeMichele and Darby [75] conclude that the ability of the structures to support the magnetic forces is a primary constraint.

Superconducting magnets are usually wound from wire or tape made from material that is only superconducting within a volume in temperature–magnetic field–current density space [76]. In practice this means operating well below critical temperatures, usually in liquid helium (4°K). Even at

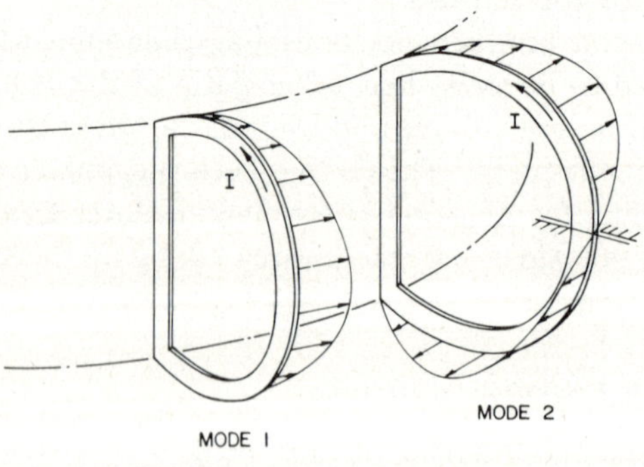

Fig. 21 Lateral bending modes of toroidal field coils for a Tokamak fusion reactor.

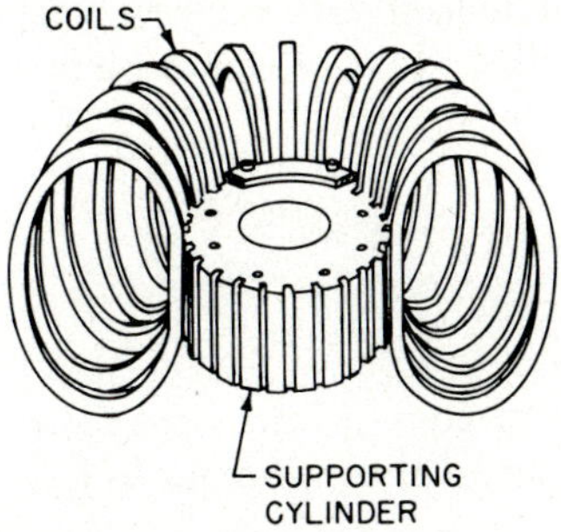

Fig. 22 Partial set of toroidal field coils for fusion reactor.

this temperature the local magnetic field and current densities cannot exceed certain critical values, or that part of the coil will become resistive—creating heat, evaporating the helium, and propagating the normal region to the rest of the coil; see, for example, [77]. Relative motion of the windings due to vibration can create heat and "quench" the coil into the normal state [78]. Successful designs are then somewhat an art based on experience as well as analysis.

A typical coil for a fusion reactor is shown in Fig. 21. Stress analysis of these large magnetic coils has been based on numerous models including a string model (no bending) [79], curved beams [80], and isotropic elasticity analyses as well as finite element methods [81].

Most designs assume that the planar configuration of the coils will remain stable for all operating currents. This assumption has been challenged by this writer and coworkers [82–84].

In the toroidal magnetic confinement fusion reactor the planar coils are arranged in a torus (Fig. 22). The mutual circumferential forces between coils cancel when the coils are symmetrically spaced but create inward radial or centering forces on the inner sides of the magnets. In addition large tension stresses along the circumferential direction of each magnet arc created by the $\mathbf{J} \times \mathbf{B}$ body forces normal to the arc of the coil.

Using a flexible arc model, File, Mills, and Sheffield [79] suggested a "D"-shaped pure tension coil similar to that in Fig. 21. If the current follows the arc, the body force $\mathbf{I} \times \mathbf{B}$ is normal to the arc, and the tension T is proportonal to the radius of curvature ρ,

$$T = IB\rho.$$

Since the toroidal magnetic field varies inversely as the distance r from the center axis of the torus, the radius of curvature is chosen to vary with r so that T remains constant over the arc, i.e.

$$B = \frac{\mu_0 NI}{2\pi r},$$

$$\rho = kr,$$

where N is the number of coils in the torus. The resulting curve for the magnet arc is given in Cartesian coordinates by

$$x = \int_1^y \frac{\pm \ln y \, dy}{(k^2 - \ln^2 y)^{1/2}}, \tag{4.10}$$

where if r_1, r_2, are inner and outer radii of the magnet arc respectively, then k is given by

$$k = \tfrac{1}{2} \ln (r_2/r_1)$$

and the tension is

$$T = \mu_0 I^2 \frac{N}{4\pi} \ln (r_2/r_1). \tag{4.11}$$

The problem with this model is that the finite cross-section of the coil results in different radii of curvature and magnetic field across the coil. Thus the tension will vary across the cross-section and bending will result.

Other "D"-shaped designs have been presented by the University of Wisconsin Fusion Feasibility Group [81] who have used a finite element model of a coil with alternate layers of stabilized superconductor and stainless-steel reinforcing band for the magnet windings.

Another fusion reactor design by the Oak Ridge National Laboratory [85] employs twenty-four circular coils, wound on a 4.5-m diameter bore and each carrying 3.8×10^6 A-turns per coil.

These three designs have in common the fact that each coil is surrounded by a helium dewar and the magnetic forces are resisted by cold structure. Any lateral constraints on the coils to prevent bending out of their planes must be kept small to minimize heat leaks to the warm structure outside the magnets. With a small number of lateral supports on each coil the coils will be susceptible to out-of-plane bending and possible out-of-plane buckling due to self-forces and mutual forces between coils.

To avoid these problems Powell and Bezler [86] have proposed transmission of the magnetic forces on the cold superconductor to a warm

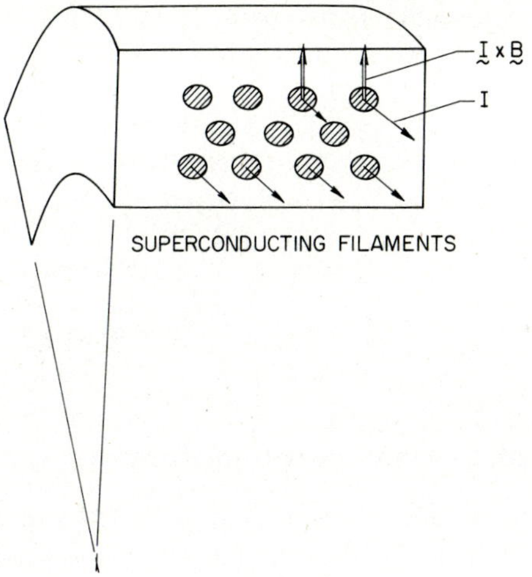

Fig. 23 Cross section of a filamentary superconductor showing stress concentration areas.

structure through insulating material. Their stress analysis of this problem again uses a given body force distribution in a finite element code.

DeMichele and Darby [87] have used a three-dimensional finite element stress code to calculate stresses in the Oak Ridge circular coil [85] and the Princeton Plasma Physics Lab "D" coil [79] under operating conditions as well as a "quench" condition in one coil.

A problem that has not been reported involves the local stresses in the superconducting wire or tape itself. Most of these analyses using finite elements obtain resultant forces at nodes averaged over areas much larger than the wire cross-section. However, in a filamentary superconductor the body forces are applied to the filaments, which are embedded in a normal matrix such as copper. This should represent a classic stress concentration problem (Fig. 23). Also the composite nature of these materials has not been fully studied or characterized from a mechanics viewpoint.

Another problem which has received very little attention is the diamagnetic forces on the superconductor. These materials partially exclude the magnetic flux from the superconductor thereby locally increasing the flux in one area of the magnets above the average calculated by neglecting this effect. Stevenson and Atherton [88] have made a heuristic analysis of this effect and attribute its neglect to the structure failure of one of their magnets. (See also [89].)

Fig. 24 Kink instability in a linear conductor.

4.6 Stability of Current-carrying Rods

Mechanicians have long been familiar with the fire-hose instability in which fluid flowing through a flexible tube can cause transverse oscillations of the hose. One is prompted then to ask whether the analog of current flowing through a conductor can produce such dynamic instability in the solid. Theoretical analysis leads to an affirmative answer and there is also some experimental evidence to back this up.

In the following only static instabilities or buckling are discussed, and one may ask whether flutter or dynamic instabilities can also occur. Suspicions of flutter-type instabilities in current-carrying systems might be reinforced by the observation that when the current is forced through a solid the forces $\mathbf{J} \times \mathbf{B}$ follow the motion in part as $\mathbf{J}$ conforms to the moving solid boundaries. However, for velocity independent current distributions a magnetic energy $W_m(\mathbf{J}, \mathbf{r})$ can be written which depends only on the positions of the currents. If the forces can be derived from W_m, then the system is conservative and only buckling can occur. This conclusion does not hold when induced or eddy currents flow in a solid.

In plasma physics a current-carrying arc is known to have a lateral "kink" instability [90]; Fig. 24. An elementary calculation of the forces on an initially straight, current-carrying cylinder when it is displaced laterally in a sinusoidal displacement $w(x) = w_0 \sin kx$, suggests a force proportional to the curvature [90],

$$F \sim \frac{\mu_0 I^2}{4\pi} \frac{\partial^2 w}{dx^2} \ln (kr_0) + 0[(ka)^2], \tag{4.12}$$

where r_0 is the radius of the conductor.

An early treatment of this problem for a flexible conductor was given by Leontovich and Shafranov [91] in 1961 who looked at the combined effect of the self-field of the current and a longitudinal magnetic field. Dolbin [92] extended this to the analysis of waves in an elastic cylinder and Dolbin and Morozov [93] examined the particular case of bending vibrations in a current-carrying rod. They concluded that the rod could buckle under the perturbed forces generated by the currents in the rod above. In these papers the current is assumed to flow on the outside of the rod. The perturbed magnetic field external to the rod is calculated using magnetostatic field theory and standard perturbation techniques. Prudnikov [94] has extended Dolbin's analysis to include an initial magnetic field along the rod.

In a more recent paper Chattopadhyay and Moon [95] have reexamined the problem and have presented experimental evidence for the destabilizing effect of currents in a rod. In their analysis the current is assumed to be uniformly distributed across a circular rod. Magnetostatic theory is again used with boundary conditions that the current near the surface be tangential to the deformed rod surface. Thus perturbed currents and fields in the rod, as well as outside the rod, are calculated. The bending vibrations of the rod must satisfy (see Fig. 25)

$$\rho A \frac{\partial^2 w}{\partial t^2} + YJ \frac{\partial^4 w}{\partial z^4} = F_y + \frac{\partial C_x}{\partial z}.$$

The transverse magnetic body force is calculated from

$$d\mathbf{F} = \iint \mathbf{J} \times \mathbf{B}\, da\, dz.$$

In this theory the integrated force across the rod cross section yields a magnetic couple distribution C_x as well as a force F_y. For sinusoidal deformation $w_0\, e^{ikx}$, in the long wavelength limit, $2\pi/k \gg a$,

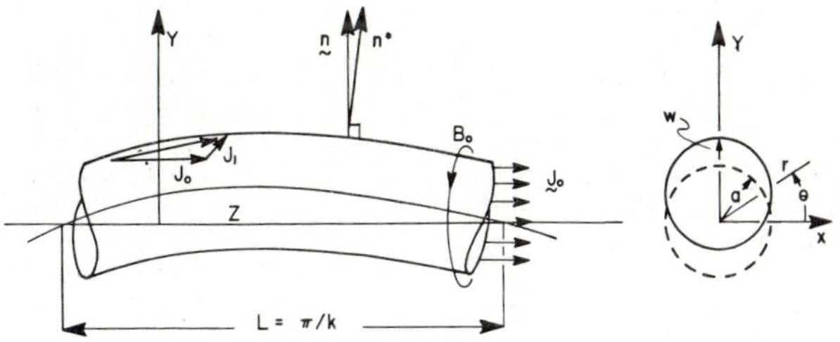

Fig. 25 Element of a bent circular wire carrying electric current.

$$C_x = \frac{\mu_0 I^2}{8\pi} ikw, \tag{4.13}$$

$$F_y = \frac{\mu_0 I^2}{4\pi} wk^2 \left[\ln \frac{2}{ka} - C - \tfrac{1}{4} \right], \tag{4.14}$$

where $C = 0.577$ is Euler's constant.

The theory predicts that the natural bending frequencies will decrease with current, i.e.

$$\omega^2 = \omega_0^2(1 - I^2/I_c^2). \tag{4.15}$$

The critical buckling current is given by

$$\mu_0 I_c^2 = \frac{4\pi(ka)^2 YJ/a^2}{[\ln(2/ka) - 1.327]}. \tag{4.16}$$

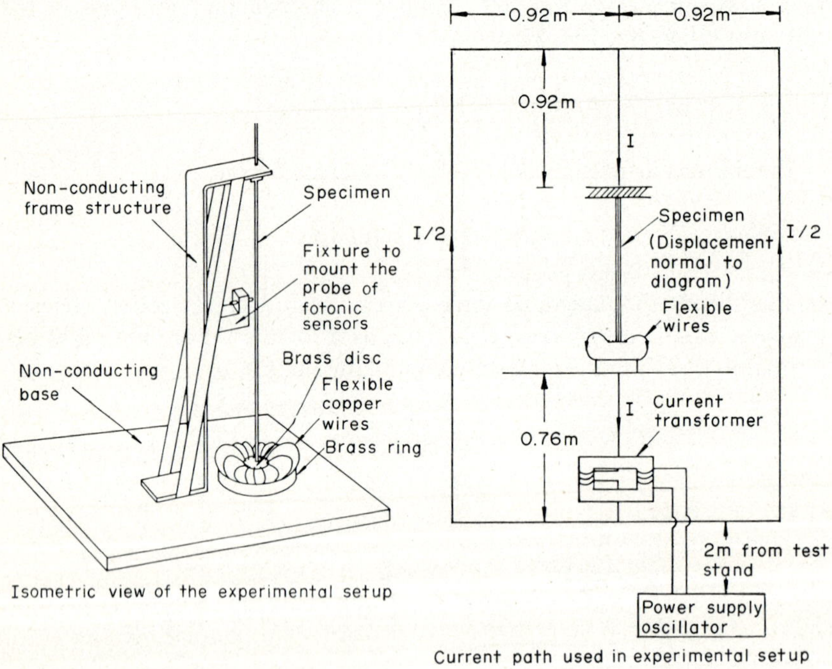

Fig. 26 Diagram of experimental apparatus for vibrations of a rod carrying electric current (from S. Chattopadhyay and F. C. Moon, "Magnetoelastic Buckling and Vibration of a Rod Carrying Electric Current," *J. Appl. Mech.*, **42** (Dec. 1975) 809–814.

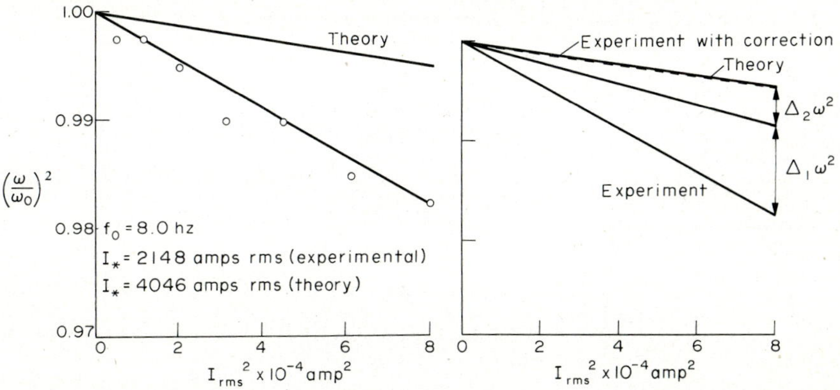

Fig. 27 Bending frequency versus current for a circular rod—comparison of theory and experiment (from S. Chattopadhyay and F. C. Moon, "Magnetoelastic Buckling and Vibration of a Rod Carrying Electric Current," *J. Appl. Mech.*, **42** (Dec. 1975) 809–814.

In the limit as $ka \to 0$ this agrees with the result of Doblin and Morozov [93]. For the case of surface current they obtain a constant 1.077 instead of 1.327 in (4.16).

Experiments on a cantilevered rod with currents up to 400 amperes at room temperature were performed as illustrated in Fig. 26. The detailed procedures and analysis of the results are discussed in [96].

The results showed a decrease in natural frequency of the rod with the square of the current; Fig. 27. The measured slope of the frequency–current dispersion curve was much higher than that predicted by the theory (4.16). Analysis of the temperature effects on the elastic modulus and length change of the rod due to Joule heating suggests that the decrease in frequency is in part due to thermoelastic effects. When these effects are subtracted from the data, the predicted and theoretical results are in good agreement; Fig. 27.

It was clear from these experiments that detection of the "kink" instability in solids using normal conductors will be difficult because of thermal effects. Ultimate confirmation of the theory will have to be done on superconducting solids. Such experiments are under way at the time of this writing.

Experimental evidence for the instability of a solid conductor carrying electric current is claimed by Abramova *et al.* [98]. In their experiments a multi-kiloamp current is passed through a 0.5-mm diameter wire of 70 mm in length. Below a certain current helical deformations are observed, while above a threshold energy the wire melts and pinch or sausage-type deformations are believed to occur and the wire disintegrates. Photographs of their results for the helical instability are shown in Fig. 28.

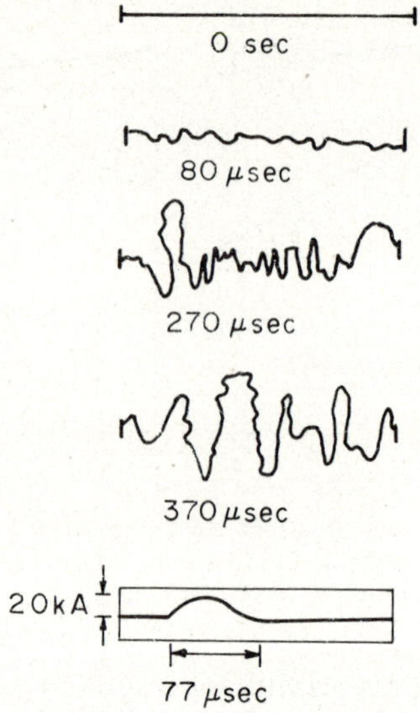

Fig. 28 Vibrations of a wire carrying a pulse of 20 kA current (from K. B. Abramova *et al.*, "Magnetohydrodynamic Instabilities during Electrical Explosion," *Sov. Phys.-Dokl.* **11**, No. 4 (Oct. 1966) 301–304).

While it is clear that the kink instability due to self-magnetic forces could explain the data, one cannot neglect thermoelastic effects which must certainly be present since the wire is heated to near melting. In fact if the wire were rigidly fixed between electrodes thermal expansion of the wire could itself cause buckling.

The experiments of Abramova *et al.* [98] inspired a paper by Vandakurov and Kolesnikova [97] using an analysis similar to that of Dolbin [92]. They assumed an elastic cylinder with uniform current distribution. They give numerical results for the critical current for both helical and pinch instabilities. They also give numerical data for the growth times for the instabilities. They do not, however, give an explicit formula for the frequency–current dispersion relation.

4.7 Conducting Rods in Magnetic Fields

The previous section described the destabilizing effects of the self-fields of the currents in the conductor. In this section we describe problems in which the current interacts with a static magnetic field produced by an

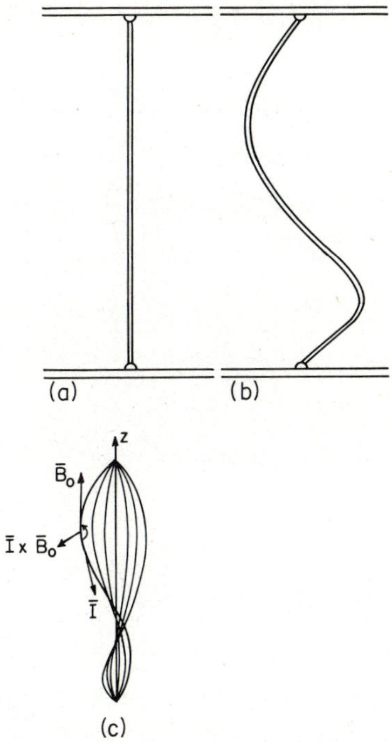

Fig. 29 Instability of a helical wire carrying current in a longitudinal magnetic field (from H. H. Woodson and J. R. Melcher, *Electromechanical Dynamics*, Parts I, II, III, Wiley (1968)).

independent source. When no initial current is present, vibrations of the rod in the magnetic field can induce current flow in the rod which results in magnetic damping forces. This problem has been investigated by Leibowitz and Ackerberg [99]. Smith and Herrmann [100] have extended this problem to investigate the effect of magnetic damping on a vibrating rod with an independent follower force.

When current flows in an initially straight flexible wire in an axial magnetic field, a helical buckling instability can occur; Fig. 29. An analysis of this elementary problem may be found in the text by Woodson and Melcher [101] who also present a photograph of such an instability; this text is an excellent collection of physical problems for the novitiate to the field of magneto-mechanics.

An extension of this problem to biaxial bending of a rod carrying helical current filaments embedded in it may have application to superconducting

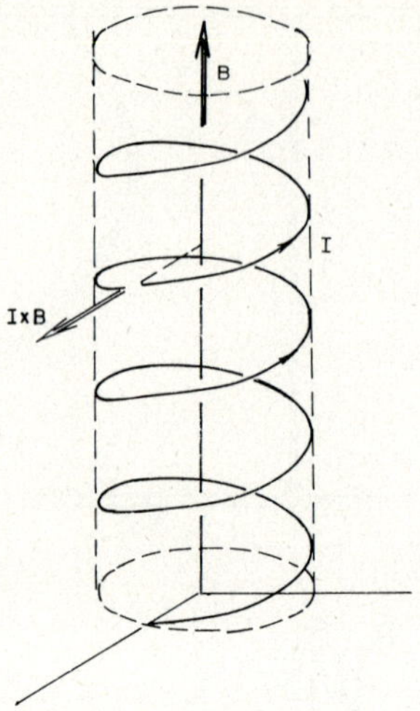

Fig. 30 Diagram of a helical current element in a longitudinal magnetic field.

wires. Superconducting wires are often made of multifilamentary superconducting strands embedded in a normal conductor; Fig. 30. The strands are sometimes twisted into helices along the wire axis to prevent the wire from going into the normal resistive state. The twisted current-carrying filaments can be represented by an axial current **I** and an effective magnetization **M**, where

$$d\mathbf{M} = \int \mathbf{r} \times \mathbf{J} \, da \, dz.$$

In a uniform external axial magnetic field **B** the rod will experience a force $\mathbf{I} \times \mathbf{B}_0$, and a couple $\mathbf{M} \times \mathbf{B}_0$. The self-fields of **I**, **M** are neglected. The vectors **I**, **M** rotate with the rod or wire where the rotation vector is given by

$$\mathbf{\Omega} = \left(-\frac{\partial v}{\partial z}, \frac{\partial u}{\partial z}, 0 \right).$$

The rotated vectors **I**, **M** are then

$$\mathbf{I} = I_0 \mathbf{e}_z + \mathbf{\Omega} \times \mathbf{I}_0,$$
$$\mathbf{M} = M_0 \mathbf{e}_z + \mathbf{\Omega} \times \mathbf{M}_0.$$

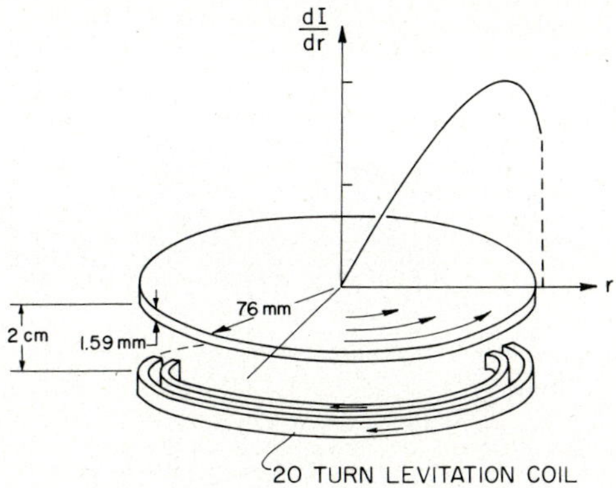

Fig. 31 Current distribution in a levitated circular aluminum plate.

The equation for the vibration and stability becomes

$$D_1 \frac{\partial^4 u}{\partial z^4} + \rho A \frac{\partial^2 u}{\partial t^2} = I_0 B_0 \frac{\partial v}{\partial z} + B_0 M_0 \frac{\partial^2 u}{\partial z^2},$$
$$D_2 \frac{\partial^4 v}{\partial z^4} + \rho A \frac{\partial^2 v}{\partial t^2} = -I_0 B_0 \frac{\partial u}{\partial z} + B_0 M_0 \frac{\partial^2 v}{\partial z^2}. \qquad (4.17)$$

The details of the analysis are not presented here but it would appear that with $M_0 = 0$ (i.e. no twisting) the rod could become unstable as in the case of the stretched wire with no bending rigidity. With twisting, the term with M_0 would either stabilize or destabilize the rod depending on the sign of $B_0 M_0$.

4.8 Effect of Currents on Plate Vibrations

To obtain quasi-static currents in a flexible structure, without using superconducting materials, the writer has conducted experiments on circular aluminum plates levitated above a neighboring coil with sinusoidal currents; Fig. 31. Circulating currents of up to 10^3 A were induced in the thin aluminum plates 15.2 cm diameter, 1.6 mm thick. The natural frequencies were observed at sublevitating currents. To minimize thermal effects, resonance and free vibration techniques were used within 10 seconds after the current was turned on.

The changes in the frequency squared were found to be linearly related

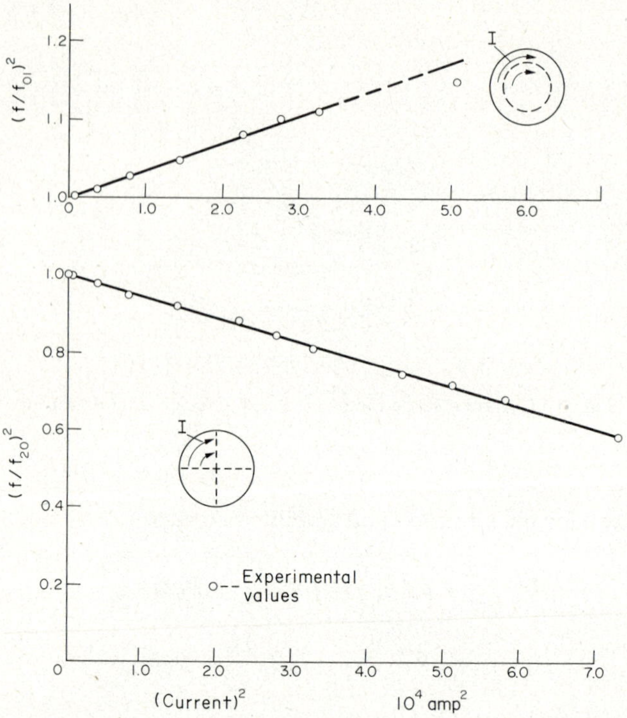

Fig. 32 Effect of circulating currents on the vibration of circular plates—frequency versus current (from F. C. Moon, "Mechanics of Structures with High Electric Currents," Princeton University Report AMS 1111).

to the squares of the exciting coil currents; Fig. 32. The 2-, 3-, 4-nodal diameter modes were all observed to suffer a decrease in natural frequency, up to 22% decrease for the two-nodal diameter mode. This behavior can be explained in terms of the "kink" instability due to current–current forces in the plate; Fig. 31.

The symmetric modal frequencies were increased with the presence of circulating currents. The behavior can be explained in terms of two concentric filaments of current in the plate, interacting to produce attractive forces between the loops and hence stiffening the plate and raising the frequency; Fig. 31.

The interpretation of the results need further explanation. First, the effect of the exciting coil, which was 2 cm below the plate, was ruled out since the plate rigid-body levitation frequency was about 3 Hz, whereas the changes in elastic mode frequencies were of the order of 20 Hz or more. Second, the effect of thermal stresses cannot be ruled out. For although the temperature rise due to Joule heating was small, less than 5°C, the thermal gradients could be large. For small times the temperature will behave like J^2; see Fig. 31 for the J distribution. This places a higher temperature on the edges than at the center. The effects of these thermal gradients on the frequencies are believed to account for up to 50% of the observed changes in frequency.

Thus again, as in the case of the rod, the attempt to observe the "kink" instability due to self current forces in a normal conductor is inevitably complicated by the Joule heating thermoelastic effects. This writer has thus abandoned the search for this effect in normal conductors and has begun experiments on elastic superconducting rings. As a preliminary to these experiments an analysis of the effect of high currents in superconducting rings was undertaken and has found application in the design of toroidal field magnets for proposed fusion reactors.

4.9 Elastic Stability of Superconducting Magnets

As shown in Subsection 4.8, the analysis of a straight elastic rod with current, leads to a critical value of the current I_c, above which the straight configuration is not stable; I_c is not to be confused with the "critical" current for superconductors to go into the normal or resistive state. One is naturally led to ask whether the large, high current superconducting coils being designed for fusion reactors (Subsection 4.5) will remain elastically stable in their planes for the rated design currents. The tentative answer, based on analysis, is that they will be unstable if not properly constrained.

To examine potential buckling of these coils self-field forces on each magnet as well as mutual attractive forces on each magnet in a torus configuration must be included. These destabilizing magnetic forces must be counteracted by the elastic bending and torsional stiffness of each coil, intercoil constraints, and the field-induced tension in the coil windings.

To estimate the self current–current forces during bending we can use the rod as an analog to see if the physical variables in the proposed superconducting magnets are large enough to warrant a magnetoelastic stability analysis. These variables are current I, Young's elastic modulus Y, the cross-sectional area $A = \pi d^2/4$ and the distance between supports which we assume is of the order of the outside diameter D for a circular coil. Then using the rod as an analog the criterion for instability can be roughly stated using the nondimensional group M_e which I call appropriately, "*magnetoelastic number*"

$$M_e = \frac{\mu_0 I^2}{YA}\left(\frac{D}{d}\right)^2 \sim 0(1). \tag{4.18}$$

Typical values for such a magnet are $I = 5 \times 10^6$A, $Y = 10^{11}$ N/m^2, $d = 30$ cm, $D = 600$ cm, and we find the magnetoelastic number, $M_e \sim 2$. Thus a detailed elastic stability analysis is justified for these devices.

(a) *Single-coil Analysis.* To study the single toroidal magnet coil for magnetoelastic instability we use the theory of curved beams, or rings [102]. The magnet can bend in its original plane or out of its plane. In this section we review only the out-of-plane case [82], Chattopadhyay [96] has looked at the inplane bending. At each cross-section of the magnet coil there will be force resultants representing coil tension, bending moments, and transverse shear. The centroidal axes will be displaced an amount $v(z)$ normal to the coil plane, where z is a coordinate along the arc of the coil. In addition, each cross-section can rotate about the z-axis an amount $\phi(z)$. If the unconstrained part of the "D"-shaped magnet (Fig. 33) is replaced by an arc of radius "a" then v, ϕ must satisfy the equations (neglecting rotary inertia)

$$A\frac{\partial^2}{\partial z^2}\left[\frac{\phi}{a} - \frac{\partial^2 v}{\partial z^2}\right] + \left(G_0 + \frac{C}{a}\right)\frac{\partial}{\partial z}\left(\frac{\partial \phi}{\partial z} + \frac{1}{a}\frac{\partial v}{\partial z}\right) + T_0\frac{\partial^2 v}{\partial z^2} + F_m = m\frac{\partial^2 v}{\partial t^2},$$

$$C\frac{\partial}{\partial z}\left[\frac{\partial \phi}{\partial z} + \frac{1}{a}\frac{\partial v}{\partial z}\right] - \left(\frac{A}{a} + G_0\right)\left(\frac{\phi}{a} - \frac{\partial^2 v}{\partial z^2}\right) = 0, \tag{4.19}$$

where A is the bending stiffness, C is the torsional stiffness, T_0 is the initial tension due to $\mathbf{I} \times \mathbf{B}$, G_0 is the initial moment due to $\mathbf{I} \times \mathbf{B}$, and F_m is the *perturbed magnetic force* due to bending.

The magnetic field creates an initial radial body force $\mathbf{I} \times \mathbf{B}$, which varies across the magnet cross-section, creating an initial tension T_0 and resultant moment G_0.

To investigate the elastic stability, one must look at the forces which arise when the structure is deformed. Here F_m represents the perturbed magnetic forces normal to the original magnet plane when the coil bends out of its plane. These forces can be characterized as stabilizing or destabilizing in so far as they increase or decrease the natural frequency of the coil. Based on an analysis of a simply supported arc, one can conclude:

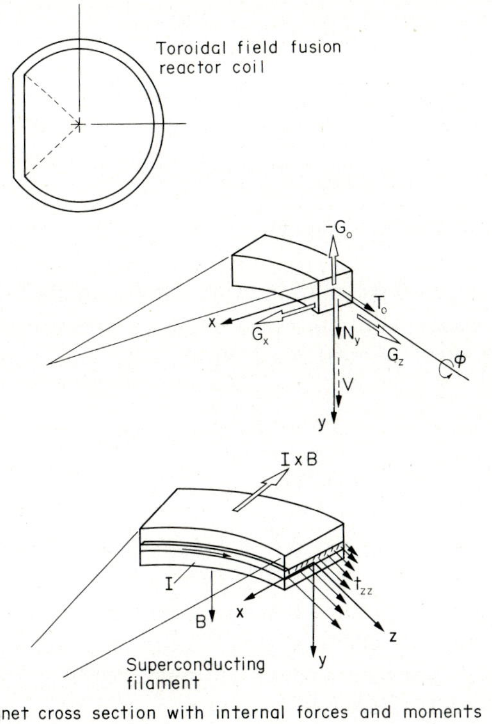

Fig. 33 Cross-section of a circular superconducting coil element showing forces, force resultants, and stresses (from F. C. Moon and S. Chattopadhyay, "Elastic Stability of Thermonuclear Reactor Coil," *Proc. 5th Symp. on Engng Problems of Fusion Research, Nov. 1973*, Princeton, N.J., IEEE Nuclear and Plasma Sci. Soc., N.Y., Publ. No. 73, CH 0843-3-NPS (Apr. 1974) 544–578).

T_0 acting alone is stabilizing;
G_0 acting alone is destabilizing; and
F_m acting alone is destabilizing.

T_0, G_0 can be found from the undeformed current and field distribution in the magnet. To find the perturbed magnetic forces F_m an energy method has been used similar to that described in Subsection 2.2. The displacements are written in terms of a complete set of orthonormal functions, i.e.,

$$\begin{aligned} v &= \alpha_n Z_n(z), \\ \phi &= \beta_n Z_n(z), \quad n \text{ summed.} \end{aligned} \tag{4.20}$$

The inductance is then expanded in a Taylor series in α_n, β_n, i.e.

$$L(\alpha_n, \beta_n) = L_0 + \frac{1}{2}\frac{\partial^2 L}{\partial \alpha_1^2}\alpha_1^2 + \cdots, \tag{4.21}$$

where we have used the fact that

$$\frac{\partial L}{\partial \alpha_n} = 0 \quad \text{and} \quad \frac{\partial L}{\partial \beta_n} = 0,$$

because the unperturbed magnetic fields create no out-of-plane magnetic forces.

Using a Galerkin-type solution, with a finite number of modes, one can find an approximate solution. For a one-term approximation in (4.20) and neglecting the effect of twist on L in (4.21) the magnetic energy has the form

$$W = \tfrac{1}{2}I^2(L_0 + L_1\alpha_1^2), \tag{4.22}$$

and the *modal magnetic force* is

$$\frac{\partial W}{\partial \alpha_1} = I^2 L_1 \alpha_1. \tag{4.23}$$

If $L_1 > 0$ then the perturbed magnetic force is in the direction of deformation α_1, and the effect is destabilizing. We call L_1 the *incremental modal inductance*. It has been calculated numerically by treating the current in the coil as a set of discrete line elements [96, 82]. The results indicate that $L_1 > 0$, or that F_m is destabilizing.

The combined effect of tension T_0, due to the radial body force, and perturbed magnetic force F_m is complicated and is currently under study both experimentally and analytically by the writer and coworkers.

(b) *Stability of a Toroidal Set of Magnets*

(i) *Theory.* A toroidal set of magnets for proposed fusion reactors is shown in Fig. 22. These magnets have their planes normal to a common base plane and intersect at a common line or toroidal axis. The coils are assumed to be constrained along part of their boundary by more rigid structures and interact with each other only through mutual magnetic forces produced by the currents in each magnet.

In the previous section the isolated coil analysis was sketched. Although the perturbed self-magnetic forces are destabilizing, the initial attractive forces between coils, due to small out-of-plane bending deformation, have been found to be of greater importance [83]. While the net magnetic force on a coil is zero when symmetrically placed between two similar magnets in the torus, small out-of-plane bending will introduce a magnetic force which tends to increase the bending. For a large enough coil current, the elastic restraining forces will not be able to restore the coil to its unbent shape and the whole set will bend to a new equilibrium position, providing the coils do not fail structurally or go normal.

Even if the structure is stiff enough to maintain the unbent equilibrium (i.e. $I < I_c$, the critical current) the current will couple the coils dynamically and affect the vibration modes of the system. Thus a vibration analysis also includes the buckling problem. That mode whose frequency goes to zero as $I \to I_c$ is the buckling mode for the set of magnets. To determine the new buckled equilibrium position requires a nonlinear analysis of the bending. This has been done for the isolated coil by Chattopadhyay [96].

To analyze the vibrations of a set of coils, each coil is assumed to vibrate in a particular coil mode. Thus if v_n is the out-of-plane displacement of the nth coil we assume that

$$v_n(z, t) = u_n(t) f(z), \tag{4.24}$$

where $f(z)$ is common to all coils. When the current is zero, the $u_n(t)$ all exhibit harmonic oscillations when perturbed, i.e.

$$\ddot{u}_n + \omega_0^2 u_n = 0, \qquad I = 0.$$

When $I \neq 0$, there are mutual forces between each magnet and its pairs of neighbors. Thus if there are N magnets in the set, the displacements u_n satisfy

$$\ddot{u}_n + \omega_0^2 u_n = \frac{1}{m_0} \sum_{m=1}^{N-1} F_{nm} = Q_n/m_0, \tag{4.25}$$

where m_0 is the modal mass and F_{nm} the mutual magnetic forces.

The calculation of F_{nm} is best done by considering the nth coil and its mth nearest neighbors. With this procedure the large initial forces cancel out and the Q_n is a linear function of the coil displacement, i.e.,

$$Q_n = \sum_{m=1}^{M} K_m(2u_n - u_{n-m} - u_{n+m}), \tag{4.26}$$

where $M = (N - 1)/2$ for N odd, and $M = N/2$ for N even..

The K_m act as negative spring constants between the nth coil and its mth nearest neighbors. The vibrations of the set of magnets can thus be represented by a linear chain of oscillators coupled by negative springs and having periodic boundary conditions. The motions of such a set are well known and are treated in many texts, e.g. Brillouin [103].

The general motion is of the form

$$u_n(t) = \sum_{\alpha=0}^{N-1} U_\alpha\, e^{i\omega_\alpha t}\, e^{i(2\pi n\alpha)/N}. \tag{4.27}$$

The frequency of each mode ω_α can then be shown to have the following form [83]:

$$\omega_\alpha^2 = \omega_0^2\left[1 - I^2 \sum_{m=1}^{M} 4C_m \sin^2 \frac{\pi m\alpha}{N}\right], \tag{4.28}$$

where $I^2 C_m = K_m/m_0\omega_0^2$ describes the magnetic forces. Buckling occurs when I^2 attains the value

$$I_c^{-2} = \sum_{m=1}^{M} C_m \sin^2 \frac{\pi m\alpha}{N}. \tag{4.29}$$

To determine the magnetic force constants C_m, the energy method is again used, i.e.

$$Q_n = \frac{\partial W(u_n, I)}{\partial u_n},$$

where W is the magnetic energy.

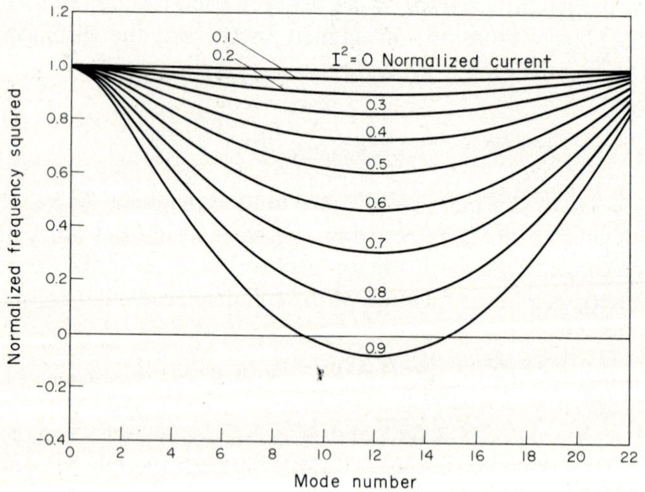

Fig. 34 The effect of current on the out of plane bending frequencies of a coupled set of toroidal field magnets (from *J. Appl. Phys.* **47**, No. 3 (Mar. 1976)).

If W is expressed in terms of the mutual inductance between all of the coils then one can show that [83]

$$2K_m u_n = I^2 \frac{\partial}{\partial u_n}(L_{n(n+m)} + L_{n(n-m)}) \tag{4.30}$$

with $u_{n-m} = u_{n+m} = 0$, and where L_{kl} is the mutual inductance between the kth and lth coils.

To determine the K_m, the change in mutual inductance between a coil and its nearest neighbors has been calculated for a typical set of coils. One can see that the force constants drop off rapidly after the first pair.

Such a procedure leads to a frequency–current dispersion relation for the set of coils. A typical plot is shown in Fig. 34. At a large enough value of current I, the frequency of one mode becomes zero and the coils buckle such that each successive pair bends toward each other.

The critical value of I depends, of course, on the original deformation mode $f(z)$ in (4.24). This function will change depending on the lateral constraints placed on the coils.

In conclusion, while the large initial tension stresses due to the $\mathbf{I} \times \mathbf{B}$ body forces must be the primary concern of the coil designer, each design should

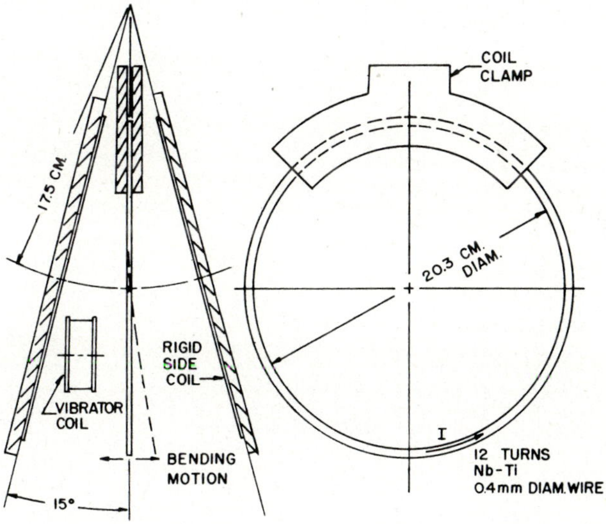

Fig. 35 Geometrical arrangement of three circular superconducting coils in a toroidal segment for lateral stability tests (from *J. Appl. Phys.* **47**, No. 3 (Mar. 1976)).

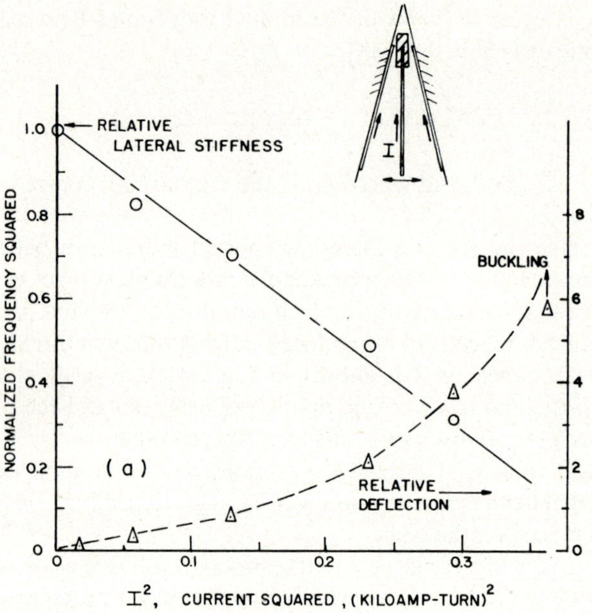

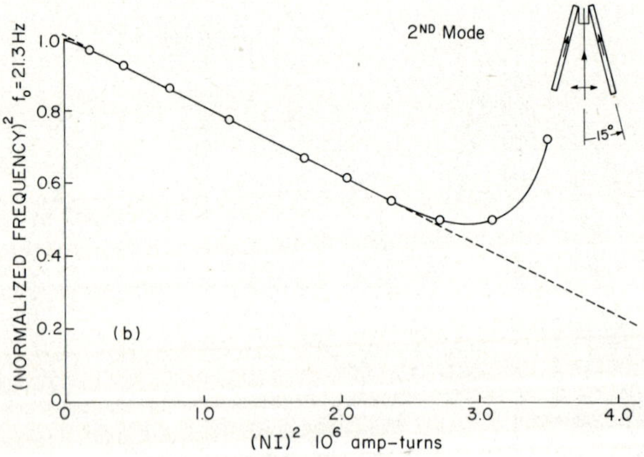

Fig. 36 Decrease in the lateral stiffness and buckling of a superconducting coil in a toroidal segment with increasing current. (a) First buckling mode (from *J. Appl. Phys.* **47**, No. 3 (Mar. 1976)); (b) second mode.

be checked for magnetoelastic stability when the magnetoelastic number M_e in (4.18) is of order unity.

(ii) *Experiments on buckling of superconducting magnets.* Experiments have recently been conducted by the writer on the stability of a flexible superconducting coil between two similar coils in a partial torus; Fig. 35. The coils were wound in the radial direction with twelve turns of niobium–titanium multifilament wire and embedded in an epoxy matrix. The results shown in Fig. 36 show the decrease in lateral stiffness (proportional to the square of the frequency) with current and subsequent buckling of the coil against one of the fixed coils, [84].

4.10 Mechanical Properties of Superconductors

While many materials will become superconducting at temperatures above absolute zero, that is they will exhibit zero resistance to the flow of electric current, most of these materials remain superconducting only for small current densities or low values of the magnetic field. However, a class of compounds known as Type II superconductors are able to conduct currents at high densities (e.g. 10^4 amps/cm^2) in high magnetic field environments (i.e. greater than 1 Wb/m^2). The commercial form of two of the most widely used Type II materials, niobium–tin (Nb_3S_n) and niobium–titanium (Nb–Ti), are actually composite materials. The superconducting material is embedded in a normal conductor matrix such as copper, sometimes cladded with stainless steel for strength. Figure 37 shows the sandwich type (Nb_3Sn) and the multifilamentary type (Nb–Ti).

Comparatively little work has been done on the effective properties of these materials from the composite point of view, compared with the purely structural composites developed in the 1960s. Mechanical properties of the sandwich or tape-type superconductor have been given by Benz and Coffin [104]. A summary of the mechanical properties of the filamentary type may be found in a paper by Easton and Koch [105].

Of greater interest perhaps is the effect of stress on the superconducting properties themselves. An early study is that of Hudson [106] who showed that the limiting superconducting current decreased as the stress in the direction of current neared the yield point. Another group, Ekin *et al.* [107], has shown that the superconducting–normal conductor boundary in the current–magnetic field plane is lowered by the application of stress. Since the normal matrix in which the superconducting material is embedded is

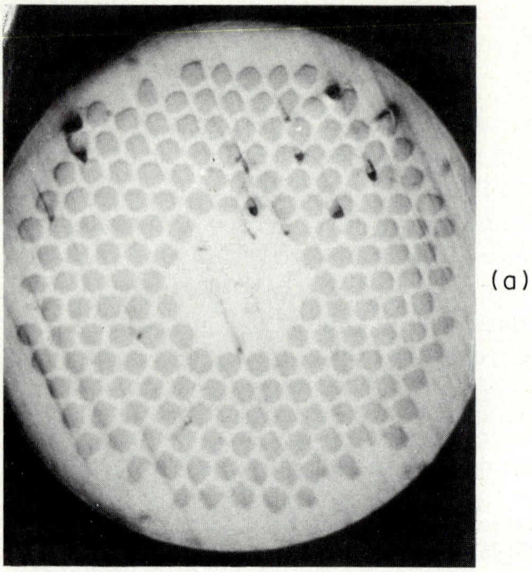

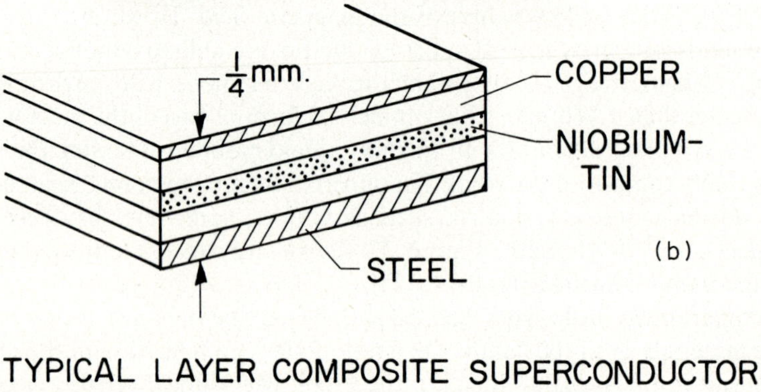

Fig. 37 (a) Cross-section of a 0.4-mm diameter multifilament superconductor of niobium–titanium in a copper matrix. (b) Cross section of a layered niobium–tin superconducting tape between copper plys.

supposed to accommodate the current if the superconductor goes into the normal state, the resistivity of the matrix at low temperatures is important. In the low-temperature region, the resistivity is determined mainly by impurities and defects in the lattice. Fisher and Linz [108] have shown that cyclic strain will increase the resistivity at 4.2°K. Thus cyclic loads on super-

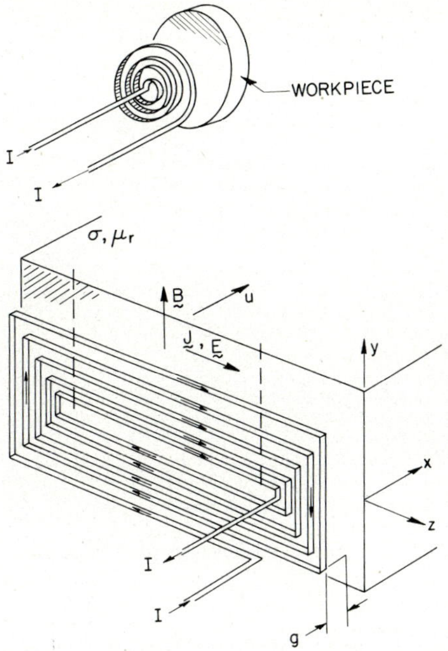

Fig. 38 Diagram of a magnetic forming coil near a conductor.

conducting magnets may limit the current-carrying ability of the material should the magnet go into the normal state.

Finally the damping and dispersion of ultrasonic waves in filamentary superconductors has been studied by Sachse [109].

5 DYNAMIC MAGNETIC FORCES IN SOLIDS

5.1 Introduction

In magnet design the stress analysis centers on equilibrating the magnetic forces which would otherwise accelerate the parts of the magnet. In the application of pulsed magnetic fields one is often concerned with producing an unbalanced magnetic force to accelerate and deform the solid. Pulsed magnetic fields were originally designed to produce high transient magnetic fields for physics research. In designing such devices it was quickly learned that these high magnetic fields could move or deform solids for some useful purpose; see, for example Furth [110]. These applications include:

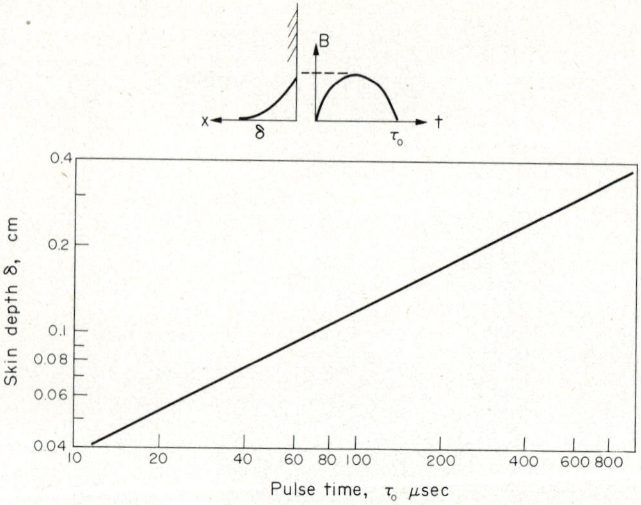

Fig. 39 Skin depth versus current pulse time.

(i) magnetic generation of stress waves;
(ii) magnetic forming of metals;
(iii) dynamic testing of materials;
(iv) impulse loading of structures; and
(v) acceleration of metals for solid-waste separation.

In all these applications a time-varying source of magnetic field induces electric currents in a nearby conducting solid. This is usually done by placing a coiled conductor near the solid and discharging a stored charge capacitance bank through the coil (Fig. 38). The interactions of the pulsed magnetic field with the currents in the metal include the following mechanical effects:

$\mathbf{J} \times \mathbf{B}$ body forces;
magnetization forces;
thermoelastic strains due to $\mathbf{J} \cdot \mathbf{J}$;
ablation due to melting; and
Rayleigh–Taylor instability at the solid–field interface.

For wave propagation of low amplitude stress waves in nonferromagnetic metals, addition of magnetic effects to the linear theory of thermoelasticity will suffice for analysis. For magnetic-forming applications, a plasticity model is required. If the magnetic forces are sufficiently high, the shear effects may be neglected and a magneto-fluid analysis can be used.

In pulsed field applications the magnetic field is excluded from the metal

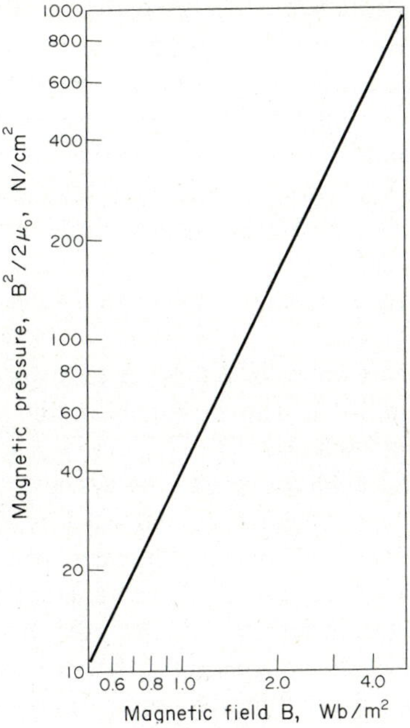

Fig. 40 Magnetic pressure versus magnetic field.

by the induced eddy currents. The depth of penetration of the field into the metal at the time of maximum field is of the order of

$$\delta = (2\tau_0/\mu_0\sigma\pi)^{1/2}, \tag{5.1}$$

where σ is the conductivity of the metal and τ_0 is duration of the pulse. This is called the skin depth and as shown in Fig. 39 can be quite small compared to structural dimensions for pulses shorter than 10^{-4} sec.

When the skin depth δ is small, the magnetic forces on the structure can sometimes be replaced by an equivalent pressure force of intensity $p_m = B_t^2/2\mu_0$ where B_t is the value of the component of the field tangential to the surface of the solid. Values of "magnetic pressure" P_m versus field are shown in Fig. 40.

5.2 Magnetic Generation of Stress Waves

The interaction of stress waves and magnetic fields has received a great deal of attention especially in the study of ferromagnetic solid-state theory [111]. Such phenomena, often treated from the viewpoint of spin waves and phonons, have resulted in applications in the electronics industry. This subject will not be treated here. Another problem which has received a great deal of attention from mechanicians is the effect of a static magnetic field on the propagation of stress waves. A review of these papers has been given by Paria [18]. This work includes the magnetic body force $\mathbf{J} \times \mathbf{B}$ and the Joule heating rate $\mathbf{J} \cdot \mathbf{J}/\sigma$ in the equations of thermoelasticity. The nonlinear terms are linearized about an initial magnetic field. The effect of the magnetic field on the frequency–wavelength dispersion equation is then found.

An early work on stress waves in a static field, neglecting thermal effects, was by Robey [112], in 1953, who examined sound waves in conducting plates. Another was that by Knopoff [14], in 1955, who examined the effect of the Earth's magnetic field on seismic waves and concluded that such effects are unimportant. Early experimental work was reported by Galkin and Koroliuk [113] in 1958, who found a linear dependence of longitudinal sound speed in tin on the square of the magnetic field. These measured increases were only of the order $10^{-3}\%$, however. Another experiment on metal single crystals in static magnetic fields was performed by Alers and Fleury [114] in 1963 who also reported small changes in sound speed for fields of the order of 1 $\mathrm{Wb/m^2}$. Additional references are found in Paria [18], and further review of this subject is not given here. Two recent papers on the subject not found in [18] are by Pao and Hutter [115], who treat soft ferromagnetic materials, and Nayfeh and Nemat Nasser [116]. Using the theory of the propagation of singular surfaces in nonlinear continuum mechanics, McCarthy [117] has examined the propagation, growth, and decay of acceleration waves in conducting materials. Another study of plane waves in magnetoelasticity is that of Recker [118]. The direct generation of stress waves by dynamic magnetic fields has received less attention. This problem has application to deforming solids and generating large transient forces in structures for dynamic testing.

One method for magnetically generating short-duration stress waves in plates is to pass current along two conducting strips each cemented to the edge of an elastic plate and separated by a thin dielectric. The repulsion between the two strips can load the surface and excite a stress wave in the plate. Use of such a technique has been reported by Snell *et al.* [119] with stress wave rise times of 2 μsec.

A precursor to the analysis of the magnetic-forming problem is the magnetic generation of stress waves in elastic conductors. Magnetic generation of one-dimensional waves in nonferromagnetic conductors was treated by Moon and Chattopadhyay [120] who also conducted experiments. The problem concerns stress waves in a half-space ($x > 0$) generated by an applied jump in tangential magnetic field at the boundary. The magnetic field diffuses into the conductor and generates heat. The main conclusions of the analysis are that *two waves are generated by the magnetic field.* One is a pressure wave due to the $\mathbf{J} \times \mathbf{B}$ body force and is proportional to $B^2/2\mu_0$ on the boundary; Fig. 41a. The other one is a thermoelastic wave generated by the sudden rise in temperature at the surface; Fig. 41b. As one can see, this wave has both tension and compression which reverse at a very high frequency. Although this wave can be of the order of the $\mathbf{J} \times \mathbf{B}$ wave, it is highly damped, which probably accounts for why it was not observed in the experiments.

Analysis of this problem is based on thermoelasticity. Let u be the longitudinal displacement, and T the temperature, and B the magnetic field normal to u, then one can derive the following governing equations [120]:

$$(\lambda + 2\mu)\frac{\partial^2 u}{\partial x^2} - \rho\frac{\partial^2 u}{\partial t^2} = -\frac{\partial(B^2/2\mu_0)}{\partial x} + \alpha(3\lambda + 2\mu)\frac{\partial T}{\partial x},$$

$$K\frac{\partial^2 T}{\partial x^2} - \rho C\frac{\partial T}{\partial t} = -\frac{1}{\sigma\mu_0^2}\left(\frac{\partial B}{\partial x}\right)^2 + \left[\alpha T_0(3\lambda + 2\mu)\frac{\partial^2 u}{\partial x\,\partial t}\right],$$

$$\frac{\partial^2 B}{\partial x^2} - \mu_0\sigma\frac{\partial B}{\partial t} = \left[\mu_0\sigma\frac{\partial}{\partial x}B\frac{\partial u}{\partial t}\right]. \tag{5.2}$$

Here λ, μ are Lamés constants, α is the thermoelastic constant, while K, C are thermal conductively and specific heat respectively.

The analysis of [120] is based on an approximation which drops the bracketed terms in these equations. Thus although the problem is still nonlinear, the equations are hierarchical, i.e. one first finds $B(x, t)$, then $T(x, t)$, and finally $u(x, t)$. Neglect of the bracketed term in the heat equation uncouples the thermoelastic equations, while dropping the right-hand side of Maxwell's equation neglects the convective effect of the motion on the field. The former approximation effectively neglects the decay in the stress waves which can be estimated independently [120]. The convective effect of the motion on the diffusion of magnetic field into the solid might be important if the structure were a thin plate or shell.

The generation of thermoelastic waves by electromagnetic radiation by such devices as lasers or electron beams has been observed experimentally and also studied analytically [121–123], but is not covered in this review.

Electric fields in conducting solids have been generated by stress waves with no external fields using the electron inertia effect [124]. This problem

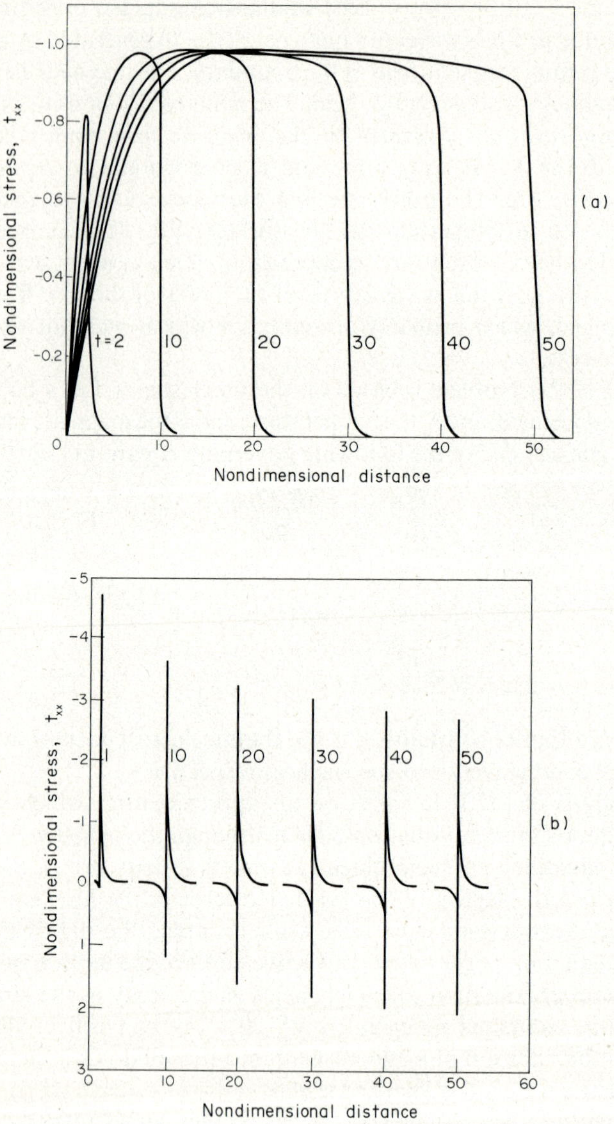

Fig. 41 (a) Eddy current pressure wave versus distance. (b) Joule heating induced thermoelastic wave versus distance (from F. C. Moon and S. Chattopadhyay, "Magnetically Induced Stress Waves in a Conducting Solid—Theory and Experiment," *J. Appl. Mech.* **41**, No. 3 (Sept. 1974) 641–646).

has been studied by this writer [52] using a two-continuum model for the electrons and ions.

5.3 Magnetic Forming of Metals

The use of pulsed magnetic fields to deform solids was developed in the 1960s in a number of countries. Its developers saw it as a special case of high-energy-rate metal forming, which includes explosive and electro-hydraulic forming. The successful application of this new forming tool encouraged many engineering groups to develop devices rather than pursue a full description of the mechanics of the deformation process.

Magnetic forming can only be applied directly to conducting solids. Transient currents in the coil of the magnetic-forming tool create a change of flux near the work piece. This generates circulating or eddy currents in the solid which in turn create heat and temperature gradients. The solid may also be magnetized if it is ferromagnetic. The *direct magnetic forces* result from $\mathbf{J} \times \mathbf{B}$ body forces as well as magnetization forces. The *indirect magnetic effects* include thermoelastic deformation and magnetostriction in ferromagnetic materials. Most analyses of magnetic forming treat only the eddy-current forces. These currents in the workpiece screen out magnetic fields, thereby allowing the field between the coil and workpiece to build up and create very high magnetic fields and forces.

The devices usually consist of a conducting multiturn coil of inductance L, which is connected to a charge capacitance bank of capacity C. When the circuit is closed the current rises and falls in a time of the order of $2\pi(LC)^{1/2}$. If V is the voltage across the capacitors, the stored energy in the bank is $CV^2/2$ which in actual devices runs from 10^4 to 2×10^5 joules. Some experimental studies have shown, however, that only a maximum of 15% of this energy is used to deform the solid. Other potential energy-storage devices for magnetic forming include large rotating motors or fly-wheels, and superconducting energy-storage devices.

Early work on magnetic forming was reported by Birdsall *et al.* [125] in 1961 and developed rapidly in the United States with support from the National Aeronautics and Space Administration; see, for example, [126]. A layman's description of the process is given by Furth [110]. A discussion of the physics of pulsed magnets and the effects of high transient magnetic fields on solid conductors has also been given by Furth, Levine, and Waniek [127] in 1957 and Furth [128].

Technical data on the coils necessary for magnetic forming may be found

Fig. 42 Pancake and solenoidal coils for magnetic forming.

in reports by Waniek *et al.* [129, 130], as well as in a NASA technology utilization report [126].

An analytical estimate of the maximum velocity of magnetically imploded tubes was given by Lippmann and Schreiner [131]. An experimental study of stress, and forming efficiency in magnetically formed flat plates and cylinders, has been given by Baines, Duncan, and Johnson [132]. For the case of a solenoidal coil in a thin conducting tube, the ratio of plastic work done to electrical energy discharged was found to be between 6–14%.

The most common magnetic-forming coils are usually flat pancake coils, for forming flat sheet solids, and the solenoidal coil for imploding shells or swagging, and for radially expanding thin shells; see Fig. 42. However, the high energy pulse forces generated by the magnetic field have also been used to compress powdered metal into a solid (see, for example, Barbarovich [133]), and for riveting (Leftheris [134]). In the latter application a pancake coil generates a stress wave in a conical bar. The stress wave intensifies as it propagates down the conical bar and compresses the head of a rivet at the narrow end of the conical bar.

Another application of pulsed magnetic forces has been in the area of impact welding. The magnetic coil accelerates a plate obliquely against another plate and the force of impact welds the two together; Meyer [135].

Further studies of magnetic forming include a dimensionless group analysis for scaling calculations by Lawrence [136]. Details of the electromagnetics including calculation of eddy currents, magnetic field diffusion, and force distributions have been given by Gerber [137], Hillier and Lal [138], and Bauer [139]. In the latter dissertation a Mises yield condition is used to examine the plastic deformation of a magnetically expanded shell. Another parametric study of the magnetic-forming process, including a number of references, is that of Al-Hassani *et al.* [140].

Magnetic implosion of conducting shells has also been used to produce megagauss magnetic fields by compressing the initial flux in the shell. A brief note on this application is given by Freeman, Cnare, and Waag [141]. They use the mutual inductance between the deformed shell and the outer primary coil to calculate the forces and the radial motion of the shell. In this simple model no buckling or wrinkling of the shell is considered. This topic is also discussed by Knoepfel [142].

5.4 Magnetic Impulse Testing of Solids and Structures

In addition to forming metals, dynamic magnetic forces have been used to generate pulsed forces for dynamic testing of structures. Al-Hassani and Johnson [143] use a pulsed current coil near cantilevered beams, and frame structures to test their transient response. Forrestal and Overmier [144] have used a single wire near a thin shell to produce a lateral dynamic force on the shell of duration in the order of microseconds.

Magnetic forces have been used for dynamic testing of materials. Using a magnetically driven piston, Waniek *et al.* [130] performed dynamic stress–strain measurements on uniaxial test specimens. Walling and Forrestal [145] have investigated the elasto-plastic behavior of aluminum rings using pulsed magnetic forces of μsec duration.

Wessenburg [146] uses a solenoidal coil with pulsed current to magnetically collapse thin shells, in dynamic buckling tests. Al-Hassani has also investigated magnetic buckling of thin shells [147].

Of particular interest to mechanicians is the instability of the interface between the imploding shell and magnetic field, producing Rayleigh–Taylor instabilities of a two-fluid interface. Analysis of this phenomenon is given by Somon [148] who treats the imploding shell as a fluid.

The strength of a single-turn conducting ring under a pulsed current flowing around the ring has been investigated by Herlach and McBroon [149].

Current from a shorted capacitor bank creates a megagauss (100 tesla) magnetic field in the center of 2-10-mm diameter copper coils in the order of microseconds. It is found that above 100 tesla the elastic restraining forces offer minimal resistance to the radial magnetic forces and the coils retain their integrity through inertia alone.

High field effects on solids in single-turn tests were discussed in the 1957 paper by Furth, Levine, and Waniek [127]. Near 100 tesla, surface melting was observed along with saw-like cuts on the inner edges of the turn.

5.5 Magnetic Forming Forces in Ferromagnetic Conductors

The dynamic magnetic forces on nonferromagnetic conductors are calculated from the Lorentz forces $\mathbf{J} \times \mathbf{B}$ on the solid. However, if the solid can be magnetized, additional forces act on the solid. As discussed above (Subsection 3.2), the distribution of the magnetization forces is not unique and depends on the chosen stress–strain–magnetization relation. The total force on the solid, however, is independent of the particular force model chosen and can be calculated from integration of the Maxwell stress vector on a surface S_0 surrounding the magnetized body, i.e.

$$\mathbf{F} = \frac{1}{\mu_0} \int_S \mathbf{n} \cdot [\mathbf{BB} - \tfrac{1}{2} B^2 \mathbf{I}] \, da, \tag{5.3}$$

where $\mathbf{n}$ is the outward surface normal and $\mathbf{I}$ the identity matrix.

Consider the case of a magnetic forming coil on one side of a plate; Fig. 43. This configuration is used to form plates against a die. Let A^-, A^+

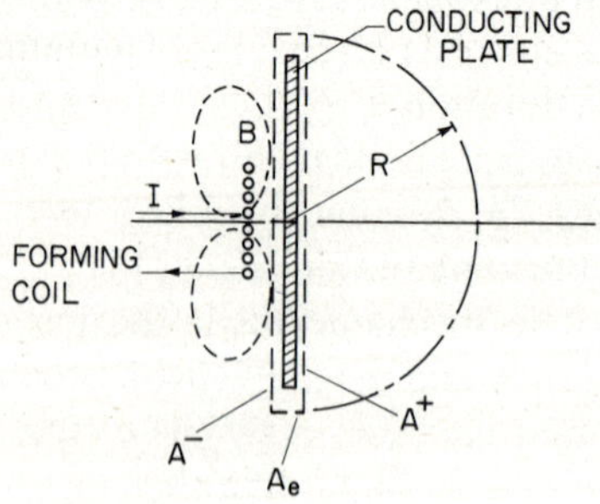

Fig. 43 Integration contour for Maxwell stresses around a conducting plate near a magnetic forming coil.

be the magnet and die sides of the plate respectively, while A_e is the edge of the plate. On A^+, A^-,

$$\begin{aligned} \mathbf{n} &= -\mathbf{e}_z \quad \text{on} \quad A^- \\ &= +\mathbf{e}_z \quad \text{on} \quad A^+. \end{aligned}$$

Then the magnetic force to accelerate the plate against the die is F_z and is given by

$$F_z = \frac{1}{\mu_0} \int_{A^+} - \int_{A^-} (B_z^2 - \tfrac{1}{2}B^2)\, da + \frac{1}{\mu_0} \int_{A_e} B_z B_e\, da \tag{5.4}$$

where B_e is the component of the magnetic field normal to the edge surface A_e. For a finite size coil, the magnetic field decreases with radial distance R from the center of the coil as a dipole, i.e.

$$\mathbf{B} \sim 0\left(\frac{1}{R^3}\right) \quad \text{as} \quad R \to \infty.$$

Thus as the area of the plate is made larger, the integral over A_e above becomes negligible. Also the integral over A^+ can be replaced by the hemispheric-like surface shown in Fig. 43, on which the fields decrease as $1/R^3$. Thus for plates of area dimensions much larger than the dimensions of magnet–plate gap, the force is found to be

$$F_z = \frac{1}{2\mu_0} \int_{A^-} (B_t^2 - B_z^2)\, da, \tag{5.5}$$

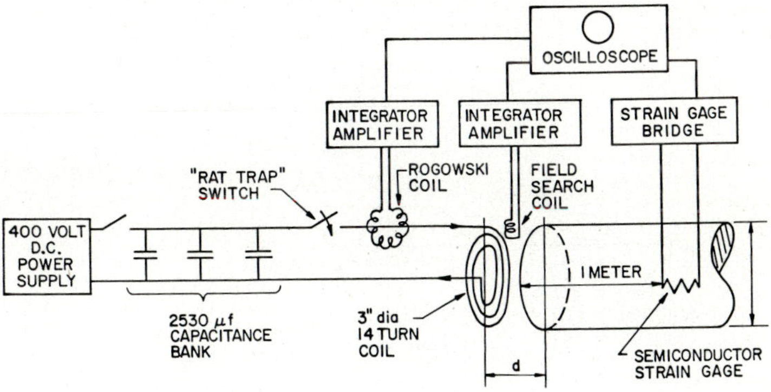

Fig. 44 Experimental apparatus for measuring magnetic forming forces.

where B_t is the tangential component of the magnetic field outside the magnet face of the plate.

For nonferromagnetic materials or small times after the coil current is switched on, the magnetic field does not penetrate the plate and hence $B_z \simeq 0$ and the force on the plate acts away from the magnet coil. For ferromagnetic materials the field will attempt to become normal to the plate surface, after sufficient time for diffusion, and the force might become negative and the plate will be attracted to magnet coil. Thus for materials such as nickel or steel the plate may not be accelerated against the forming die.

This effect can be demonstrated experimentally as shown in Figs. 44, 45. A pulsed current from a capacitance bank is sent into a coil placed against a 2-m long, 7-cm diameter steel bar. The bar is used to store the magnetic force history in a stress wave, which is measured with strain gages at the midpoint of the bar. The delay in the force or strain signal caused by the finite wave speed in the bar allows the strain signal to be separated in time from the electrical noise due to switching the current.

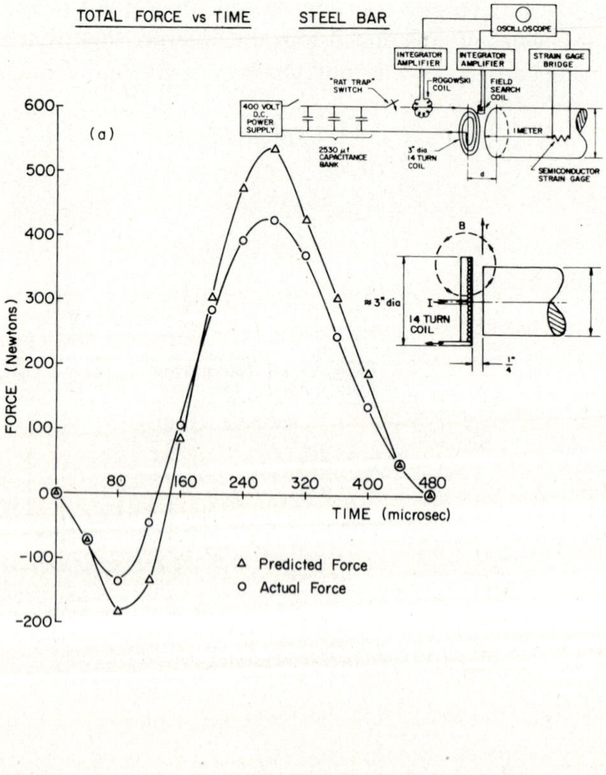

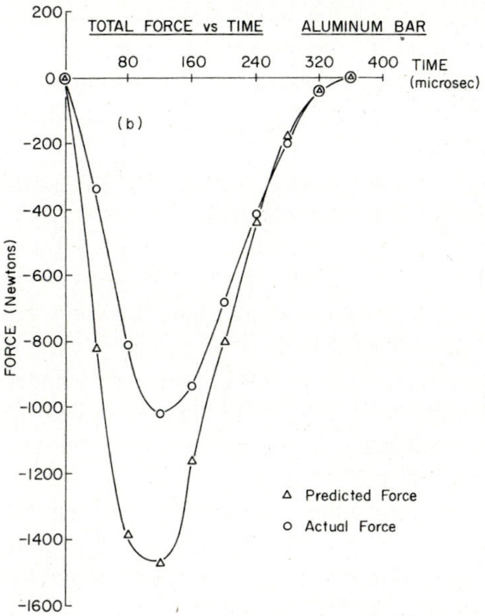

Fig. 45 (a) Force history in a steel bar induced by a pulsed magnetic forming coil near the end of the bar. (b) Force history in an aluminum bar. (Predicted stresses based on measured magnetic fields and the Maxwell stress tensor.

For a half sine pulse of current (Fig. 45) the magnetic force in the steel is initially repulsive (away from the coil) but soon becomes attractive. When an aluminum cap is epoxied to the end of the bar at the coil end, the force becomes completely repulsive.

Measurements of the magnetic field components at the end of the bar were made and numerical integration of the Maxwell stress tensor over the coil end of the bar was performed. Comparison of the measured force from the strain gage and the predicted force using the measured fields in (5.5) is shown in Figs. 45a, b for steel and aluminum. The difference between the measured and predicted forces is believed due to contributions in the integral over A_e in (5.5) which were neglected.

6 EPILOG

This review has touched on a broad range of problems. It is not the intent of this writer to convey to the reader that all these problems have been treated fully or that most of the interesting problems have been solved in magneto-solid mechanics. To the contrary, as long as electric power is central to an industrial world, then new and ever fascinating devices will continue to demand an ever-increasing understanding of the interaction of electromagnetic fields and solids.

As examples the interaction of electric arcs, electron beams, and lasers with solid materials should receive increasing attention. Also new methods in applied mathematics will be needed to solve inherently nonlinear problems resulting from forces of the type $\mathbf{J} \times \mathbf{B}$, $\mathbf{M} \cdot \nabla \mathbf{B}$ or Joule heating terms $\mathbf{J} \cdot \mathbf{J}$ or motion-induced currents $\sigma \mathbf{v} \times \mathbf{B}$, let alone the conventional constitutive and geometric nonlinearities usually encountered in mechanics.

While for many applications recourse to numerical techniques will certainly be required, finite difference and finite element codes coupling magneto, thermal and stress fields have only recently been developed (e.g. [150]) and further work in this area will be required.

Since the time of the original writing of this article the proceedings of the Sixth Symposium on Engineering Problems of Fusion Research has been published [151] which contains many papers or summaries of research on the stress analysis of superconducting magnets.

Acknowledgment. Supported in part by grants from the National Science Foundation (ENG75-09079) and the Energy Research and Development Administration (AEC No. AT(11-1)-2493 to Princeton University).

7 REFERENCES

1. Todhunter, I. and Pearson, K., *A History of the Theory of Elasticity and of Strength of Materials*, Dover Publ., Inc., N.Y.
2. Maxwell, J. C., *A Treatise on Electricity and Magnetism*, Dover Publ., Vols. I, II.
3. Jackson, J. D., *Classical Electrodynamics*, J. Wiley & Sons, New York (1962).
4. Stratton, J. A., *Electromagnetic Theory*, McGraw-Hill, N.Y. (1941).
5. Brown, W. F., *Magnetoelastic Interactions*, Tracts in Natural Philosophy, No. 9, C. A. Truesdell (ed.), Springer, Berlin (1966).
6. Penfield, P. and Haus, H. A., *Electrodynamics of Moving Media*, Research Monograph No. 40, M.I.T. Press, Cambridge, Mass (1967).
7. Pao, Y. H., Review Article in *Mechanics Today*, S. Nemat Nasser, Editor (this volume).
8. Toupin, R. A., "A Dynamical Theory of Elastic Dielectrics," *Int. J. Engng Sci.* **1**, 89–112.
9. Eringen, A. C., "On the Foundations of Electro-elastodynamics," *Int. J. Engng Sci.* **1** (1961) 127–153.
10. Hague, B., *The Principles of Electromagnetism Applied to Electrical Machines* (orig. Oxford Univ. Press, 1929), Dover Publ., N.Y. (1962).
11. Dwight, H. B., "Calculation of Magnetic Force on Disconnecting Switches," *Amer. Inst. E.E.* **39**, pt. II (1920) 1337–1355.
12. Higgins, T. J., "Formulas for Calculating Short-circuit Forces Between Channels Located Back to Back," *Trans. AIEE* **63** (Oct. 1944) 710–711.

13. Holm, Ragnar, *Electric Contacts*, 4th edition, Springer-Verlag, N.Y. (1967).
14. Knopoff, L., "The Interaction Between Elastic Wave Motions and a Magnetic Field in Electrical Conductors," *J. Geophys. Res.* **60** (1955) 441–456.
15. Tiersten, H. F., "Coupled Magnetomechanical Equations for Magnetically Saturated Insulators," *J. Math. Phys.*, **5** (1964) 1–21.
16. Strauss, W., "Magnetoelastic Properties of Yttrium–Iron Garnet," *Phys. Acoust.*, W. P. Mason (ed.), **IV** (1968), pt. B, Academic Press.
17. Kaliski, S. and Petykiewicz, J., "Dynamical Equations of Motion and Solving Functions for Elastic and Inelastic Anisotropic Bodies in the Magnetic Field," *Proc. Vibr. Publ.* **1**(2)(1959/1960) 17–35.
18. Paria, G., "Magneto-elasticity and Magneto-thermoelasticity," *In. Adv. Appl. Mech.* **10** (1967) 73–112, Academic Press.
19. Snowdon, A. C., "Studies of Electromagnetic Forces Occurring at Electrical Contacts," *Proc. AIEE Appl. Ind.* **24** (Mar. 1961).
20. Williams, J. E. C., *Superconductivity and its Applications*, Pion Ltd., London (1970).
21. Silvester, P. and Popovic, B. D., "The Integral Equations of Superconductive Levitation Systems," *Digests of the Intermag. Conf., Int. Magnetics Conf. IEEE, Toronto, Canada* (May 1974) 19.9.
22. Moon, F. C., "Mechanics of Structures with High Electric Currents," Princeton University Report AMS 1111, available NTIS (Order No. PB234253/AS) (July 1973).
23. Crandal, S. H., Karnopp, D. C., Kurtz, E. F., Jr. and Pridmore-Brown, D. C., *Dynamics of Mechanical and Electromechanical Systems*, McGraw-Hill Book Co., N.Y. (1968).
24. Reitz, J. R., "Forces on Moving Magnets due to Eddy Currents," *J. Appl. Phys.* **41**, No. 5 (April 1970) 2067–2071.
25. Frei, E. H., Shtrikman, S. and Treves, D., *Phys. Rev.* **106** (1957) 446.
26. Aharoni, A., "Complete Eigenvalue Spectrum for the Nucleation in a Ferromagnetic Prolate Spheroid," *Phys. Rev.* **131**, No. 4 (Aug. 15, 1963) 1478–1482.
27. DeBlois, R. W., "Ferromagnetic Properties of Single-crystal Nickel Platelets and Sub-micron Whiskers," *J. Appl. Phys.* **38**, No. 3 (Mar. 1967) 1291.
28. Mozniker, R. A., "Effect of a Uniform Magnetic Field on the Free Oscillations of Mechanical Systems" (in Russian), *Dopovidi Acad. Nauk Ukr. SSR* (1959) 847–852.
29. Panovko, Y. C. and Gubanova, I. I., *Stability and Oscillations of Elastic Systems* (Engl. transl.), Consultants Bureau (1965).
30. Moon, F. C., "Magnetoelastic Stability and Vibration," Ph.D. Thesis, Cornell University, Ithaca, N.Y. (Feb. 1967).
31. Moon, F. C. and Pao, Y.-H., "Magnetoelastic Buckling of a Thin Plate," *J. Appl. Mech.* **35**, No. 1 (1968) 53–58.
32. Moon, F. C. and Pao, Y.-H., "Vibration and Dynamic Instability of a Beam-plate in a Transverse Magnetic Field," *J. Appl. Mech.* **36**, No. 1 (1969) 92–100.
33. Moon, F. C., "The Mechanics of Ferroelastic Plates in a Uniform Magnetic Field," *J. Appl. Mech.* **37**, No. 1 (1970) 153–158.
34. Kaliski, S., "Quasi-static Approximation to the Equation of Elastic Vibrations in a Ferromagnetic Plate under the Action of a Transverse Magnetic Field," *Bull. de l'Academie Polonaise des Sci.* **XVII**, No. 9 (1969) 411–418.
35. Dunkin, J. W. and Eringen, A. C., "Propagation of Waves in an Electromagnetic Elastic Solid," *Int. J. Engng Sci.* **1** (1963) 461–495.
36. Pao, Y.-H. and Yeh, C.-S., "A Linear Theory for Soft Ferromagnetic Elastic Solids," *Int. J. Engng Sci.* **11**, No. 4 (Apr. 1973) 415.
37. Yeh, C.-S., "Linear Theory of Magnetoelasticity for Soft Ferromagnetic Materials and Magnetoelastic Buckling," Ph.D. Thesis, Cornell University, Ithaca, N.Y. (Jan. 1971).
38. Wallerstein, D. V. and Peach, M. O., "Magnetoelastic Buckling of Beams and Thin Plates of Magnetically Soft Material," *J. Appl. Mech.* **39**, No. 2 (1972) 451–455.
39. Dalrymple, J. M., Peach, M. O. and Viegelahn, G. L., "Magnetoelastic Buckling of Thin

Magnetically Soft Plates in the Cylindrical Mode," *J. Appl. Mech.* **41**, Series E, No. 1 (Mar. 1974) 145–150.
40. Dalrymple, J. M., Peach, M. O. and Viegelanh, G. L., "Magnetoelastic Buckling: Theory vs. Experiment," *Soc. Expl. Stress Anal.*, No. 2357 (May 1975).
41. Popelar, C. H., "Postbuckling Analysis of a Magnetoelastic Beam," *J. Appl. Mech.* **39** (Mar. 1972) 207–211.
42. Popelar, C. H. and Bast, C. O., "An Experimental Study of the Magnetoelastic Postbuckling Behavior of a Beam," *Exp. Mech.* **12**, No. 12 (Dec. 1972) 537–542.
43. Moon, F. C., "The Buckling of a Dielectric Fiber in an Electric Field," *Letters in Appl. Engng Sci.* **1** (July 1973) 327–336.
44. Love, A. E. H., *The Mathematical Theory of Elasticity*, Dover Publ. edition, New York.
45. Matteucci, C., *Annales de Chimie*, **XXVIII**, 1850, 493–499.
46. Birr, M. P. and Koryu Ishii, T., "Dynamic Stiffness of Magnetically Biased Laminated Iron Core," *Digests of the 1972 Intermag. Conf., Kyoto, Japan, IEEE Magnetics Soc.* (Apr. 1972) 8.5.
47. Srinivasan, S., "Vibration and Stability of Beams and Plates in the Presence of Electromagnetic Fields," Ph.D. Thesis, Ohio State University, Columbus, Ohio (1970).
48. Yuan, K., "Magneto-thermo-elastic Stresses in an Infinitely Long Cylindrical Conductor Carrying a Uniformly Distributed Axial Current," *Appl. Sci. Res.* **26** (1972) 307–314.
49. Roberts, B. W., "Oscillatory Magneto-acoustic Phenomena in Metals," *Phys. Acoustics*, W. P. Mason (ed.), **IV**, Part B (1968).
50. Steele, M. C. and Vural, B., *Wave interactions in Solid State Plasmas*, McGraw-Hill (1969).
51. Moon, F. C., "A Theory for High Current Elastic Conductors," *Recent Adv. in Engng Sci.* **5** (1970) 87–103.
52. Moon, F. C., "The Generation of Electromagnetic Radiation by Elastic Waves," *Dev. in Mech., Proc. of 12th Midwest Mech. Conf.* **6**, 623–631.
53. Green, A. E. and Naghdi, P. M., *Int. J. Engng Sci.* **3** (1965) 231.
54. Demiray, H., Ph.D. Thesis, Princeton University (1971).
55. Eringen, A. C., "Micromagnetism and Superconductivity," *J. Math. Phys.* **12**, No. 7 (July 1971) 1353–1358.
56. Tiersten, H. F., "An Extension of the London Equations of Superconductivity," *Physica* **37** (1967) 504–538.
57. Pipkin, A. C. and Rivlin, R. S., "Electrical Conduction in Deformed Isotropic Materials," *J. Math. Phys.* **1**, No. 2 (Mar.–Apr. 1960) 127–130.
58. Pipkin, A. C. and Rivlin, R. S., "Electrical Conduction in a Noncircular Rod," *J. Math. Phys.* **2**, No. 6 (Nov.–Dec. 1961) 865–868.
59. Pipkin, A. C. and Rivlin, R. S., "Non-rectilinear Current Flow in a Straight Conductor," *J. Math. Phys.* **3**, No. 2 (Mar.–Apr. 1962) 368–371.
60. Brechna, H., *Superconducting Magnet Systems*, Springer-Verlag, N.Y. (1973), pp. 131–150.
61. Montgomery, D. B., *Solenoid Magnet Design*, Wiley, New York (1969).
62. Landau, E. and Lifshitz, M., *Electrodynamics of Continuous Media*, Pergamon Press (1960).
63. Cockroff, J. D., "The Design of Coils for the Production of Strong Magnetic Fields," *Phil. Trans. Roy. Soc.* A **227** (1928) 317.
64. Middleton, A. J. and Trowbridge, C. W., "Mechanical Stresses in Large High Field Magnet Coils," *2nd Int. Conf. on Magnet Technology, Oxford* (July 1967), pp. 140–149.
65. Kilb, R. W. and Westendorp, W. F., *Stress Calculations for Cylindrical Magnetic Field Coils*, General Electric Co. Report 67-C-440, Schenectady, New York (Nov. 1967).
66. Westendorp, W. F., *Balancing of Magnetic Stresses and Winding Stresses in Superconducting Coils*, General Electric Co. Report 71-C-161 (June 1971).
67. Garden, P. O., "Mechanical Stresses in Bonded Plane Helical Solenoids with Arbitrary External Field," *J. Sci. Inst.*, Series 2, **I** (April 1968) 437.
68. Furth, H. P., Levine, M. A., and Waniek, R. W., "Production and Use of High Transient Magnetic Fields II," *Rev. Sci. Instr.* **28** (1957) 949–958.

69. Wakefield, K. E., *Design of Force-free Toroidal Magnets*, Plasma Physics Lab., Princeton Univ., N.J. (Mar. 1964), Report No. MATT-208.
70. Wells, D. R. and Mills, R. G., "Force Reduced Toroidal Systems," *High Magnetic Fields*, Kolm, H., Lax, B., Bitter, F. and Mills, R. (eds.), M.I.T. Press and J. Wiley & Sons, Inc., N.Y. (1962), Chap. 5, pp. 44–47.
71. Goldstein, H., *Classical Mechanics*, Addison-Wesley Publ. Co., Reading, Mass. (1950), p. 69.
72. Parker, E. N., "Reaction of Laboratory Magnetic Fields Against Their Current Coils," *Phys. Rev.* **109** (1958) 1440.
73. Levy, R. H., "Author's reply to Willinski's Comment on Radiation Shielding of Space Vehicles by Means of Superconducting Coils'," *Amer. Rocket Soc. J.* **32** (May 1962) 787.
74. Hein, R. A., "Superconductivity: Large Scale Applications," *Science* **185**, No. 4147 (July 1974) 211–222.
75. DeMichele, D. W. and Darby, J. B., Jr., "An Analysis of the Material Constraints on Superconducting C.T.R. Magnets," *Proceedings of Texas Symposium on Technology of Controlled Thermonuclear Fusion Experiments and Engineering Aspects of Fusion Reactors* (Nov. 20–23, 1972), Austin, Texas.
76. Meyer, G. and Maix, R., "Superconductors and Superconducting Magnets," *Brown Boveri Review*, No. 8/9 (1970) 1–8.
77. Greene, W. J. and Saibel, E., "Stability of Internally Cooled Superconductors," *Adv. Cryogenic Engng* **14**, Proc. 1968 Cryo Engng Conf. Case West Res. Univ., Cleveland, Plenum Press, N.Y. (1969).
78. Sarafi, Z. W. and Meyn, E. H., "Effect of Mechanical Vibrations on the Performance of Superconducting Magnets," NASA-TN-D3954 (May 1967).
79. File, J., Mills, R. G. and Sheffield, G. V., "Large Superconducting Magnet Designs for Fusion Reactors," *4th Symp. on Engng Problems of Fusion Research-Naval Res. Lab., Wash., D.C.* (Apr. 1971).
80. Frankenberg, J., Citrolo, and Bonanos, P., "PLT Confining Field Coil-loading and Stresses," *Proc. 5th Sym. on Engng Problems of Fusion Res., Princeton Univ., Nov. 1973*, IEEE Publ. No. 73, CH0843-3-NPS, 332–334.
81. Badger, B., et al., "UWMAK-1: A Wisconsin Toroidal Fusion Reactor Design," University of Wisconsin, UWFDM-6, Nov. 20, 1973, revised: Mar. 15, 1974; see also Moses, R. W. Jr. and Young, W. C., "Analytical Expressions for Magnetic Forces on Sectored Toroidal Coils," *Proc. 6th Symp. Engng Problems of Fusion Research* (Nov. 1975), IEEE Pub., N.Y. (1976), pp. 917–921.
82. Moon, F. C. and Chattopadhyay, S., "Elastic Stability of a Thermonuclear Reactor Coil," *Proc. 5th Symp. on Engng Problems of Fusion Research, Nov. 1973*, Princeton, N.J. IEEE Nuclear and Plasma Sci. Soc., N.Y., Publ. No. 73, CH 0843-3-NPS (Apr. 1974), pp. 574–578.
83. Moon, F. C. and Swanson, C., *Vibration and Stability of a Set of Superconducting Toroidal Magnets, J. Appl. Phys.* **47**, No. 3 (March 1976) 914–919.
84. Moon, F. C., "Buckling of a Superconducting Coil Nested in a Three-Coil Toroidal Segment," *J. Appl. Phys.* **47**, No. 3 (March 1976) 920–921.
85. Lubell, M. S., Cannon, D. D., Burn, P. B., Shannon, T. E. and Anderson, J. L., "Toroidal Field System," Appendix D.7, pp. D-59-D85 in ORMAK F/BX; *A Tokamak Fusion Test Reactor*, Hanbenreich, P. N. and Roberts, M. (eds.), Oak Ridge Nat. Lab., U.S. AEC Report ORNL-TM-4634.
86. Powell, J. R. and Bezler, P., *Warm Reinforcement and Cold Reinforcement Magnet Systems for Tokamak Fusion Power Reactors; A Comparison*, Brookhaven National Laboratory Report BNL-17434 (Nov. 1972).
87. DeMichele, D. W. and Darby, J. B., "Three-dimensional Mechanical Stresses in Toroidal Magnets for Controlled Thermonuclear Reactors," *Proc. 5th Symp. on Engng Problems of Fusion Research*, IEEE Publ. No. 73 CH-0843-3-NPS, pp. 558–569.
88. Stevenson, R. and Atherton, D., "Diamagnetic Forces in Superconducting Magnets," *IEEE Trans. Magnetics*, Vol. MAG-11, No. 2 (Mar. 1975) 528–531.

89. Walstrom, P. L., "Diamagnetic Effects in a Tape Wound Superconducting Solenoid," *J. Appl. Phy.* **45**, No. 5 (May 1974) 2293–2295.
90. Thompson, W. B., *An Introduction to Plasma Physics*, Pergamon Press, N.Y., pp. 106–108 (1962).
91. Leontovich, M. A. and Shafranov, V. D., "The Stability of a Flexible Conductor in a Magnetic Field," *Plasma Physics and the Problem of Controlled Thermonuclear Reactions*, **1** (1961), Pergamon Press.
92. Dolbin, N. I., "Propagation of Elastic Waves in a Current-carrying Rod," *PMTF*, No. 2 (1962) (in Russian).
93. Dolbin, N. I. and Morozov, A. I., "Elastic Bending Vibrations of a Rod Carrying Electric Current," *Zh. Prik. Mekh. i Techni. Fiziki* (in translation) **3** (1966) 97–103.
94. Prudnikov, V. V., "Elastic Oscillations of a Current-carrying Rod in a Longitudinal Magnetic Field," *Zh. Prik. Mekh. i Tekhni. Fiziki* **9**, No. 1 (1968) 168–172.
95. Chattopadhyay, S. and Moon, F. C., "Magnetoelastic Buckling and Vibration of a Rod Carrying Electric Current," *J. Appl. Mech.* **42** (Dec. 1975) 809–814.
96. Chattopadhyay, S., "Vibration and Stability of Structures Carrying High Electric Currents," Doctoral Dissertation, Princeton University (1973).
97. Vandakurov, Yu-V. and Kolesnikova, E. N., "Stability of a Solid Conducting Cylinder in the Magnetic Field of its Self Current," *Soc. Phys.-Tech. Phys.* **12**, No. 11 (May 1968) 1458–1464.
98. Abramova, K. B., Valitskii, V. P., Vanderkurov, Yu. V., Zlatin, N. A. and Peregud, B. P., "Magnetohydrodynamic Instabilities During Electrical Explosion," *Sov. Phys.-Doklady* **11** No. 4 (Oct. 1966) 301–304.
99. Leibowitz, M. A. and Ackerberg, R. C., "The Vibration of a Conducting Wire in a Magnetic Field," *Quart. J. Mech. Appl. Math.* **XVI**, pt. 4 (1963) 507–519.
100. Smith, T. E. and Herrmann, G., "Stability of Circulatory Elastic Systems in the Presence of Magnetic Damping," *Acta Mech.* **12**, (1971) 175–188.
101. Woodson, H. H. and Melcher, J. R., *Electromechanical Dynamics*, Parts I, II, III, Wiley (1968).
102. Culver, C. G., "Natural Frequencies of Horizontally Curved Beams," *Proc. Amer. Soc. Civil Engineers, J. of the Structural Div.* (Apr. 1967) 189–203.
103. Brillouin, L., *Wave Propagation in Periodic Structures*, Dover Publ., N.Y. (1953).
104. Benz, M. G. and Coffin, L. F., Jr., "Mechanical and Electrical Properties of Non-symmetric Copper Nb_3–Sn Stainless Steel Composite Superconductors," *2nd Int. Conf. on Magnet Technology, Oxford* (1967), pp. 513–518.
105. Easton, D. S. and Koch, C. C., "Mechanical Properties of Superconducting Nb–T_i Composites," *Proc. Cryogenic Engng Conf., Kingston, Ont., Canada* (July 1975).
106. Hudson, W. R., "Effect of Tensile Stress on Current-carrying Capacity of Commercial Superconductors," NASA Technical Note TN-D-3745 (1966).
107. Ekin, J. W., Fickett, F. R. and Clark, A. F., "Effect of Stress on the Critical Current of Nb–T_i Multifilimentary Composite Wire," *Proc. of Cryogenic Engng Conf., Kingston, Ont., Canada* (July 1975), Paper J-3.
108. Fisher, E. S., Kim, S. H. and Linz, R. J., "Effect of Cyclic Strain on Electrical Resistivity of Cu at 4.2°K," *Proc. Int. Cryogenic Materials Conf., Kingston, Ont., Canada* (July 1975).
109. Sachse, W., "Group Velocity of R.F. Bursts and Attenuation of Ultrasonic Pulses in Nb–Ti-Filamentary Composite Superconductor Wires," *Ultrasonics Symposium Proc.*, IEEE Cat. no. 75CHO-994-4SU (1975).
110. Furth, H. P., "Magnetic Pressure," *Int. Sci. Tech.* (Sept. 1966), pp. 32–40.
111. Schlömann, E., "Generation of Phonons in High Power Ferromagnetic Resonance Experiments," *J. Appl. Phy.* **31**, No. 9 (Sept. 1960) 1647–1656.
112. Robey, D. H., "Magnetic Dispersion of Sound in Electrically Conducting Plates," *J. Acoust. Soc. Amer.* **25**, No. 4 (July 1953) 603–609.

113. Galkin, A. A. and Koroliuk, A. P., "Dispersion of Sound in Metals in a Magnetic Field," *J. Expl Theoret. Phys.* (USSR) **34** (Apr. 1958) 1025–1026.
114. Alers, G. A. and Fleury, P. A., "Modification of the Velocity of Sound in Metals by Magnetic Fields," *Phys. Rev.* **129**, No. 6 (Mar. 1963) 2425–2429
115. Pao, Y.-H. and Hutter, K., "Magnetoelastic Waves in Soft Ferromagnetic Solids."
116. Nayfeh, A. H. and Nemat-Nasser, S., "Electromagneto-thermoelastic Plane Waves in Solids with Thermal Relaxation," *J. Appl. Mech.* (Mar. 1972) 108–113.
117. McCarthy, M. F., "The Propagation and Growth of Plane Acceleration Waves in a Perfectly Electrically Conducting Elastic Material in a Magnetic Field," *Int. J. Engng Sci.* **4** (1966) 361–381.
118. Recker, W. W., "A Difference Method for Plane Problems in Magnetoelastodynamics," *J. Appl. Mech.* **39**, No. 3 (Sept. 1972) 689–695.
119. Snell, MacKallor, and Guernsey, "An Electromagnetic Plane Stress Wave Generator," *Expl. Mech.* (Nov. 1973) 472–479.
120. Moon, F. C. and Chattopadhyay, S., "Magnetically Induced Stress Waves in a Conducting Solid—Theory and Experiment," *J. Appl. Mech.* **41**, No. 3 (Sept. 1974) 641–646.
121. Percival, C. M., "Laser-generated Stress Waves in a Dispersive Elastic Rod," *J. Appl. Phys.* **38**, No. 13 (Dec. 1967) 5313–5315.
122. Richter, B. J., "Stress Waves in Conductors Induced by the Incompletely Neutralized Electromagnetic Fields of Relativistic Electron Beams," Ph.D. Thesis, Stanford Univ. (1971).
123. Morland, L. W., "Generation of Thermoelastic Stress Waves by Impulsive Electromagnetic Radiation," *AIAA J.* **6**, No. 6 (June 1968) 1063–1066.
124. Kennedy, J. D. and Curtis, C. W., "Transient Electron-inertia Field Produced by a Strain Pulse," *J. Acoust. Soc. Amer.* **41** (2) (Feb. 1967) 328–325.
125. Birdsall, D. H., Ford, F. C., Furth, H. D. and Riley, R. E., "Magnetic Forming," *Amer. Mach.* (Mar. 1961), pp. 117–121.
126. Anon., *The Electromagnetic Hammer*, National Aeronautics and Space Administration Technology Utilization Report, NASA SP-5034 (Dec. 1965).
127. Furth, H. P., Levine, M. A. and Waniek, R. W., *Rev. Sci. Instr.* **28** (1957) 949.
128. Furth, H. P., *Pulsed Magnets*, Chap. 22 "High Magnetic Fields," Kolm, H. *et al.* (eds.), M.I.T. Press, Proc. of Int. Conf. on High Magnetic Fields (Nov. 1961).
129. Waniek, R. W. *et al.*, "Research, Development and Manufacture of Magnetic Forming Coils," Advanced Kinetics Inc., NASA Contract NAS-8-11757 (Mar. 1965).
130. Waniek, R. W., "New Concepts for High Energy Rate Forming System Summary Report," Advanced Kinetics Inc., Costa Mesa, Calif. Avail. NTIS (Aug. 1965), NASA-CR-74918.
131. Lippmann, H. J. and Schreiner, H., "Zur Physik der Metallumforming mit Hohen Magnetfeldimpulsen," *Zeit. fur Metallkind* **55** (1964) 737–740.
132. Baines, K., Duncan, J. L., and Johnson, W., "Electromagnetic Metal Forming," *Proc. Instn. Mech. Engrs* **180** (1965–66) 348–362 (pt. 3I).
133. Barbarovich, Yu. K., "Use of the Energy of a Strong Pulsed Magnetic Field for Powder Compaction," Engl. trans. of *Poroshkovaya Metallurgiya*, No. 10 (82) (Oct. 1969) 798–803 (orig. pp. 24–31).
134. Leftheris, B., "Design, Development, Manufacture, and Deliver New Concepts for High Energy Rate Forming System Quarterly Report No. 11," Republic Aviation Corp., Mineola, N.Y., NASA-CR-60848 (Jan. 1965).
135. Meyer, M. D., *Impact Welding Using Magnetically Driven Flyer Plates,* Sandia Laboratories Report SLL-73-5006, Mar. 1973, Livermore, Calif., U.S.A.
136. Lawrence, W. N., "Scale Modelling Calculations for Electromagnetic Forming," *2nd Int. Conf. on High Energy Forming, Estes Park, Colorado* (June 1969).
137. Gerber, H. L., "Diffusion Effects in Magnetic Forming," Ph.D. Thesis Illinois Inst. of Tech., Chicago, Univ. Microfilms No. 73-3623.

138. Lal, G. K. and Hillier, M. J., "The Electrodynamics of Electromagnetic Forming," *Int. J. Mech. Sci.* **10**, 491–500.
139. Bauer, Dietrich, "New Method for Measuring Forces, Work Processes, Form Changes, Rates of Form Changes and Strength After Form Changes During Widening of Cylindrical Work Pieces by Fast Changing Magnetic Fields," (in German) from Technischen Hochschule, Hannover (1967).
140. Al-Hassani, S. T. S., Duncan, J. L. and Johnson, W., "On the Parameters of the Magnetic Forming Process," *J. Mech. Engng Sci.* **16**, No. 1 (1974) 1–9.
141. Freeman, J. R., Cnare, E. C. and Waag, R. C., "Magnetically Imploded Metal Foils," *Appl. Phys. Letters*, **10**, No. 4 (15 Feb. 1967) 11–113.
142. Knoepfel, Heinz, *Pulsed High Magnetic Fields*, North-Holland Publ., London (1970).
143. Al-Hassani, S. T. S. and Johnson, W., "The Magnetomotive Loading of Cantilevers, Beams and Frames," *Int. J. Mech. Sci.* **12** (1970) 711–722.
144. Forrestal, M. J. and Overmier, D. K., "An Experiment on an Impulse Loaded Elastic Ring," *AIAA J.* **12**, No. 5 (May 1974) 722–724.
145. Walling, H. C. and Forrestal, M. J., "Elastic-plastic Expansion of 6061-76 Aluminum Rings," *AIAA J.* **11**, No. 8 (Aug. 1973) 1196–1197.
146. Wessenburg, D. L., *Dynamic Buckling of Thin Aluminum, Circular Cylindrical Shells Subjected to Uniform Impulse Loading*, Sandia Lab. Report SLL 73-0045 (1973).
147. Al-Hassani, S. T. S., "Plastic Buckling of Thin Walled Tubes Subject to Magnetomotive Force," *J. Mech. Engng Sci.* **16** (Apr. 1974).
148. Somon, J. P., "The Dynamical Instabilities of Cylindrical Shells," *J. Fluid Mech.* **38** (1969) 769–791.
149. Herlach, F. and McBroon, R., "Megagauss Fields in Single Turn Coils," *J. Physics E.: Sci. Instr.* **6** (1973) 652.
150. Miya, K., Aw, S., Ando, Y., Ohta, M. and Suzuki, Y., "Application of Finite Element Method to Electro-mechanical Dynamics of Superconducting Magnet Coil and Vacuum Vessel," *Proc. 6th Symp. of Engng Problems of Fusion Research, Nov. 1975*, IEEE Publ., N.Y. (1976).
151. *Proceedings of the Sixth Symposium on Engineering Problems of Fusion Research, Nov. 1975, San Diego*, IEEE Publ. No. 75CH1097-5-NPS, N.Y. (1976).

VI

On Nonequilibrium Thermodynamics of Continua: Addendum

S. Nemat-Nasser

Northwestern University, Evanston, Illinois

1 INTRODUCTION

This article is supplementary to the article which appeared as Chapter II of Volume 2 of this series [1]. Here, the implications of the second law of thermodynamics, as stated by Eqs. (4.4) and (4.5) in [1], are considered, and by means of an example, it is shown that these statements are too restrictive, and should be relaxed, in order to apply to many commonly used models for the macroscopic behavior of materials.† In addition, the consequences of (thermodynamic) *material stability* are studied, and the corresponding restrictions that this may place on the evolutionary equations for internal variables are explored. In particular, it is shown that when the material is thermodynamically stable (as will be defined later on), and if there are only eight scalar-valued internal variables, then, under additional rather mild assumptions, the rate of change of the internal variables (i.e. the fluxes) must necessarily be proportional to the gradient of a *scalar function of state*, taken with respect to the internal forces. This result is then used to show that the *inelastic* rate of change of strain, the (inelastic) rate of stress relaxation, the rate of entropy production due to internal dissipation, and the rate of change of temperature due to internal relaxation are proportional to the gradient of a suitable potential function (an *inelastic* potential depending on *state variables* only) taken with respect to the stress, strain, temperature, and entropy, respectively (i.e. a general set of normality rules).

† Except for the theorem on p. 123 of [1], no other result presented in [1] is affected by this modification.

2 PRELIMINARIES

Let $\mathbf{C}$, θ, e, and $\mathbf{S}$ denote, respectively, Green's deformation tensor relative to a suitably chosen reference configuration $\mathscr{C}$, the empirical temperature ($\theta > 0$), the internal energy density per unit mass ($e > 0$), and the second Piola–Kirchhoff stress† (measured and taken relative to $\mathscr{C}$). Because the thermomechanical properties of inelastic solids depend to a large extent on the history of the deformation and temperature changes that they have undergone, it appears a straightforward and logical conclusion that the stress and internal energy at a given particle of a continuum are functionals of the entire history up to and including the present time, of the deformation-temperature of that particle, i.e.

$$\mathbf{S} = \underset{s=-\infty}{\overset{t}{\mathbf{S}}} (\mathbf{C}(s), \theta(s)), \qquad e = \underset{s=-\infty}{\overset{t}{e}} (\mathbf{C}(s), \theta(s)). \tag{2.1}$$

Since these functionals can be of very general form, it may appear that (2.1) is a very general description of the thermomechanics of inelastic solids. However, unless (2.1) is given a more explicit structure, it is no more than a mark on paper. One such explicit structure is obtained by a nonlinear multiple integral representation [2] of the functionals in (2.1). In practice, one has to be content with only the first few integrals, and even then it is a formidable task to evaluate the corresponding kernels. Hence, the method has so far met with limited success; for a summary of the state of the art in viscoelasticity, and references, see [3].

An alternative to the above *explicit* approach is to have an *implicit* representation of constitutive relations; see [1] for discussion and references. In this approach, one envisages a *material system* which is homogeneous and has unit mass, and undergoes homogeneous deformation and temperature changes. Then one introduces a set of parameters, collectively denoted by ξ, and one writes the stress and internal energy as *functions* of the *observable* variables $\mathbf{C}$ and θ, as well as the additional parameters ξ, to obtain

$$\mathbf{S} = \hat{\mathbf{S}}(\mathbf{C}, \theta; \xi), \qquad e = \hat{e}(\mathbf{C}, \theta; \xi). \tag{2.2}$$

The set ξ may consist of finite, or possibly infinite, elements, each element being a tensor of a suitable order. In fact, this set may be divided into subsets, each subset having a finite, or possibly an infinite, number of elements of a given tensorial order. Some, or all, of these subsets may be dense or

† This stress is conjugate to the Lagrangian strain $\mathbf{E} = \frac{1}{2}(\mathbf{C} - \mathbf{I})$, where $\mathbf{I}$ is the identity tensor, in the sense that $\mathbf{S} \cdot \dot{\mathbf{E}} = \frac{1}{2}\mathbf{S} \cdot \dot{\mathbf{C}}$ is the rate of work per unit mass; superposed dot denotes material time derivative. Note that here the rate of work is per unit mass, whereas usually it is defined per unit initial volume.

they may be discrete in their mathematical character. In practice, however, one first considers the simplest possibility before proceeding to more complex cases. Indeed, it is for this very fact that this *implicit* representation has the advantage over the *explicit* representation (2.1).

In the sequel it will be assumed that there are m real numbers ξ_i, $i = 1, 2, \ldots, m$, which define the set ξ completely. The extension to more general cases is only of technical mathematical nature, and therefore does not relate to the physics of the theory.

Within the framework of a macroscopic nonequilibrium thermodynamics it is unnecessary to associate the parameters ξ (which will also be called hidden or internal variables) with any individual microscopic structural changes that are in reality instrumental in the inelastic behavior of solids. *With this viewpoint, ξ's are regarded as average global parameters which in an average approximate manner manifest the most macroscopically dominant effect of, generally, several microscopic structural changes. The accuracy of this representation, therefore, will clearly depend on the extent to which ξ's measure the relevant microscopic rearrangements in a given solid.*

The first law, i.e. the conservation of energy, in the present context and for a material system, takes on the form†

$$dQ = de - \tfrac{1}{2}\hat{\mathbf{S}} \cdot d\mathbf{C}, \tag{2.3}$$

where dQ is the increment of heat added to the material system, and $\mathbf{S} = \hat{\mathbf{S}}(\mathbf{C}, e; \xi)$; i.e. $\mathbf{C}$, e, and ξ are used as state variables. Equation (2.3) states that the change in internal energy equals the work done by the external forces plus the heat supplied.

3 SECOND LAW

Since the internal variables ξ are essentially unobservable, and hence uncontrollable, it is natural to seek a statement for the second law, which does not involve these parameters directly. In [1], p. 119, statements (4.4) and (4.5) are of this kind. In particular, statement (4.5) of [1] can be written as:

> A material system in a state $\{\mathbf{C}_0, \theta_0, \xi_0\}$, having internal energy e_0, cannot adiabatically attain nearby states which have the same deformation measure $\mathbf{C}_0$, and internal energy $e < e_0$, no matter what the corresponding initial, intermediate, and *final* values of the internal variables ξ may be. (3.1)

† The notation used here is consistent with that of [1].

This statement implies that in an adiabatic cycle of deformation the internal energy of the material system cannot decrease, no matter what the initial and the final values of the internal variables may be. In this sense, it is extremely restrictive. For example, it excludes material behavior which is modelled by linear viscoelasticity.

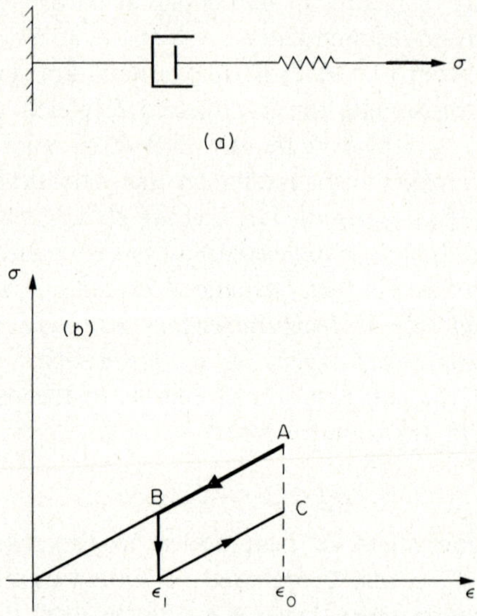

Fig. 1

To show this, consider the Maxwell model which consists of a linear spring in series with a linear dashpot; Fig. 1a. In terms of the state variables, this model has one deformation measure, i.e. the total strain ε, and one scalar internal variable. For the initial state consider the spring extended to strain ε_0. At this state allow the spring to relax *rapidly* to strain $\varepsilon_1 < \varepsilon_0$, doing work equal to the area of trapezoid $AB\varepsilon_1\varepsilon_0$; see Fig. 1b. At this state hold the total strain ε_1 constant, allowing the dashpot to move and the spring to relax from stress σ_1 to zero stress. This will involve no external work. Now, change the strain instantaneously from ε_1 to its initial value ε_0, which requires work equal to the area of triangle $C\varepsilon_1\varepsilon_0$. The net work in this strain cycle is negative (positive work is done by the material system). The total internal energy is reduced from its initial value by an amount equal to the area of rhomboid $AB\varepsilon_1C$. Therefore, if one does not restrict the final value of the internal variable, then statement (3.1) excludes models of the above kind. A similar problem exists in classical plasticity. (A counter-example in this case has been suggested by Prof. J. Rice; this will not be discussed here.)

In order to eliminate difficulties of this kind, one is forced to modify statement (3.1) to include restrictions on the *final values* of the internal variables. An acceptable statement is the following:

> Second Law: A material system in a state $\{\mathbf{C}_0, \theta_0, \xi_0\}$ having internal energy e_0, cannot adiabatically attain nearby states which have the same deformation measure $\mathbf{C}_0$, *the same internal variables* ξ_0, and have internal energy $e < e_0$, no matter what the intermediate values of the deformation measure and the internal variables may be. (3.2)

In terms of Carathéodory's adiabatic inaccessibility, (3.2) is generalized to:

> Second Law: Every ε-neighborhood of a given state of a material system M, contains a (connected) set of states that cannot be attained by M in adiabatic transitions. (3.3)

One may object to statement (3.2) on the grounds that it is not possible to control the final values of the internal variables. However, since statement (3.2) does not contradict this, the objection cannot carry much force.

The second law in the form of (3.2) or (3.3) can now be used to establish the existence of the entropy function, $\eta = \hat{\eta}(\mathbf{C}, e; \xi_0)$, and the absolute temperature, in exactly the same manner as discussed in [1]. There is no change in the basic logic, because to establish the fact that there exists an integrating factor, $\theta = \hat{\hat{\theta}}(\mathbf{C}, e; \xi_0)$, for the energy balance equation (2.3) which does not explicitly involve the increments of the internal variables, one fixes the internal variables at $\xi = \xi_0$, and then employs statement (3.3) or (3.2) in this special context; see the discussion which follows Eq. (4.6) in [1]. In fact, for this purpose, the following restrictive statement is amply sufficient: in every ε-neighborhood of the given state, $\{\mathbf{C}_0, \theta_0, \xi_0\}$, of a material system there is a connected set of states in the hyperplane, $\xi = \xi_0$, which cannot be adiabetically attained from the given state; see [4], pp. 304–316.

Whatever argument may be employed, the final result as far as the integrability of the first law (2.3) is concerned, will remain the same. Hence, one obtains from (2.3),

$$\hat{\hat{\theta}} = \frac{1}{(\partial\hat{\hat{\eta}}/\partial e)}, \qquad \hat{\hat{\mathbf{S}}} = -2\,\frac{(\partial\hat{\hat{\eta}}/\partial \mathbf{C})}{(\partial\hat{\hat{\eta}}/\partial e)}. \tag{3.4}$$

3.1 On the Principle of Nondecreasing Entropy

There are a number of ways by which one can prove from a suitable statement of the second law that in any adiabatic transition of a material system from a given state to a neighboring one, the entropy must not decrease. If the internal variables ξ are given the same status as that of the deformation variables $\mathbf{C}$, in the sense that they can be controlled (i.e. if one assumes that from a given state of a material system all neighboring states with prescribed deformation and internal variables are attainable, provided that the temperature or the internal energy is allowed to take on suitable values), then the principle of adiabatic inaccessibility (3.3) can be used to prove that the entropy function must be monotonic for adiabatic transitions; see [4]. In a similar manner, if one uses a very restrictive statement of the second law, which states the adiabatic inaccessibility principle totally in terms of the observable variables, allowing the internal variables to attain suitable values, one can prove the monotone nature of entropy for adiabatic transitions; this was done in [1], Subsection 4.5. Other alternatives can also be pursued. However, as in the case of classical thermodynamics, one may choose to state the monotone character of the entropy function for adiabatic transitions as "the second part of the second law". We shall briefly summarize this below (see [4], Sect. 6, pp. 317–322).

To this end one assumes that the entropy function, $\eta = \hat{\hat{\eta}}(\mathbf{C}, e; \xi)$, is a continuous and continuously differentiable function of *all* its arguments in a suitably chosen neighborhood of a given state. The rate of change of entropy then becomes

$$\dot{\eta} = \frac{\partial \hat{\hat{\eta}}}{\partial \mathbf{C}} \cdot \dot{\mathbf{C}} + \frac{\partial \hat{\hat{\eta}}}{\partial e} \dot{e} + \frac{\partial \hat{\hat{\eta}}}{\partial \xi} \cdot \dot{\xi}$$

$$= \frac{1}{\hat{\hat{\theta}}} \{\dot{e} - \tfrac{1}{2}\hat{\hat{\mathbf{S}}} \cdot \dot{\mathbf{C}}\} + \frac{\partial \hat{\hat{\eta}}}{\partial \xi} \cdot \dot{\xi} = \frac{\dot{Q}}{\hat{\hat{\theta}}} + \frac{\partial \hat{\hat{\eta}}}{\partial \xi} \cdot \dot{\xi} \tag{3.5}$$

For adiabatic transitions, $\dot{Q} = 0$. The quantity $\partial \hat{\hat{\eta}} / \partial \xi$ is a state function, and therefore has a fixed value at a given state. To require that the rate of entropy must not decrease in any adiabatic transition is to require that the rate of change of internal variables, $\dot{\xi}$, must be restricted in such a manner that

$$\frac{\partial \hat{\hat{\eta}}}{\partial \xi} \cdot \dot{\xi} \geq 0, \tag{3.6}$$

which, in turn, restricts the evolution of the internal variables. In [4], transitions which are in compliance with (3.6) were called *admissible*. Statement (3.6) may be postulated as the second part of the second law for inelastic materials.

3.2 Internal Forces

Having established the existence of an entropy function, one may choose $\mathbf{C}$, θ, and ξ as state variables, introduce Helmholtz free energy

$$\psi = \hat{\psi}(\mathbf{C}, \theta; \xi), \qquad \hat{\psi} = \hat{e} - \theta\hat{\eta}, \tag{3.7}$$

where $\eta = \hat{\eta}(\mathbf{C}, \theta; \xi)$, and obtain for the stress, $\mathbf{S} = \hat{\mathbf{S}}(\mathbf{C}, \theta; \xi)$, and entropy, $\eta = \hat{\eta}(\mathbf{C}, \theta; \xi)$,

$$\hat{\mathbf{S}} \equiv 2\frac{\partial\hat{\psi}}{\partial\mathbf{C}}, \qquad \hat{\eta} \equiv -\frac{\partial\hat{\psi}}{\partial\theta}. \tag{3.8}$$

Since $\theta > 0$, one also obtains from (3.6),

$$\frac{\partial\hat{\psi}}{\partial\xi}\cdot\dot{\xi} \leq 0. \tag{3.9}$$

One can introduce internal forces, $\mathbf{\Lambda} = \hat{\mathbf{\Lambda}}(\mathbf{C}, \theta; \xi)$, conjugate to the internal variables as

$$\mathbf{\Lambda} \equiv -\frac{\partial\hat{\psi}}{\partial\xi}. \tag{3.10}$$

Then (3.9) becomes

$$\hat{\mathbf{\Lambda}}\cdot\dot{\xi} \geq 0, \tag{3.11}$$

which must be regarded as an alternative expression of the second part of the second law of thermodynamics. It states that the rate of work (i.e. energy dissipation) of the internal forces must be nonnegative. Note carefully that *the internal forces are state functions, hence* (3.11) *restricts the rate of change of internal variables. In particular, it states that one cannot reverse, at a given state, the rate of change of internal variables.* This is important, since it brings into focus the difference between internal variables and the observable variables $\mathbf{C}$ and θ: at a given state it is assumed that one can prescribe arbitrarily the rate of change of the observable variables, i.e. *externally* one can control and prescribe $\dot{\mathbf{C}}$ and $\dot{\theta}$ at a given state; in contrast, one cannot do the same for $\dot{\xi}$. For example, at every state one can assign either $\{\dot{\mathbf{C}}_0, \dot{\theta}_0\}$ or $\{-\dot{\mathbf{C}}_0, -\dot{\theta}_0\}$ to the deformation and temperature rates. On the other hand, the second law states that if $\dot{\xi}_0$ is an *admissible* rate at a given state, then $-\dot{\xi}_0$ is not; note that this is a macroscopic observation; for some discussion of this matter, see pp. 223–227 of [5].

4 ON THE NATURE OF EVOLUTIONARY EQUATIONS

It is reasonable and essential to require that the internal variables should be independent parameters, that is, one should select the ξ's in such a manner that no one of these variables can be expressed as a function of the others; otherwise this variable should be eliminated in favor of the others. Hence, it is essential to require that the vector equation (3.10) is (at least locally) invertible. Considering a finite number of internal variables ξ_i, $i = 1, 2, \ldots, m$, a necessary and sufficient condition for this local invertibility is that the Jacobian determinant, $J = \det |\partial^2\hat{\psi}/\partial\xi_i\xi_j|$, be sign-definite, either positive- or negative-definite.

The material will be called *thermodynamically stable*, if at constant $\mathbf{C}$ and θ,

$$\frac{\partial^2\hat{\psi}}{\partial\xi_i\,\partial\xi_j}\dot{\xi}_i\dot{\xi}_j > 0 \qquad \text{for} \quad \dot{\xi} \neq \mathbf{0}, \tag{4.1}$$

i.e. the Jacobian determinant J is positive-definite for all admissible non-equilibrium states; this implies that the eigenvalues of the symmetric matrix $[\partial^2\hat{\psi}/\partial\xi_i\,\partial\xi_j]$ are all positive functions of state. Among other things, material stability condition (4.1) implies that, at constant deformation and temperature (that is, if we suddenly fixed and held constant the strains and the temperature of a material system), the vector of the rate of change of the internal forces, $\dot{\mathbf{\Lambda}}$, makes an obtuse angle with the corresponding fluxes.† This in turn implies that the internal forces "relax" if the deformation and temperature of a material system are suddenly fixed and held constant thereafter. Clearly, this is a minimum that one must require for thermodynamic material stability.

Condition (4.1) implies that by means of a Legendre transformation,

$$\Psi = \Lambda_i\xi_i + \psi = \hat{\Psi}(\mathbf{C}, \theta; \mathbf{\Lambda}), \tag{4.2}$$

where $\mathbf{C}$ and θ are regarded as parameters, one can obtain

$$\xi_i = \partial\hat{\Psi}/\partial\Lambda_i. \tag{4.3}$$

From this relation it follows that

$$\partial\xi_i/\partial\Lambda_j = \partial\xi_j/\partial\Lambda_i \quad \text{and} \quad \frac{d}{dt}(\partial\xi_i/\partial\Lambda_j) = \frac{d}{dt}(\partial\xi_j/\partial\Lambda_i). \tag{4.4}$$

However, these relations *do not imply that* $\partial\dot{\xi}_i/\partial\Lambda_j = \partial\dot{\xi}_j/\partial\Lambda_i$, i.e. such a reciprocity in general may not exist, and more importantly, that in general

† That is, $\dot{\mathbf{\Lambda}} \cdot \dot{\xi} < 0$.

such a reciprocity does not follow from the reciprocal relation $(4.4)_1$. Under some restrictive assumptions such relations, however, will apply; see [6–10]. For example, if one assumes that each flux is only a function of the corresponding force and no other internal forces, then for $i \neq j$, $\partial\dot{\xi}_i/\partial\Lambda_j = 0$, and the reciprocal relation for fluxes applies. Also, when the fluxes are written as linear functions of the forces, $\dot{\xi}_i = L_{ij}\Lambda_j$, where L_{ij} is a symmetric positive-definite matrix (Onsager's assumption), then the reciprocal relations apply by assumption.

A major question to explore is finding out the extent to which the material stability assumption (4.1) restricts the fluxes $\dot{\xi}$. It will be shown that *if the fluxes are expressed as*†

$$\dot{\xi} = \mathbf{H}(\boldsymbol{\Lambda}) \tag{4.5}$$

and if there are no more than 8 *scalar-valued internal parameters, i.e.* $m \leq 8$, *then* (4.1) *implies that there exists a positive scalar function* $\lambda = \lambda(\boldsymbol{\Lambda})$, *and a potential function* $\varphi = \varphi(\boldsymbol{\Lambda})$, *such that*

$$\dot{\xi} = \mathbf{H} = \lambda \frac{\partial \varphi}{\partial \boldsymbol{\Lambda}}. \tag{4.6}$$

In a more general setting, however, no such relation may exist, unless restrictive assumptions are imposed.

To sketch the proof of the above assertion, first consider

$$\begin{aligned} \dot{\mathbf{S}} &= \left\{ \frac{\partial \hat{\mathbf{S}}}{\partial C_{AB}} \dot{C}_{AB} + \frac{\partial \hat{\mathbf{S}}}{\partial \theta} \dot{\theta} \right\} + \frac{\partial \hat{\mathbf{S}}}{\partial \xi_i} \dot{\xi}_i \\ &= \dot{\mathbf{S}}^{(e)} + \dot{\mathbf{S}}^{(in)}, \end{aligned} \tag{4.7}$$

where the quantity in the braces (denoted by $\dot{\mathbf{S}}^{(e)}$) in the first line of this equation represents the *elastic* rate of change of $\mathbf{S}$, and the last term represents the corresponding *inelastic* rate of change. This division is compatible with the basic character of the corresponding state variables. Since the ξ's represent inelastic structural changes, then if they are held fixed, no such structural changes occur, the corresponding change therefore being purely elastic.

In a similar manner, at constant deformation and temperature, the inelastic rate of change of entropy is defined by

$$\dot{\eta}^{(in)} = \frac{\partial \hat{\eta}}{\partial \xi_i} \dot{\xi}_i. \tag{4.8}$$

† This means that the fluxes are (implicit) functions of the state through the state functions $\boldsymbol{\Lambda} = \hat{\boldsymbol{\Lambda}}(\mathbf{C}, \theta; \xi) \equiv -\partial\hat{\psi}/\partial\xi$; i.e. they are *explicit* functions of internal forces only.

Next consider the quantity

$$F = \dot{\mathbf{\Lambda}} \cdot \dot{\xi} = \left\{ \frac{\partial \hat{\Lambda}_i}{\partial C_{AB}} \dot{C}_{AB} + \frac{\partial \hat{\Lambda}_i}{\partial \theta} \dot{\theta} \right\} \dot{\xi}_i + \frac{\partial \hat{\Lambda}_i}{\partial \xi_j} \dot{\xi}_j \dot{\xi}_i$$
$$= \left\{ \frac{\partial}{\partial C_{AB}} \left(-\frac{\partial \hat{\psi}}{\partial \xi_i} \right) \dot{C}_{AB} + \frac{\partial}{\partial \theta} \left(-\frac{\partial \hat{\psi}}{\partial \xi_i} \right) \dot{\theta} \right\} \dot{\xi}_i - \frac{\partial^2 \hat{\psi}}{\partial \xi_j \partial \xi_i} \dot{\xi}_j \dot{\xi}_i \tag{4.9}$$
$$= -\dot{\mathbf{S}}^{(\text{in})} \cdot \dot{\mathbf{C}} + \dot{\eta}^{(\text{in})} \dot{\theta} - \frac{\partial^2 \hat{\psi}}{\partial \xi_j \partial \xi_i} \dot{\xi}_j \dot{\xi}_i.$$

Equation (3.10) shows that the internal forces,

$$\mathbf{\Lambda} = \hat{\mathbf{\Lambda}}(\mathbf{C}, \theta; \xi) = \hat{\mathbf{\Lambda}}(P), \qquad P \equiv \{\mathbf{C}, \theta; \xi\}, \tag{4.10}$$

are functions of state P. Substitution for $\mathbf{\Lambda}$ into (4.5) yields

$$\dot{\xi} = \mathbf{H}(\hat{\mathbf{\Lambda}}(P)) = \mathbf{G}(P); \tag{4.11}$$

i.e. fluxes are also functions of state. Hence, (4.9) can be written as

$$F = -\boldsymbol{\sigma}(P) \cdot \dot{\mathbf{C}} + f(P)\dot{\theta} - g(P), \qquad g(P) > 0, \tag{4.12}$$

so that *at a given state only the rates* $\dot{\mathbf{C}}$ *and* $\dot{\theta}$ *can be prescribed arbitrarily.*

Let the material system M be in a given nonequilibrium state P_0. The corresponding internal forces and fluxes then are $\mathbf{\Lambda}_0 = \hat{\mathbf{\Lambda}}(P_0)$ and $\mathbf{G}_0 = \mathbf{G}(P_0)$, with the restriction that $\dot{\xi}_0 \cdot \mathbf{\Lambda}_0 > 0$. Let the rates $\dot{\mathbf{C}} = \mathbf{F}(t)$ and $\dot{\theta} = \tau(t)$ be given as suitably smooth functions of time in an interval $0 \le t < \Delta t$. Then the transition of M is the solution of the system of differential equations

$$\dot{\xi} = \mathbf{G}(P), \qquad \dot{\theta} = \tau(t), \qquad \dot{\mathbf{C}} = \mathbf{F}(t), \qquad 0 \le t < \Delta t, \tag{4.13}$$

with the initial conditions $\xi(0) = \xi_0$, $\theta(0) = \theta_0$, and $\mathbf{C}(0) = \mathbf{C}_0$. This is a curve, $P = P(t)$, in the state space, which originates at point P_0. Since $\mathbf{\Lambda}$ is a function of state, substitution into (4.10) gives $\mathbf{\Lambda} = \mathbf{\Lambda}(t)$ which is the corresponding curve in the $\mathbf{\Lambda}$-space that originates from the point $\mathbf{\Lambda}_0$.

The functions $\tau(t)$ and $\mathbf{F}(t)$ can be prescribed arbitrarily. Since all involved state functions will be assumed to be suitably smooth, one can choose $\tau(t)$ and $\mathbf{F}(t)$ in such a manner that the quantity

$$F = -\boldsymbol{\sigma}(P) \cdot \mathbf{F}(t) + f(P)\tau(t) - g(P) \tag{4.14}$$

be strictly negative. This can always be done. One choice is $\tau(t) = 0$ and $\mathbf{F}(t) = \mathbf{0}$, which yields $F = -g(P) < 0$.

Consider now a very small interval of time, Δt. During this interval the solution curve of (4.13) can be approximated by

$$\xi(t) = \xi_0 + \mathbf{G}_0 t, \qquad \theta(t) = \theta_0 + \tau_0 t, \qquad \mathbf{C}(t) = \mathbf{C}_0 + \mathbf{F}_0 t, \qquad 0 \le t < \Delta t, \tag{4.15}$$

where $\mathbf{G}_0$, τ_0, and $\mathbf{F}_0$ are the initial values of the rate of change of the corresponding state variables; τ_0 and $\mathbf{F}_0$ can, of course, be prescribed arbitrarily, but $\mathbf{G}_0 = \mathbf{G}(P_0)$ is fixed. Substitution into (4.14) yields

$$F = \{-\boldsymbol{\sigma}(P_0) \cdot \mathbf{F}_0 + f(P_0)\tau_0 - g(P_0)\} + 0(t), \tag{4.16}$$

where $0(t)$ denotes terms linear and higher order in t. Since $t < \Delta t$ is very small, the sign of F is given by that of the quantity in braces in (4.16). Hence one can choose τ_0 and $\mathbf{F}_0$ such that, in this small interval of time, F is strictly negative. Corresponding to this choice one obtains in the $\boldsymbol{\Lambda}$-space

$$\boldsymbol{\Lambda} = \hat{\boldsymbol{\Lambda}}(\mathbf{C}_0 + \mathbf{F}_0 t, \theta_0 + \tau_0 t; \xi_0 + \mathbf{G}_0 t), \qquad 0 \le t < \Delta t, \tag{4.17}$$

which is an elementary curve emanating from $\boldsymbol{\Lambda}_0$, on which F is strictly negative.

Consider now the special case where the number of internal variables is not more than 8, say $m = 8$. Then the $\boldsymbol{\Lambda}$-space is 8-dimensional Euclidean. From the above consideration it is clear that one can choose the seven parameters τ_0 and $\mathbf{F}_0$, such that as they range over all their admissible values, the elementary curve (4.17) marks a *solid*† cone (with vertex at $\boldsymbol{\Lambda}_0$) in the $\boldsymbol{\Lambda}$-space, in and on which the function F is strictly negative. Hence, in every suitably small neighborhood of point $\boldsymbol{\Lambda}_0$ there exists a connected set of points in which F is strictly negative. Since the functions $\tau(t)$ and $\mathbf{F}(t)$ can be chosen arbitrarily, the above-mentioned neighborhood can be further extended. This implies that if one confines attention to a suitably small neighborhood of $\boldsymbol{\Lambda}_0$, then there is a connected set of points close to $\boldsymbol{\Lambda}_0$, and includes $\boldsymbol{\Lambda}_0$ as a boundary point, which cannot be connected to $\boldsymbol{\Lambda}_0$ by the solution curves of $F\,dt = \dot{\xi} \cdot d\boldsymbol{\Lambda} = \dot{\xi} \cdot \dot{\boldsymbol{\Lambda}}\,dt = 0$. This is precisely the condition under which the differential form $\dot{\xi} \cdot d\boldsymbol{\Lambda}$ admits an integrating factor‡ $\lambda = \lambda(\boldsymbol{\Lambda})$, such that $\dot{\xi} \cdot d\boldsymbol{\Lambda}/\lambda = d\varphi$ is a perfect differential, where $\varphi = \varphi(\boldsymbol{\Lambda})$.

The integrating factor λ can be strictly positive or strictly negative, but clearly it cannot change sign; here λ will be taken to be positive. The function φ will be called the *inelastic potential.* It is defined to within an additive constant.

Let the material system M be in the *nonequilibrium* state P_0, with internal

† That is, a region in the $\boldsymbol{\Lambda}$-space which has an *interior* point.

‡ See, for example, [11].

forces $\mathbf{\Lambda}_0$. If the strain and temperature are held fixed, then the internal forces relax into a new value which corresponds to that of the local *equilibrium* state. Choose the origin of the coordinate system in the $\mathbf{\Lambda}$-space to coincide with this equilibrium state. Then at the origin the corresponding internal forces are assigned zero values. The inelastic potential can now be defined by

$$\varphi(\mathbf{\Lambda}_0) = \int_0^{\mathbf{\Lambda}_0} \frac{\mathbf{H}(\mathbf{x})}{\lambda(\mathbf{x})} \cdot d\mathbf{x}. \tag{4.18}$$

Since the integral is path-independent, the integration may be carried out along the radial line $\mathbf{x} = \mathbf{\Lambda}_0 \tau$, $0 \leq \tau \leq 1$, i.e.

$$\varphi(\mathbf{\Lambda}_0) = \int_0^1 \left[\frac{\mathbf{H}(\mathbf{\Lambda}_0 \tau)}{\lambda(\mathbf{\Lambda}_0 \tau)} \cdot \mathbf{\Lambda}_0 \tau \right] \frac{d\tau}{\tau}. \tag{4.19}$$

The quantity in the brackets in (4.19) is nonnegative, since λ is positive, $\dot{\xi} \cdot \mathbf{\Lambda} \geq 0$. Therefore φ is an increasing function along any radial line emanating from the origin in the $\mathbf{\Lambda}$-space. φ attains its minimum value in the $\mathbf{\Lambda}$-space at the local equilibrium position; see [9], [10], and other references cited in [10]. When the fluxes are linearly dependent on the forces, and when the Onsager relations hold, φ becomes proportional to the rate of entropy production. In general, however, it is the inelastic potential φ, and not the rate of entropy production, which attains a minimum value at the equilibrium state.

5 NORMALITY RULES

Let the fluxes be defined in terms of the inelastic potential φ by (4.6). By substitution for the internal forces in terms of the state variables, obtain

$$\varphi = \varphi(\hat{\mathbf{\Lambda}}(\mathbf{C}, \theta; \xi)) \equiv \hat{\Omega}(\mathbf{C}, \theta; \xi), \qquad \lambda = \lambda(\hat{\mathbf{\Lambda}}(\mathbf{C}, \theta; \xi)) \equiv \hat{\lambda}(\mathbf{C}, \theta; \xi). \tag{5.1}$$

The inelastic rate of stress relaxation can then be written as

$$\begin{aligned} -\dot{\mathbf{S}}^{(\text{in})} &= -\frac{\partial \hat{\mathbf{S}}}{\partial \xi_i} \dot{\xi}_i = -\frac{\partial^2 \hat{\psi}}{\partial \xi_i \, \partial \mathbf{C}} \dot{\xi}_i \\ &= \hat{\lambda} \frac{\partial \hat{\Lambda}_i}{\partial \mathbf{C}} \frac{\partial \varphi}{\partial \Lambda_i} = \hat{\lambda} \frac{\partial \hat{\Omega}}{\partial \mathbf{C}}. \end{aligned} \tag{5.2}$$

Hence the gradient of the inelastic potential in the $\mathbf{C}$-subspace is proportional to the inelastic rate of stress relaxation.

In the same manner one obtains†

$$\dot{\eta}^{(\text{in})} = -\frac{\partial^2 \hat{\psi}}{\partial \xi_i \, \partial \theta} \dot{\xi}_i = \hat{\lambda} \frac{\partial \hat{\Omega}}{\partial \theta}. \tag{5.3}$$

Since $\hat{\lambda} > 0$, (5.3) shows that the inelastic potential cannot be a decreasing function of temperature.

One may use **S**, θ, and ξ as state variables. To this end one writes

$$\mathbf{C} = \overline{\mathbf{C}}(\mathbf{S}, \theta; \xi), \tag{5.4}$$

introduces the Legendre transformation

$$\chi = \mathbf{S} \cdot \mathbf{C} - \psi \equiv \overline{\chi}(\mathbf{S}, \theta; \xi), \tag{5.5}$$

and obtains

$$\mathbf{C} = \frac{\partial \overline{\chi}}{\partial \mathbf{S}}, \qquad \eta = \frac{\partial \overline{\chi}}{\partial \theta} \equiv \overline{\eta}(\mathbf{S}, \theta; \xi), \qquad \mathbf{\Lambda} = \frac{\partial \overline{\chi}}{\partial \xi} \equiv \overline{\mathbf{\Lambda}}(\mathbf{S}, \theta; \xi). \tag{5.6}$$

The inelastic strain rate is then given by

$$\dot{\mathbf{C}}^{(\text{in})} = \frac{\partial \overline{\mathbf{C}}}{\partial \xi_i} \dot{\xi}_i \equiv \overline{\lambda}(\mathbf{S}, \theta; \xi) \frac{\partial \overline{\Omega}(\mathbf{S}, \theta; \xi)}{\partial \mathbf{S}}, \tag{5.7}$$

as can easily be verified. In (5.7), $\overline{\lambda}$ and $\overline{\Omega}$ represent the same integrating factor λ and the inelastic potential φ, expressed in terms of the new state variables.

Again, if a Legendre transformation is applicable, one may use **C**, η, and ξ as state variables, and obtain

$$e = \psi + \theta \eta \equiv \tilde{e}(\mathbf{C}, \eta; \xi), \tag{5.8}$$

† In general, $\dot{\eta}^{(\text{in})} \geq 0$. However, since $\dot{\eta}^{(\text{in})}$ is the rate of entropy change due to internal structural changes at instantaneously fixed deformation and temperature (rather than at fixed deformation and internal energy), its nonnegativeness is not a restatement of the second law. To establish $\dot{\eta}^{(\text{in})} \geq 0$, note that at fixed deformation and temperature, one has, from Eq. $(3.7)_2$,

$$\theta \dot{\eta}^{(\text{in})} = \dot{e} + \mathbf{\Lambda} \cdot \dot{\xi} = -\dot{Q} + \mathbf{\Lambda} \cdot \dot{\xi},$$

where $\dot{Q}$ is the rate at which heat must be removed from the material system in order to maintain the instantaneously fixed temperature. The rate of dissipation, $\mathbf{\Lambda} \cdot \dot{\xi}$, may be divided into two parts: (1) the rate of energy dissipated into heating, which must be equal to $\dot{Q}$; and (2) the rate of energy dissipation due to irreversible processes other than those directly generating heat. This latter dissipation includes energy loss because of the formation of internal cracks, voids, entanglement of dislocations, breakage of polymer bonds, and other irreversible processes which dissipate energy but do not cause a temperature increase. For example, in metal plasticity only a portion of the plastic work is converted into the heat energy which causes a temperature increase. Hence $\mathbf{\Lambda} \cdot \dot{\xi} - \dot{Q} \geq 0$, i.e. $\dot{\eta}^{(\text{in})} \geq 0$.

so that

$$\mathbf{S} = \frac{\partial \tilde{e}}{\partial \mathbf{C}} \equiv \mathbf{S}(\mathbf{C}, \eta; \xi), \qquad \theta = \frac{\partial \tilde{e}}{\partial \eta} \equiv \tilde{\theta}(\mathbf{C}, \eta; \xi), \qquad \mathbf{\Lambda} = -\frac{\partial \tilde{e}}{\partial \xi} \equiv \tilde{\mathbf{\Lambda}}(\mathbf{C}, \eta; \xi). \tag{5.9}$$

Then it can easily be shown that the rate of change of temperature caused by internal relaxation is given by

$$\dot{\theta}^{(\text{in})} = \frac{\partial \tilde{\theta}}{\partial \xi_i} \dot{\xi}_i \equiv -\tilde{\lambda} \frac{\partial \tilde{\Omega}}{\partial \eta}, \tag{5.10}$$

where the integrating factor $\tilde{\lambda}$, and the inelastic potential $\tilde{\Omega}$, are the *same* quantities as before, but expressed in terms of the new state variables.

Equations (5.2), (5.3), (5.7), and (5.10) define a complete set of normality rules. It is also easy to see that

$$\dot{\mathbf{\Lambda}}^{(\text{in})} = \frac{\partial \hat{\mathbf{\Lambda}}}{\partial \xi_i} \dot{\xi}_i \equiv \hat{\lambda} \frac{\partial \hat{\Omega}}{\partial \xi}. \tag{5.11}$$

The validity of these normality rules rests on the assumptions (4.5) and (4.6); see Mandel [8] for a careful discussion of this point in viscoplasticity. Neither of these assumptions constitutes a law of nature. Therefore these normality rules do not have general validity. Note that in the linearized case (still for large deformations), when the fluxes are linear functions of the forces, and when the Onsager assumption holds, then $\varphi = \frac{1}{2} L_{ij} \Lambda_i \Lambda_j$, and hence the normality rules apply. This remark remains valid even when a continuous spectrum for the internal variables is assumed. Thus, in linear viscoelasticity and linear creep, the inelastic strain rate is always a gradient of the inelastic potential taken with respect to the conjugate stress.

Acknowledgments. This work has been supported by the National Science Foundation under Grant GK-31352 to Northwestern University. The author wishes to thank his wife Éva and Mrs. Erika Ivansons for their help in preparing the manuscript. Also, many stimulating discussions and private communications with Prof. J. Rice of Brown University, which were very helpful in clarifying the role of the internal variables in a proper statement of the second law, are gratefully acknowledged.

6 REFERENCES

1. Nemat-Nasser, S., "On Nonequilibrium Thermodynamics of Continua," in *Mechanics Today*, **2**, S. Nemat-Nasser (ed.), Pergamon Press, New York (1975) 94–158.
2. Green A. E. and Rivlin, R. S., "The Mechanics of Nonlinear Materials with Memory, Part I," *Arch. Rational Mech. Anal.* **1** (1957) 1–21.
3. *Mechanics of Visco-Elastic Media and Bodies*, Proceedings, IUTAM Symposium, J. Hult (ed.), Springer-Verlag (1975).
4. Nemat-Nasser, S., "On Nonlinear Thermoelasticity and Nonequilibrium Thermodynamics," in *Nonlinear Elasticity*, R. W. Dickey (ed.), Academic Press, New York (1973) 289–338.
5. *Foundations of Continuum Thermodynamics*, J. J. D. Domingos, M. N. R. Nina, and J. H. Whitelaw (eds.), The Macmillan Press (1974) 223–227.
6. Kestin, J. and Rice, J. R., "Paradoxes in the Application of Thermodynamics to Strained Solids," in *A Critical Review of Thermodynamics*, E. B. Stuart, B. Gal-Or, and A. J. Brainard (eds.), Mono Book Corp., Baltimore (1970) 275–298.
7. Rice, J. R., "Inelastic Constitutive Relations for Solids: An Internal-Variable Theory and Its Application to Metal Plasticity," *J. Mech. Phys. Solids* **19** (1971) 433–455.
8. Mandel, J., "Thermodynamics and Plasticity," in *Foundations of Continuum Thermodynamics*, J. J. D. Domingos, M. N. R. Nina and J. H. Whitelaw (eds.), The Macmillan Press, London (1974) 283–315.
9. Edelen, D. G. B., "On the Characterization of Fluxes in Nonlinear Irreversible Thermodynamics," *Int. J. Engng Sci.* **12** (1974) 397–411.
10. Nemat-Nasser, S., "On Nonequilibrium Thermodynamics of Viscoelasticity: Inelastic Potentials and Normality Conditions," in *Mechanics of Visco-Elastic Media and Bodies*, Proceedings, IUTAM Symposium, J. Hult (ed.), Springer-Verlag (1975) 375–391.
11. Buchdahl, H. A., *The Concepts of Classical Thermodynamics*, Cambridge University Press, Cambridge (1966).

Author Index†

† Italic numerals denote full references. Other numerals denote page(s) where the author is referred to by name or reference number or both.

Subject Index